Michel Deville (Ed.)

Proceedings of the Seventh GAMM-Conference on Numerical Methods in Fluid Mechanics

Notes on Numerical Fluid Mechanics
Volume 20

Series Editors: Ernst Heinrich Hirschel, München
Keith William Morton, Oxford
Earll M. Murman, M.I.T., Cambridge
Maurizio Pandolfi, Torino
Arthur Rizzi, Stockholm
Bernard Roux, Marseille

(Addresses of the Editors: see inner back cover)

Volume 1 Boundary Algorithms for Multidimensional Inviscid Hyperbolic Flows (K. Förster, Ed.)

Volume 2 Proceedings of the Third GAMM-Conference on Numerical Methods in Fluid Mechanics (E. H. Hirschel, Ed.) (out of print)

Volume 3 Numerical Methods for the Computation of Inviscid Transonic Flows with Shock Waves (A. Rizzi / H. Viviand, Eds.)

Volume 4 Shear Flow in Surface-Oriented Coordinates (E. H. Hirschel / W. Kordulla)

Volume 5 Proceedings of the Fourth GAMM-Conference on Numerical Methods in Fluid Mechanics (H. Viviand, Ed.) (out of print)

Volume 6 Numerical Methods in Laminar Flame Propagation (N. Peters / J. Warnatz, Eds.)

Volume 7 Proceedings of the Fifth GAMM-Conference on Numerical Methods in Fluid Mechanics (M. Pandolfi / R. Piva, Eds.)

Volume 8 Vectorization of Computer Programs with Applications to Computational Fluid Dynamics (W. Gentzsch)

Volume 9 Analysis of Laminar Flow over a Backward Facing Step (Ken Morgan / J. Periaux / F. Thomasset, Eds.)

Volume 10 Efficient Solutions of Elliptic Systems (W. Hackbusch, Ed.)

Volume 11 Advances in Multi-Grid Methods (D. Braess / W. Hackbusch / U. Trottenberg, Eds.)

Volume 12 The Efficient Use of Vector Computers with Emphasis on Computational Fluid Dynamics (W. Schönauer / W. Gentzsch, Eds.)

Volume 13 Proceedings of the Sixth GAMM-Conference on Numerical Methods in Fluid Mechanics (D. Rues / W. Kordulla, Eds.) (out of print)

Volume 14 Finite Approximations in Fluid Mechanics (E. H. Hirschel, Ed.)

Volume 15 Direct and Large Eddy Simulation of Turbulence (U. Schumann / R. Friedrich, Eds.)

Volume 16 Numerical Techniques in Continuum Mechanics (W. Hackbusch / K. Witsch, Eds.)

Volume 17 Research in Numerical Fluid Dynamics (P. Wesseling, Ed.)

Volume 18 Numerical Simulation of Compressible Navier-Stokes Flows (M. O. Bristeau / R. Glowinski / J. Periaux / H. Viviand, Eds.)

Volume 19 Three-Dimensional Turbulent Boundary Layers — Calculations and Experiments (B. van den Berg / D. A. Humphreys / E. Krause / J. P. F. Lindhout)

Volume 20 Proceedings of the Seventh GAMM-Conference on Numerical Methods in Fluid Mechanics (M. Deville, Ed.)

Volume 21 Panel Methods in Fluid Mechanics with Emphasis on Aerodynamics (J. Ballmann / R. Eppler / W. Hackbusch, Eds.)

Michel Deville (Ed.)

Proceedings of the Seventh GAMM-Conference on Numerical Methods in Fluid Mechanics

Friedr. Vieweg & Sohn Braunschweig / Wiesbaden

Manuscripts should have well over 100 pages. As they will be reproduced photomechanically they should be typed with utmost care on special stationary which will be supplied on request. In print, the size will be reduced linearly to approximately 75 %. Figures and diagramms should be lettered accordingly so as to produce letters not smaller than 2 mm in print. The same is valid for handwritten formulae. Manuscripts (in English) or proposals should be sent to the general editor Prof. Dr. E. H. Hirschel, Herzog-Heinrich-Weg 6, D-8011 Zorneding.

Vieweg is a subsidiary company of the Bertelsmann Publishing Group.

Produced by W. Langelüddecke, Braunschweig
Printed in Germany

ISSN 0179-9614

ISBN 3-528-08094-9

FOREWORD

The GAMM-Committee for Numerical Methods in Fluid Mechanics (GAMM-Fachausschuss für Numerische Methoden in der Strömungsmechanik) organizes every odd year the GAMM-CONFERENCE ON NUMERICAL METHODS IN FLUID MECHANICS.

The conferences started in Germany at the DFVLR in Cologne (1975-77-79) and then moved around Europe : at ENSTA in Paris (1981), the University of Rome (1983) and again the DFVLR in Göttingen (1985). The Seventh Conference was held at the "Faculté des Sciences Appliquées", Université Catholique de Louvain, Louvain-La-Neuve, Belgium, from September 9 to September 11, 1987.

The GAMM-Conference is intended to gather scientists, engineers and researchers who are working in the field of computational fluid mechanics. The main objective consists in exchanges and stimulating discussions between the various fields of CFD such as Aerodynamics, Combustion, Propulsion, Fluidmachinery , Meteorology, Biofluidmechanics, Heat transfer and convection, etc. The subjects covered in the Conference are essentially related to theoretical and algorithmical aspects of numerical methods in fluid mechanics (finite difference methods, finite element methods, spectral methods,etc) or to particular applications to fluid flow problems where novelties of the methods are emphasized. Moreover reports are given on GAMM-WORKSHOPS sponsored by the Committee where specialized topics are investigated by scientists involved in these particular fields. The 1987 Conference was attended by about 110 scientists from 15 different countries There were 52 contributed papers and activities of two workshops were reported. The contributions are presented in these proceedings by alphabetical order according to the first author.

Current trends in CFD come out from these papers. Three-dimensional computations become the norm. Related questions like mesh generation, domain decomposition, adaptive meshes are treated deeply in accordance with the impact of vector and parallel processing.

From the algebraic point of view, conjugate gradients with pre-conditioning and multigrid techniques made major break throughs for the solution of very large problems. Among the applications, hypersonic flows with real gas constitute a large area of renewed interest. In fact, these proceedings give a picture of the state of the art in numerical fluid mechanics and shed some light on the prospects for future directions in research and industrial applications.

The chairman of the conference (Prof. M. Deville) would like to acknowledge the support from : Gesellschaft für Angewandte Mathematik und Mechanik (GAMM), Fonds National de la Recherche Scientifique (FNRS), Ministère de l'Education Nationale et de la Culture Française, Office of Naval Research, London Branch Office (ONRL), IBM Belgium S.A. (IBM), Data General Belgium S.A. (DG).

Finally, I would like to express my gratitude to my colleagues and personnel of the Unité de Mécanique Appliquée who helped me to organize successfully this conference.

January 28, 1988. MICHEL DEVILLE.

CONTENTS

REPORTS ON GAMM-WORKSHOPS

APPLICATION OF MULTIGRID METHODS FOR THE COUPLED AND DECOUPLED SOLUTION OF THE INCOMPRESSIBLE NAVIER-STOKES EQUATIONS

Ch. Arakawa
University of Tokyo
Tokyo, Japan

A.O. Demuren
University of Lagos
Lagos, Nigeria

W. Rodi B. Schönung
University of Karlsruhe
Karlsruhe, F.R. Germany

ABSTRACT

This paper presents the application of multigrid methods to the solution of the Navier-Stokes equations for incompressible, two-dimensional, laminar flow problems. The Full Approximation Storage - Full Multigrid Algorithm (FAS - FMG) is utilised with relaxation schemes which solve the primitive variables in either a coupled or a decoupled manner. The performance of the multigrid methods is compared with those of several single-grid methods, using two flow problems as test cases. The results show that the multigrid methods always converged faster than equivalent single-grid methods. More importantly, however, the best multigrid method was found to converge 10 - 80 times faster than the best single-grid method.

INTRODUCTION

Multigrid techniques, which were initially developed for the efficient solution of linear, elliptic differential equations, are used more and more also in the field of numerical fluid dynamics. They became quite common for the calculation of compressible flows, whereas up to now only few attempts have been reported on their application to the solution of the incompressible Navier-Stokes equations with primitive variables. Most applications to the Navier-Stokes equations have employed a stream function-vorticity formulation (see Ghia et al. [1]). The reason for this lies in the solution technique for the pressure field. With the compressible solution procedures, the velocity components are obtained from the momentum equations, the density from the continuity equation and the pressure via the enthalpy equation or from the equation of state. Thus, for each variable to be calculated, a specific equation is given. This is not the case for incompressible flows for which there is no explicit equation for the pressure. In the primitive variable formulation, two principally different methods have been developed in the past for determining the pressure field in incompressible flows:

(i) A special pressure equation or pressure-correction equation is derived with the aid of the momentum equations and the continuity equation. The momentum equations and the pressure-determining equation are solved sequentially (decoupled method). A typical example of this is the SIMPLE algorithm of Patankar and Spalding [2].

(ii) The momentum equations and the continuity equation are solved in a coupled way (coupled method). Caretto et al. [3] proposed such an algorithm called SIVA.

Most of the calculation procedures, based on the decoupled method, determine the pressure field via a pressure-correction equation and use a predictor method for obtaining the velocity and pressure field. More specifically, after solving the momentum equations with a guessed pressure to yield the velocity components and after the solution of the pressure-correction equation, the pressure and velocity fields are updated to satisfy continuity. The application of multigrid techniques in connection with this procedure is problematic because, after smoothing the momentum equations, errors of higher frequencies can be introduced via the velocity corrections. This problem has already been discussed by Brandt and Dinar [4] as well as by Fuchs [5]. These authors suggest specific pressure-correction equations, which result however in smooth velocity corrections only for a specific range of Reynolds numbers.

More recently, Vanka [6] proposed a multigrid method for a coupled solution procedure. Coupled solution methods have not been used very often in the past because of the following two reasons:

a) If the coupling is performed in the whole solution domain (direct method) huge matrices result, which are difficult to solve.
b) If the coupling is performed only in small subdomains, the resulting matrices are reasonable to handle, but poor convergence rates are obtained, especially for fine grids, because of the weak coupling of the different subdomains. It was for this reason that Caretto et al. [3] found the SIVA algorithm generally inferior to the SIMPLE algorithm in many incompressible flow problems. But with multigrid techniques this problem can be overcome once the asymptotic convergence rate is approached which is independent of the total number of grid points.

In the present paper, the multigrid technique is applied to the coupled and decoupled solution of the incompressible Navier-Stokes equations in primitive variables. In the decoupled solution method, the multigrid technique is used for the whole system of equations, that is, the momentum equations and the pressure-correction equation are smoothed on one grid level and then the residuals of the velocity and the pressure corrections are transferred to the next one. The performance of the following pressure-correction methods is analysed: SIMPLE, SIMPLEC, PISO and SIMPLEST. The latter three are all improvements proposed for accelerating the convergence rate of the popular SIMPLE method (see e.g. Patankar [7]). For the coupled solution procedure the method of Vanka [6] is adopted and both point-relaxation (SCGS) and line relaxation (CL-SOR) schemes are utilised.

MATHEMATICAL MODEL

The Navier-Stokes equations may be written for incompressible, steady, laminar plane flow as:

$$\frac{\partial(UU)}{\partial x} + \frac{\partial(UV)}{\partial y} = -\frac{1}{\rho}\frac{\partial P}{\partial x} + \nu\left(\frac{\partial^2 U}{\partial x^2} + \frac{\partial^2 U}{\partial y^2}\right), \qquad (1)$$

$$\frac{\partial(UV)}{\partial x} + \frac{\partial(VV)}{\partial y} = -\frac{1}{\rho}\frac{\partial P}{\partial y} + \nu\left(\frac{\partial^2 V}{\partial x} \quad \frac{\partial^2 V}{\partial y}\right), \qquad (2)$$

and the continuity equation as:

$$\frac{\partial U}{\partial x} + \frac{\partial V}{\partial y} = 0\,, \qquad (3)$$

where the symbols have their usual meaning.

Relaxation Methods

The finite-difference approximation of the set of equations (1) - (3) is solved by relaxation methods either in a coupled or decoupled manner. All the methods use a staggered-grid system with the control volumes for the velocities and the pressure displaced as shown in Figure 1. The differences lie in the pressure-velocity coupling.

The coupled methods solve for the velocities and the pressure at each grid node, simultaneously. Two variants are considered in the present study. The *first is a point-relaxation method in which the pressure at the centre of the control volume, shown in Figure 2 and the four velocities along its faces are determined simultaneously. Through some simplification (see Vanka [6]), the resulting 5 x 5 matrix equation at each node can be solved algebraically. This method is called symmetric coupled Gauss Seidel (SCGS). The solution is obtained for all grid nodes in a lexicographical order. The second variant employs a line-relaxation technique to obtain the nodal pressure and the face velocities, simultaneously, along a complete line of grid nodes. Algebraic solution is not now possible, but the efficient Thomas algorithm is utilised. The method is called coupled line successive over relaxation (CLSOR).*

The decoupled methods solve for the velocities and the pressure distribution in the whole flow domain in a sequential manner. All the four methods considered in the present study are based on the SIMPLE algorithm of Patankar and Spalding [2]. In this, the velocities are first solved with a guessed pressure field, these are then applied to a pressure-correction equation derived from the continuity and the momentum equations. Solution of the former yields the pressure corrections and subsequently the velocity corrections with which the computed pressure and velocity fields are updated. In the original formulation some terms linking the velocity corrections at neighbouring grid nodes are neglected. This enables the solution to be obtained in a line-by-line procedure or through an ADI (alternate direction implicit) scheme. However, this neglect resulted in rather poor convergence characteristics. The other three methods employ various proposed improvements to enhance the pressure-velocity coupling, thereby increasing the rate of convergence. SIMPLEC (SIMPLE-Consistent), proposed by van Doormal and Raithby [8] assumes that the velocity corrections at neighbouring nodes are of the same order, so that the terms neglected are related to the differences between these values and not the values themselves. Only a slight modification of coefficients is necessary to implement this. The PISO (Pressure Implicit Split Operator) method proposed by Issa [9] has, in addition to the predictor and corrector stages in SIMPLE, a second corrector stage in which the influences of the neighbouring points are not neglected. Thus, each iteration requires more computational work (typically 40 - 50%), and more memory is required to store additional coefficients. The SIMPLEST method proposed by Spalding [10] neglects only the diffusive contribution to the velocity corrections at the neighbouring nodes. As may be expected, this method performs well in lower Reynolds number flows.

All these relaxation methods require many iterations for convergence of the solution, but, as will be shown later, these increase at an unbounded rate as the computational domain is refined. Hence, the need for methods, such as the multigrid ones which have convergence characteristics which are independent of the degree of grid refinement.

Multi-grid Methods

The multigrid concept is based on the observation that most relaxation methods are efficient in smoothing out error components whose wavelengths are comparable to the grid mesh size. Error components with longer wavelengths are smoothed out at comparatively slower rates. Thus, the idea of the multigrid technique is to smooth out high frequency components by performing a few iterations of the relaxation process on the fine grid. The remaining errors are then transferred to a coarser grid where the corresponding lower frequency error components are again smoothed out in a few iterations. Further transfers are made to even coarser grids and the process is repeated until all the error components have been smoothed out. The results are then progressively transmitted back to the finer grids. The overall effect of this procedure is that the various Fourier components of the error are removed on grid meshes most efficient for the purpose, thereby accelerating the convergence rate on the fine grid on which the solution of the finite difference equation is sought.

In the present study, the Full Approximation Storage - Full Multi-Grid (FAS - FMG) algorithm due to Brandt [11] is employed for both the coupled and the decoupled solution schemes, because of the non-linear nature of the system of equations. FMG implies that converged solutions are first obtained on coarser grids, which are then prolongated to finer grids to act as first approximations to the fine grid solutions. FAS requires that for a finite-difference equation, which is written for the kth grid as:

$$L^k U^k = F^k, \tag{4}$$

and which has U^k as the fine grid approximation, the full approximation U^{k-1} is stored on the coarse grid (k-1) rather than its correction δU^{k-1}. U^{k-1} is given by:

$$U^{k-1} = I_k^{k-1} U^k + \delta U^{k-1} \tag{5}$$

The full approximation is obtained on the coarse grid by solving the corresponding FAS equation:

$$L^{k-1} U^{k-1} = L^{k-1}\left(I_k^{k-1} U^k\right) + I_k^{k-1}\left(F^k - L^k U^k\right). \tag{6}$$

The fine grid approximation is then corrected as:

$$U^k_{new} = U^k_{old} + I^k_{k-1}\left(U^{k-1} - I_k^{k-1} U^k_{old}\right). \tag{7}$$

In equations (4) - (7), I_k^{k-1}, and I_{k-1}^k, denote the restriction and prolongation operators, respectively. In the present study, bilinear interpolation is employed everywhere for these.

The multigrid codes differ in their cycling procedures. The decoupled code uses fixed V cycles for switching between grids. 5 iterations are performed on each of the finer grids before transferring the residuals to the next coarser grid. During the prolongation half of the cycle, when the results are transferred back to the finer grids, 10 iterations are performed on each of the coarser grids. The coupled code, on the other hand, uses adaptive cycles in which transfers to coarser grids are controlled solely by the rate of smoothing of the residuals. If the rate of reduction of the average residual on any but the coarsest grid falls below 0.8 the residuals are transferred to the next coarser grid for smoothing, otherwise iteration continues on the current grid. Once the corresponding residuals on the coarser grid have been reduced to less than 20% of their original value, the results are transferred back to the finer grid, and the multigrid cycle continues until a converged solution is obtained on the finest grid. The convergence criterion applied is that the average normalized residual $\bar{R}$ should be less than 10^{-5}, where

$$\bar{R} = \left[\left(\sum (R^u_{i,j})^2 + \sum (R^v_{i,j})^2 + \sum (R^c_{i,j})^2\right) / \left(3\cdot(NI-2)(NJ-2)\right)\right]^{1/2} \tag{8}$$

(R^u, R^V and R^C are residuals of equations (1), (2), (3), respectively, and NI and NJ are number of grid points in the x- and y-directions.) The multigrid cycle is applied on a maximum of 4 grid levels in the decoupled procedure, but is completely flexible in the coupled one. Up to 6 levels are used in the present study.

RESULTS AND DISCUSSION

The performance of the prediction methods using both single-grid and multigrid techniques is evaluated with two test cases: (a) laminar flow in a lid-driven square cavity, at Reynolds numbers (= U_oH/ν) of 100 and 1000; (b) laminar flow over a backward-facing step at Reynolds number (= U_oh/ν) of 100 and 500, with aspect ratios (L/H) of 3 and 6, respectively. These test cases are illustrated in Figure 3.

Figures 4 and 5 show the evolution of the CPU times per grid point, required for convergence of the single-grid methods as the computational grid is refined. These are seen to increase approximately linearly with the number of grid points, so that the actual CPU times would increase quadratically. Figures 6 and 7 show similar comparisons for the multigrid methods. The CPU times per grid point are now mostly of the same order of magnitude, no matter the total number of grid points. Some methods show slight increases, whereas others show decreases. Thus, the total CPU time would, in general, increase linearly with increase in the number of computational grid points, so that grid refinement becomes a feasible means of obtaining better numerical accuracy.

Table 1 show a comparison of the CPU times for convergence and the number of iterations on the finest grid required for the various multigrid methods and the best of the single-grid methods, applied to test case (a). The latter is usually the PISO or SIMPLEC method. The PISO method, which is actually very similar to the SIMPLER method proposed by Patankar [7], converges in the least number of iterations, of all single grid methods, but it requires 40 - 50% more CPU time per iteration, and is thus not always the most efficient. Further, it requires more memory storage than the other methods, so, SIMPLEC is clearly the preferred method for single-grid calculations. Typically, it converges about 2.5 times faster than the SIMPLE method. The point relaxation method SCGS has by far the slowest convergence rate of all single-grid methods. However, with multigrid, it leads to the fastest convergence rate for case (a). Considerable savings in CPU times are achieved by the application of the multigrid procedures with both coupled and decoupled relaxation schemes, especially on very fine computational grids. The pertinent measure of multigrid performance is the ratio CPU time of best single-grid method over CPU time of best multigrid method. This ratio has values of 80 and 20, on the finest grid, for the two problems of test case (a). Figures 4 and 5 suggest that the ratio will increase with further grid refinement. Thus, multigrid methods are particularly useful when very fine grid calculations are desired.

The results for test case (b), the flow over a backward-facing step, are presented in Table 2. The multigrid methods again show considerable savings in CPU times for the very fine grids. The point-relaxation method with multigrid also performs best for the lower Reynolds number and aspect ratio (L/H = 3), but is not so good for the higher Reynolds number and aspect ratio (L/H = 6). The reason for this is the large difference in magnitudes of the directional coefficients that would result from the large cell aspect ratio. Brandt [11] had pointed out that point-relaxation methods would not perform well under such conditions, and line-relaxation should be preferred. The coupled line relaxation method with multigrid converges fastest for this problem. In this case, CPU times are reduced by the ratio of 20 and 13, on the finest grid, through the use of multigrid procedures.

CONCLUSIONS

It has been shown that multigrid techniques can be applied with both coupled and decoupled approaches for solving the Navier-Stokes equations, leading to considerable savings in CPU time. The savings would increase with grid refinement. In contrast, proposed improvements to solution procedures with single grids result in only moderate savings in computer time. The best choice of the relaxation method to apply with the multigrid technique appears to depend on the particular problem, but the coupled procedures show slight advantages over the decoupled ones.

ACKNOWLEDGEMENT

The calculations were performed on the Siemens 7881 computer of the University of Karlsruhe, when the first two authors were on leave from their institutions. Sponsorship from the Deutsche Forschungsgemeinschaft and the Alexander von Humboldt Foundation is gratefully acknowledged.

REFERENCES

[1]. GHIA U., GHIA K.N., SHIN C.T.: "High-Re solutions for incompressible flow using the Navier-Stokes equations and a multigrid method", J.Comp. Physics, 48 (1982), pp. 387-411.

[2]. PATANKAR, S.V., SPALDING, D.B.: "A calculation procedure for heat, mass and momentum transfer in three-dimensional parabolic flows", Int. J. Heat Mass Transfer, 15 (1972), pp. 1878-1806.

[3]. CARETTO, L.S., CURR, R.M., SPALDING, D.B.: "Two numerical methods for three-dimensional boundary layers", Comp. Methods in Appl. Mech. and Eng., 1 (1972).

[4]. BRANDT, A., DINAR, N.: "Multigrid solution to elliptic flow problems", in Numerical Methods in PDE, ed. by S.V. Parter, Academic Press, (1979).

[5]. FUCHS, L.: "New relaxation methods for incompressible flow problems", in Numerical Methods in Laminar and Turbulent Flow, edited by C. Taylor, J.A. Johnson and W.R. Smith, Pineridge Press, (1983), pp. 606-616.

[6]. VANKA, S.P.: "Block-implicit multigrid solution of Navier-Stokes equations in primitive variables", J. Comput. Phys., 65 (1986), pp. 138-158.

[7]. PATANKAR, S.V.: "Numerical heat transfer and fluid flow", McGraw Hill, New York (1980).

[8]. VAN DOORMAL, J.P., RAITHBY, G.D.: "Enhancements of the SIMPLE method for predicting incompressible fluid flows", Num. Heat Trans., 7 (1984).

[9]. ISSA, R.I.: "Solution of implicitly discretized fluid flow equations by operator-splitting", Internal Report, Dept. of Min. Resources Eng., Imperial College, London (1982).

[10]. SPALDING, D.B.: "Mathematical modelling of fluid mechanics, heat transfer and mass transfer processes", Mech. Eng. Rept., No. HTS/80/1, Imperial College, London (1980).

[11]. BRANDT, A.: "Multi-level adaptive solutions to boundary-value problems", Math. of Comp., 31, No. 138 (1977).

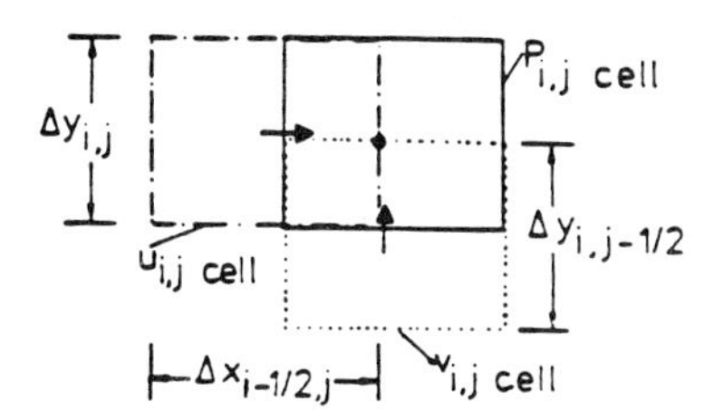

Fig 1: Control volumes for staggered grid

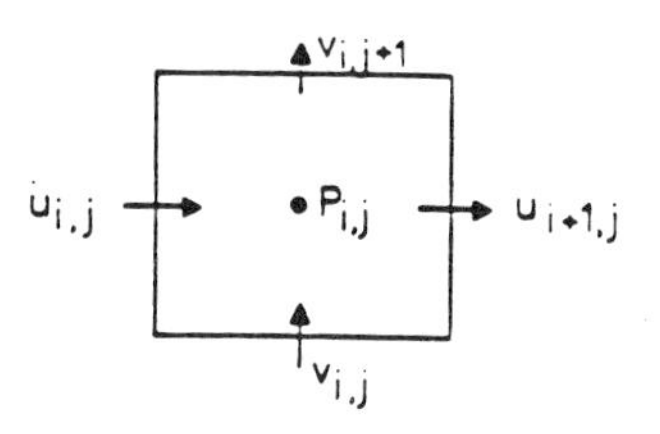

Fig. 2: Dependent variables for central node

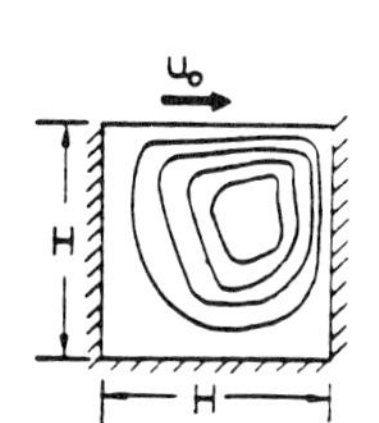

Fig. 3(a): Lid-driven cavity flow

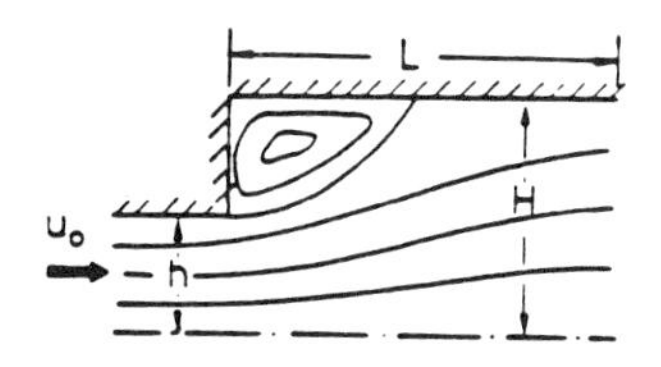

Fig. 3(b): Flow over a backward-facing step

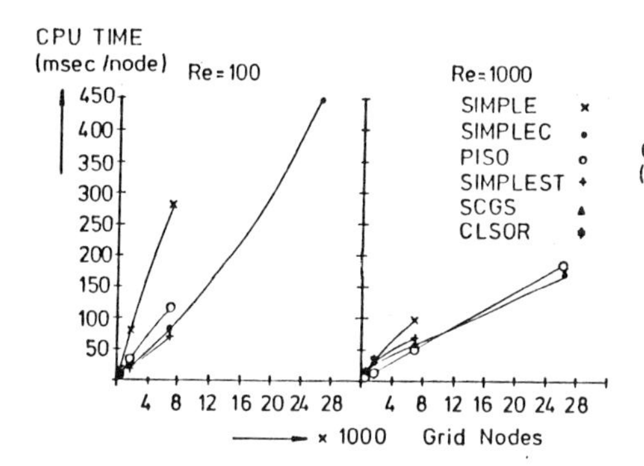

Fig. 4: CPU Times for convergence; case (a), single grid methods

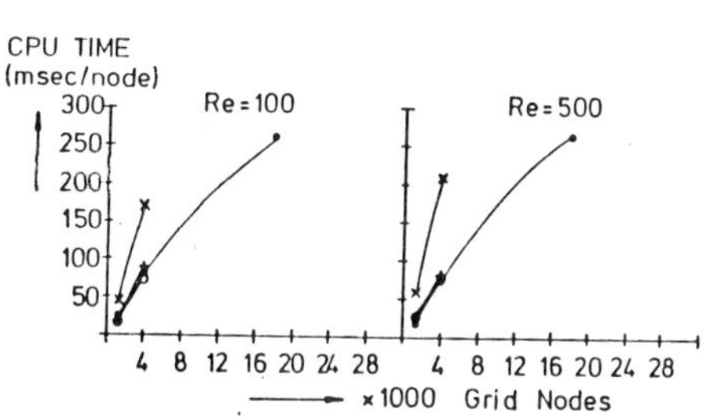

Fig. 5: CPU Times for convergence; case (b), single grid methods

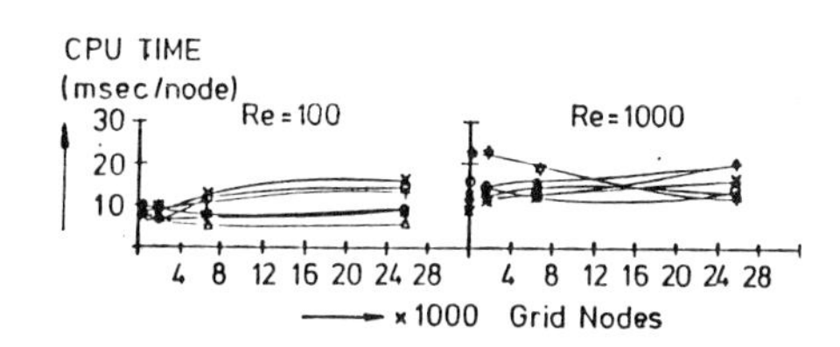

Fig. 6: CPU Times for convergence; case (a), multigrid methods

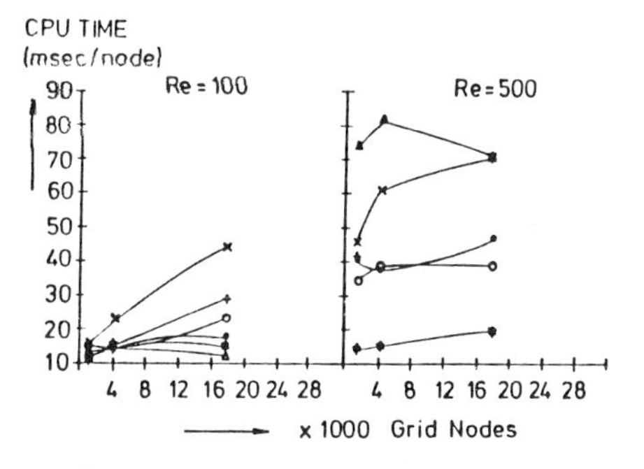

Fig. 7: CPU Times for convergence; case (b), multigrid methods

TABLE 1: CPU time (secs)/No. of iterations on finest grid; for lid-driven square cavity flow

GRID	BEST SINGLE GRID		MULTIGRID					
			Decoupled				Coupled	
	METHOD	RESULT	SIMPLE	SIMPLEC	PISO	SIMPLEST	SCGS	CLSOR
Re = 100								
22x22	SIMPLEC	6/61	4/17	5/21	4/15	5/21	4/19	4/20
42x42	SIMPLEC	65/187	23/27	18/20	18/15	23/25	13/14	15/14
82x82	SIMPLEC	885/625	98/26	64/17	84/17	85/21	38/12	55/15
162x 162	SIMPLEC	11800/2000	371/22	227/15	376/17	299/17	145/11	219/15
Re = 1000								
22x22	PISO	5/36	5/23	6/29	8/23	6/27	5/20	11/42
42x42	PISO	34/69	28/32	28/31	33/22	26/27	26/23	42/33
82x82	SIMPLEC	348/248	103/27	102/26	113/21	92/22	102/18	131/22
162x 162	SIMPLEC	4496/750	388/16	294/15	303/12	363/17	325/13	315/12

TABLE 2: CPU time (secs)/No. of iterations on finest grid; for laminar flow over a backward-facing step

GRID	BEST SINGLE GRID		MULTIGRID					
			Decoupled				Coupled	
	METHOD	RESULT	SIMPLE	SIMPLEC	PISO	SIMPLEST	SCGS	CLSOR
Re = 100, L/H = 3								
34x34	SIMPLEC	26/120	19/36	17/30	14/20	18/32	16/22	13/28
66x66	PISO	340/312	101/46	62/27	64/21	72/32	63/22	61/30
130x 130	SIMPLEC	4400/1180	754/76	315/30	390/26	496/46	214/21	260/31
Re = 500, L/H = 6								
34x34	SIMPLEC	28/123	42/93	36/69	28/43	38/80	87/55	18/38
66x66	SIMPLEC	371/404	267/13	121/58	127/43	N/A	358/51	70/27
130x 130	SIMPLEC	4490/1200	N/A	639/51	494/36	N/A	1200/41	339/34
N/A = Not available due to excessive CPU time.								

A CONTROL VOLUME BASED FULL MULTIGRID PROCEDURE FOR THE PREDICTION OF TWO-DIMENSIONAL, LAMINAR, INCOMPRESSIBLE FLOWS

M. Barcus*, M. Peric**, G. Scheuerer**

* Institut für Mathem. Maschinen und Datenverarbeitung III, Universität Erlangen-Nürnberg, Martensstr. 3, D-8520 Erlangen

** Lehrstuhl für Strömungsmechanik, Universität Erlangen-Nürnberg, Egerlandstr. 13, D-8520 Erlangen

SUMMARY

A Full Multigrid procedure has been incorporated in a finite volume solution method for the two-dimensional Navier-Stokes equations. The novelties of the approach are a colocated variable arrangement and the use of a Correction Scheme instead of the more often used Full Approximation Scheme. The SIMPLE algorithm is employed on all grid levels for pressure-velocity coupling. Results are presented for two test cases, the flow in a lid driven cavity and over a backward facing step, showing considerable savings in computer time.

INTRODUCTION

Accurate numerical solutions of practical flow problems require very fine computational grids, especially for high Reynolds numbers. On the other hand, it is well known that the efficiency of traditional iterative solution methods decreases when the grid is refined. The emerging computational demands exceed therefore very often the available computer resources. As a consequence, the interest in rapidly converging algorithms is very high. These considerations have lead to a more intensive application of multigrid techniques as one method to improve the convergence properties of iterative schemes.

The present paper describes a Full Multigrid (FMG) algorithm for the calculation of two-dimensional, incompressible, laminar flows in rectangular geometries, based on a conservative finite volume discretization. The continuity and momentum equations on each grid level are solved with the SIMPLE method of Patankar and Spalding [1]. In contrast to existing multigrid procedures (e.g. Vanka [2]), a colocated arrangement of the variables on the numerical grid is used, minimizing the computational effort for restriction and prolongation within the FMG-cycles. While most other methods use the so-called Full Approximation Scheme (FAS), see Brandt [3], the simpler Correction Scheme (CS) is employed, here. It demands only the transfer of residuals to the coarser grids, thus reducing computer storage and time relative to the FAS method.

The paper proceeds by presenting the base numerical method on which the FMG procedure is built, followed by a description of the multigrid technique. Results are presented for two test cases, the flow in a lid driven cavity and over a backward facing step.

NUMERICAL BASE METHOD

The flows considered here are described by the two-dimensional, steady Navier-Stokes equations for incompressible fluids. The multigrid scheme developed for their solution is based on a conservative finite volume method with colocated variable arrangement, discussed in detail by Peric et al. [4]. Only a short summary will be provided here.

Fig. 1 shows the colocated variable arrangement. The velocities in the x- and y-directions are denoted by U and V, the pressure by P. The mass fluxes C_x and C_y are stored at the control volume faces. The colocated arrangement has several advantages over the staggered approach: i) the finite volume coefficients representing the convective and diffusive momentum fluxes are identical for U and V, ii) all control volumes coincide with the boundaries of the integration domain, facilitating the treatment of the boundary conditions, and iii) it can be used for general non-orthogonal grids in conjunction with Cartesian velocity components. Extensions to multigrid methods require refinement and coarsening of only one control volume set.

The discretization of the flow equations follows the finite volume practice, outlined by Patankar [5]. The first order upwind scheme (UDS) is used to discretize the convective fluxes. The diffusive terms are approximated by central differences. The resulting finite volume equations can be written as:

$$\Sigma\ A_m\ U_m = -\delta y\ \delta P_x, \qquad (1)$$

$$\Sigma\ A_m\ V_m = -\delta x\ \delta P_y. \qquad (2)$$

$\delta P_x = P_e - P_w$ and $\delta P_y = P_n - P_s$ are the pressure differences across a control volume. The index m runs over the grid points P, W, E, S, N, cf. Fig. 1. A_m represent the finite volume coefficients arising from the discretization of the convection and diffusion terms.

The velocity and pressure fields are iteratively calculated with the SIMPLE algorithm of Patankar and Spalding [1]. The momentum equations are initially solved with a guessed pressure field, followed by the solution of a pressure correction equation, derived from the continuity equation. The velocities and the pressure are then updated to satisfy the continuity equation after each SIMPLE iteration.

It is usually necessary to underrelax the solution of the U-V-P system to obtain convergent solutions, see Patankar [5]. In fact, the convergence rates depend quite strongly on the choice of the underrelaxation factors. In many cases, nearly optimum convergence can be achieved by relating the underrelaxation parameters for the velocities α_u and the pressure correction α_p via $\alpha_p = 1 - \alpha_u$, with $\alpha_u \simeq 0.8$, see Peric et al. [4]. This will be discussed in more detail when presenting the results.

The systems of linear equations resulting from discretization of the momentum and pressure correction equations are solved with the strongly implicit method of Stone [6]. This algorithm showed better convergence properties than the point- or line-iterative Gauss-Seidel methods often used in connection with multigrid schemes. For the momentum equations only one inner iteration is performed per outer iteration of the SIMPLE algorithm. Up to six inner iterations are made for the pressure correction equation, unless the sum of the absolute residuals has fallen by a factor of 5.

Convergence of the method is controlled by checking the normalized sums of the absolute residuals of the momentum and continuity equations. If all three sums are below a certain criterion (here 0.001), the iterations are stopped.

FULL MULTIGRID ALGORITHM

The above method is used to set up a FMG algorithm. The FMG procedure starts by solving the flow equations with the base method on the coarsest grid. The velocity and pressure fields are then extrapolated (prolongated) to the next finer grid, and used there as starting values. The grid is refined by halving the control volumes in both coordinate directions. Thus, four fine grid control volumes result out of one coarse grid control volume, as exemplified in Fig. 2. Bilinear interpolation is used to transfer the variables to the finer grid.

After ν_1 SIMPLE iterations with the base method on the fine grid, approximate solutions U^1, V^1, P^1 result, satisfying the discretized momentum equations up to residuals R^u and R^v, respectively:

$$\Sigma\, A_m^1\, U_m^1 = -\delta y\ \delta P_x^1 - R^u, \tag{3}$$

$$\Sigma\, A_m^1\, V_m^1 = -\delta x\ \delta P_y^1 - R^v. \tag{4}$$

The notation A_m^1 indicates that the coefficients are assembled with the velocity fields U^1 and V^1. In order to obtain the exact solution of Eqs. (1) and (2), corrections must be added to the velocity and pressure fields, i.e.:

$$\Sigma\, (A_m^1 + A_m')(U_m^1 + U_m') = -\delta y\ (\delta P_x^1 + \delta P_x'), \tag{5}$$

$$\Sigma\, (A_m^1 + A_m')(V_m^1 + V_m') = -\delta x\ (\delta P_y^1 + \delta P_y'). \tag{6}$$

By subtracting Eqs. (3) and (4) from Eqs. (5) and (6), the defect (or error) equations are obtained:

$$\Sigma\, (A_m^1 + A_m')\, U_m' = -\delta y\ \delta P_x' + R^u - \underline{\Sigma\, A_m'\, U_m^1}, \tag{7}$$

$$\Sigma\, (A_m^1 + A_m')\, V_m' = -\delta y\ \delta P_y' + R^v - \underline{\Sigma\, A_m'\, V_m^1}. \tag{8}$$

These are restricted to the coarse grid; the underlined terms are neglected, after test calculations showed a negligible effect on the overall convergence rates.

The defect equations have the same structure as the discretized momentum equations (1) and (2). They represent the conservation of the errors U' and V'. Since the coarse and fine grid control volumes have common boundaries, and a conservative finite volume discretization is employed, the coarse grid equations can be assembled without any interpolation. The mass fluxes through the control volume faces, appearing in the coefficents A_m, are calculated by adding the two fine grid mass fluxes of the corresponding coarse grid control volume face, see Fig. 2. The diffusion fluxes are recalculated in the standard central difference fashion. The residuals, representing source terms in the error equations, are restricted by summing up the contributions of the four related fine grid control volumes.

The SIMPLE algorithm is also employed on the coarse grid, but now in terms of errors instead of the original variables. In the first iteration the pressure errors are estimated as $P' = 0$, and Eqs. (7) and (8) are solved for U' and V'. A pressure correction equation, identical to the one of the base method, is then solved to provide corrections to U', V', and P', in order to satisfy continuity. The whole process on the coarse grid is repeated for a specified number of iterations. After this, the errors are bilinearly interpolated to the fine grid and added to U^1, V^1 and P^1. The two-grid procedure is completed by performing ν_2 smoothing iterations on the fine grid.

The two-grid V-cycle just described is repeated until the converged solution on the fine grid is obtained; it is then prolongated to the next finer grid. The multigrid cycle is restarted, but now with two coarser grids. The procedure can be extended to an arbitrary number of grids. Fig. 3 shows this schematically for a five-grid system. Typically, only one SIMPLE iteration is made on the finest grid, before switching to the next coarser grid. There, an increasing number of iterations is performed (see the numbers shown in Fig. 3). With this strategy less work is done on the fine and more on the coarser grids. This sequence turned out to be nearly optimum in terms of computing times.

RESULTS

Two test cases were considered:

- the flow in a lid driven square cavity (LDC) at various Reynolds numbers $Re = \rho U_L H/\mu$, and
- the flow over a backward facing step (BFS) at a Reynolds number $Re = \rho U_b H/\mu = 100$.

The geometry of the solution domains and the boundary conditions are shown in Fig. 4. Uniform grids (non-square for the BFS) were employed in both cases. The residuals in the continuity and momentum equations were normalized with representative momentum and mass fluxes. For the LDC, these fluxes were computed with $0.1U_L$ and H, and for the BFS with U_b and H. In addition to the single grid (SGM) calculations starting with zero initial estimates for all variables, a prolongation scheme (PGM) was also employed. Here, extrapolated coarse grid solutions were used as starting values for the iterations on the fine grid. Since PGM is a component of the FMG procedure, the results allow to distinguish the accelerations due to the multigrid method and to the PGM.

Prior to a more detailed analysis of the test cases, the influence of the underrelaxation parameters on the convergence rates is discussed. Fig. 5 shows the number of iterations necessary for convergence on the finest grid as a function of the underrelaxation factor α_u for the velocities. Three grid levels were used in the FMG procedure. α_p was fixed at 0.2, except where a different value is indicated in the diagram. The convergence rates of both single grid methods are a strong function of α_u. The optimum value of α_u is around 0.9, and the number of iterations almost doubles, when it is reduced to 0.8 or raised above 0.95. In contrast, the convergence rate of the FMG method depends very little on α_u.

Table 1 shows the computing times and the number of iterations spent on each grid for the calculations of the cavity flow. The Reynolds number was varied between 100 and 5000. Underrelaxation factors of α_u = 0.8 and α_p = 0.4 were employed, unless otherwise indicated. Due to the excessive computing costs, solutions on the finest grids were only obtained with the FMG scheme. The corresponding SGM and PGM computing times are extrapolated from the values of the coarser grids.

The largest savings in computer time relative to the base method are noted for Re = 100. PGM needs about three times less computing time on the finest grid at Re =100, and 1.3 times less at Re = 5000. The FMG procedure is much more efficient: computing times are reduced by factors of 220 and 20 at Re = 100 and 5000, respectively. One reason for the smaller savings at higher Reynolds numbers is that the base method converges increasingly faster.

Table 1 Results for the lid driven cavity flow

	Computing Time					Number of Iterations				
	10x10	20x20	40x40	80x80	160x160	10x10	20x20	40x40	80x80	160x160
	Re = 100, α_u = 0.8, α_p = 0.4									
SGM	0.22	1.42	24.23	449.09	8000.*	20	57	206	796	-
PGM	0.22	0.79	14.02	197.61	2600.*	20	33	119	386	-
FMG	0.22	0.68	2.50	9.52	35.34	20	17	15	15	15
	Re = 1000, α_u = 0.8, α_p = 0.4									
SGM	0.24	1.48	14.02	219.64	3000.*	25	59	123	389	-
PGM	0.24	1.17	11.81	156.81	1800.*	25	49	94	299	>855
FMG	0.24	0.75	2.90	13.56	53.26	25	19	17	17	17
	Re = 5000, α_u = 0.8, α_p = 0.4									
SGM	0.38	1.47	14.29	166.03	2000.*	46	60	121	294	-
PGM	0.38	1.10	12.15	137.44	1310.*	46	47	100	262	>624
FMG	0.28	0.98	3.85	23.28	114.80	32	25	23	27	31

As an additional test case, the flow over a backward facing step is considered. Table 2 shows the results. The FMG procedure is also for this case the most efficient scheme. It requires 80 times less computing time than the SGM on the finest grid. The prolongation scheme saves a factor of four relative to the base procedure.

Table 2 Results for the backward facing step

	Computing Time					Number of Iterations				
	20x10	40x20	80x40	160x80	320x160	20x10	40x20	80x40	160x80	320x160
	Re = 100, α_u = 0.8, α_p = 0.3									
SGM	0.62	4.41	84.18	1190.9	18000.*	32	82	248	841	-
PGM	0.62	2.84	34.68	381.30	4000.*	32	61	153	414	-
FMG	0.62	0.68	2.33	10.38	53.38	32	23	23	27	29

CONCLUSIONS

A Full Multigrid procedure based on a Correction Scheme (CS) was developed to solve the 2D-Navier-Stokes equations. The SIMPLE algorithm of Patankar and Spalding [1] was used for the pressure-velocity coupling on all grid levels. The results presented above comply with the multigrid theory developed for linear equations. The computational effort increases almost linearly with the total number of control volumes, and the number of fine grid iterations remains about constant, irrespective of grid refinement. These features were verified over a wide range of Reynolds numbers for the lid driven cavity flow. In contrast to the base method, the convergence rate of the FMG procedure is only weakly dependent on the choice of the underrelaxation factors. The same values could be used for all grids and for all Reynolds numbers.

The present method has several advantages when compared to other proposals (e.g. [2]). It requires less storage and operations, because no interpolation is involved in the restriction to coarser grids, and the errors transferred back to the fine grids are readily available. Another advantage of the present method is the colocated variable arrangement. It uses only one control volume set on each grid level. Therefore, the same restriction and prolongation operators apply to all variables, and special near-boundary interpolation techniques, as in the staggered approach, are unnecessary. Finally the CS scheme presented here, can be easily extended to treat larger systems of coupled equations.

ACKNOWLEDGEMENTS

The present work was sponsored by the Deutsche Forschungsgemeinschaft within the program "Finite Approximationen in der Strömungsmechanik". This support is gratefully acknowledged. The authors are also grateful to Prof. J. H. Ferziger from Stanford University for many helpful discussions.

REFERENCES

[1] PATANKAR, S. V., SPALDING, D. B.: "A Calculation Procedure for Heat, Mass and Momentum Transfer in Three-Dimensional Parabolic Flows", Int. J. Heat Mass Transfer, 15 (1972) pp. 1787-1806.

[2] VANKA, S. P.: "Block-Implicit Multigrid Solution of Navier-Stokes Equations in Primitive Variables", J. Comput. Phys., 65 (1986) pp. 138-158.

[3] BRANDT, A.: "Multigrid Techniques: 1984 Guide with Applications to Fluid Dynamics", GMD Bonn, GMD-Studien Nr. 85 (1984).

[4] PERIC, M., KESSLER, R., SCHEUERER, G.: "Comparison of Finite-Volume Numerical Methods with Staggered and Non-Staggered Grids", University of Erlangen-Nbg., Lehrst. f. Strömungsmechanik, Rept.-Nr. LSTM 163/T/87 (1987), (acc. for publ. in Computers and Fluids).

[5] PATANKAR, S. V.: "Numerical Heat Transfer and Fluid Flow", McGraw Hill, New York, 1980.

[6] STONE, H. L.: "Iterative Solution of Implicit Approximations of Multi-Dimensional Partial Differential Equations", SIAM J. Num. Anal., 5 (1968) pp. 530-558.

FIGURES

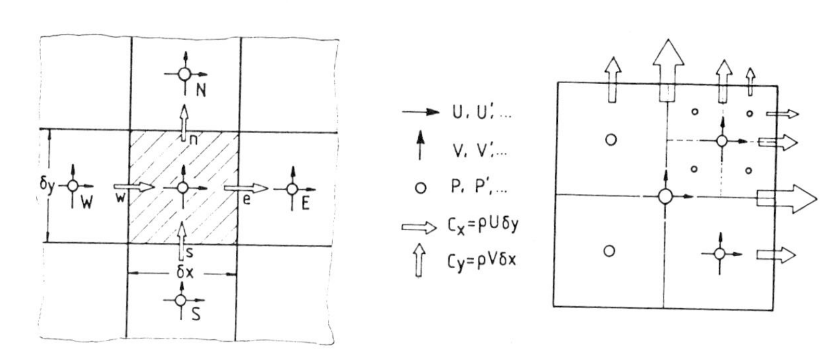

Fig. 1 Colocated variable arrangement

Fig.2 Location of coarse and fine grids

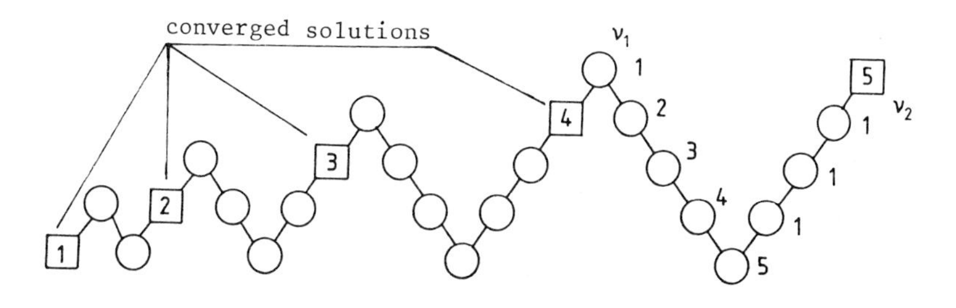

Fig. 3 Schematic Full Multigrid sequence

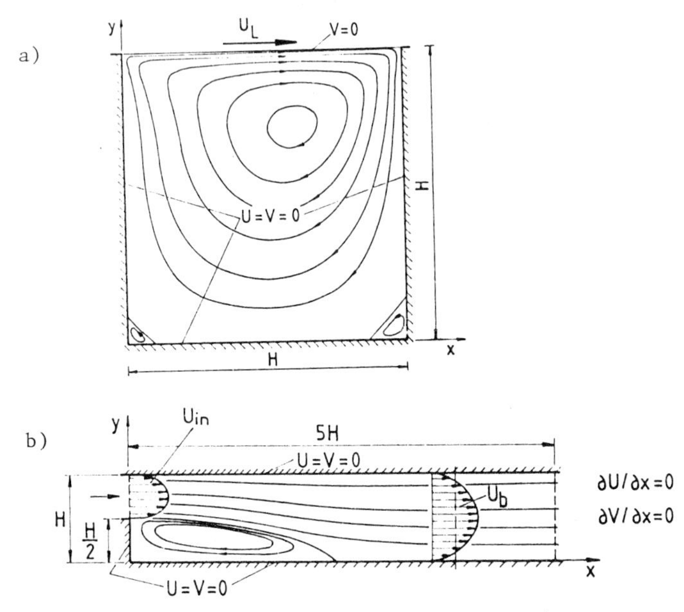

Fig. 4 Geometry of test cases: a) lid driven cavity, b) backward facing step flow

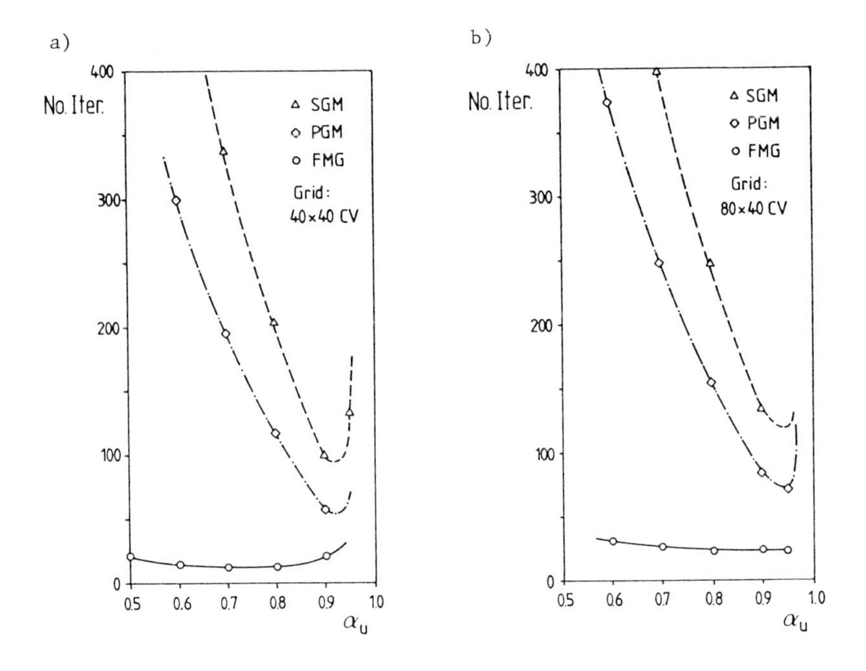

Fig. 5 Influence of underrelaxation parameter α_u onconvergence rate: a) lid driven cavity, b) backward facing step flow

A LOCAL MULTIGRID STRATEGY FOR VISCOUS TRANSONIC FLOWS AROUND AIRFOILS

F. Bassi*, F. Grasso**, M. Savini***

* Politecnico di Milano, Dip. di Energetica, Piazza L. da Vinci 32, 20133 Milano, Italia

** Universita' di Roma "La Sapienza", Dip. di Mecc. e Aeronautica, via Eudossiana 18, 00100 Roma, Italia

*** CNR/CNPM, Viale Baracca 69, 20068 Peschiera Borromeo (Mi), Italia

SUMMARY

A grid embedding technique for viscous flow computations has been developed. The method exploits a local multigrid strategy and the mesh embedding procedure is on a cell by cell basis. Second order interpolation is used to obtain interface boundary conditions. Computations of viscous flows around NACA0012 and RAE2822 airfoils are presented. The method yields accurate results with the use of a rather coarse grid, and the implemented interface treatment does not produce spurious oscillations.

INTRODUCTION

The present paper deals with the development of an accurate and robust explicit finite volume algorithm that uses adaptive embedded meshes for the solution of viscous transonic flows about airfoils.

In a recent paper that summarizes the results of a Viscous Transonic Airfoil Workshop [1], it has been assessed that the most important numerical errors, that arise with Navier Stokes solvers, are due to inappropriate grid clustering and refinement. Such errors can be drastically reduced by use of adaptive techniques such as dynamic clustering, zonal, and embedding.

In the dynamic clustering technique mesh points are displaced toward regions of high gradients. This technique [2] has the advantage that the number of grid points is kept fixed and the domain is logically rectangular. However its use can produce excessive skewing and stretching of the grid at the expense of the accuracy of the results. Moreover it cannot easily handle complex geometries.

In the zonal approach [3] the computational domain is partitioned into subregions, and different grids are generated for each zone. Different sets of equations can be used in the different subregions. The technique is well suited to handle complex three dimensional geometries. However zonal interfaces must be properly treated to ensure a conservative transport across them.

In the adaptive embedded grid approach the accuracy is

enhanced in regions that need greater resolution by enriching the mesh locally, and the new grid points are organized either in a single fine grid or in fine grid patches, embedded in the coarse grid. Different grids are kept independently, each one with its own solution vector. As for the zonal approach, grid interfaces must be treated carefully to properly transfer fluxes computed on the different grids [4-7].

In the present work an embedded grid approach that exploits a local multigrid strategy is developed for the solution of viscous compressible flows about airfoils [7]. Embedded (fine) grid points are defined on a cell by cell basis in regions where a criterion based on a cumulative distribution function of pressure gradient is not satisfied. Correct coupling of the grids at interfaces is ensured by high order interpolation to obtain fine grid points boundary conditions, and by transfering the fine grid fluxes to the coarse grid, so as to satisfy conservation.

The method has been applied to compute two test cases of the Viscous Transonic Airfoil Workshop [1], and a laminar supersonic test case of the GAMM Workshop on the Numerical Simulation of Compressible Navier Stokes Flows [8]. The computed results obtained on a rather coarse grid (100 x 28) are in very good agreement with the experiments and with other numerical results obtained on much finer grids.

In the next sections the numerical solution is described, the results presented and concluding remarks given.

NUMERICAL SOLUTION

A fully conservative formulation of the Navier Stokes equations is employed. The equations are:

$$\frac{\partial}{\partial t}\int_V W\,dV + \oint_{\partial V}\left(f n_x + g n_y\right)dS = 0 \tag{1}$$

where

$$W = [\rho, \rho u, \rho v, \rho E]^T$$

$$f = \left[\rho u, \rho u u + p - \sigma_{xx}, \rho u v - \sigma_{xy}, \rho u E + u p - u\sigma_{xx} - v\sigma_{xy} + q_x\right]^T$$

$$g = \left[\rho v, \rho u v - \sigma_{xy}, \rho v v + p - \sigma_{yy}, \rho v E + v p - u\sigma_{xy} - v\sigma_{yy} + q_y\right]^T$$

$$p = (\gamma - 1)\rho\left(E - \frac{1}{2}\left(u^2 + v^2\right)\right)$$

$$\overleftrightarrow{\sigma} = \mu\left(\left(\nabla\vec{u} + \nabla\vec{u}^T\right) - \frac{2}{3}\nabla\cdot\vec{u}\overleftrightarrow{U}\right)$$

$$\vec{q} = -\frac{\gamma}{Pr}\mu\nabla e\,.$$

The Baldwin-Lomax turbulence model is employed for the turbulent test case computations, while constant viscosity coefficients are used in the laminar case.

The computational solution is obtained in two or more phases, depending on the levels of grid embedding.

Phase I

First the solution on the standard (coarse) grid is obtained by using an explicit finite volume method. The system of governing equations is reduced to a system of ordinary differential equations and time integration is performed by a three stage Runge Kutta algorithm. A cell centered formulation is employed and the solution at each stage is obtained as follows

$$W_C^{(0)} = W_C^n$$

$$W_C^{(k)} = W_C^{(0)} + \alpha_k \frac{\Delta t_C}{V_C}\left[-C\left(W_C^{(k-1)}\right) + D\left(W_C^{(0)}\right) + AD\left(W_C^{(0)}\right)\right] \quad (2)$$

$$W_C^{n+1} = W_C^{(2)}$$

where C, D, AD are respectively the net inviscid, diffusion and adaptive dissipation contributions and they are evaluated as in Refs. [7-9] and $\alpha_1 = \alpha_2 = .6; \alpha_3 = 1$.

Phase II

In this phase the solution is advanced on the mesh with embedded grids. Within Phase I a quasi converged solution is obtained, then an adaptation criterion is checked and embedded grids are constructed.

The procedure followed in the present work generates embedded grids by refining automatically coarse cells, and a cumulative distribution function, based on the modulus of pressure gradient weighted by cell volume, is used for the adaptation criterion. The cells where the adaptation criterion is not satisfied are flagged, and four embedded cells are created for each flagged coarse cell by halving the mesh spacing in both directions. The new grid points are organized so as to define a fine grid that is overlaid on the coarse one. The grids are kept independently in a tree-like structure and interact through interface boundaries. Different solution vectors are defined on different grids and time stepping is performed separately on each grid.

It is worth mentioning that an important issue in embedded grid approaches is that of data structure [4,7]. For the sake of brevity the problem is not addressed here. However it must be observed that with the proposed technique, in the presence of embedded grids, the overhead per cell is approximately 15%.

For computational purposes fictitious embedded cells are generated along the boundary of the adaptation regions. The advantage of using fictitious cells is two-fold: 1) it allows to properly evaluate interface fluxes; 2) it is computationally efficient since the same finite difference formulas can be used on every embedded cell.

The computational strategy exploits a multigrid strategy [7,10], and proceeds as follows

Initialization

At the beginning of Phase II the solution on the embedded grid is initialized by interpolating the coarse grid solution by using Legendre polynomials [11].

Embedded grid solution

$$W_E^{(0)} = W_E^n$$

$$W_E^{(k)} = W_E^{(0)} + \alpha_k \frac{\Delta t_E}{V_E}\left[-C\left(W_E^{(k-1)}\right) + D\left(W_E^{(0)}\right) + AD\left(W_E^{(0)}\right)\right] \tag{3}$$

$$W_E^{n+1} = W_E^{(3)}$$

Injection

In the regions of adaptation the fine grid solution is injected on to the coarse grid, and coarse grid solution is updated as follows

$$W_C = \overset{C}{\underset{E}{I}} W_E = \sum V_E W_E / V_C$$

Coarse grid solution

In this phase the coarse grid solution is obtained as in Phase I. However in the regions of adaptation the residuals, collected from the embedded cells, are transfered on to the coarse ones via a forcing function that is added to Eqn. (1) as is done in multigrid. The solution is evaluated as follows

$$W_C^{(0)} = W_C^n$$

$$W_C^{(k)} = W_C^{(0)} + \alpha_k \frac{\Delta t_C}{V_C}\left[-C\left(W_C^{(k-1)}\right) + D\left(W_C^{(0)}\right) + AD\left(W_C^{(0)}\right) + \sum_E R_E^{n+1} - R_C^n\right] \tag{4}$$

$$W_C^{n+1} = W_C^{(3)}$$

Correction stage

In the regions of adaptation the embedded grid solution is updated by distributing the coarse grid residuals as follows

$$W_E = W_E + \overset{E}{\underset{C}{I}}\left(W_C^{n+1} - W_C^n\right)$$

Interface treatment

Embedded grids introduce boundaries which are internal to the computational domain and which require a correct choice of boundary conditions to ensure conservation, accuracy and stability. Two approaches are generally implemented, when a cell centered formulation is used.

Allmaras and Baron [12] concentrate on finding an accurate and stable definition of interface fluxes by interpolating coarse and fine grid fluxes, and they reach the conclusion that stability can be obtained at the expense of accuracy. Berger [4] solves the problem by first defining fine grid boundary conditions by using high order interpolation, and then enforcing conservation by transfering the fine grid fluxes on to the coarse cell.

In the present work an approach similar to that of Refs. [4,7] is employed. It is found that an accurate definition of the fine grid boundary values is obtained by use of Legendre polynomial interpolation of the coarse grid solution. Across the interfaces, the coarse grid flux is evaluated as the sum of the fine grid fluxes.

If finer levels of grid embedding are introduced, the solution proceeds exactly in the same way as described above.

RESULTS AND DISCUSSION

The method here developed has been applied to compute viscous transonic flows around NACA0012 and RAE2822 airfoils. The selection of these airfoils relies on the availability of experimental and numerical results presented at the GAMM Workshop on Numerical Simulation of Compressible Navier Stokes Flows, held in Nice in 1985, and at the Viscous Transonic Airfoil Workshop held in Reno in 1987.

The objective of the computations is two-fold: 1) for code validation; 2) to show that having to rely on limited computer resources, it is still possible to obtain an accurate solution when one uses the proposed mesh embedding technique.

The computations have been performed on a Gould-3267 computer and use a rather coarse grid (100 x 28 for the turbulent test cases, and 80 x 24 for the laminar case); for the RAE2822 airfoil a computation has also been performed on a (160 x 48) standard grid to study the effect of global mesh refinement versus embedding.

The grid is generated by solving the incompressible inviscid equations in natural coordinates around the airfoil by multigrid technique, and orthogonal cells are obtained. Clustering of grid lines, at leading and trailing edges, and at guessed shock location, is achieved by using control points properly distributed near the airfoil surface. For the turbulent test cases, Y^+ , the minimum (nondimensional) cell distance from the wall, varies between 1 and 15. The farfield boundary is set at 12 chords away from the profile for the turbulent cases, and 6 chords for the laminar case.

RAE2822 : turbulent; $Mach = .75$; $\alpha = 2.70$; $Re = 6.2 \times 10^6$

The computed results (see Fig. (1)) show the same flow features, i.e. Cp, Cf, separation extent, iso-p, iso-M, etc., as found by other authors, and the same discrepancies in comparison with experiments, due to limitations of the physical submodels, in particular on account of turbulence [1]. However it is important to stress that the present results are obtained on a (100 x 28) coarse grid with only 338 coarse cells being refined, while most of the computed results discussed in Ref. [1] are obtained on grids having four times as many cells. The computed (experimental) values of the aerodynamic coefficients are: CL = .762 (.743); CD = .247 (.242). Note that the results obtained on the more refined grid (160 x 48) are practically coincident with those obtained on the (100 x 28) grid with embedding, and are not reported.

NACA0012 : turbulent; $Mach = .55$; $\alpha = 8.34$; $Re = 9.0 \times 10^6$

Fig. (2) show a comparison of the computed and measured Cp vs x/c. In this case there is a small supersonic region near the leading edge that is adequately resolved, and the values of CL and CD are: CL = .981 (.983); CD = .376 (.253). A small separation bubble appears at the foot of the shock, and an incipient separation is noticed at the trailing edge, whilst both phenomena are not observed without embedding.

NACA0012 : laminar; $Mach = 2.0$; $\alpha = 10.$; $Re = 106$

This test case is very useful for evaluating the proposed method: 1) being the flow laminar, uncertainties due to turbulence modeling are removed; 2) a bow shock stands off the leading edge.

The results of the computations are shown in Fig. (3), and indicate that the method does not give spurious oscillations, even if the boundaries of the embedded cells are close to high gradient regions. The computed values of CL and CD are: CL = .327; CD = .483.

CONCLUSIONS

A grid embedding technique for viscous flow computations has been developed. The method exploits a local multigrid strategy in order to enhance the accuracy of the coarse grid solution. The mesh embedding procedure is on a cell by cell basis so as to refine only those cells where a higher resolution is required. The interface treatment, based on second order interpolation to obtain fine grid boundary conditions, does not give rise to spurious oscillations.

ACKNOWLEDGEMENTS

This research was supported by CNR-PFE2 86.0075.859 and MPI 40%.

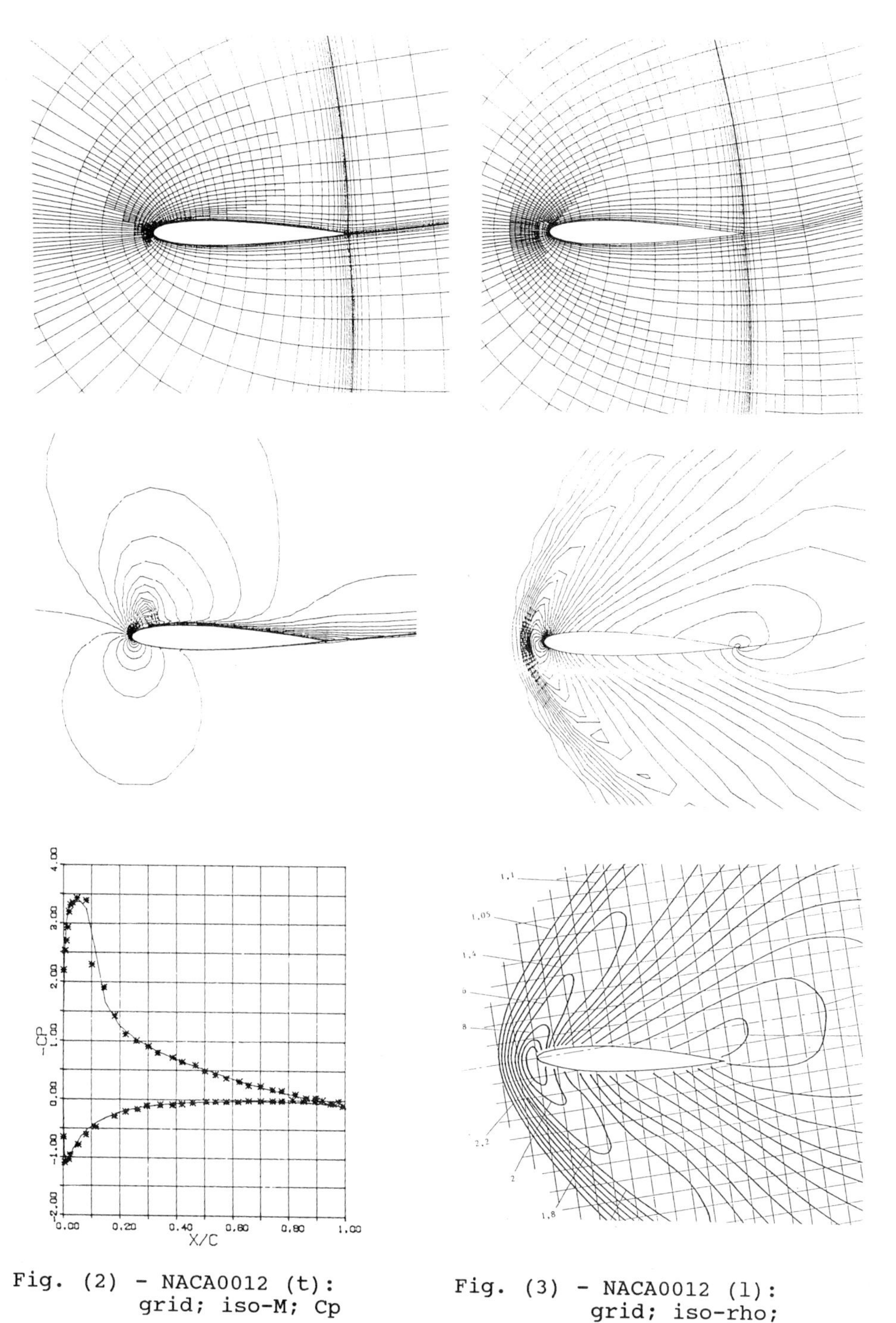

Fig. (2) - NACA0012 (t):
grid; iso-M; Cp

Fig. (3) - NACA0012 (l):
grid; iso-rho;
iso-rho(exp)

REFERENCES

[1] Holst, T.L., AIAA paper no. 87-1460, 1987.
[2] Dwyer, H.A., AIAA J., 22, 12, 1984, pp 1705-1712.
[3] Rai, M.M., AIAA paper no. 85-0488, 1985.
[4] Berger, M.J., Von Karman Institute, Lecture Series 1987-04, 1987.
[5] Berger, M.J., Jameson, A., AIAA J., 23, 4, 1985, pp 561-568.
[6] Dannenhoffer, J.F., Baron, J.R., AIAA paper no. 86-0495, 1986.
[7] Bassi, F., Grasso, F., Savini, M., Springer Verlag, Lecture Notes in Physics, 264, 1986.
[8] Grasso, F., Jameson, A., Martinelli, L., GAMM Workshop on Numerical Solution of the Compressible Navier Stokes Equations, Nice, France 1985.
[9] Martinelli, L., Jameson, A., Grasso, F., AIAA paper no 86-0208, 1986.
[10] Angrand, F., Leyland, P., private communication, 1987.
[11] van Leer, B., J. of Comp. Phys., 23, 1977, pp 276-299.
[12] Allmaras, S.R., Baron, J.R., AIAA paper no. 86-0509, 1986.

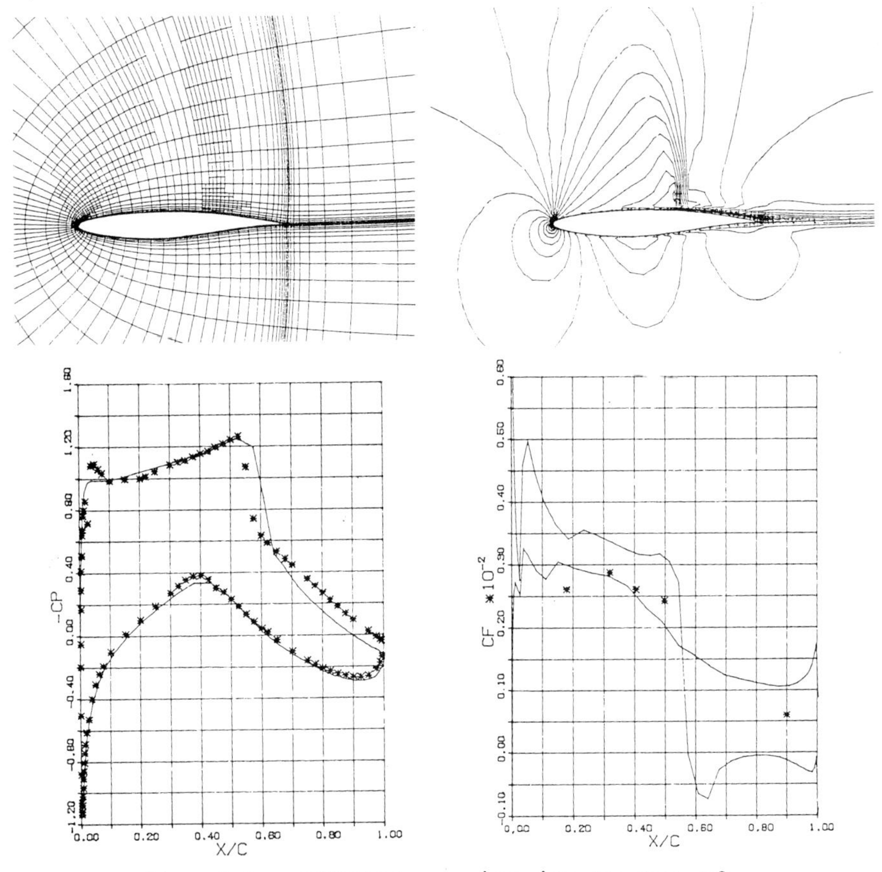

Fig. (1) - RAE2822 : grid; iso-M; Cp; Cf

THE OPTIMAL TIME STEP FOR THE IMPLICIT APPROXIMATE FACTORIZATION SCHEME - THEORY AND APPLICATIONS

M. BORSBOOM, A.K. STUBOS, P.-H. THEUNISSEN
Von Karman Institute for Fluid Dynamics
Ch. de Waterloo 72, 1640 Rhode-St-Genese, Belgium

An analytical prediction is derived for the value of the optimal time step for the implicit approximate factorization scheme, when used as an iterative solver for centrally discretized, steady-state convection-diffusion equations. It has been verified that for practical applications this prediction optimizes indeed very well the convergence properties of this relaxation scheme, both for convection-dominated and for diffusion-dominated flow problems.

INTRODUCTION

A series of physical and numerical arguments leads to the conclusion that a solution method for the discretized steady-state compressible Navier-Stokes equations that combines robustness, general applicability and good convergence properties is to be sought in the class of first-order implicit relaxation schemes [1]. Explicit methods, on the contrary, must be expected to be intrinsically slow, due to their rather severe stability limits [1, 2].

The most convenient way of constructing such an implicit relaxation scheme for this nonlinear multi-dimensional problem is by means of a backward Euler discretization in time of the unsteady equations. The steady-state perturbations introduced in the system by the initial guess can be relaxed through a finite "time step" Δt, while the implicit part of the scheme is normally linearized in order to be solvable. This would allow the use of an infinitely large time step, leading to quadratic convergence (Newton's method). However, such a method will be inefficient due to the sparseness of the matrix to be inverted [3,p167].

The amount of computational effort per iteration step can be reduced significantly by approximately factorizing the implicit part of the scheme [4]. Unfortunately, the latter simplification has to be paid for by a strong reduction in the convergence speed of the relaxation scheme, due to the factorization error that is proportional to Δt^2. The choice of the relaxation parameter Δt becomes therefore very critical. The purpose of the present paper is to develop a theoretical basis for the estimation of the optimal time step for this implicit approximate factorization scheme (AF).

THEORY

The following linear model equation is considered:

$$\frac{\partial\Phi}{\partial t} + u\frac{\partial\Phi}{\partial x} + v\frac{\partial\Phi}{\partial y} = \nu_x\frac{\partial\Phi^2}{\partial x^2} + \nu_y\frac{\partial\Phi^2}{\partial y^2} + f \ , \quad \Phi = \Phi(t,x,y) \ , \quad f = f(x,y) \ , \qquad (1)$$

with constant convective speeds u, v and constant diffusion coefficients ν_x and ν_y. The steady-state solution $\Phi^*(x,y)$ of this unsteady, convection-diffusion equation with appropriate boundary conditions is sought. To this end, the equation is discretized implicitly in time, while the two space dimensions in the implicit part are uncoupled through the AF:

$$[1 + \Delta t(u\frac{\partial}{\partial x} - \nu_x \frac{\partial}{\partial x^2})]\,[1 + \Delta t(v\frac{\partial}{\partial y} - \nu_x \frac{\partial}{\partial y^2})]\,\Delta\Phi^t = \Delta t RHS(\Phi^{t-1})\,, \qquad (2)$$

$$t = 1, 2, 3, \ldots .$$

The unknown $\Delta\Phi^t$ of this equation denotes the *correction* or *increment* $\Phi^t - \Phi^{t-1}$. The iterative process is initialized by means of some suitably chosen Φ^0. The symbol RHS at the right-hand side of (2) stands for the differential operator:

$$RHS(\Phi) = -u\frac{\partial\Phi}{\partial x} - v\frac{\partial\Phi}{\partial y} + \nu_x\frac{\partial^2\Phi}{\partial x^2} + \nu_y\frac{\partial^2\Phi}{\partial y^2} + f\,.$$

The factorization error of (2) is $O(\Delta t^2)$, indicating that for large time steps, the factorization error becomes very large. This causes the convergence speed of the scheme to tend to zero, when Δt tends to infinity [5]. For very small Δt, on the contrary, the factorization error will be very small, but the speed of convergence will be small also. It is clear that there must be an optimal time step, for which value the positive effect of increasing Δt for still better convergence properties, is balanced by the negative effect of an increasing factorization error. The subject of this paper is to determine that value, by means of a Fourier mode analysis.

For that purpose, the recursive expression needs to be derived that is satisfied at every iteration step by the discretized *steady-state error* $\Delta\underline{\Phi}^{*t}$. Here, $\Delta\underline{\Phi}^{*t}$ denotes the difference between the approximation $\underline{\Phi}^t$ of the vector of nodal-point values $\underline{\Phi}^*$, and $\underline{\Phi}^*$ itself, where the vector $\underline{\Phi}^*$ is the solution of the system of algebraic equations $RHS^h(\underline{\Phi}) = \underline{0}$, obtained upon the discretization in space of the differential equation $RHS(\Phi) = 0$. Using classical centered finite differences on a uniform cartesian grid of size Δx times Δy, the expression that relates $\Delta\underline{\Phi}^{*t}$ to $\Delta\underline{\Phi}^{*t-1}$ becomes:

$$[\frac{1}{\Delta t} + u\frac{(\mu\delta)_x}{2\Delta x} - \nu_x\frac{\delta_x^2}{\Delta x^2}]\Delta\tilde{\Phi}^t_{ij} = RHS^h(\Phi^{t-1}_{ij}) - RHS^h(\Phi^*_{ij}) = \qquad (3a)$$

$$= [-u\frac{(\mu\delta)_x}{2\Delta x} - v\frac{(\mu\delta)_y}{2\Delta y} + \nu_x\frac{\delta_x^2}{\Delta x^2} + \nu_y\frac{\delta_y^2}{\Delta y^2}]\Delta\Phi^{*t-1}_{ij}\,,$$

$$[\frac{1}{\Delta t} + v\frac{(\mu\delta)_y}{2\Delta y} - \nu_y\frac{\delta_y^2}{\Delta y^2}](\Delta\Phi^{*t}_{ij} - \Delta\Phi^{*t-1}_{ij}) = \Delta\tilde{\Phi}^t_{ij}\,, \qquad (3b)$$

where δ_x and μ_x denote the central difference operators $\delta_x\Phi_{ij} = \Phi_{i+1/2\,j} - \Phi_{i-1/2\,j}$ and $\mu_x\Phi_{ij} = \Phi_{i+1/2\,j} + \Phi_{i-1/2\,j}$, and where δ_y and μ_y denote similar operators in the y-direction, with i and j the indices of the nodal points.

Expression (3a) represents a system of equations, where the unknowns are uncoupled in j-direction. In this way, the intermediate correction $\Delta\tilde{\Phi}^t_{ij}$ can indeed be calculated very efficiently, by solving one tridiagonal system of equations per j-line. The calculation of the correction $\Delta\Phi^t_{ij}$ to the approximation Φ^{t-1}_{ij} becomes equally efficient, by solving tridiagonal systems of equations that are associated with the i-lines.

For convergence of scheme (3), the sequence of successive corrections $\{\Delta\underline{\Phi}^{*t}\}$ should satisfy the relation:

$$|\Delta\underline{\Phi}^{*t}| \leqslant Ke^{-\epsilon t}\ |\Delta\underline{\Phi}^{*0}|\ , \quad t = 1, 2, 3, \ldots\ , \tag{4}$$

where ϵ is a positive constant, and K a polynomial in t of finite degree with positive coefficients.

The maximum possible value of ϵ in (4), ϵ_{max}, indicates the asymptotic *speed of convergence*. In order to verify whether inequality (4) holds for relaxation scheme (3), we will derive the expression of its amplification matrix G^h, defined as $\Delta\underline{\Phi}^{*t}/\Delta\underline{\Phi}^{*t-1}$, by means of a Fourier mode analysis.

Such an analysis does not allow to include many important features, such as the effect of boundary conditions and irregular grids, in the study of the convergence properties of the scheme, but it is the only practical way that exists to optimize (although to a rather limited extend) the relaxation parameter Δt. We will suppose therefore that the steady-state error of (1) and of (2) is periodic in space, over a length L_x in x-direction, and over a length L_y in y-direction. Normally, L_x and L_y are multiples of the size of the domain that we consider, and depend on the type of boundary conditions, as we will see later with the applications. The prolongation $\Delta\Phi^{*ht}$ of $\Delta\underline{\Phi}^{*t}$ can then be written as the infinite sum of trigonometric functions:

$$\Delta\Phi^{*ht}(x,y) = \sum_{p,q>0} \Delta\Phi^{*ht}_{pq} \ ,$$

where the $\Delta\Phi^{*ht}_{pq}$ denote the Fourier modes:

$$\Delta\Phi^{*ht}_{pq} = [A^t\cos(\omega_p\frac{x}{\Delta x}) + B^t\sin(\omega_p\frac{x}{\Delta x})]\ [C^t\cos(\omega_q\frac{y}{\Delta y}) + D^t\sin(\omega_q\frac{y}{\Delta y})]\ , \tag{5}$$

with the frequencies $\omega_p = p\frac{2\pi\Delta x}{L_x}$ and $\omega_q = q\frac{2\pi\Delta y}{L_y}$, $p, q = 1, 2, \ldots$.

For the convergence analysis it is sufficient to consider only the frequencies lower than or equal to π, since this is the highest frequency that can be represented unambiguously on the grid.

The amplification factor per Fourier mode is found immediately from the relation (3) between $\Delta\underline{\Phi}^{*t}$ and $\Delta\underline{\Phi}^{*t-1}$, by using expression (5) to replace the value of the unknowns at the nodal points:

$$G^h_{pq} = \frac{[(1+\gamma_p\gamma_q)^2 + \beta_p{}^2\gamma_q{}^2 + \beta_q{}^2\gamma_p{}^2 + \beta_p{}^2\beta_q{}^2]^{1/2}}{[(1+\gamma_p+\gamma_q+\gamma_p\gamma_q)^2 + \beta_p{}^2(1+\gamma_q)^2 + \beta_q{}^2(1+\gamma_p)^2 + \beta_p{}^2\beta_q{}^2]^{1/2}}\ , \tag{6}$$

with: $\beta_p = u\frac{\Delta t}{\Delta x}\sin\omega_p$, $\qquad \beta_q = v\frac{\Delta t}{\Delta y}\sin\omega_q$,

$$\gamma_p = \nu_x\frac{\Delta t}{\Delta x^2}(2-2\cos\omega_p)\ , \quad \gamma_q = \nu_y\frac{\Delta t}{\Delta y^2}(2-2\cos\omega_q)\ .$$

Unconditional or A-stability of the scheme can easily be shown; for every $\Delta t \geqslant 0$, and for all Fourier modes, we have $G^h_{pq} \leqslant 1$.

For optimal convergence, Δt should minimize the maximum of G^h_{pq} over p and q, since it is this value that determines the asymptotic speed of convergence of the scheme. Because of the complexity of expression (6), it is impossible to determine this value analytically, as a function of u, v, ν_x, ν_y, Δx, Δy, L_x and L_y. Instead, we have used the computer in order to gain some basic insight in the maximum of the function G^h_{pq}. This has led to some indications about the simplifications that can be made, after which an analytical approach became feasible.

First of all we have restricted ourselves to the case that the convection and the diffusion is the same in both spatial directions, by introducing $u = v$, and $\nu_x = \nu_y$. We will present the maximum of G^h_{pq} as a function of the CFL-number CFL and the cell Reynolds number Re_c, that we have defined as:

$$CFL = \frac{u\Delta t}{\sqrt{(\Delta x \Delta y)}}, \quad Re_c = \frac{u\sqrt{(\Delta x \Delta y)}}{\nu_x}.$$

Actually, we will give the value of the natural logarithm of the inverse of $\max(G^h_{pq})$, since this corresponds to the asymptotic speed of convergence ϵ_{max}. It is very convenient to consider ϵ_{max} as a function of non-dimensionalized variables only. Hence:

$$\epsilon_{max} = \epsilon_{max}(CFL, Re_c, \frac{\Delta x}{\Delta y}, \frac{L_x}{\Delta x}, \frac{L_y}{\Delta y}), \tag{7}$$

where $\Delta x/\Delta y$ denotes the grid-cell ratio, and where $L_x/\Delta x$ and $L_y/\Delta y$ are the dimensions of the periodicity domain, expressed in number of grid cells. The first variable in (7) should be optimized in terms of the last four variables, in the sense that ϵ_{max} should be as large as possible.

With $\Delta x/\Delta y$, $L_x/\Delta x$ and $L_y/\Delta y$ equal to 4, 100 and 25 respectively, ϵ_{max} has been determined as a function of CFL and Re_c, by calculating for every set of parameters the maximum of G^h_{pq} with respect to ω_p and ω_q. The result is presented in Figure 1, together with the two frequencies for which the maximum of G^h_{pq} is reached. Similar results are obtained for other values of $\Delta x/\Delta y$, $L_x/\Delta x$ and $L_y/\Delta y$ [1].

All results show the same trend, in the sense that for any set of parameters, and for any Re_c, the optimal CFL-number is *finite*, and that when $CFL \to \infty$ the speed of convergence tends to zero. This asymptotic behavior can also be deduced directly from (6) and is in agreement with the discussion in the introduction. Another observation to be made is that the optimum of the CFL-number which maximizes ϵ_{max} for a certain value of Re_c, is rather pronounced. This phenomenon can be explained as follows.

For increasing CFL, starting from zero, each mode is damped out better and better, until the value is reached for which that particular mode is damped out optimally. After this point, the damping starts deteriorating again, due to the increasing factorization error. Of course, the optimal damping of different modes occurs in general at different CFL. Hence for each CFL-number, error modes are damped out with different speeds.

For a specific set of parameter values, the asymptotic speed of convergence is determined by the worst damped mode, or *critical mode*, that maximizes G^h_{pq}. Furthermore, for different parameter values, the critical mode will in general be different. So leaving all the other parameters constant, there must be at least one value of CFL, that we will indicate by CFL^*, for which the critical mode for $CFL < CFL^*$ is equally well damped as the critical mode for the range $CFL > CFL^*$. In other words, for $CFL = CFL^*$, *two* modes are critical.

Our observations have indicated that for each set of values of Re_c, $\Delta x/\Delta y$, $L_x/\Delta x$ and $L_y/\Delta y$, there is only one such CFL^*, for which value ϵ_{max} always has its maximum. Therefore CFL^* is refered to as the *optimal CFL-number*. The two critical modes involved in its determination are essentially a function of Re_c.

Figure 1 shows that at low Re_c, the two critical modes that are damped out with a speed ϵ_{max} when the CFL-number is equal to CFL^*, are the modes (π,π) and $(2\pi\Delta x/L_x, 2\pi\Delta y/L_y)$. For medium high Re_c, we have that by approximation the modes $(\pi/2,\pi/2)$ and $(2\pi\Delta x/L_x, 2\pi\Delta y/L_y)$ become equally well damped at $CFL = CFL^*$, while for high Re_c the critical modes at CFL^* are $(\pi/2,\pi/2)$ and (π,π). We observed that this general behavior of CFL^* does not change substantially with $\Delta x/\Delta y$, $L_x/\Delta x$ and $L_y/\Delta y$. The optimal CFL-number for each of the three ranges can therefore easily be determined from (6), since for that value the amplification factor of both critical modes should be equal. This yields:

$$CFL_l^* \approx \frac{Re_c}{4\pi}\left[\frac{\Delta y/\Delta x + \Delta x/\Delta y}{\Delta x\Delta y/L_x^2 + \Delta x\Delta y/L_y^2}\right]^{1/2},$$

$$CFL_m^* \approx \frac{1}{\sqrt{2}\pi}\left[\frac{\Delta y/\Delta x + \Delta x/\Delta y}{\Delta x\Delta y/L_x^2 + \Delta x\Delta y/L_y^2}\right]^{1/4}, \quad \text{and} \tag{8}$$

$$CFL_h^* \approx .5(Re_c)^{1/3},$$

where CFL_l^*, CFL_m^* and CFL_h^* denote the "analytically" predicted optimal value for the CFL-number, for low, medium high and high cell Reynolds number respectively.

This approximation of the function $CFL^*(Re_c, \Delta x/\Delta y, L_x/\Delta x, L_y/\Delta y)$ is shown in Figure 1, by means of the dashed lines. We see that the prediction is excellent, and certainly good enough to be used in practice. It shows that for optimal convergence the time step at low Re_c should be taken proportional to the number of points per direction in the grid, while at medium high cell Reynolds numbers Δt should be proportional to $\sqrt{(\Delta x\Delta y)}$. The former result has been derived already before in the literature [6], while the latter has been established empirically by Pulliam [7], who obtained a considerable increase in convergence speed in this way. However, the expressions (8) are more complete, and take into account not only the size $\Delta x\Delta y$ of the grid cell, but also the grid cell ratio and the length and width of the periodicity domain.

It should be mentioned that the expressions in (8) have been obtained by considering the full system of Fourier modes (5), assuming that all modes are to be damped out by the relaxation scheme (3). However, in practice the computational domain is not periodic but bounded, and different (relaxation) schemes are applied at the boundaries due to the discretization of the physical and numerical boundary conditions. This may strongly enhance the damping of certain error modes, which as a consequence do not have to be considered anymore in the derivation of the optimal CFL-number. Also the dissipative properties of physical boundary conditions may improve substantially the convergence properties of the scheme as a whole. As these effects cannot be included in a Fourier mode analysis, only heuristic arguments and numerous testing may decide about how a good analytical prediction of CFL^* should be obtained in practice.

APPLICATION TO A NAVIER-STOKES CALCULATION

The first author has used the prediction of CFL^* for the optimization of the speed of convergence of the AF for the calculation of the steady viscous flow through a blade-to-blade passage. To this end, the compressible Navier-Stokes equations were discretized by means of an accurate finite volume technique on a highly irregular, structured grid of type H, using computational molecules of three times three points [1]. Similar expressions as in (8) were derived in order to take this extension of the molecule into account.

Total conditions were applied at the inlet, while at the outlet the static pressure was imposed. For a quasi-uniform flow, this allows for a system of waves propagating at the speeds $u \pm a$ that are reflected back and forth between inlet and outlet, where u is the velocity component normal to inlet and outlet and a the speed of sound. The largest wave length compatible with this system equals four times the length of the domain [1]. The approximation is especially good at low Mach number ($u \ll a$), when also the dissipative properties of the inlet boundary conditions are small. So when this flow is to be calculated on a uniform rectangular grid, $L_x/\Delta x$ should be taken equal to four times the number of grid cells in x-direction. Because of the imposed periodicity, $L_y/\Delta y$ should be taken equal to the total number of grid cells in y-direction.

The error mode of highest frequency that has been considered is the mode $(\pi/2,\pi/2)$. The implicit discretization of the numerical boundary conditions was found to damp out very efficiently all higher frequency modes, due to the strong coupling it provides between the unknowns at even- and odd-numbered grid points [1]. Hence only two expressions predicting the optimal CFL-number were used, obtained by equating $G^h(2\pi\Delta x/L_x, 2\pi\Delta y/L_y)$ and $G^h(\pi/2,\pi/2)$: one for low Re_c by neglecting the terms with coefficients u and v, and one for medium high and high Re_c by neglecting the terms with ν_x and ν_y.

The expressions were generalized in a straightforward manner, in order to determine for each molecule of the irregular grid the optimal time step (local time stepping). So on each molecule Δx and Δy were taken equal to the average local size of the grid in i- and j-direction, $L_x/\Delta x$ and $L_y/\Delta y$ were taken equal to four times the number of grid cells in i-direction and to the number of grid cells in j-direction, while in the formulas we used the maximum speed of propagation $|u|+a$ and the physical kinematic viscosity ν as typical values of the "convection" speed and diffusion coefficient of the system. The steady compressible viscous flow through a cascade of NACA0012 profiles with solidity 1 was calculated at a Reynolds number of 200 and an inlet Mach number of 0.21, on a grid of 78 times 39 points. Grid and results are shown in Figure 2, as well as the convergence history of the relaxation scheme using three different local time steps. It is clear that the analytical prediction of the optimal time step optimizes indeed very well the speed of convergence, notwithstanding the very limited applicability of the Fourier mode stability analysis for this case.

Remark that this result could *only* be obtained when the *full* characteristic equations were used as numerical boundary conditions at inlet and outlet, discretized in time by the same approximate factorization scheme. *Any* simpler boundary treatment seems to be incompatible with the relaxation scheme that is used inside the domain, and may therefore change considerably the convergence properties of the scheme as a whole. This effect is unpredictable with a Fourier mode analysis, making the derived predictions for the optimal time step completely useless.

HEAT DISSIPATION IN A LIQUID SATURATED DEBRIS BED

The next application deals with a two-dimensional model describing the steady-state thermohydraulic behavior of a liquid saturated, self heated particulate bed, the formation of which is postulated after severe accident sequences in liquid cooled nuclear reactors. The heat generated is removed by conduction, natural convection and phase change processes. The onset and evolution of a dry zone in the boiling region of the bed mark a sharp change in the coolability of the system and therefore their prediction is of special interest in this safety analysis problem.

The mass and energy balance, the equation of motion for each phase (liquid, vapor) in a porous bed as well as an expression for capillarity, which has an important driving effect due to the tininess of the debris particles (100-2000 μm), are combined to produce after non-dimensionalization a system of two coupled nonlinear partial differential equations for the temperature T and the liquid pressure P_ℓ in the bed [8]:

$$\nabla\cdot\frac{K_\ell}{\gamma}[\nabla P_\ell-\rho_\ell g] + \nabla K_v[\nabla(P_\ell+A_2 J)-\rho_v g] = 0 \tag{9a}$$

$$-\nabla\cdot[A_1 K_e \nabla T] - \nabla K_v[\nabla(P_\ell+A_2 J)-\rho_v g] = Q(x,y) \tag{9b}$$

where the relative permeabilities K_ℓ, K_v and the non-dimensional capillary pressure J are strongly nonlinear functions of the liquid fraction (also called saturation) s in the bed, which in turn can be related to the temperature through a suitable smoothing function [8]. ρ_ℓ, ρ_v denote the density of the liquid and vapor phase respectively, K_e is the effective thermal conductivity of the bed, Q is the power dissipated in it and A_1, A_2, γ are non-dimensional constants.

The bed rests in a vertical cylindrical container and the computational domain consists of a rectangle with the axis of symmetry forming one of its sides. The boundary conditions are specified as follows:

- Top: imposed T_{top} and $P_{\ell,top}$.
- Bottom: impermeable wall and imposed T_{bot}, or impermeable wall and heat losses through a finite heat transfer coefficient h and a given external cooling temperature T_c:

$$[-A_1 K_e \nabla T - K_v \nabla(P_\ell + A_2 J)]\cdot\underline{n} = A_1 h(T_c - T) \tag{10}$$

- Left: axisymmetry,
- Right: similar to the bottom one.

Using successive substitution of the two unknowns the equations can be solved separately and the numerical problem one is faced with becomes how to solve a nonlinear Poisson equation throughout the bed. The implicit approximate factorization scheme described above is employed to find its solution on a uniform rectangular grid, to which end artificial time derivatives have been introduced in the equations (9). The axisymmetric boundary condition is implemented implicitly while the others are treated explicitly in the code.

Several cases were studied in order to test the validity and applicability of the predicted value of the optimal time step Δt^* for the present diffusion dominated, nonlinear problem. Since the pressure solution is almost hydrostatic it is the temperature equation (9b) that dictates the convergence properties of the system. Therefore the asymptotic speed of convergence of this equation will be considered for three different steady-state cases:

Case I: The converged solution on a 20x50 grid, with the temperature fixed at three boundaries, is shown in Figure 3. Most of the bed is boiling, the minimum saturation value being 0.11. To predict the optimal time step the possible critical modes should be identified taking into account the type of the boundary conditions which dictate the range of frequencies of the perturbation modes to be considered. It can be easily seen (Figure 4) that the highest frequency mode in x-direction compatible with the requirement of axisymmetry on one and fixed value on the other boundary is $\pi/3$ and the lowest frequency mode is $2\pi\Delta x/4l_x$, with l_x the length of the domain in x-direction (i.e. $L_x=4l_x$). Turning to the y-direction, the corresponding extreme modes (for fixed values at both boundaries) become π and $2\pi\Delta y/2l_y$ respectively, with $L_y=2l_y$. For the optimal time step both extreme modes become critical and hence its value for this case is found by solving:

$$G(\frac{\pi}{3},\pi) = G(\frac{2\pi\Delta x}{4l_x},\frac{2\pi\Delta y}{2l_y}) \tag{11}$$

where this time different values of the diffusion coefficients ν_x, ν_y are considered. An expression for CFL_I^*, similar to the one in (8) is thus obtained.

In Figure 5 the convergence histories of equation (9b) for Δt's around the predicted optimum are drawn. The asymptotic speeds of convergence resulting from such curves are listed in Table 1. It is evident that for the present case the real optimum lies within 10% of the predicted one. Notice that although in Figure 5 the convergence seems to be faster for $\Delta t=1.35\Delta t^*$, the *asymptotic* speed of convergence ϵ_{max} is actually smaller than for $\Delta t=\Delta t^*$. This is because the amplitude of the critical mode for $\Delta t=1.35\Delta t^*$ that is introduced in the system by the initial condition is relatively small and therefore it appears only at low error values.

Case II: As Figure 6 shows, the top part of the bed is now subcooled, while the bottom one undergoes boiling, and incipient dryout is identified at a small region close to the axis of symmetry. The right and bottom walls are characterized by the "hybrid" boundary condition (10) which can be used to obtain an estimation for the periodicity length of the steady-state error in each direction and subsequently for the corresponding critical mode to be considered. After linearization (10) gives for $\Delta\Phi^{*t}$, in the y-direction:

$$\frac{\partial\Delta\Phi^{*t}}{\partial y} + c_y\Delta\Phi^{*t} = 0 \quad \text{at} \quad y = 0. \tag{12}$$

The lowest frequency mode $2\pi\Delta y/L_y$ should satisfy the condition of $\Delta\Phi^{*t}=0$ at $y=l_y$ as well as (12), while the corresponding mode in the x-direction should satisfy similar conditions. This, for the present values of c_x and c_y, can happen only if $L_x=2l_x$ and $L_y=3l_y$. The computations done using the above estimations for the critical modes show that the optimal Δt prediction based on this analysis is quite acceptable (Table 1).

Case III: In this last boiling case (Figure 7) the derivative of the temperature is fixed on the right wall. Though one would expect $L_x=2l_x$, the best prediction of Δt^* is still obtained assuming $L_x=4l_x$ for the determination of the critical mode (Table 1). In other words the system behaves as if the right boundary has fixed temperature (see Figure 4). This is explained by the explicit treatment of the right boundary condition where temporarily the condition $\Delta\Phi^{*t}=0$ (i.e. fixed value) is applied for the implicit solution of the inner points of the computational domain.

CONCLUSIONS

An analytical prediction of the optimal time step for the AF has been derived, with which (nearly) optimal convergence properties were obtained for two difficult numerical problems of practical interest. Such a prediction can only be obtained by means of a *correctly performed* Fourier mode analysis, where the type of boundary conditions determine which range of error modes are to be considered. Otherwise, the effect of boundary conditions is not taken into account in the prediction. This makes that the prediction may fail completely when the implementation of the numerical boundary conditions is too much different from the scheme used inside the domain, as this may change completely the convergence properties of the scheme as a whole.

Notice that Δt can also be optimized for the damping of only a part of the error modes, e.g. for the scheme to be primarily dissipative for the error modes of high frequency. This would make the AF a very useful smoother for a multigrid technique.

REFERENCES

[1] BORSBOOM M. (1987), *A Numerical Solution Method for the Steady-State Compressible Navier-Stokes Equations with Application to Channel and Blade-to-Blade Flow*, Doctoral Thesis, Université de Liege, Belgium.

[2] JESPERSEN D.C. (1984), "Recent Developments in Multigrid Methods for the Steady Euler Equations", in: *Computational Fluid Dynamics*, Von Karman Institute LS1984-04.

[3] DAHLQUIST G., BJÖRCK A., ANDERSON N. (1974), *Numerical Methods*, Prentice-Hall, Englewood Cliffs, New Jersey.

[4] WARMING R.F., BEAM R.M. (1978), "On the Construction and Application of Implicit Factored Schemes for Conservation Laws", SIAM-AMS Proc. 11, 85-129.

[5] VAN LEER B., MULDER W.A. (1985), "Relaxation Methods for Hyperbolic Conservation Laws", in *Numerical Methods for the Euler Equations of Fluid Dynamics*, Angrand F. et al., eds., SIAM, Philadelphia.

[6] ABARBANEL S.S., DWOYER D.L., GOTTLIEB D. (1986), "Improving the Convergence Rate to Steady State of Parabolic ADI Methods", J.Comput.Phys. 67, 1-18.

[7] PULLIAM T.H., STEGER J.L (1985), "Recent Improvements in Efficiency, Accuracy, and Convergence for Implicit Approximate Factorization Algorithms", AIAA Paper 85-0360.

[8] THEUNISSEN P.-H., BUCHLIN J.M., STUBOS A.K. (1985), "A Two-Dimensional Code Modeling the Thermohydraulic Behaviour of Liquid Saturated Self Heated Debris Beds", VKI Preprint 1986-05.

TABLE 1:
ϵ_{max}-values for different Δt's.

Cases	Time Step				
	$0.75\Delta t^*$	$0.9\Delta t^*$	Δt^*	$1.2\Delta t^*$	$1.35\Delta t^*$
I	0.127	0.150	0.138	0.106	0.076
II	0.023	0.025	0.026	0.024	0.023
III	0.028	0.034	0.037	0.029	0.025

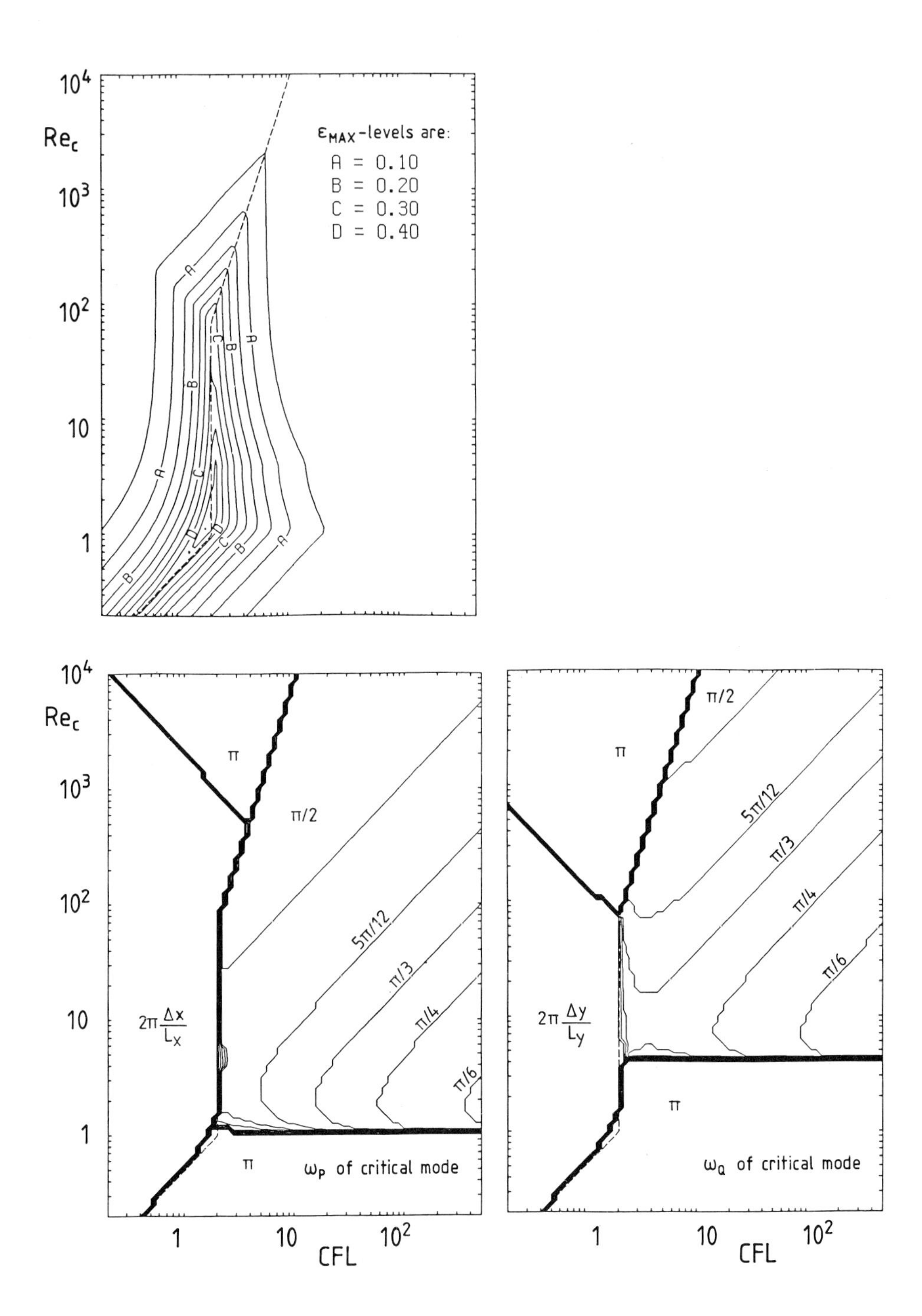

FIGURE 1: ϵ_{max} and the two frequencies of the critical mode as function of Re_c and CFL; $\Delta x/\Delta y = 4$, $L_x/\Delta x = 100$, and $L_y/\Delta y = 25$. Prediction of optimal CFL-number indicated by dashed line.

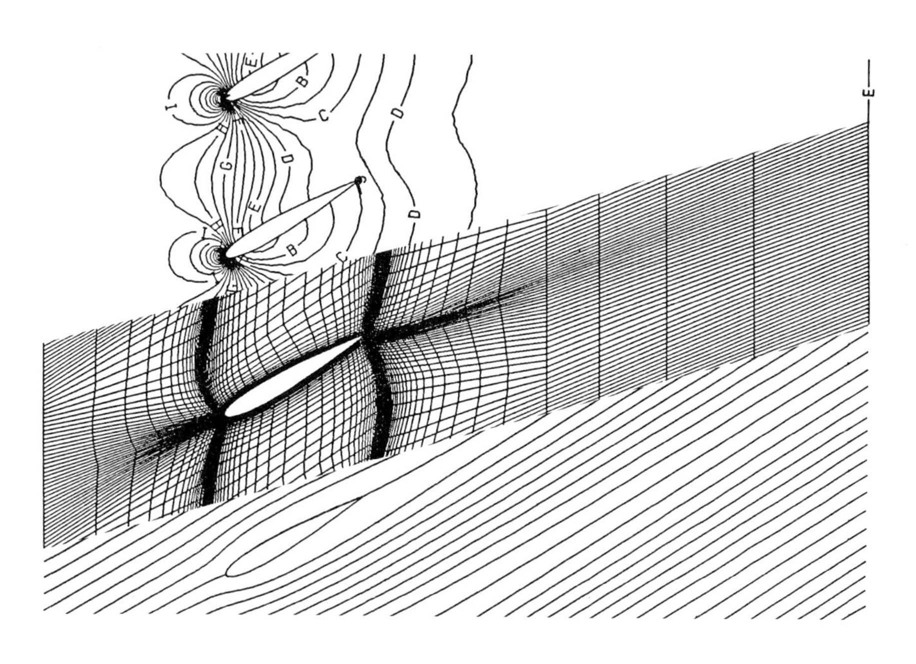

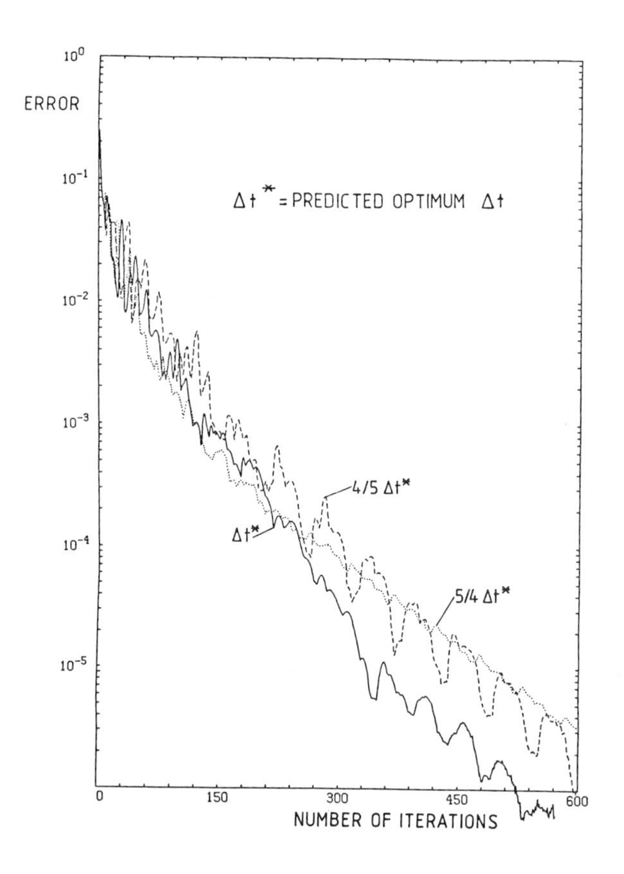

FIGURE 2:
Static pressure contours, 78 times 39 point grid, streamlines and convergence histories of the steady-state flow calculation through a blade-to-blade passage at Re = 200 and M_{in} = 0.21.

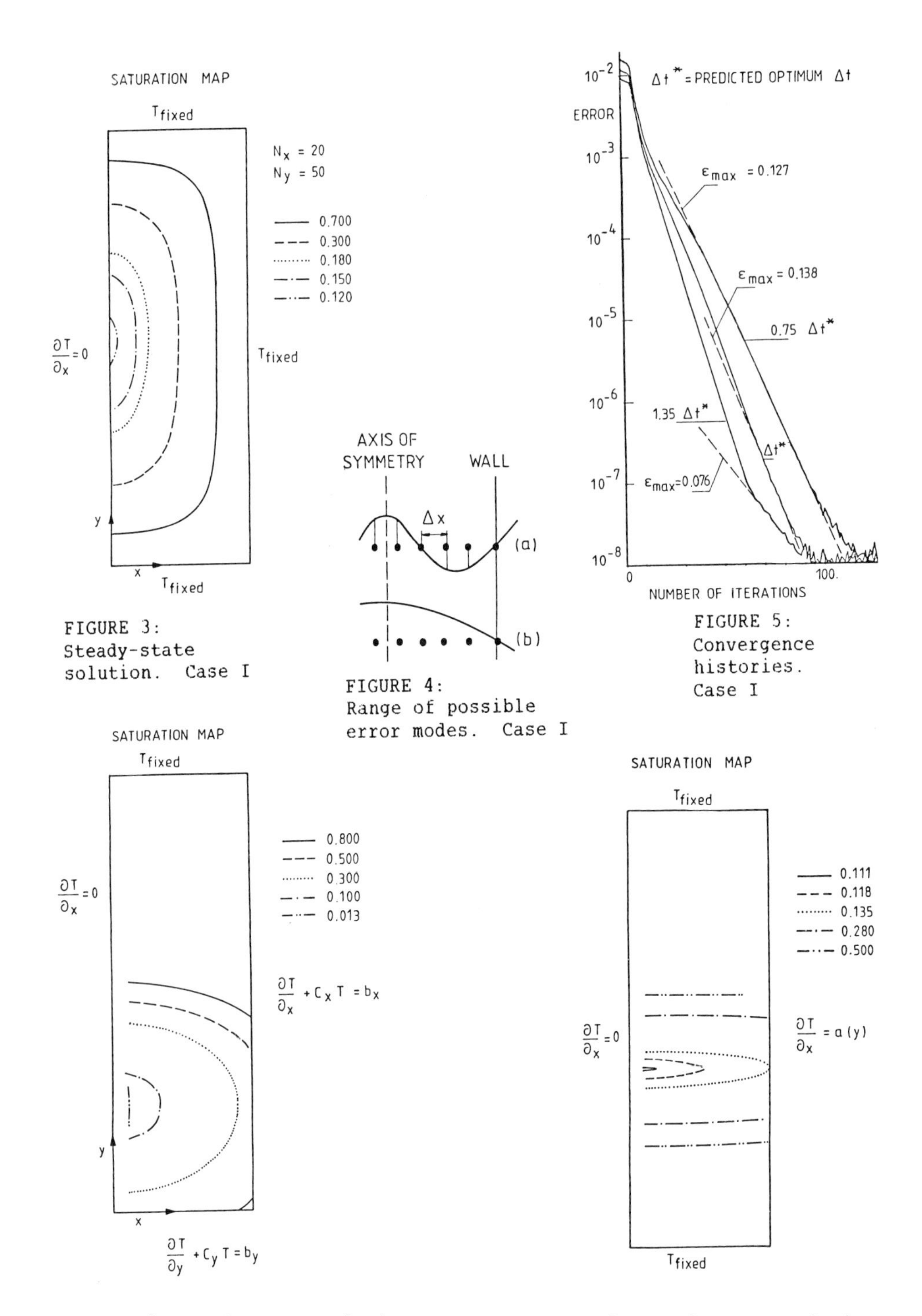

FIGURE 3: Steady-state solution. Case I

FIGURE 4: Range of possible error modes. Case I

FIGURE 5: Convergence histories. Case I

FIGURE 6: Steady-state solution. Case II

FIGURE 7: Steady-state solution. Case III

UNSTEADY TRANSONIC FLOWS PAST AIRFOILS USING A FAST IMPLICIT GODUNOV TYPE EULER SOLVER

A. Brenneis, A. Eberle
Messerschmitt-Bölkow-Blohm GmbH, LKE122
Postfach 801160, D-8000 München 80, Germany

SUMMARY

Inviscid, transonic flow over a harmonically oscillating airfoil with solid body motion in a deformable mesh is computed by solving the two-dimensional Euler equations in integral form. The applied finite volume scheme uses a Godunov type averaging procedure for the evaluation of the flow quantities at the volume cell faces which separate constant left and right states on either side. A relaxation method employed to the unfactored implicit Euler equations that will reduce significantly the computational time is described. The calculated results for the flow over the NACA 64A010 airfoil pitching about its one-quarter chord are compared with experimental data. Good agreement between experiment and theory are obtained both for low and high reduced frequencies. Reduction of the implicit time steps per cycle leads to substantial saving of CPU time without fundamental worsening the results.

INTRODUCTION

The proper computation of unsteady, transonic flows around oscillating airfoils remains an outstanding and important problem in fluid dynamics. Efficient and complete analytical methods with the ability to accurately predict the occurrence of aircraft aeroelastic phenomena would find immediate applications in the treatment of such problems as flutter and buffet. With the introduction of supercomputers, high speed and large memory facilities, the numerical calculation of the rotational, compressible flow over complex configurations has become feasible.

The theoretical analysis of transonic flows is complicated by the presence of mixed subsonic and supersonic flow regions within the flowfield and moving shock waves of varying strength. During the last decade, quite a number of computational methods for solving the unsteady problems have been developed which take into account interaction between the steady and the unsteady flow fields, and if shock waves are present also the effect of the periodic shock-wave motion. Many of the numerical schemes are based on the transonic small-perturbation (TSP) equation, either linearized or non-linearized, and the full potential equation. The TSP-method which is based on the assumption that the steady and unsteady perturbations of the flow field remain small is therefore limited to thin airfoils and small amplitudes of oscillation. In addition, in the linearized TSP-method the effect of schock waves can be included only, if the shocks do exist already in the steady flow field. The potential equation which has not these drawbacks suffers from the limitation to weak and moderately strong shock waves. However, in order to calculate the unsteady flow over the next generation of aircraft higher order solver based on the Euler and Navier-Stokes equations are needed. By assuming the flow to be inviscid, which implies that the effect of the boundary layer is not included which may lead to a change in effective airfoil contour and thus, to a change in circulation, shock strength and shock position, the Euler equations provide the correct description in the transonic and all other flow regions.

The unsteady Euler equations have been solved by a number of authors, /1,2,3,4,5/, using an explicit scheme /5,6,7,8/ or an approximate-factorization (AF) method /3,4/. The explicit time stepping schemes suffer from the limitation of the global time step needed in time-accurate unsteady problems, leading to a great number of time steps per cycle and hence to large CPU time. The principal disadvantage of the AF scheme in two dimensions is that optimal convergence is obtained at a value of the time step that is not a priori known. The optimal time step is found by trial and error and varies from problem to problem, although usually it corresponds to a Courant number of order ten. In three dimensions, the stability restriction is even more severe.

The Euler code developed in this investigation solves the unfactored implicit Euler equations by means of a relaxation method. The advantages of this iterative scheme in the application to both steady and truly unsteady flows are the dramatically increased speed of convergence to steady state and the disappearance of the transients respectively, that is unconditionally stable in a linear stability analysis and the CFL number can be arbitrarily high. Further chief attributes are the straightforward extension to three dimensions and the high degree of vectorization.

The new code was tested on the flow over the NACA 64A010 airfoil oscillating in pitch studing the effect of grids, CFL number and time steps per cycle. The results obtained are in good agreement with experiments /9/.

NOTES ON THE METHOD OF SOLUTION

- Governing Equations

The conservation law vector form of the dimensionless unsteady Euler equations in time dependent curvilinear coordinates

$$\begin{aligned} \xi &= \xi(t,x,z) \ , \\ \zeta &= \zeta(t,x,z) \end{aligned} \qquad (1)$$

can be written as

$$\frac{\partial \Phi}{\partial t} + \frac{\partial E}{\partial \xi} + \frac{\partial F}{\partial \zeta} = 0 \ , \qquad (2)$$

where:

$$\Phi = J \begin{pmatrix} \rho \\ \rho u \\ \rho w \\ e \end{pmatrix} ; \quad E = J \begin{pmatrix} \rho U \\ \rho u U + p\xi_x \\ \rho w U + p\xi_z \\ U(e+p) - p\xi_t \end{pmatrix} ; \quad F = J \begin{pmatrix} \rho W \\ \rho u W + p\zeta_x \\ \rho w W + p\zeta_z \\ W(e+p) - p\zeta_t \end{pmatrix} \qquad (3)$$

and

$$\begin{aligned} U &= (u-\dot{x})\xi_x + (w-\dot{z})\xi_z \ , \\ W &= (u-\dot{x})\zeta_x + (w-\dot{z})\zeta_z \ , \\ J &= x_\xi z_\zeta - x_\zeta z_\xi \ , \\ \xi_t &= -\dot{x}\xi_x - \dot{z}\xi_z \ , \\ \zeta_t &= -\dot{x}\zeta_x - \dot{z}\zeta_z \ . \end{aligned} \qquad (4)$$

In the above equations, ρ is the density, u and w are the velocity components in x and z directions, p is the pressure, e is the total energy, J is the Jacobian of the inverse mapping and $\dot{x}$ and $\dot{z}$ are the appropriate velocity components of the moving cell. For a perfect gas the relation

$$e = \frac{p}{\gamma-1} + \frac{1}{2}\,\rho(u^2 + w^2) \tag{5}$$

holds.

- Time Step - Explicit Scheme

For the treatment of real unsteady problems the time consistency is required. Therefore the unsteady evolution must be computed with a uniform time step Δt in the whole computational domain. This global time step is defined from a fixed CFL number:

$$\Delta t \leq \Delta t_{max} = \min_{i,k} \frac{J_{i,k} \cdot CFL}{\left[|\vec{V}| + a\right]_{i,k}} \quad . \tag{6}$$

$\vec{V}$ represents the local velocity vector and a the velocity of sound.

The explicit solution using the largest allowable time increment reads

$$\Phi^{n+1} = \Phi^n - \Delta t\,(E_\xi + F_\zeta) \;, \tag{7}$$

where Φ^{n+1} denotes the conservative solution vector at the new time level n+1, Φ^n at the old time level n. The flux differences E_ξ and F_ζ are approximated via a third order characteristic flux extrapolation scheme, using sensor functions to detect shocks, where the scheme reduces to first order accuracy /10/. The principal feature of the characteristic flux extrapolation scheme is a Godunov type averaging procedure based on an eigenvalue analysis of the Euler equations by means of which the fluxes are evaluated at the finite-volume face which separate constant sets of flow variables on either side.

- Implicit Scheme - Relaxation Method

The implicit scheme

$$\frac{\Phi^{n+1} - \Phi^n}{\Delta t} + \delta_\xi E^{n+1} + \delta_\zeta F^{n+1} = 0 \tag{8}$$

with the difference operaters δ_ξ and δ_ζ, which are assumed to be based on suitable upwind schemes of desired accuracy, is not directly solvable for the dependent variables Φ^{n+1} by virtue of its nonlinearity. Therefore a sequence of approximations denoted by q^ν such that

$$\lim_{\nu > 1} q^\nu \rightarrow \Phi^{n+1} \tag{9}$$

is constructed.

A Newton method /11/ for $q^{\nu+1}$ by linearizing Eq. (8) about the known subiteration state denoted by superscript ν is introduced as follows:

$$(\frac{1}{\Delta t} I + \delta_\xi A^\xi + \delta_\zeta A^\zeta)\ \delta q^{\nu+1} = -\ (\frac{q^\nu - \Phi^n}{\Delta t} + E_\xi^\nu + F_\zeta^\nu) = RHS\ . \tag{10}$$

q^ν and $q^{\nu+1}$ are the solution vectors at the subiteration states ν and $\nu+1$. $\delta_i (i=\xi,\zeta)$ are the first order upwind difference operators. I is the identity matrix and A^i are the true Jacobians of the terms on the right hand side.

As the numerical solution of the full Newton method in two or more dimensions is too time consuming a pointwise Gauss-Seidel (GS) relaxation method is applied to solve the unfactored implicit Euler equations, where the RHS is evaluated using the latest available values of the dependent variables. Further an underrelaxation factor ω is introduced on the RHS to ensure convergence. By application of the so-called checkerboard scheme, in which the points are divided into black and white ones, a high degree of vectorization is achieved.

- Boundary Conditions

For inviscid flow computations a tangency condition is imposed on the airfoil surface. Due to the structure of the used grid, either the ξ or the ζ direction is the tangential direction. Being ξ the tangency coordinate then along the boby surface $\zeta(t,x,z) = 0$, for the finite solid body element, which is moved with $\dot{x}$ and $\dot{z}$ the no through flow condition

$$\rho(u-\dot{x})\zeta_x + \rho(w-\dot{z})\zeta_z = 0 \tag{13}$$

must hold. This means that the normal mass flux across the boundary is set to zero. The remaining variables at the cell face are extrapolated from the interior of the computational domain. No explicit Kutta condition is specified and in the farfield nonreflecting boundary conditions are used.

- Grid Generation

The mesh used is of type H with fixed farfield frame and was generated using Poisson's equation /12/. The fourth-order (biharmonic) system

$$\nabla^4 x^i = 0 \quad (i=1,2) \tag{14}$$

is implemented as a set of two second-order equations (Poisson's and Laplace's equation):

$$\begin{aligned} x_{\xi\xi} + x_{\zeta\zeta} &= P\ , \\ z_{\xi\xi} + z_{\zeta\zeta} &= Q\ , \end{aligned} \tag{15}$$

with

$$\begin{aligned} P_{\xi\xi} + P_{\zeta\zeta} &= 0\ , \\ Q_{\xi\xi} + Q_{\zeta\zeta} &= 0\ . \end{aligned} \tag{16}$$

The discretization of the derivatives by central differences leads to linear algebraic equations in physical space and yields by rearranging of the equations the grid point coordinates $(x,z)_{i,k}$ and the source terms

$(P,Q)_{i,k}$. The resulting grid is stored and used for the flowfield computation, whereby after each time step the airfoil is moved and the mesh is rearranged by one to two grid iterations.

RESULTS AND DISCUSSION

The relaxation scheme was tested on the flow past the NACA 64A010 configuration with harmonically varying angle of attack

$$\alpha_{(t)} = \alpha_0 + \alpha_1 \cos(\omega t) \tag{17}$$

at the Mach number 0.8. Calculations have been made for the mean value of the angle of attack α_0 = 0 deg, the amplitude of the airfoil oscillation α_1 = 1 deg and the reduced frequncy

$$k = \frac{\omega \, c}{2 \, U_\infty} \tag{18}$$

of 0.025 and 0.202, pitch axis located at 0.25 chord. This standard computational test cases are denoted as CT3 and CT6 and were compared with the experimental data from /9/. The unsteady computations, always started with parallel flowfield, were carried out for three cycles to ensure that the surface pressure and the integrated loads repeat from cycle to cycle. The unsteady pressures are calculated as real (in-phase) and imaginary (out-of-phase) parts of the first harmonic component of the pressure. The harmonic components are normalized by the nondimensional amplitude of motion in radians and calculated using a Fourier analysis. The airfoil geometry was taken from /13/ and the computational grids have common outer boundary located 5 chord lengths above and below the airfoil, and 5 chord lengths upstream and downstream from the airfoil leading edge and trailing edge respectively.

Fig. 1 shows the coarse mesh with 88 points in x and 48 points in z direction. Each surface is covered by 50 nodes and in the region where the shock is expected to be located the mesh is clustered. The first calculation done with a CFL number of 400 requires 200 time steps per cycle (Δt/cycle). The comparison of the computational results with free stream boundary condition and the experimental measurements, Fig. 2a, shows that the computed mean pressures ahead of the shock are underpredicted. Computations made with the experimentally measured wall-pressure boundary-condition, /14/, indicate that this underprediction of the pressure is in part a result of the wind-tunnel wall interference. The components of the first harmonic analysis, Fig. 2b, agree very well with the experimental data execpt in the shock-wave region. The overprediction of the shock strength and the shock location aft of the experimental shock position indicates that viscous effects are present in the real flow.

In order to assess the influence of the grid refinement on the accuracy of the results, calculations for the CT6 were repeated with a fine (170 x 92 points) and a very fine (390 x 240 points) mesh which has 106 and 210 nodes respectively located on each surface. The grid with 390 x 240 points has its outer boundary located 10 chord lengths above, below, upstream and downstream of the airfoil. The employed grids around the airfoil and the corresponding real and imaginary parts are shown in Fig. 3. The obtained results (Δt/cycle=200) are identical; only the suction peak at the leading edge is better predicted by using finer meshes. The oscillations up to about x/c=0.4 are due to poor geometry definition given by /13/ and can be eliminated by smoothing the airfoil surface.

As the CPU time depends principally on the number of time steps per cycle computations were made varying Δt/cycle on the coarse mesh. The results shown in Fig. 4 are identical, to within two digits. By decrease of Δt/cycle (50, 25) upstream of the shock the in-phase component increases slightly as the out-of-phase component decreases. Downstream of the shock both real and imaginary part tend to oscillate at the very low number of Δt/cycle =25. This effect can be avoided by increase of the GS iterations and/or the non-linear subiterations. However this rises the CPU time of the order of the case with Δt/cycle=50. The computational efforts, compared also with the explicit solution, are listed in Table 1. It can be seen that without loss of accuracy considerable CPU time saving is possible by reduction of Δt/cycle.

Finally the test case CT3, reduced frequency k=0.025, coarse mesh, is computed with the explicit and implicit code. The unsteady results depicted in Fig. 5 show good agreement with the experiment. Table 2 emphasizes the advantages of the implicit algorithm with regard to the computational time.

CONCLUSIONS

A relaxation method for solving the unfactored implicit Euler equations was used to calculate a transonic flow over the NACA 64A010 airfoil pitching about its one-quarter chord. The computational results were correlated with experimental data from the NASA Ames 11 x 11 ft transonic wind tunnel. Good agreement with the experimental measurements was achieved both for high and low frequencies except in the shock-wave region, where the effects of shock-wave/boundary-layer interaction must be taken into account. No influence of grid refinement to accuracy of the unsteady pressures was observed. The implicit algorithm reduces the CPU time up to one tenth or more of the explicit one. This depends on the refinement of the grid, reduced frequency and time steps per cycle. Computations with about 50 to 100 time steps per cycle yield results which cannot be improved by rise of them. Extension to three dimensions is straightforward and first calculations will be done on a rectangular supercritical wing and the LANN-wing.

REFERENCES

/1/ Magnus R., Yoshihara H.: Unsteady Transonic Flows over an Airfoil. AIAA Journal, Vol. 13, Dec. 1975, pp. 1622-1628

/2/ Lerat A., Sides J.: Numerical Simulation of Unsteady Transonic Flows Using the Euler Equations in Integral Form. Israel Journal of Technology, Vol. 17, 1979, pp. 302-310

/3/ Steger J.: Implicit Finite-Difference Simulation of Flow About Arbitrary Two-Dimensional Geometries. AIAA Journal, Vol. 16, July 1978, pp. 679-686

/4/ Chyu W. J., Davis S. S., Chang K. S.: Calculation of Unsteady Transonic Flow over an Airfoil. AIAA Journal, Vol. 19, June 1981, pp. 684-690

/5/ Jameson A., Venkatakrishnan V.: Transonic Flows About Oscillating Airfoils Using the Euler Equations. AIAA-85-1514, July 1985

/6/ Smith G. E., Whitlow W. Jr., Hassan H. A.: Unsteady Transonic Flows Past Airfoils Using the Euler Equations. AIAA-86-1764, 1986

/7/ Deslandes R.: Eine explizite Methode zur Lösung der Eulergleichungen angewandt auf Instationäre Ebene Strömungen. MBB/S/PUB/257, Juni 1986

/8/ Jameson A., Schmidt W., Turkel E.: Numerical Solutions of the Euler Equations by Finite Volume Methods Using Runge Kutta Time Stepping Schemes. AIAA-81-1259, June 1981

/9/ Davis S. S.: NACA 64A010 Oscillatory Pitching. Data Set 2, Compendium of Unsteady Aerodynamics Measurements, AGARD-702, 1982

/10/ Eberle A.: 3 D Euler Calculation Using Characteristic Flux Extrapolation. AIAA-85-0119, 1985

/11/ Chakravarthy S. R.: Relaxation Methods for Unfactored Implicit Upwind Schemes. AIAA-84-0165, 1984

/12/ Schwarz W.: Elliptic Grid Generation System for Three-Dimensional Configurations Using Poisson's Equation. 1st Intern. Conf. on Numerical Grid Generation in Comp. Fluid Dynamics. Pineridge Press, 1986

/13/ Abbott I. H., v. Doenhoff A. E.: Theory of Wing Sections. Dover Publications, Inc., New York, 1958

/14/ King L. S., Johnson D. A.: Calculations of Transonic Flow About an Airfoil in a Wind Tunnel. AIAA-80-1366, 1980

Table 1 Comparison of the CPU time of the explicit and implicit algorithm for the NACA 64A010, CT6

	explicit*	implicit			
CFL	0.4	400	800	1600	3200
Δt/cycle	17545	200	100	50	25
CPU [min/cycle]	97.3	50.6	26.1	18.8	10.9

Table 2 Comparison of the CPU time of the explicit and implicit algorithm for the NACA 64A010, CT3

	explicit*	implicit		
CFL	0.4	1600	3200	7400
Δt/cycle	152000	400	200	100
CPU [min/cycle]	836	152	80	54

* The explicit algorithm was applied to a modified mesh with the wake cells being discarded, seen in Fig. 5. This causes an acceleration of the explicit code by a factor of 10.

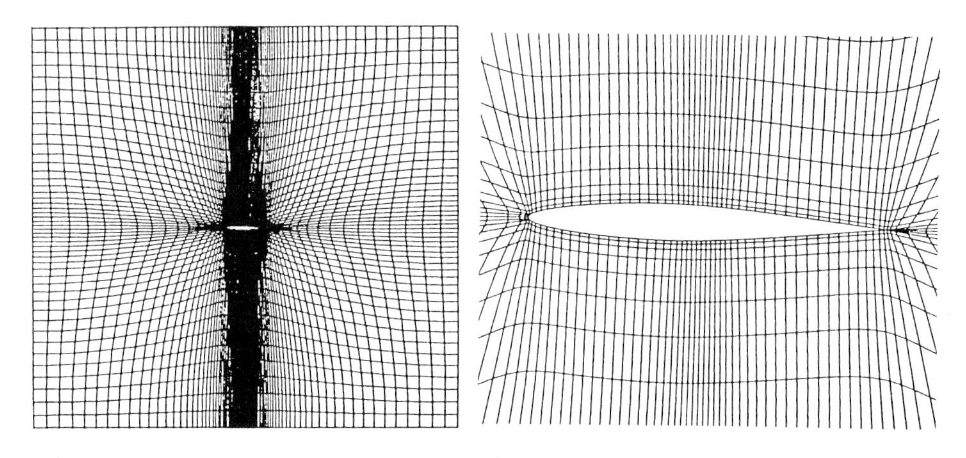

Fig. 1 Computational mesh for NACA 64A010 (88 x 48 points)

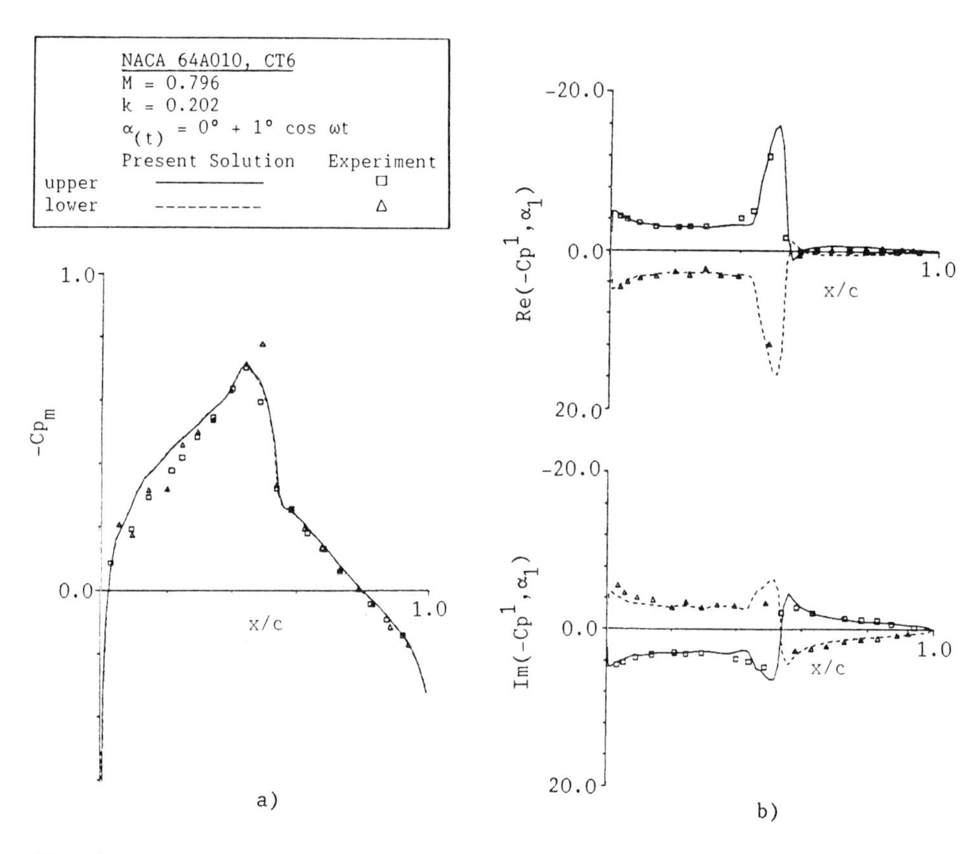

Fig. 2 Mean pressure distribution (a) and first harmonic analysis of the pressure coefficient (b)

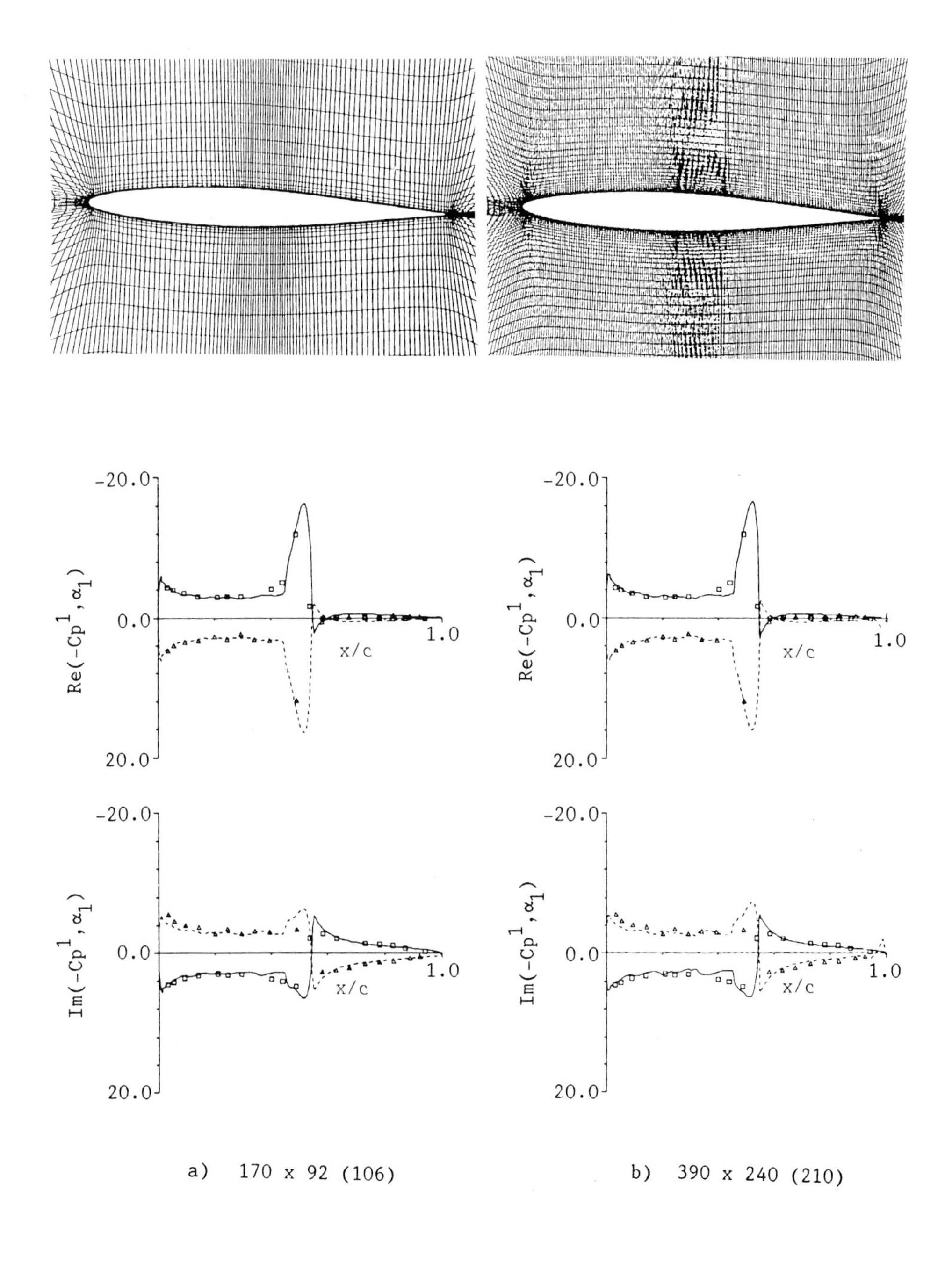

Fig. 3 Partial view of the computational meshes and unsteady pressure distribution for the NACA 64A010, CT6

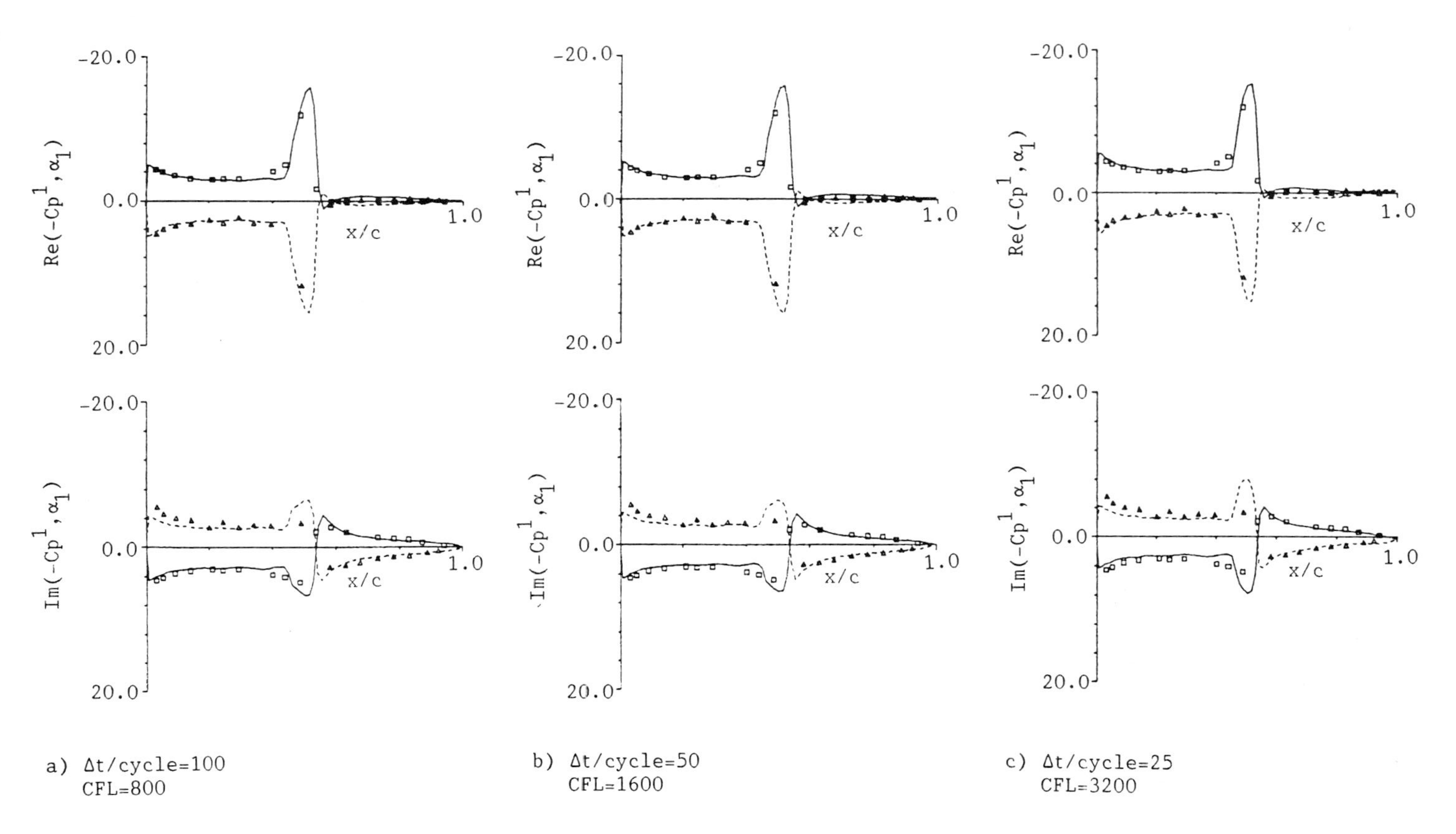

Fig. 4 Comparison of the unsteady pressure distribution: Variation of time steps per cycle for NACA 64A010, CT6

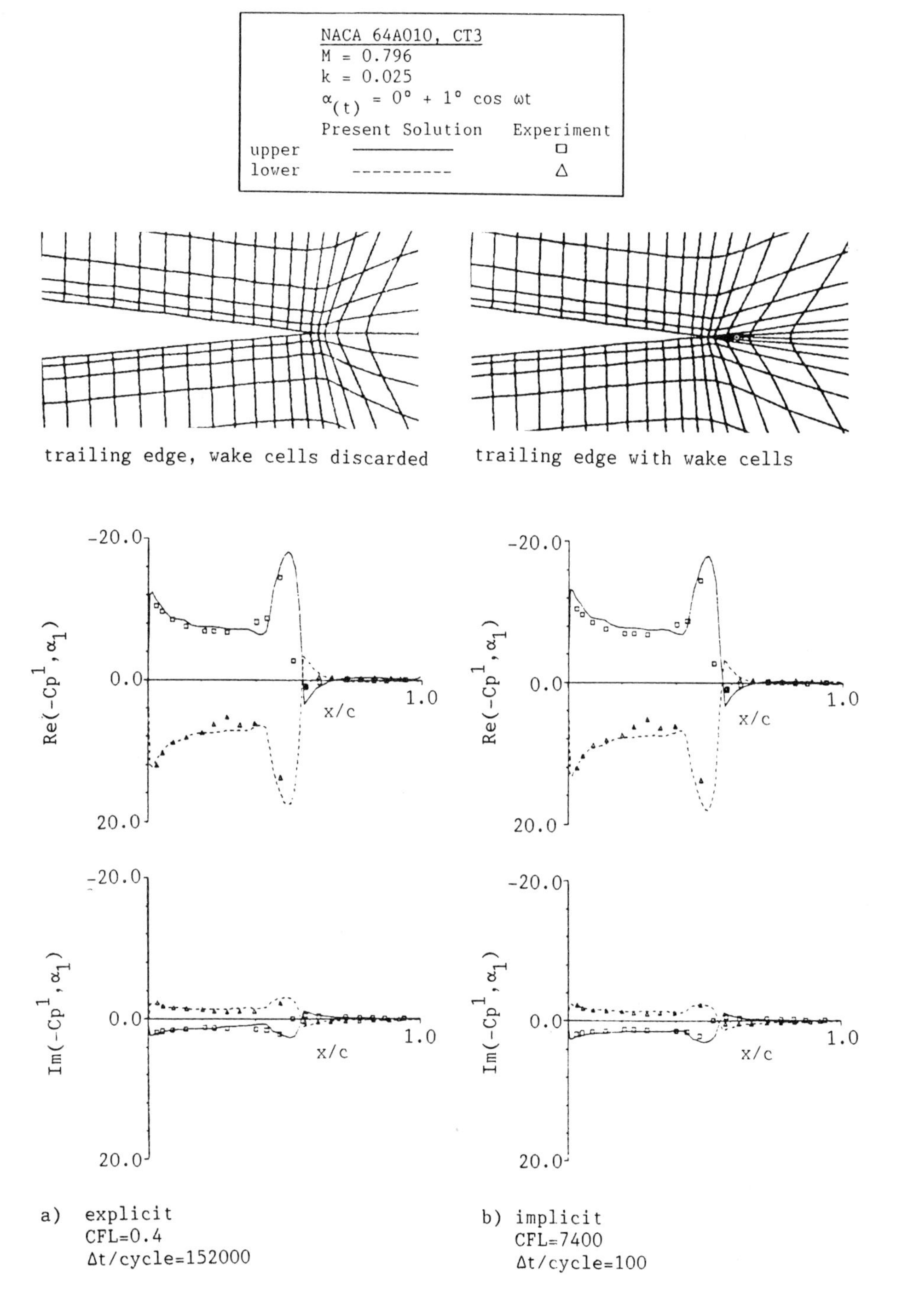

Fig. 5 Unsteady pressure distribution of the NACA 64A010, CT3

NAVIER-STOKES COMPUTATIONS OF LAMINAR COMPRESSSIBLE AND INCOMPRESSIBLE VORTEX FLOWS IN A CHANNEL

U. Brockmeier, N. K. Mitra, M. Fiebig
Institut für Thermo- und Fluiddynamik
Ruhr-Universität Bochum, Germany

SUMMARY

To investigate the structure of compressible and incompressible vortices behind a small delta wing in a channel at low Reynolds and Mach numbers, computer programs have been developed in order to solve complete three dimensional Navier-Stokes and energy equations with the help of a marker and cell method. Results show qualitatively similar vortex formation, flattening of the vortex core and movement of the core away from the channel center and towards the bottom wall for both incompressible and compressible flows. However, compressible fluid results into larger velocities, particularly in azimuthal components than the incompressible fluid.

INTRODUCTION

Turbulators in form of vortex generators increase the convective heat transfer in the gas side of a plate fin heat exchanger (see fig. 1). The gas flow between two plate fins can be modeled as a flow in a rectangular channel around built-in obstacles in form of circular or elliptic tubes. This flowfield is extremely complicated and consists of locally separated and dead water zones. Typical Reynolds numbers of these flows lie between 500 and 2000. The vortex generators in form of delta wings or delta winglets are punched out of the plate fin. The longitudinal vortices behind these vortex generators disturb the boundary layer growth on the plate and interfere with the dead water zones behind cylinders.

The computation of flow and temperature fields in plate-fin heat exchangers with built-in tube bank and vortex generatores is a tremendous task and is possibly beyond the capacity of even supercomputers. The present work addresses to the investigation of the structure of the flowfield in a threedimensional channel behind a single vortex generator.

Although delta wing has been subjected to many theoretical and experimental investigations at high Reynolds and Mach numbers because of aerodynamic interests, the structure of flow behind a delta wing inside a channel has never been reported. It is expected that because of the favorable pressure gradient in channel flows, the vortices behind the wing will be stable and vortex breakdown will not occur. On the other hand, at low Reynolds number flows, the viscous effect from the walls should damp, possibly distort and influence the vortex motion.

The characteristic velocity, consequently the Mach number for delta wing in a channel (or plate fins) for application in heat exchanger will be very small ($M \simeq 0.01$), however the effect of compressibility in flow can manifest itself through large temperature difference between the gas and the heat exchanger.

The purpose of the present work is to investigate vortex generation and transport in a channel behind a delta wing at low Reynolds numbers for both incompressible and compressible (low Mach number) flows. To this end computer programs are developed in order to solve threedimensional Navier-Stokes and energy equations for compressible and incompressible flows and results are compared.

BASIC EQUATIONS AND METHOD OF SOLUTION

The computational domain (see fig. 2a) consists of a rectangular channel with a vortex generator in form of a delta wing on one of the plate. Since symmetry prevails in the vertical central plane of the channel, flowfield only in the half of the channel has to be computed. The equations to be solved are the threedimensional unsteady compressible and incompressible continuity, momentum (Navier-Stokes) and energy equations. These equations are written in cartesian coordinates. The dissipation terms in the energy equations have been neglected and the density for the compressible flow has beeen assumed to be independent of pressure and a function of temperature only ($\rho \sim 1/T$). These assumptions are justified since the Mach number anywhere in the flowfield is extremely small. The basic equations will not be presented here since they can be found in any standard text book on fluid mechanics.

The basic equations have been nondimensionalized. It has been assumed that the viscosity is a function of temperature ($\mu \sim T^{\omega}$ where $0.5 < \omega < 1$). The dimensionless equations contain as parameters Reynolds number $Re = \bar{u}\ H / \nu_0$ and Prandtl number Pr where $\bar{u}$ is the average velocity at the channel inlet, H is the channel height and ν_0 is the dynamic viscosity at the channel inlet. The Prandtl number has been assumed to be constant and equal to 0.7.

The basic equations are solved numerically with the help of a modified marker and cell (MAC) method [1]. MAC is a semi-implicit time-dependent method in which second upwind scheme is used to discretize the convective and diffusive terms of momentum and energy equations in a staggered mesh (fig. 2b). The pressure, temperature and density are defined in the center of a cell, the velocity components on the faces of the cell on which the components are orthogonal [1]. The solution of the finite difference equations proceeds in two steps. In the first step, the unsteady momentum equations (in finite difference form) are solved explicitly for the velocity field in a new time level. In the second step an SOR-like pressure-velocity correction method is used to calculate iteratively the presure and velocity fields which satisfy the continuity equation. This pressure-velocity iteration corresponds to the solution of Poisson equation for pressure [2]. For the incompressible case, the energy equation is solved at the end after a steady state solution of momentum and continuity equation is obtained. For the compressible flow, the energy equation is solved at each step along with the pressure-velocity iteration.

The pressure-velocity iteration takes nearly 90% of the computational time. This computational time has been reduced by implementing a multigrid technique in the SOR-like iterations. The details of the computational scheme can be seen in ref. [3].

RESULTS AND DISCUSSION

For the initial condition, uniform temperature and fully developed flow of incompressible medium for both compressible and incompressible cases have been assumed. Furthermore it has been assumed that the channel walls (plate fins) and the vortex generator possess a temperature which is 30% larger than the incoming gas temperature. For boundary conditions at the exit the second derivatives of variables have been assumed to be zero.

Fig. 3 shows velocity vectors on nine cross cross sections in the channel for incompressible flow. One notices the formation of the vortices and deformation of the vortex core to an elliptical cross section as the vortices move along in the channel. Such deformation of vortex cores in a channel has also been observed in experiments [4]. Computations also show, as has been observed experimentally [4] that vortices behind the delta wing move outward away from the axis and also towards the bottom plate of the channel.

Qualitatively, the flowfield for compressible case is similar to incompressible flow. Figures 4a, 4b and 4c show isolines of velocity differences Δu, Δv, and Δw (u, v and w are usual velocity components) between incompressible and compressible flows at a channel cross section one wing length dowmstream from the wing base. Large differences (isolines 1 or 2) appear under the vortex axis as well as on the heated bottom plate and on the left and right boundaries. Figure 4 makes it clear that axial (u) as well as azimuthal (v and w) components of velocity near the wall become much larger for compressible than for imcompressible flows. The largest differences appear in the lateral component w. As cold fluid comes in contact with the hot plate, it quickly expands and is accelerated.

In figure 4 only density has been assumed to be temperature dependent for compressible flows, and viscosity has been assumed to be constant. If viscosity is also taken to be temperature dependent for compressible flow, the differences between velocities for compressible and incompressible flows become smaller. Figure 5a shows isolines of absolute velocity differences for incompressible flow with constant properties and compressible flow with variable properties. Figure 5b shows corresponding isolines for compressible flows with variable properties and constant properties. The difference in velocities in fig. 5a is larger than in 5b

ACKNOWLEDGEMENT

This work has been supported by the Deutsche Forschungsgemeinschaft

REFERENCES

[1] Hirt, C.W., Nicols, B.D., Romero, N.C., "SOLA - A Numerical Solution Algithm for Transient Flows", Los Alamos Scientific Report LA-5652 (1975).

[2] Brandt, A., Dendy. Jr., J.E., Ruppel, H.J., "The Multigrid Method for Semi-Implicit Hydrodynamic Codes", J. Comp. Phys. 34, (1980).

[3] Brockmeier, U., "Numerische Verfahren zur Berechnung dreidimensionaler Strömungs- und Temperaturfelder in Kanälen mit Längswirbelerzeugern und Untersuchung von Wärmeübergang und Strömungsverlust", Dissertation Ruhr-Universität Bochum (1987).

[4] Kallweit, P., "Längswirbelerzeuger für den Einsatz in Lamellenwärmetauschern", Dissertation, Ruhr-Universität Bochum (1986).

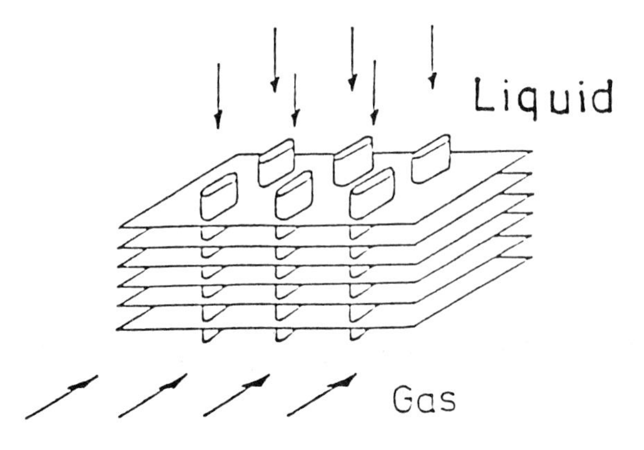

Fig. 1 Schematic of a plate-fin heat exchanger

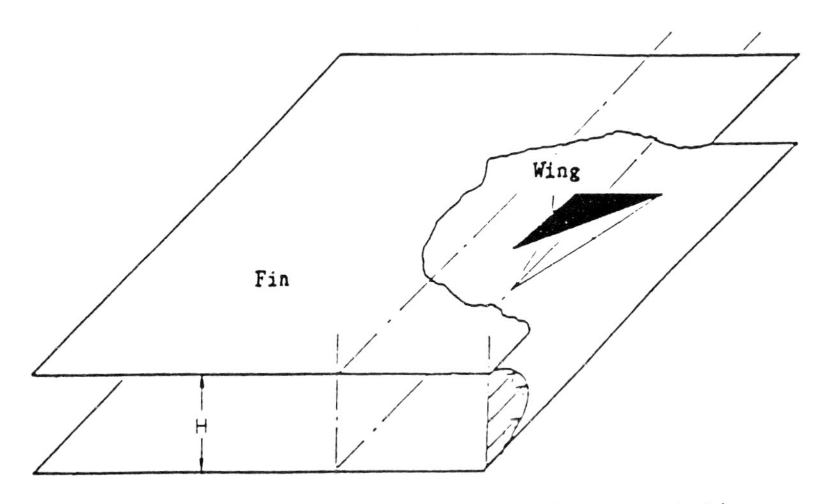

Fig. 2a Schematic of the model geometry for computation

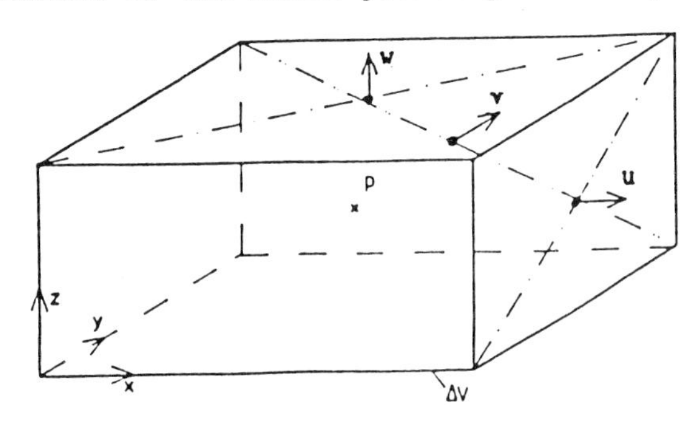

Fig. 2b Three dimensional staggered grid showing definitions for discrete variables

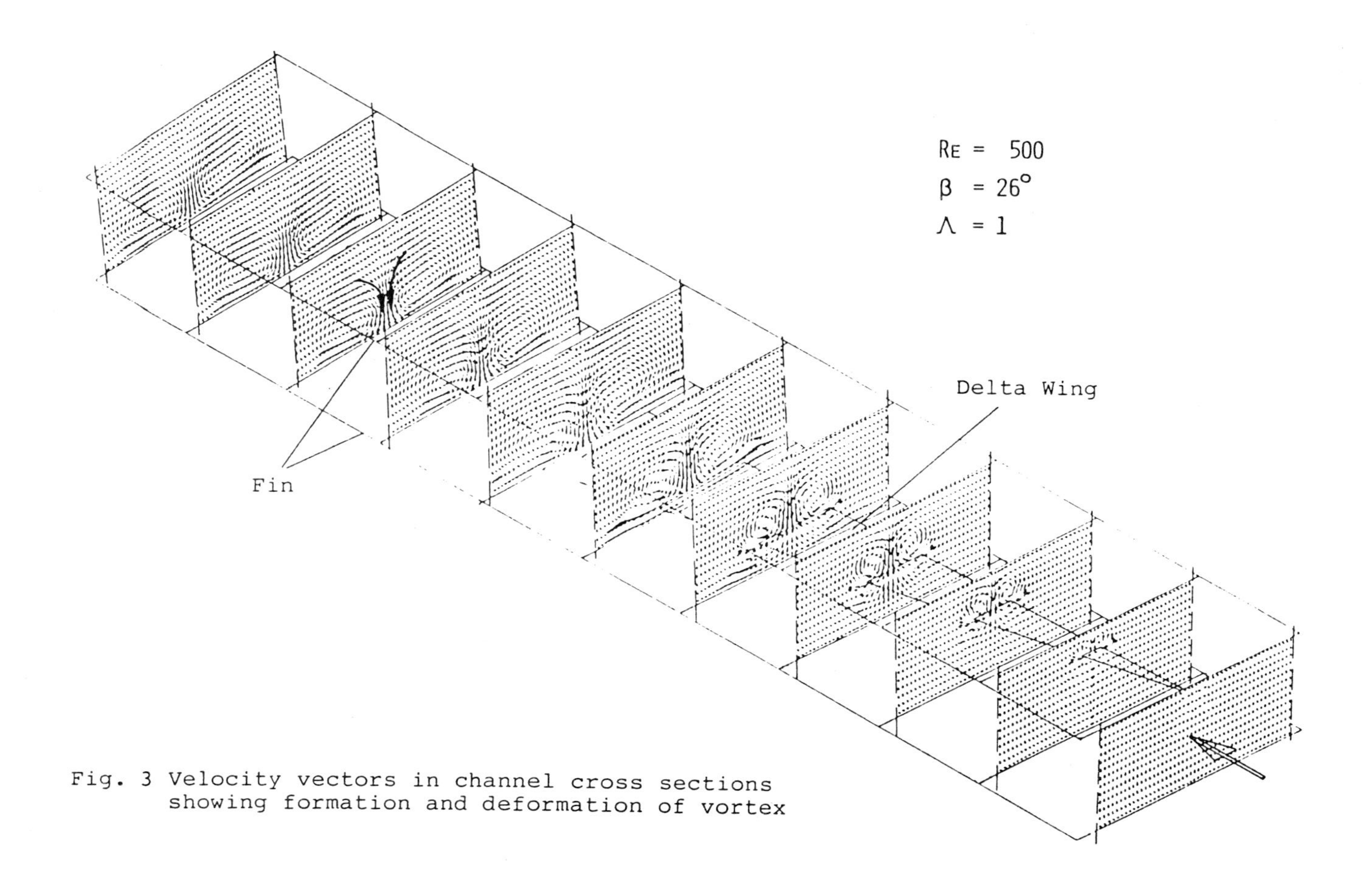

Fig. 3 Velocity vectors in channel cross sections showing formation and deformation of vortex

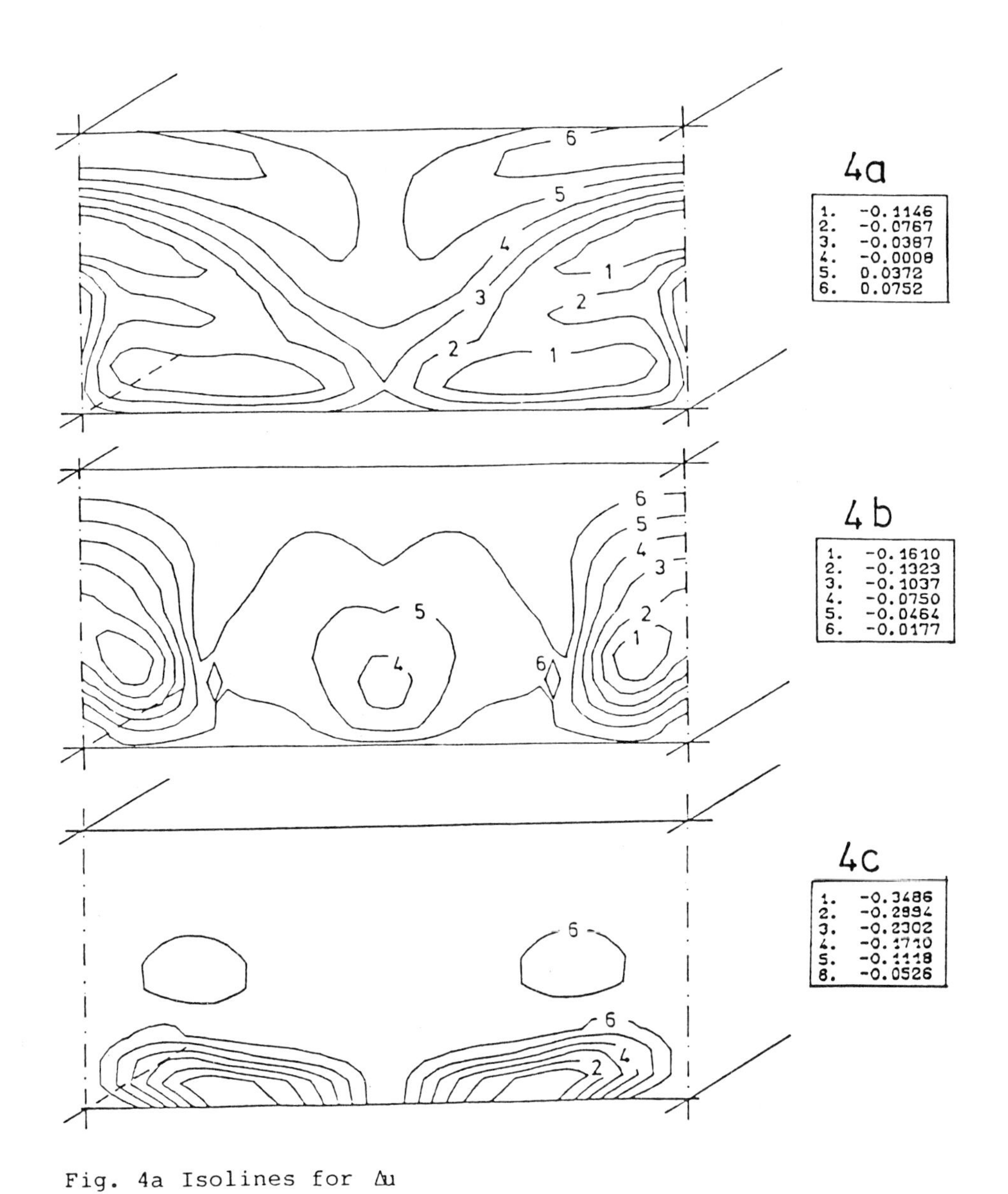

Fig. 4a Isolines for Δu
Fig. 4b Isolines for Δv
Fig. 4c Isolines for Δw
Isolines for differences of velocity components between incompressible and compressible flows at a channel cross section one wing length downstream of wing base. Re = 500, Wing geomerty is same as in fig. 3.

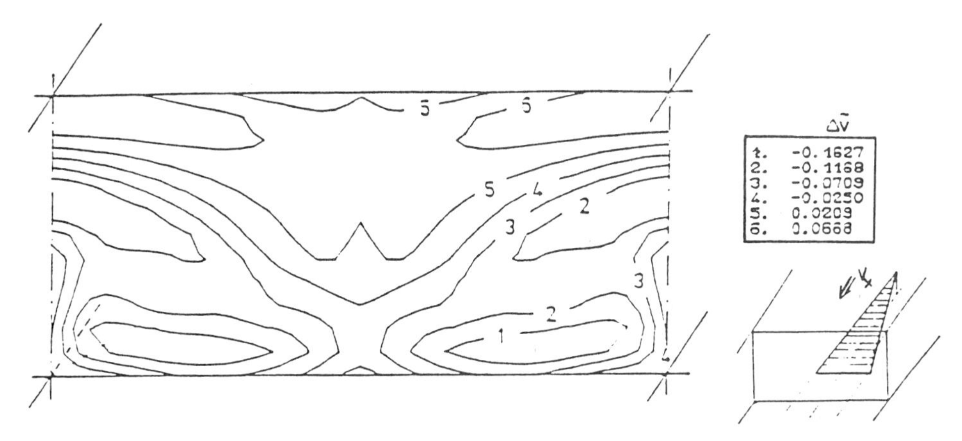

Fig. 5a Isolines of differences of absolute velocities for incompressible flow with constant properties and compressible flow with variable properties in a channel cross section: one wing length downstream behind a delta wing with $\beta = 26^{\circ}$, $\Lambda = 1$.

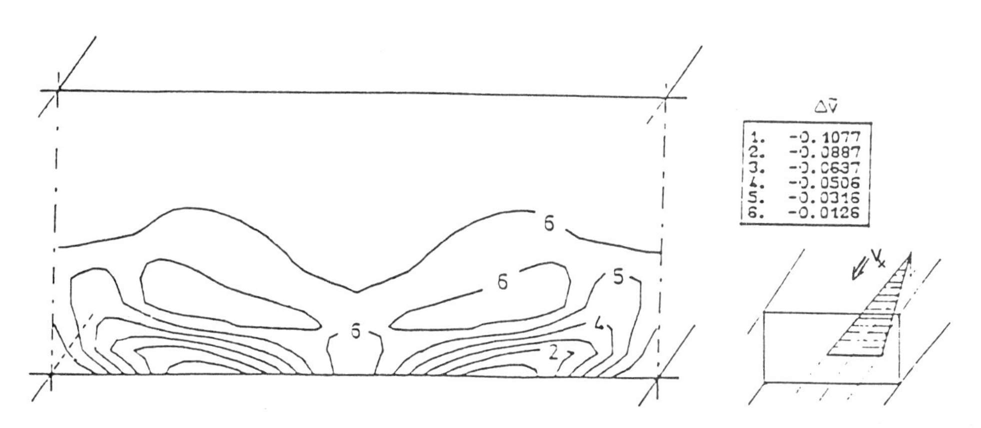

Fig. 5b Isolines of differences of absolute velocities for compressible flow with variable properties and compressible flow with constant properties for the case of fig. 5a.

NUMERICAL COMPUTATION OF EXTERNAL TRANSONIC FLOWS

D M Causon
Department of Mathematics & Physics
Manchester Polytechnic
Manchester M1 5GD
UK

SUMMARY

An improved Euler solver is presented for computing steady axisymmetric transonic flows around projectiles. The method is pseudo time-dependent, split and uses a finite volume formulation. Shock waves are captured crisply by means of a total variation diminishing (TVD) MacCormack scheme which has no problem-dependent adjustable parameters. The method is robust, versatile and holds promise for treating complex three-dimensional projectile geometries, including vortical base flow regions, accurately and economically. A Navier-Stokes solver based on the method described is currently being developed.

INTRODUCTION

Computational aerodynamics is playing an increasingly valuable role in practical aerodynamic analyses. The increasing number of solutions of compressible flow problems around complex configurations testifies to the significant contribution which theoretical analysis can make in aircraft design. A major area of concern in artillery shell design is the total aerodynamic drag. The designer, seeking maximum range and terminal velocity is looking for ways of reducing the drag. The total drag of a projectile can be divided into three components: (1) pressure (wave) drag; (2) viscous (skin friction) drag and (3) base drag. For a typical shell geometry at Mach 0.9, the respective components account for (1) 20%; (2) 30% and (3) 50% of the total drag. The pressure and viscous drag cannot generally be reduced significantly without adversely affecting the stability of the shell. Recent attempts to reduce the total drag have been directed at reducing the base drag. The use of boat-tailing has been found very effective in reducing the drag; however at transonic speeds it has a negative effect on the stability of the shell. Another effective means of reducing the drag is that of base injection wherein a small amount of mass is injected into the base region to increase the base pressure and thereby reduce the drag. Studies of mass injection have been quite extensive in supersonic flow, but less widely studied in the transonic range. Most of the latter work has been largely empirical or experimental, and only recently have sophisticated numerical techniques been employed [1,2].

The present approach is based on a finite-volume method described at previous GAMM meetings [3,4]. The method uses a total-variation diminishing (TVD) MacCormack scheme in which it is possible selectively to add artificial compression to improve the resolution of computed discontinuities. A multi-domain treatment supports the computation of the flow around projectile geometries, including the recirculatory base-flow region.

FORMULATION

Since a body-fitted mesh will be required, the equations of motion are written in integral form as a prelude to discretisation by the finite volume method.

$$\frac{\partial}{\partial t}\iiint_{vol} \underset{\sim}{U}\, d\,vol + \oiint_s \underline{\underline{H}} \cdot \underline{\hat{n}}\, ds = 0 \tag{1}$$

where $\underset{\sim}{U} = [\rho,\rho w_1,\rho w_2,\rho w_3,e]^T$ $\qquad \underline{\underline{H}} = \begin{bmatrix} \rho \underline{q} \\ \rho w_1\ \underline{q} + p\underline{a}_1 \\ \rho w_2\ \underline{q} + p\underline{a}_2 \\ \rho w_3\ \underline{q} + p\underline{a}_3 \\ (e+p)\underline{q} \end{bmatrix}$,

and the flow velocity $\underline{q} = w_\ell a_\ell$, where a_ℓ are the Cartesian unit base vectors. We solve equations (1) using a factored sequence of one-dimensional difference operators, where each component operator relates to its respective co-ordinate direction. Further details may be found in [5].

TVD MACCORMACK FINITE VOLUME SCHEME (TVDM)

The MacCormack finite volume operator $L_1(\Delta t)$ is:

$$\underset{\sim}{U}\,\overline{^{n+1}_{ijk}} = \underset{\sim}{U}^n_{ijk} - \Delta t \left[\underline{\underline{H}}^n_{ijk}\underline{S}_{i+½} + \underline{\underline{H}}^n_{i-1jk}\underline{S}_{i-½} \right] \tag{2a}$$

$$\underset{\sim}{U}^{n+1}_{ijk} = ½ \left[\underset{\sim}{U}^n_{ijk} + \underset{\sim}{U}^{\overline{n+1}}_{ijk} - \Delta t \left[\underline{\underline{H}}^{\overline{n+1}}_{i+1jk}\underline{S}_{i+½} + \underline{\underline{H}}^{\overline{n+1}}_{ijk}\underline{S}_{i-½} \right]\right] \tag{2b}$$

where $\underset{\sim}{U}_{ijk} = vol_{ijk}[\rho,\rho w_1,\rho w_2 \rho w_3,e]^T_{ijk}$ and $\underline{S}_{i\pm½}$ are the area vectors on opposite faces on the cell, corresponding to the surface x^1 = constant. Scheme (2) can be updated easily to total variation diminishing (TVD) form by appending to the right hand side of the "corrector" step (2b) the term:

$$+ \left[K^+_{i+½}\left(r^+_i\right) + K^-_{i+½}\left(r^-_{i+1}\right)\right] \Delta \underset{\sim}{U}^n_{i+½} - \left[K^+_{i-½}\left(r^+_{i-1}\right) + K^-_{i-½}\left(r^-_i\right)\right] \Delta \underset{\sim}{U}^n_{i-½}\ , \tag{3}$$

where for clarity we have suppressed subscripts j,k and

$$K^{\pm}(r^{\pm}) = 0.5C(\nu)[1-\Phi(r^{\pm})]\ ,$$

$$\nu = \max_i |\lambda_i| \frac{\Delta t}{\Delta x^1}\ ,$$

$$c(\nu) = \begin{cases} \nu(1-\nu)\ , & \nu \leqslant 0.5 \\ 0.25\ , & \nu > 0.5 \end{cases}\ ,$$

$$r^+_i = \frac{\left[\Delta \underset{\sim}{U}^n_{i-½}\ ,\ \Delta \underset{\sim}{U}^n_{i+½}\right]}{\left[\Delta \underset{\sim}{U}^n_{i+½}\ ,\ \Delta \underset{\sim}{U}^n_{i+½}\right]}\ , \qquad r^-_i = \frac{\left[\Delta \underset{\sim}{U}^n_{i-½}\ ,\ \Delta \underset{\sim}{U}^n_{i+½}\right]}{\left[\Delta \underset{\sim}{U}^n_{i-½}\ ,\ \Delta \underset{\sim}{U}^n_{i-½}\right]}$$

$$\Phi(r) = \begin{cases} \min(2r,1)\ , & r > 0 \\ 0\ , & r \leqslant 0\ . \end{cases} \tag{4}$$

TVD MACCORMACK SCHEME WITH ARTIFICIAL COMPRESSION (TVDMAC)

In order selectively to add artificial compression in regions where the numerical solution changes abruptly, we append to the right hand side of the corrector step of the TVDM scheme the term:

$$-\frac{\lambda}{2}\left[\,\theta_{i+\frac{1}{2}}G_{i+\frac{1}{2}}-\theta_{i-\frac{1}{2}}G_{i-\frac{1}{2}}\,\right], \tag{5}$$

where again for clarity we have suppressed subscripts j,k and:

$$\theta_{i+\frac{1}{2}} = \max\left[\,\hat{\theta}_i,\hat{\theta}_{i+1}\,\right],$$

$$\hat{\theta}_i = \begin{cases} \left|\dfrac{|\Delta_{i+\frac{1}{2}}\rho| - |\Delta_{i-\frac{1}{2}}\rho|}{|\Delta_{i+\frac{1}{2}}\rho| + |\Delta_{i-\frac{1}{2}}\rho|}\right|, & |\Delta_{i+\frac{1}{2}}\rho| + |\Delta_{i-\frac{1}{2}}\rho| > \varepsilon \\ 0, & |\Delta_{i+\frac{1}{2}}\rho| + |\Delta_{i-\frac{1}{2}}\rho| \leqslant \varepsilon \end{cases}$$

$$\varepsilon = 0.01 \max_i |\rho^n_{i+1} - \rho^n_i|,$$

$$G^m_{i+\frac{1}{2}} = g^m_i+g^m_{i+1} - |g^m_{i+1}-g^m_i|\ \mathrm{sgn}\left[\,U^m_{i+1}-U^m_i\,\right],$$

$$g^m_i = \alpha_i\,\left(U^m_{i+1}-U^m_{i-1}\right),$$

$$\alpha_i = \max\left\{0,\ \min_{1\leqslant m\leqslant M}\frac{\min\left[\left(U^m_{i+1}-U^m_i\right),\ \left(U^m_i-U^m_{i-1}\right)\ \mathrm{sgn}\left(U^m_{i+1}-U^m_i\right)\right]}{|U^m_{i+1}-U^m_i| + |U^m_i-U^m_{i-1}|}\right\}$$

and where m = 1(1)M, M being the number of components of the solution vector $\underset{\sim}{U}$.

The M, TVDM and TVDMAC schemes have been used to solve the one-dimensional Riemann problem. At time t=0, the left and right states are:

$$\underset{\sim}{U}_L = \begin{bmatrix} 0.445 \\ 0.311 \\ 8.928 \end{bmatrix}, \quad \underset{\sim}{U}_R = \begin{bmatrix} 0.5 \\ 0.0 \\ 1.4275 \end{bmatrix}.$$

With increasing time, a mixing process takes place such that a rarefaction wave moves to the left and a contact discontinuity and a shock wave move to the right. The results shown in Fig 1 were obtained after 100 time steps with 140 cells and λ in (5) set to 1.0. Other computed solutions are given in [6]. It can be seen that scheme TVDM exhibits none of the oscillations of the M scheme and that the resolution of the contact discontinuity is improved dramatically with artificial compression. The results shown in Fig 1 cannot be distinguished from those obtained from computationally more expensive TVD schemes based on Riemann solvers (see Harten [6]). Clearly, any production computer code employing the MacCormack method can be modified quickly and simply, as described above, to yield a high-resolution scheme.

The computational overhead associated with artificial compression term (5) is not as high as one might expect since (5) is turned off in regions where the solution is smooth. The automatic switch functions $\theta_{i\pm\frac{1}{2}}$ appearing in (5) can be replaced by the flux limiter functions given in square brackets in (3). This saves a small amount of computing time without significantly degrading the results. Yee [7] gives various flux limiters which could be employed in place of (4). We intend to study these in the future. Further discussion of the use of artificial compression techniques can be found in Harten [6].

MULTI-DOMAIN OPERATOR-SPLITTING ALGORITHM (MDOSA)

Fig 2a,b illustrates a secant-ogive cylinder boat-tail (SOCBT) geometry, typical of an artillery shell and the flow pattern in the base region. The solution procedure for this problem is shown in Fig 3 and requires a multi-domain approach with two mesh systems. The algorithm is implemented through the following four steps:

1. In Mesh 2 (base region) L_1 is applied along all rows (requiring a "base" boundary condition on the left and extrapolation of variables on the right).

2. In Mesh 1 (field region) L_2 is applied along all columns from the inflow boundary to radial line XX. (A flow tangency condition is used on the left and a "radiation" boundary procedure on the right.)

3. In Mesh 2 and Mesh 1 (base and field together) L_2 is applied along all columns from XX+1 to the downstream boundary (with a symmetry boundary condition applied on the centre-line of the wake and a right-hand boundary procedure as 2).

4 In Mesh 1 (field region) L_1 is applied along all rows (requiring an "inflow" boundary condition on the left and extrapolation of variables at downstream boundary).

The initial conditions in the field region are that the flow variables are assumed uniformly to be those consistent with the freestream Mach number. In the base region the fluid is assumed to be at rest. The TVDM operators $L_1(\Delta t)$ and $L_2(\Delta t)$ are used (in steps 1,3) whenever the solution in the base region is advanced, in order to preclude any spurious oscillations when mixing begins to occur. Elsewhere, the Operator-Switching MacCormack-Upwind operators $L_1(\Delta t)$ and $L_2(\Delta t)$ are used (see [5]).

In order to simulate mass injection, the non-porous boundary condition applied along the base ($\underline{q}\cdot\hat{\underline{n}}= 0$) is replaced by:

$$w_1 = u_J$$
$$w_2 = 0.0$$
$$\rho = \rho_J = \rho_{st} ,$$

where u_J is the velocity of the injected air and the stagnation density ρ_{st} is given by:

$$\frac{\rho_{st}}{\rho_\infty} = \left[1 + \frac{\gamma-1}{2} M_\infty^2 \right]^{\frac{1}{\gamma-1}} .$$

The amount of air injected into the base region can be specified by the mass flow rate $\dot{m}_J$. Then, since ρ_J and A_J (the area of the jet) are known, u_J can be calculated for any specified $\dot{m}_J$. However, it is customary to specify a mass injection parameter, $I = \dot{m}_J/\rho_\infty U_\infty A$ where A is area of the base.

RESULTS

The results shown in Figs 4a-d have been obtained using the Multi-Domain Operator-Splitting Algorithm (MDOSA) in axysymmetric form and implemented in accordance with Steps 1-4. Treatment of cases involving mass injection require a modification of the "base" boundary condition in Step 1, as outlined above. The results relate to the SOCBT geometry at M=0.9. The Cp distribution (Fig 4a) and recirculatory base flow pattern (Fig 4b) agree closely with the results of Sturek [2] who used a Navier-Stokes solver. As can be seen in Figs 4c,d, mass injection can reduce recirculation in the base region significantly.

It is thought that whilst the present Euler solver is a useful tool, accurate predictions of the flow around a projectile, particularly over aft sections, will require a Navier-Stokes solver. This would allow pressure fluctuations to feed back from the base region through the boundary layer along the side of the body, thus changing the shock pattern and, hence, the forces and moments on the projectile. To this end, the present high-resolution TVD method will be used to replace the explicit part of MacCormack's 1982 [8] explicit-implicit solver. The absence of problem-dependent adjustable parameters for artificial viscosities should lead to a more robust algorithm. We hope to report on this in the near future.

REFERENCES

[1] J Sahu & C J Nietubicz, "Navier-Stokes computations of projectile base flow with and without base injection", AIAA Paper 83-0224, 1983.

[2] W B Sturek, "Applications of CFD to the aerodynamics of a spinning shell", AIAA paper 84-0323, 1984.

[3] D M Causon & P J Ford, "Computations in external transonic flow", in Proc 5th GAMM Conf on Num Meth in Fl Mech, Vieweg, 1984.

[4] D M Causon & C M Kwong, "Numerical experiments with a TVD MacCormack scheme", in Proc 6th GAMM Conf on Num Meth in Fl Mech, Vieweg, 1986.

[5] D M Causon & P J Ford, "An improved Euler method for computing steady transonic flows", in Proc 9th Int Conf on Num Meth in Fl Dyn, Springer-Verlag, 1985.

[6] A Harten, "High resolution schemes for hyperbolic conservation laws", J Comp Phys, 49, pp 357-393, 1983.

[7] H Yee, "Construction of explicit and implicit symmetric TVD schemes and their applications", J Comp Phys, 68, pp 151-179, 1987.

[8] R W MacCormack, "A numerical method for solving the equations of compressible viscous flow", JAIAA, 20, pp 1275-1281, 1982.

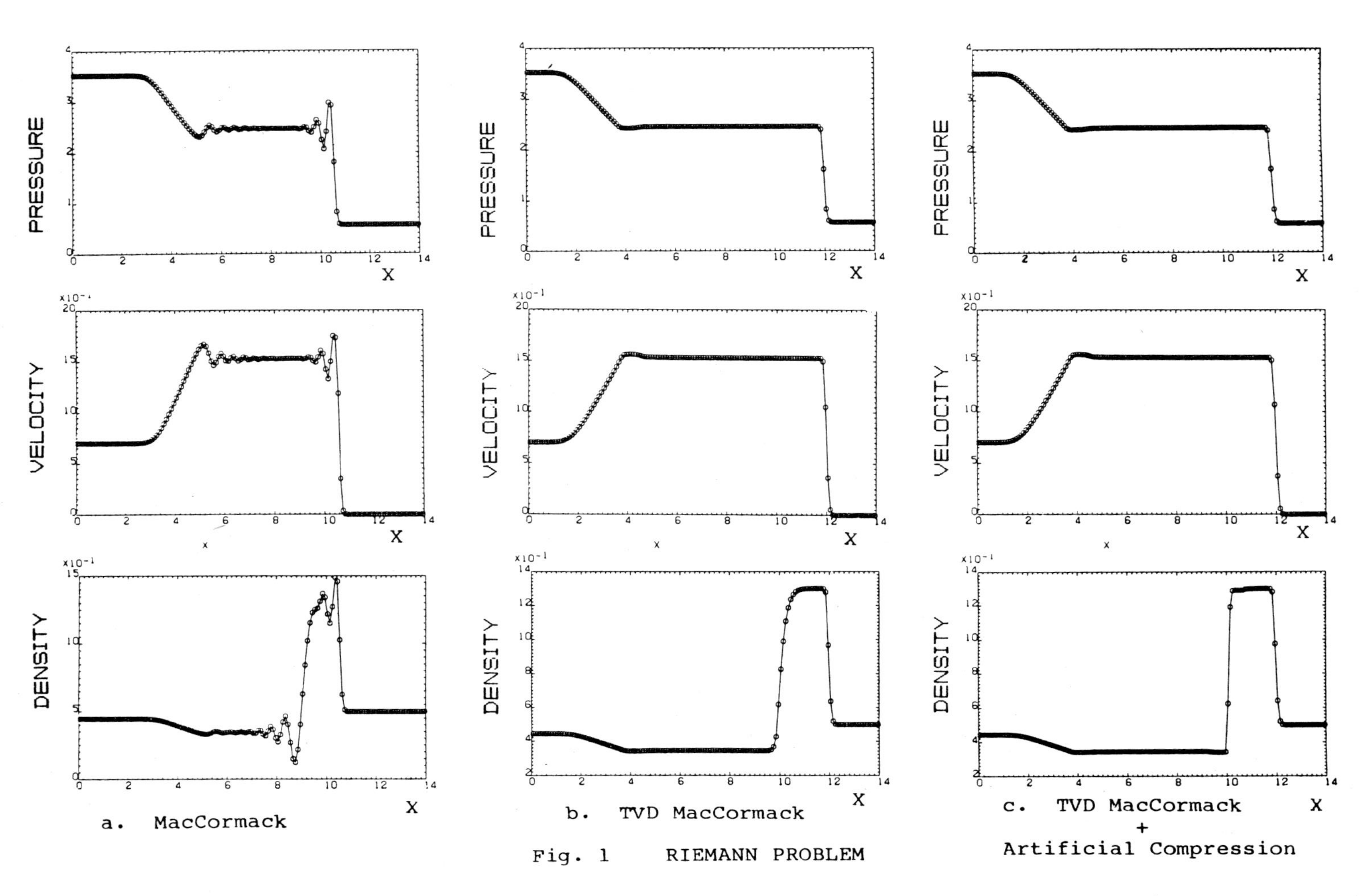

Fig. 1 RIEMANN PROBLEM

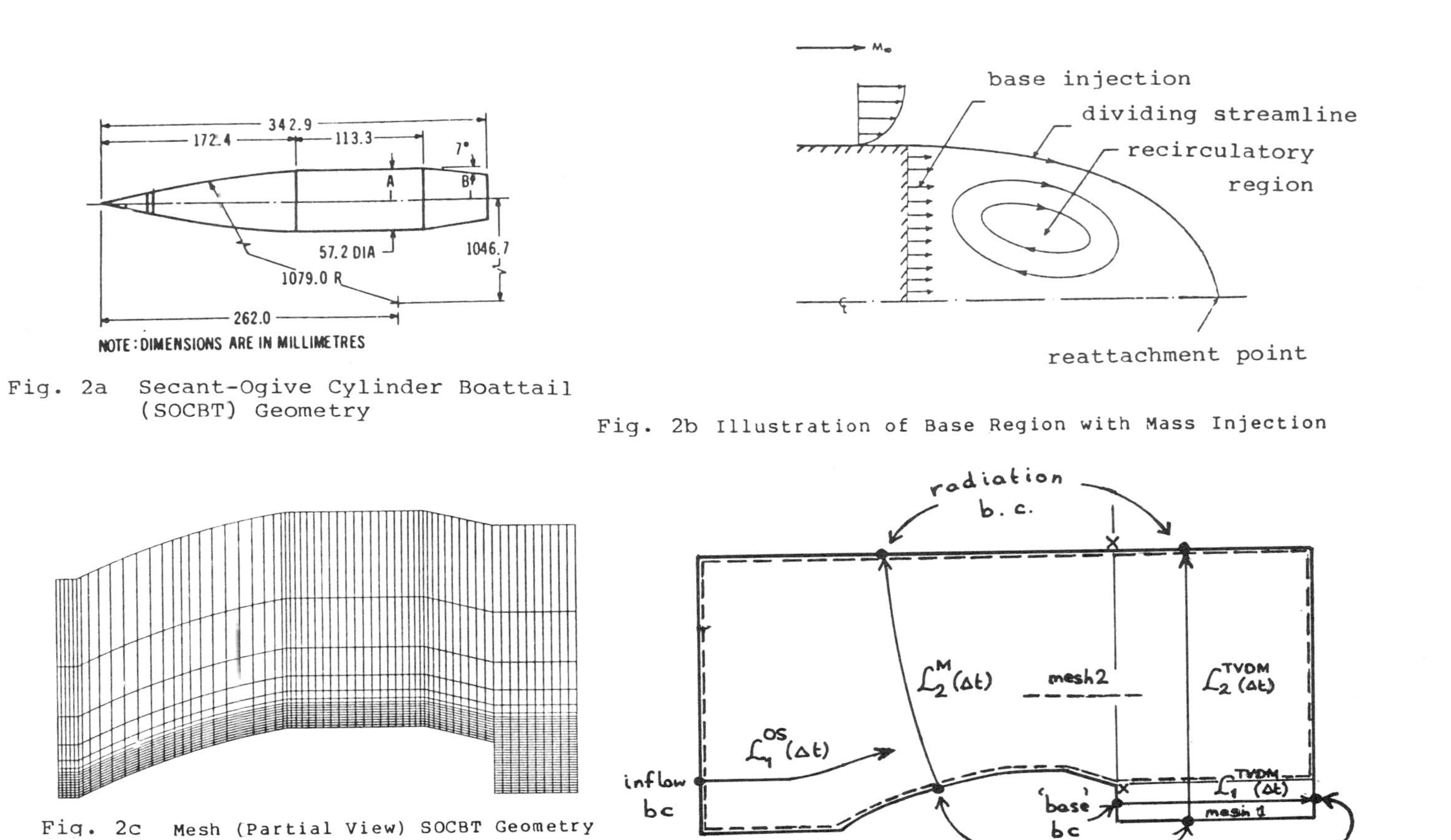

Fig. 2a Secant-Ogive Cylinder Boattail (SOCBT) Geometry

Fig. 2b Illustration of Base Region with Mass Injection

Fig. 2c Mesh (Partial View) SOCBT Geometry

Fig. 3 Solution Procedure

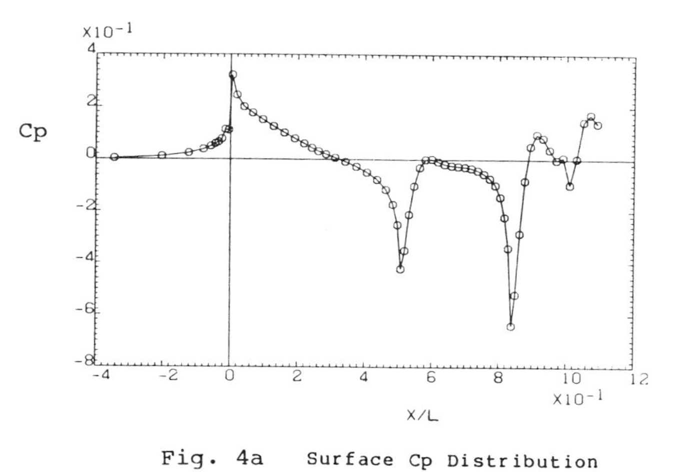

Fig. 4a Surface Cp Distribution

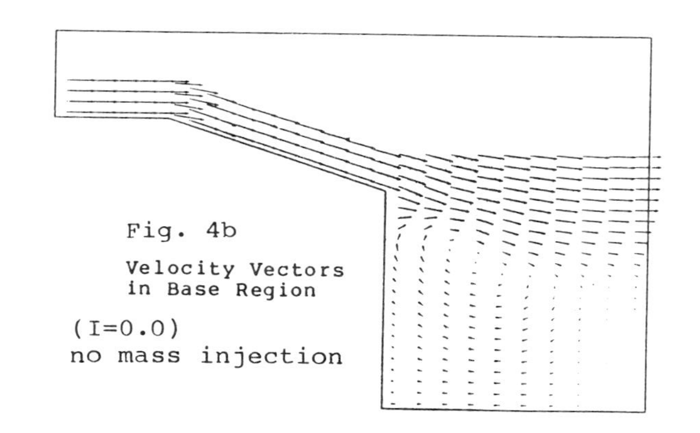

Fig. 4b

Velocity Vectors in Base Region

(I=0.0)
no mass injection

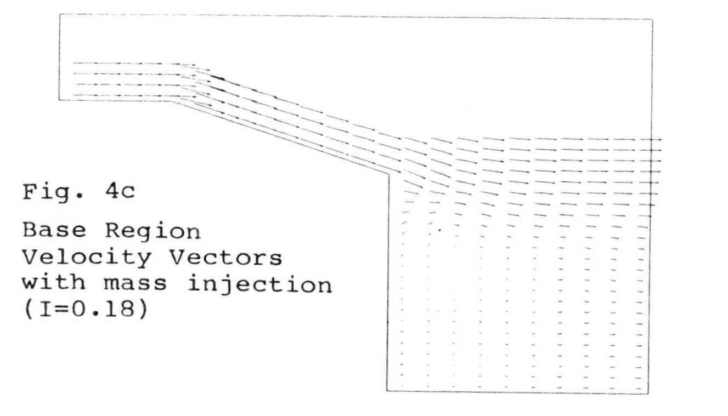

Fig. 4c

Base Region
Velocity Vectors
with mass injection
(I=0.18)

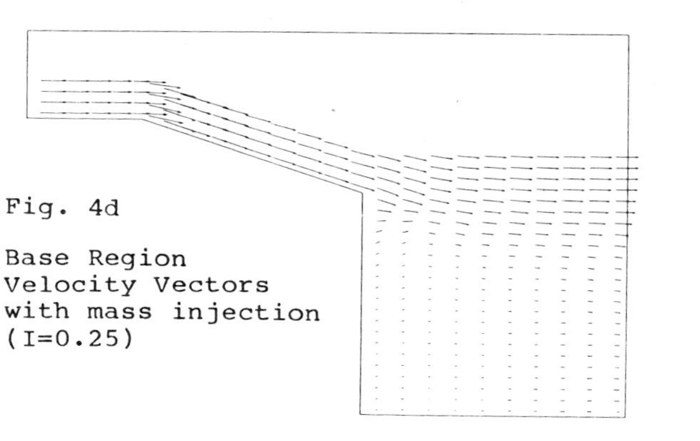

Fig. 4d

Base Region
Velocity Vectors
with mass injection
(I=0.25)

A MULTIGRID METHOD FOR STEADY INCOMPRESSIBLE NAVIER-STOKES EQUATIONS IN PRIMITIVE VARIABLES FORM

E. DICK

Department of Machinery, State University of Ghent

Sint Pietersnieuwstraat 41, B-9000 Gent, Belgium

ABSTRACT

The flux-splitting method is applied to the convective part of the steady Navier-Stokes equations, for incompressible flow. Partial upwind differences are introduced in this split first order part while central differences are used in the second order part. The set of discrete equations, obtained in this way has the property of positiveness, so that it can be solved by collective variants of relaxation methods. The partial upwinding is optimized in the same way as for a scalar convection-diffusion equation, involving, however, several Peclet numbers. It is shown that with the optimum partial upwinding accurate results can be obtained. A full multigrid method in W-cycle form, using red-black successive underrelaxation, half-injection and bilinear interpolation is described. The efficiency of this method is demonstrated.

INTRODUCTION

The flux-vector splitting method was introduced by Steger and Warming [1] to solve unsteady Euler equations. Further, it was shown by Jespersen [2] that the flux-vector splitting method can also be used on the steady Euler equations, to generate discrete equations which form a positive set so that a solution by relaxation methods, in multigrid form, is possible.

In this paper, the flux-vector splitting method is applied to the convective (i.e. Euler-) part of the Navier-Stokes equations for incompressible flow. The fundamentals of this flux-vector splitting method were already outlined by the author in [3] . It was shown there that accurate results can be obtained. In this paper, a multigrid version of the algorithm is described.

UPWIND DIFFERENCING FOR SYSTEMS OF EQUATIONS

A system of steady convective equations, such as Euler equations, can be written in linearized form as :

$$A \frac{\partial \xi}{\partial x} + B \frac{\partial \xi}{\partial y} = 0 \qquad (1)$$

where ξ is the vector of dependent variables.

For a convective system, the matrices A and B have real eigenvalues and a complete set of eigenvectors. As a consequence, it is always possible to split the matrices into a sum of a matrix with non-negative eigenvalues and a matrix with non-positive eigenvalues :

$$A = A^+ + A^- \qquad\qquad B = B^+ + B^- .$$

Equation (1) then can be written in split form as :

$$A^+ \frac{\partial^+ \xi}{\partial x} + A^- \frac{\partial^- \xi}{\partial x} + B^+ \frac{\partial^+ \xi}{\partial y} + B^- \frac{\partial^- \xi}{\partial y} = 0 . \tag{2}$$

An upwind discretization of (2) then is obtained when the +terms are discretized by backward differences and the -terms by forward differences :

$$A^+(\xi_{i,j} - \xi_{i-1,j}) + A^-(\xi_{i+1,j} - \xi_{i,j}) + B^+(\xi_{i,j} - \xi_{i,j-1}) + B^-(\xi_{i,j+1} - \xi_{i,j}) = 0 .$$

or

$$(A^+ + B^+ - A^- - B^-)\xi_{i,j} = A^+\xi_{i-1,j} + (-A^-)\xi_{i+1,j} + B^+\xi_{i,j-1} + (-B^-)\xi_{i,j+1} . \tag{3}$$

Although it is not a general rule, for a large class of systems, the coefficient matrix $C = A^+ + B^+ - A^- - B^-$ has positive eigenvalues. In this case, the system of equations (3) forms a positive set of equations, all matrix coefficients having non-negative eigenvalues. It is clear that collective variants of relaxation methods can be used on positive sets of equations.

A systematic way to split the matrices A and B in (1) is the flux-vector splitting technique of Steger and Warming [1] , based on the splitting of the eigenvalue matrices.

By denoting the eigenvalue matrices of A and B by Λ_A and Λ_B and the left eigenvector matrices by X_A and X_B, obviously :

$$A = X_A^{-1}\Lambda_A X_A \qquad\qquad B = X_B^{-1}\Lambda_B X_B .$$

The eigenvalue matrices can be split into :

$$\Lambda_A = \Lambda_A^+ + \Lambda_A^- \qquad\qquad \Lambda_B = \Lambda_B^+ + \Lambda_B^-$$

where :

$$\Lambda_A^{\pm} = \mathrm{diag}(\lambda_{iA}^{\pm}) \;,\quad \lambda_{iA}^{+} = \max(\lambda_{iA},0) \;,\quad \lambda_{iA}^{-} = \min(\lambda_{iA},0) \qquad \ldots$$

The split matrices then are obtained by :

$$A^{\pm} = X_A^{-1}\Lambda_A^{\pm} X_A \qquad\qquad B^{\pm} = X_B^{-1}\Lambda_B^{\pm} X_B .$$

FLUX-VECTOR SPLITTING FOR STEADY NAVIER-STOKES EQUATIONS

The steady Navier-Stokes equations for an incompressible fluid are :

$$u\frac{\partial u}{\partial x} + v\frac{\partial u}{\partial y} + \frac{\partial p}{\partial x} = \nu\left(\frac{\partial^2 u}{\partial x^2} + \frac{\partial^2 u}{\partial y^2}\right) ,$$

$$u\frac{\partial v}{\partial x} + v\frac{\partial v}{\partial y} + \frac{\partial p}{\partial y} = \nu\left(\frac{\partial^2 v}{\partial x^2} + \frac{\partial^2 v}{\partial y^2}\right) ,$$

$$c^2(\frac{\partial u}{\partial x} + \frac{\partial v}{\partial y}) = 0 \ . \tag{4}$$

u and v are the Cartesian components of velocity, c is a reference velocity introduced to homogenize the eigenvalues of the system matrices, ν is kinematic viscosity and p is pressure divided by density.

In system form, the set of equations (4) becomes :

$$\begin{pmatrix} u & 0 & 1 \\ 0 & u & 0 \\ c^2 & 0 & 0 \end{pmatrix} \frac{\partial}{\partial x} \begin{pmatrix} u \\ v \\ p \end{pmatrix} + \begin{pmatrix} v & 0 & 0 \\ 0 & v & 1 \\ 0 & c^2 & 0 \end{pmatrix} \frac{\partial}{\partial y} \begin{pmatrix} u \\ v \\ p \end{pmatrix} = \begin{pmatrix} \nu & 0 & 0 \\ 0 & \nu & 0 \\ 0 & 0 & 0 \end{pmatrix} \Delta \begin{pmatrix} u \\ v \\ p \end{pmatrix} \tag{5}$$

or symbolically :

$$A \frac{\partial \xi}{\partial x} + B \frac{\partial \xi}{\partial y} = D(\frac{\partial^2 \xi}{\partial x^2} + \frac{\partial^2 \xi}{\partial y^2}) \ . \tag{6}$$

The eigenvalues of the system matrices A and B are :

$$\lambda_{1A} = u \qquad \lambda_{2A} = \frac{u + \sqrt{u^2+4c^2}}{2} \qquad \lambda_{3A} = \frac{u - \sqrt{u^2+4c^2}}{2}$$

$$\lambda_{1B} = v \qquad \lambda_{2B} = \frac{v + \sqrt{v^2+4c^2}}{2} \qquad \lambda_{3B} = \frac{v - \sqrt{v^2+4c^2}}{2} \ .$$

Obviously, λ_{2A} and λ_{2B} are always positive, λ_{3A} and λ_{3B} are always negative, while λ_{1A} and λ_{1B} change sign with u and v.

Hence :

$$\Lambda_A^+ = \mathrm{diag}(u^+, \lambda_{2A}, 0) \qquad \Lambda_A^- = \mathrm{diag}(u^-, 0, \lambda_{3A})$$

$$\Lambda_B^+ = \mathrm{diag}(v^+, \lambda_{2B}, 0) \qquad \Lambda_B^- = \mathrm{diag}(v^-, 0, \lambda_{3B})$$

with :

$u^+ = \max(u,0) \qquad u^- = \min(u,0) \qquad v^+ = \max(v,0) \qquad v^- = \min(v,0)$

According to the procedure of Steger and Warming, the split matrices become :

$$A^+ = X_A^{-1}\Lambda_A^+ X_A = \begin{pmatrix} \hat{u}^+ & 0 & \alpha_1 \\ 0 & u^+ & 0 \\ \alpha_1 c^2 & 0 & a \end{pmatrix} \qquad A^- = X_A^{-1}\Lambda_A^- X_A = \begin{pmatrix} \hat{u}^- & 0 & \alpha_2 \\ 0 & u^- & 0 \\ \alpha_2 c^2 & 0 & -a \end{pmatrix}$$

$$B^+ = X_B^{-1}\Lambda_B^+ X_B = \begin{pmatrix} v^+ & 0 & 0 \\ 0 & \hat{v}^+ & \beta_1 \\ 0 & \beta_1 c^2 & b \end{pmatrix} \qquad B^- = X_B^{-1}\Lambda_B^- X_B = \begin{pmatrix} v^- & 0 & 0 \\ 0 & \hat{v}^- & \beta_2 \\ 0 & \beta_2 c^2 & -b \end{pmatrix}$$

where $\quad \hat{u}^+ = \alpha_1 u + a \qquad \hat{u}^- = \alpha_2 u - a \qquad \hat{v}^+ = \beta_1 v + b \qquad \hat{v}^- = \beta_2 v - b$

with $\quad \alpha_1 = .5(1+\alpha) \qquad \alpha_2 = .5(1-\alpha) \qquad \beta_1 = .5(1+\beta) \qquad \beta_2 = .5(1-\beta)$

$a = c^2/\sqrt{u^2+4c^2} \qquad b = c^2/\sqrt{v^2+4c^2} \qquad \alpha = u/\sqrt{u^2+4c^2} \qquad \beta = v/\sqrt{v^2+4c^2}$.

The split form of the system (6) becomes :

$$A^+ \frac{\partial \xi}{\partial x} + A^- \frac{\partial \xi}{\partial x} + B^+ \frac{\partial \xi}{\partial y} + B^- \frac{\partial \xi}{\partial y} = D\left(\frac{\partial^2 \xi}{\partial x^2} + \frac{\partial^2 \xi}{\partial y^2}\right). \tag{7}$$

On a rectangular grid, using upwind differences in the first order part and central differences in the second order part, a positive set of equations is obtained.

However, since the momentum equations in (7) have terms in the velocity differences from the convective part and the diffusive part, a partial upwind formulation is possible for these equations, retaining the positiveness.

For example, the momentum-x equation can be discretized as :

$$\hat{u}^+\left(\theta_{xx} \frac{\delta^+ u}{\delta x} + (1-\theta_{xx}) \frac{\delta u}{\delta x}\right) + \hat{u}^-\left(\theta_{xx} \frac{\delta^- u}{\delta x} + (1-\theta_{xx}) \frac{\delta u}{\delta x}\right)$$

$$+ v^+\left(\theta_{xy} \frac{\delta^+ u}{\delta y} + (1-\theta_{xy}) \frac{\delta u}{\delta y}\right) + v^-\left(\theta_{xy} \frac{\delta^- u}{\delta y} + (1-\theta_{xy}) \frac{\delta u}{\delta y}\right)$$

$$+ \alpha_1 \frac{\delta^+ p}{\delta x} + \alpha_2 \frac{\delta^- p}{\delta x} = \nu\left(\frac{\delta^2 u}{\delta x^2} + \frac{\delta^2 u}{\delta y^2}\right) .$$

where

$$\frac{\delta^+ u}{\delta x} = (u_{i,j} - u_{i-1,j})/\Delta x_w \qquad \frac{\delta^- u}{\delta x} = (u_{i+1,j} - u_{i,j})/\Delta x_e$$

$$\frac{\delta u}{\delta x} = \frac{\Delta x_e}{2\Delta x} \frac{\delta^+ u}{\delta x} + \frac{\Delta x_w}{2\Delta x} \frac{\delta^- u}{\delta x}$$

$$\frac{\delta^+ u}{\delta y} = (u_{i,j} - u_{i,j-1})/\Delta y_s \qquad \frac{\delta^- u}{\delta y} = (u_{i,j+1} - u_{i,j})/\Delta y_n$$

$$\frac{\delta u}{\delta y} = \frac{\Delta y_n}{2\Delta y} \frac{\delta^+ u}{\delta y} + \frac{\Delta y_s}{2\Delta y} \frac{\delta^- u}{\delta y}$$

$$\frac{\delta^2 u}{\delta x^2} = \left(\frac{\delta^+ u}{\delta x} - \frac{\delta^- u}{\delta x}\right)/\Delta x \qquad \frac{\delta^2 u}{\delta y^2} = \left(\frac{\delta^+ u}{\delta y} - \frac{\delta^- u}{\delta y}\right)/\Delta y$$

with $\Delta x_w = x_{i,j} - x_{i-1,j} \qquad \Delta x_e = x_{i+1,j} - x_{i,j} \qquad \Delta x = .5(\Delta x_w + \Delta x_e)$

$\Delta y_s = y_{i,j} - y_{i,j-1} \qquad \Delta y_n = y_{i,j+1} - y_{i,j} \qquad \Delta y = .5(\Delta y_s + \Delta y_n).$

A similar discretization can be used on the momentum-y equation, involving θ_{yx} and θ_{yy}. The mass-equation in (7) is to be discretized in a full upwind way.

The optimum values for the partial upwind coefficients θ_{xx}, θ_{xy}, θ_{yx} and θ_{yy} can be determined by expressing that the linearized form of the discrete equations, i.e. coefficients like $\hat{u}^+$, $\hat{u}^-$, v^+, v^-, ..., considered as being constant, is of the form :

$$e^{ux/\nu} \, e^{vy/\nu}.$$

In this way, the optimum value of θ_{xx} is found to be given by :

$$\theta_{xx} = \frac{\dfrac{u(\Delta x_w \sigma_e + \Delta x_e \sigma_w)}{\sigma_e - \sigma_w} - 2\nu}{\hat{u}^+ \Delta x_w - \hat{u}^- \Delta x_e} \qquad (8)$$

where $\sigma_w = \dfrac{1 - e^{-u\Delta x_w/\nu}}{\Delta x_w}$ $\qquad \sigma_e = \dfrac{e^{u\Delta x_e/\nu} - 1}{\Delta x_e}$.

Similar expressions are found for θ_{xy}, θ_{yx} and θ_{yy}.

As is common practice for scalar equations, the expressions for the optimum partial upwind coefficients can be replaced by their expansions for small values of velocity, with a maximum of 1. Expansion of (8) leads to :

$$\theta_{xx} = \min\{(Pe_{xx}/6),1\} \qquad (9)$$

where the Peclet-number is :

$$Pe_{xx} = (\hat{u}^+\Delta x_w - \hat{u}^-\Delta x_e)/\nu . \qquad (10)$$

NUMERICAL EXAMPLE

Figure 1 shows a well known backward facing step problem from [4] , discretized with a coarse grid with 42 elements. This grid is the coarsest of a series of four, the finest grid having 2688 elements. In the construction of finer grids, the same stretching law is applied, as used in the coarse grid.

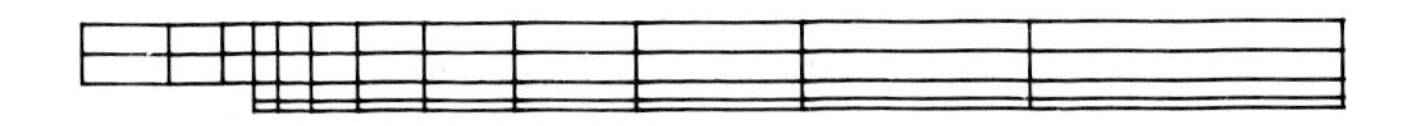

Fig. 1. Backward facing step problem discretized with a coarse grid

The following boundary conditions are imposed. At inlet $u = u_o(y)$, in which $u_o(y)$ is a parabolic profile with a mean velocity c, $v = 0$, and p from an assumption of fully developed flow :

$$\frac{\delta^- p}{\delta x} = \nu \frac{\delta^2 u}{\delta y^2} .$$

At outlet : $p = 0$, $v = 0$, $\delta^+ u/\delta x = 0$.

At solid boundaries : $u = 0$, $v = 0$ and p from an inward discretization of the normal momentum equation.

At horizontal parts of the boundaries this is :

$$\frac{\partial p}{\partial y} = \nu \frac{\partial^2 v}{\partial y^2} . \qquad (11)$$

At vertical parts of the boundaries this is :

$$\frac{\partial p}{\partial x} = \nu \frac{\partial^2 u}{\partial x^2} . \tag{12}$$

At corners, the sum of equations (11) and (12) is taken.

Figure 2 shows the solution obtained with a successive underrelaxation method (relaxation factor 0.95) in red-black ordering for

$$Re = U_{max} h/\nu = 150$$

where U_{max} is the maximum value of the velocity at the inlet section and where h is the step height. The streamlines shown in figure 2 were obtained by integration of the calculated velocity profiles. The reattachment length to step height ratio is about 6. This result is in accordance with the experimental value [4].

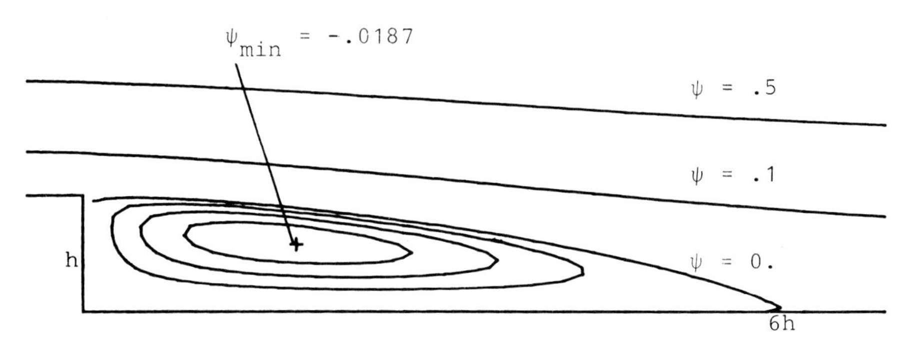

Fig. 2. Streamline pattern for the backward facing step problem, obtained at the finest grid

MULTIGRID FORMULATION

A description of the basics of the multigrid method is not given here. The reader not familiar with the terminology used in this section is referred to [5].

All equations are normalized by bringing the coefficient of u, v and p in the central node, for the momentum-x, momentum-y and pressure (mass) equation respectively, on the value 1. As a result, field equations and boundary equations take a similar form. This allows the use of full weighting as restriction operator for defects at the boundaries, taking a weighted mean of the defects of boundary equations and field equations.

Successive underrelaxation in red-black form was chosen as relaxation algorithm. For a system of first order equation the maximum relaxation factor for stability is 1 (not 2). Maximum convergence rate for a single grid calculation was found to be obtained for a relaxation factor 0.95. Although it is well known that red-black relaxation does not have optimum smoothing properties, this algorithm was chosen for its ease in vectorizing the code.

A full approximation scheme was used. For the restriction operator, experiments were done with full weighting, both on function values and on defects, injection for function values and full weighting for defects and injection for both. In the full weighting versions, experiments were done with full weighting at the boundaries and with weighting restricted to boundary points.

Due to the normalization process, the sum of the weight factors in the full weighting for the defect is to be 2. As is well known, with the red-black ordering, the weight factor using injection for the defect is not to be 2 but 1, i.e. half-injection is to be used. Bilinear interpolation was used as prolongation operator.

The classical cycle configurations were tried : V-cycle, F-cycle and W-cycle. It was found that the W-cycle performs the best. Also the full multigrid method, i.e. using a nested iteration as starting cycle was used. The cycle geometry is shown in figure 3. Each dot represents a relaxation operation. The nested iteration also has a W-form.

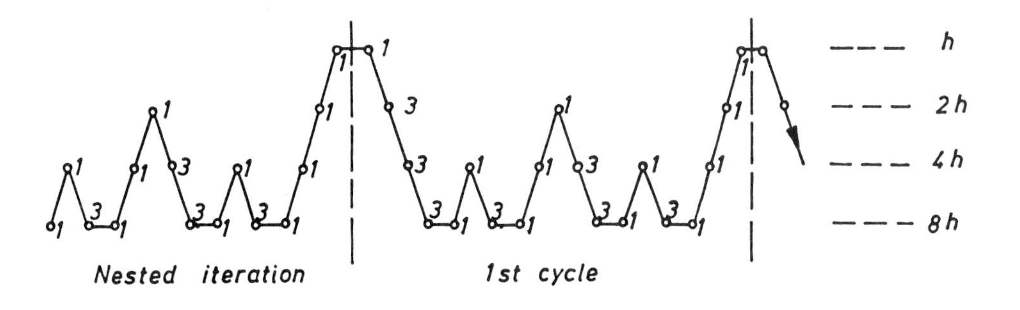

Fig. 3. Geometry of the multigrid cycle

The best efficiency of the multigrid cycle was found to be reached for the same relaxation factor as for single grid calculations : $\omega = 0.95$.

It was found that the performance is rather insensitive to the choice of the restriction operator. Using full weighting for defects is slightly more efficient than using injection, in terms of required number of cycles. Since however injection requires less residue evaluations, in terms of work units the performance is about the same. The performance is also not sensitive to the precise weighting formula : algebraic weighting (i.e. weighting factors 1/2, 1/4, 1/8) or geometric weighting (i.e. weighting factors taking into account the distances between nodes). Therefore, due to its simplicity injection both for functions and for defects was retained for further use.

Figure 4 shows the convergence history for a single grid calculation and a multigrid calculation. The initial condition is a flow with $v = 0$ and $p = 0$ everywhere and with u equal to the inlet profile in the upper part of the flowfield and $u = 0$ in the lower part of the flowfield. In the evaluation of the work of a cycle, on the finest grid, a relaxation and a residue calculation with the associated grid transfer are counted as one work unit.

The work done in each node of the cycle is indicated in figure 3. In going down in the cycle, the 3 represents a residue evaluation in the coarse grid points of the next finer grid and its injection, a residue evaluation on the coarse grid to form the right hand side in the coarse grid equations and one relaxation. The work spent in a cycle is 4.125 work units. The work spent in the nested iteration is 2.141 work units.

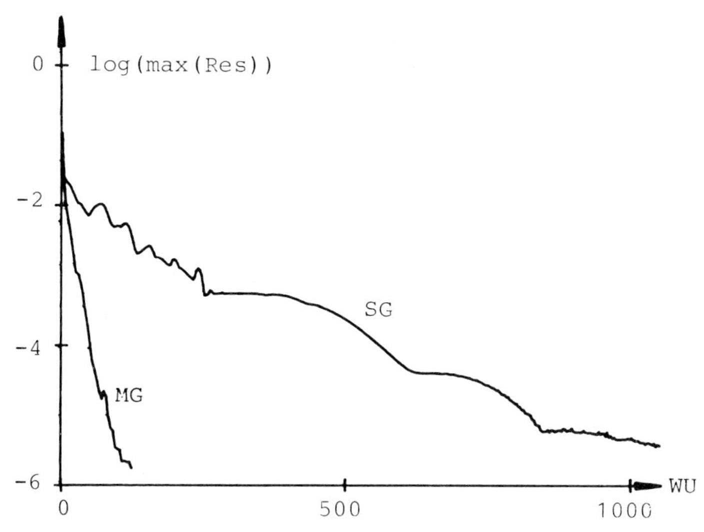

Fig. 4. Convergence history for single grid and multigrid red-black relaxation

CONCLUSION

It was shown that the flux-vector splitting technique can be applied to steady Navier-Stokes equations in incompressible flow, leading to discrete equations which can be solved by vector variants of relaxation schemes. Due to the partial upwinding an accurate solution can be obtained. By the use of the multigrid formulation a very efficient solution technique is realized.

ACKNOWLEDGEMENT

The research reported in this paper was supported by the Belgian National Science Foundation (N.F.W.O.).

REFERENCES

[1]. J.L. Steger and R.F. Warming, "Flux-vector splitting of the inviscid gas-dynamic equations with application to finite difference methods", J. Comp. Phys. 40, 263-293 (1981).

[2]. D.C. Jespersen, "A multigrid method for the Euler equations", AIAA paper 83-0124 (1983).

[3]. E. Dick, "A partial flux-splitting method for steady incompressible Navier-Stokes equations", Proc. Fifth Int. Conf. Num. Meth. Laminar and Turbulent Flow, Montreal, Pineridge Press (1987). 549-559.

[4]. K. Morgan, J. Periaux and F. Thomasset (Eds), "Analysis of laminar flow over a backward facing step", Notes on Numerical Fluid Mechanics, Vol. 9, Vieweg Verlag (1984).

[5]. K. Stüben and U. Trottenberg, "Multigrid methods : fundamental algorithms, model problem analysis and applications", Proc. Conf. on Multigrid Methods, Lecture Notes in Mathematics, Vol. 960, Springer Verlag (1981).

THE MOBILE ELEMENT METHOD FOR SYSTEMS OF CONSERVATION LAWS

M.G. Edwards *
University of Reading
Whiteknights, Reading, U.K.

SUMMARY

In this paper a novel adaptive method is presented for the solution of hyperbolic conservation laws, called the Mobile Element Method (MEM). The underlying philosophy of MEM is to move the nodes of the grid such that the total change of physical variables is minimised. The properties of the method can best be illustrated in the continuous case. The discrete approximation is derived from a functional formulation of MEM which exploits the Moving Finite Element (MFE) framework of Miller [1] and Baines [2] and is effectively a new variant of MFE. The method is applied to a standard shock tube problem in gas dynamics [7] and the Inviscid Burgers system in two dimensions.

§ 1 INTRODUCTION

We shall deal mainly with the continuous form of MEM although a parallel is drawn with the discrete approximation.

Consider the scalar hyperbolic conservation law

$$\frac{\partial u}{\partial t} + \frac{\partial f}{\partial x} = o. \tag{1.1}$$

With some initial data $u(x,o) = u_o(x)$. By applying an adaptive method to solve (1.1) a moving frame of reference is introduced into the problem. We shall choose to work in the moving frame and denote the local rate of change of the variable u with respect to the moving frame by the "Mobile derivative" of u i.e.

$$\frac{Du}{Dt} = \frac{\partial u}{\partial t} + \frac{Dx}{Dt}\frac{\partial u}{\partial x}. \tag{1.2}$$

From (1.2) (1.1) may be rewritten as

$$\frac{Du}{Dt} = -\frac{\partial f}{\partial x} + \frac{Dx}{Dt}\frac{\partial u}{\partial x}. \tag{1.3}$$

Where the grid velocity Dx/Dt is to be specified. A natural choice is to let the grid nodes move along the characteristics so that

$$\frac{Du}{Dt} = o\ , \quad \frac{Dx}{Dt} = \frac{\partial f}{\partial x}\bigg/\frac{\partial u}{\partial x}. \tag{1.4}$$

The advantages of this approach are clear from (1.4), any scheme which renders the first of (1.4) exactly will be "naturally" or "physically" stable provided that the solution gradients are non-zero ($\partial u/\partial x \neq o$). For a non-linear flux f, the formation of a discontinuity will be accompanied by a collision of the nodes which move along the intersecting characteristics and only the physically relevant solution will be selected, [3]. The numerical treatment at a shock is based on the approach of Wathen and Baines [8], full details may be found in [4].

---

*Address from 1/10/87 - College of Engineering, University of Texas at Austin, Texas, U.S.A.

§ 2 SYSTEMS OF CONSERVATION LAWS

To avoid the complexity of using one grid per characteristic component (especially in higher dimensions) the above approach is generalised to systems using a single adaptive grid.

Consider the system of hyperbolic conservation laws

$$\frac{\partial \underline{u}}{\partial t} + \frac{\partial \underline{f}}{\partial x} = o \tag{2.1}$$

with same initial data $\underline{u}(x,o) = \underline{u}_o(x)$. As in §1 we shall introduce the mobile operator and rewrite (2.1) as

$$\frac{D\underline{\underline{u}}}{Dt} = - \frac{\partial \underline{\underline{f}}}{\partial x} + \frac{Dx}{Dt}\frac{\partial \underline{\underline{u}}}{\partial x} \tag{2.2}$$

where Dx/Dt is to be specified. Although we cannot cancel the forcing term on the right hand side of (2.2) we can define Dx/Dt to be the grid velocity which minimises the ℓ_2 norm of the Mobile derivative of $\underline{u}$, [4] i.e.

$$\underset{\frac{Dx}{Dt}}{\text{Min}} \left\| \frac{D\underline{\underline{u}}}{Dt} \right\|_{\ell_2} = \underset{\frac{Dx}{Dt}}{\text{Min}} \left\| \frac{\partial \underline{\underline{f}}}{\partial x} - \frac{Dx}{Dt}\frac{\partial \underline{\underline{u}}}{\partial x} \right\|_{\ell_2}. \tag{2.3}$$

Performing the minimisation of (2.3) we obtain

$$\frac{Dx}{Dt} = \frac{\frac{\partial \underline{\underline{f}}}{\partial x} \cdot \frac{\partial \underline{\underline{u}}}{\partial x}}{\frac{\partial \underline{\underline{u}}}{\partial x} \cdot \frac{\partial \underline{\underline{u}}}{\partial x}}. \tag{2.4}$$

Note that for a single component (2.4) produces the grid velocity of (1.4). By definition (provided $\left\| \frac{\partial \underline{\underline{u}}}{\partial x} \right\|_{\ell_2} \neq o$) (2.4) is the most stable choice of grid velocity in the sense described above. It follows immediately from (2.2) and (2.4) that

$$\frac{D\underline{\underline{u}}}{Dt} \cdot \frac{\partial \underline{\underline{u}}}{\partial x} = o. \tag{2.5}$$

From (2.3) and (2.2) it may be deduced that

$$\left\| \frac{D\underline{\underline{u}}}{Dt} \right\|_{\ell_2} \leqslant \left\| \frac{\partial \underline{\underline{u}}}{\partial t} \right\|_{\ell_2} \tag{2.6}$$

which implies that in a steady state $\left\| \frac{D\underline{\underline{u}}}{Dt} \right\|_{\ell_2} \to o$ and from (2.2) $Dx/Dt \to o$.

Dominant Waves

We shall now examine (2.4) in more detail. Let A denote the Jacobian matrix $\partial \underline{f}/\partial \underline{u}$ then (2.4) may be rewritten as

$$\frac{Dx}{Dt} = \frac{\frac{\partial \underline{\underline{u}}^T}{\partial x} A \frac{\partial \underline{\underline{u}}}{\partial x}}{\frac{\partial \underline{\underline{u}}^T}{\partial x}\frac{\partial \underline{\underline{u}}}{\partial x}}. \tag{2.7}$$

Assuming a locally linear behaviour in $\underline{f}$ so that A is constant then (2.7) may be expressed in terms of the characteristic variables $\underline{v}$ as

$$\frac{Dx}{Dt} = \frac{\dfrac{\partial \underline{v}^T}{\partial x} R^T R \Lambda \dfrac{\partial \underline{v}}{\partial x}}{\dfrac{\partial \underline{v}^T}{\partial x} R^T R \dfrac{\partial \underline{v}}{\partial x}} \tag{2.8}$$

where we have assumed that A is diagonisable and employed the transformation

$$\underline{u} = R\underline{v} \qquad \text{where} \quad R^{-1} A R = \Lambda = \text{diag}(\lambda) \tag{2.9}$$

where R and Λ are the matrices of the right eigen vectors and eigen values of A respectively.

If there exists a <u>dominant wave</u> defined such that the Ith characteristic component has a gradient much larger in magnitude than the other component gradients and

$$\varepsilon_1 = \left|\frac{\partial v^{(j)}}{\partial x}\right| \Bigg/ \left|\frac{\partial v^{(I)}}{\partial x}\right| << 1 \qquad \text{for all } j \neq I \tag{2.10}$$

then it follows from (2.8) that

$$\frac{Dx}{Dt} = \lambda^{(I)} + O(\varepsilon_1) . \tag{2.11}$$

Assuming a dominant wave, (2.11) shows that the nodes of the grid will move along the corresponding characteristics and as in the scalar case the collision of nodes will signal a shock. Also note that (2.11) follows independently of the decomposition.

<u>Physical Stability</u>

Under decomposition (2.2) may be written as

$$\frac{D\underline{v}}{Dt} = - \Lambda \frac{\partial \underline{v}}{\partial x} + \frac{Dx}{Dt}\frac{\partial \underline{v}}{\partial x} . \tag{2.12}$$

Taking the ℓ_2 norm of both sides of (2.12) we obtain

$$\left\|\frac{D\underline{v}}{Dt}\right\|^2_{\ell_2} = \left\|(\Lambda - \frac{Dx}{Dt} I) \frac{\partial \underline{v}}{\partial x}\right\|^2 = \sum_p (\lambda^{(p)} - \frac{Dx}{Dt})^2 (\frac{\partial v^{(p)}}{\partial x})^2 . \tag{2.13}$$

Assuming there exists a dominant wave and (2.10) holds then

$$\left\|\frac{D\underline{v}}{Dt}\right\|^2_{\ell_2} = (\frac{\partial v^{(I)}}{\partial x})^2 \sum_p (\lambda^{(p)} - \frac{Dx}{Dt})^2 (\frac{\partial v^{(p)}}{\partial x})^2 \Big/ (\frac{\partial v^{(I)}}{\partial x})^2$$

$$= (\frac{\partial v^{(I)}}{\partial x})^2 (\lambda^{(I)} - \frac{Dx}{Dt})^2 + o(\varepsilon_1^2)$$

and by (2.11) $\quad \left\|\frac{D\underline{v}}{Dt}\right\|^2_{\ell_2} = o \quad$ to order (ε_1^2)

moreover

$$\left\|\frac{D\underline{u}}{Dt}\right\|^2_{\ell_2} = \left\|R\frac{D\underline{v}}{Dt}\right\|^2_{\ell_2} = o. \tag{2.14}$$

Therefore in the case of a dominant wave as defined by (2.10), the nodes of the grid will obtain the corresponding dominant wave velocity and move such that the system is "naturally" stable. Finally note that these observations hold for both the characteristic and the conservative variables.

§ 3 THE DISCRETE APPROXIMATION

As in [4] the above procedure (2.3) may be more generally defined by

$$\underset{\frac{Dx}{Dt}}{\mathrm{Min}} \sum_p \int_{-\infty}^{\infty} \left(\frac{Du^{(p)}}{Dt}\right)^2 dx = \underset{\frac{Dx}{Dt}}{\mathrm{Min}} \sum_p \int_{-\infty}^{\infty} \left(\frac{\partial f^{(p)}}{\partial x} - \frac{Dx}{Dt}\frac{\partial u^{(p)}}{\partial x}\right)^2 dx . \tag{3.1}$$

In the continuous case the procedure (3.1) leads directly to the grid velocity (2.4) and therefore the above properties of §2 are retained. However (3.1) now involves the minimisation of an L_2 integral norm which can be approximated within the MFE framework of Miller [1] and Baines [2]. Full details of the derivations of various discrete forms of MEM may be found in [4]. A summary of the explicit local element formulations of MEM (derived in [4]) and MFE (of Baines [2]) are given below. The only difference between the discrete approximations generated by the two schemes for the system (2.2) is in the grid velocity.

$$\dot{a}_K^{(p)} = \frac{(W_{2K}^{(p)}\Delta S_K + W_{1K+1}^{(p)}\Delta S_{K+1} + \dot{x}_K(\mu_K^{(p)}\Delta S_K + \mu_{K+1}^{(p)}\Delta S_{K+1}))}{(\Delta S_K + \Delta S_{K+1})} . \tag{3.2}$$

The grid velocities produced by MEM and MFE are given by

$$\dot{x}_{MEM} = -\frac{\sum_p(\mu_K^{(p)} W_{2K}^{(p)}\Delta S_K + \mu_{K+1}^{(p)} W_{1K+1}^{(p)}\Delta S_{K+1})}{\sum_p((\mu_K^{(p)})^2\Delta S_K + (\mu_{K+1}^{(p)})^2\Delta S_{K+1})} \tag{3.3}$$

$$\dot{x}_{MFE} = \sum_p(W_{2K}^{(p)} - W_{1K+1}^{(p)})(\mu_{K+1}^{(p)} - \mu_K^{(p)}) / \sum_p(\mu_{K+1}^{(p)} - \mu_K^{(p)})^2 \tag{3.4}$$

respectively, where

$$\Delta S_K = S_K - S_{K-1} , \quad \mu_K^{(p)} = (a_K^{(p)} - a_{K-1}^{(p)})/(S_K - S_{K-1}) . \tag{3.5}$$

$a_K^{(p)}$ are the nodal values (amplitudes) of $u^{(p)}$, S_K are the nodal positions along the x axis and $W_{iK}^{(p)}$ are element variables involving integration of the flux $\underline{f}$.

for trapezoidal rule (A)

$$W_{2K}^{(p)} = -(f_K^{(p)} - f_{K-1}^{(p)})/\Delta S_K$$

$$W_{1K+1}^{(p)} = -(f_{K+1}^{(p)} - f_K^{(p)})/\Delta S_{K+1}$$

for Simpsons rule (B)

$$W_{2K}^{(p)} = -(3f_K^{(p)} + 4f_{K-\frac{1}{2}}^{(p)} - f_{K-1}^{(p)})/\Delta S_K$$

$$W_{1K+1}^{(p)} = (3f_K^{(p)} - 4f_{K+\frac{1}{2}}^{(p)} + f_{K+1}^{(p)})/\Delta S_{K+1} \tag{3.6}$$

and $f_{K+\frac{1}{2}} = f((u_K + u_{K+1})/2)$.

From (3.6) we see that W_{2K} and W_{1K+1} are effectively backward and forward difference approximations to the derivative $-\partial f/\partial x$ at K respectively. The area weighting (ΔS_K) in (3.2) ensures that the discrete approximation is conservative on a non uniform grid, [4]. In all calculations so far only forward explicit Euler time stepping has been used to approximate the time derivatives in (3.2), (3.3) and (3.4).

§ 4 ANALYSIS OF MFE

The crucial difference between MEM and MFE is in the choice of grid velocity. Substitution of (3.5) and (3.6) into (3.3) confirms that (3.3) is a second order approximation to (2.4). Substituting (3.5) and (3.6A) into (3.4) and assuming a locally linear flux $\underline{f}$ we find that (3.4) is a second order approximation to

$$\frac{Dx}{Dt} = \left(\frac{\partial^2 \underline{f}}{\partial x^2} \cdot \frac{\partial^2 \underline{u}}{\partial x^2}\right) / \left(\frac{\partial^2 \underline{u}}{\partial x^2} \cdot \frac{\partial^2 \underline{u}}{\partial x^2}\right). \tag{4.1}$$

Making the observation that

$$\frac{D}{Dt}\frac{\partial \underline{u}}{\partial x} = -\frac{\partial^2 \underline{f}}{\partial x^2} + \frac{Dx}{Dt}\frac{\partial^2 \underline{u}}{\partial x^2}. \tag{4.2}$$

We conclude that MFE moves the nodes of the grid such that

$$\left\| \frac{D}{Dt}\frac{\partial \underline{u}}{\partial x} \right\|^2_{\ell_2} = \left\| \frac{\partial^2 \underline{f}}{\partial x^2} - \frac{Dx}{Dt}\frac{\partial^2 \underline{u}}{\partial x^2} \right\|^2_{\ell_2} \tag{4.3}$$

is minimised over Dx/Dt. Replacing first derivatives with second derivatives in (2.8) to (2.15) an analysis of MFE (similar to that of MEM [4]) shows that if there exists a dominant wave such that

$$\varepsilon_2 = \left| \frac{\partial^2 v^{(j)}}{\partial x^2} \right| \Big/ \left| \frac{\partial^2 v^{(I)}}{\partial x^2} \right| << 1 \qquad \text{for all } j \neq I \tag{4.4}$$

then it follows that

$$\frac{Dx}{Dt} = \lambda^{(I)} \quad \text{and} \quad \left\| \frac{D}{Dt}\frac{\partial \underline{u}}{\partial x} \right\|^2_{\ell_2} = \left\| R\,\frac{D}{Dt}\frac{\partial \dot{\underline{v}}}{\partial x} \right\|^2_{\ell_2} = o \quad \text{to } O(\varepsilon_2^2). \tag{4.5}$$

Therefore if (4.4) holds MFE will move the grid nodes such that the gradients of all the variables are constant. Consequently if an element expands the nodal amplitudes are forced further apart and if an element contracts the nodal amplitudes are forced closer together. In general this kind of grid movement can lead to the positions of the nodes becoming unstable.

§ 5 RESULTS OF EXPLICIT MEM AND EXPLICIT MFE

The first test case is the shock tube Riemann problem, see Sod [7]. An accurate (non-singular) initial data is obtained from the exact solution at time t = 0.1 fig (1). The result produced by MFE is shown in fig (2) at time t = 0.25, using a time step Δt = 0.001. A plausible explanation of the nodal clustering along the rarefaction may be deduced from the analysis of §4. The result produced by MEM is shown in fig (3) at time t = 0.25 and was computed in one time step, confirming the good physical stability property of the scheme. Note that each sequence of results shows the velocity u, pressure p, density ρ and energy E.

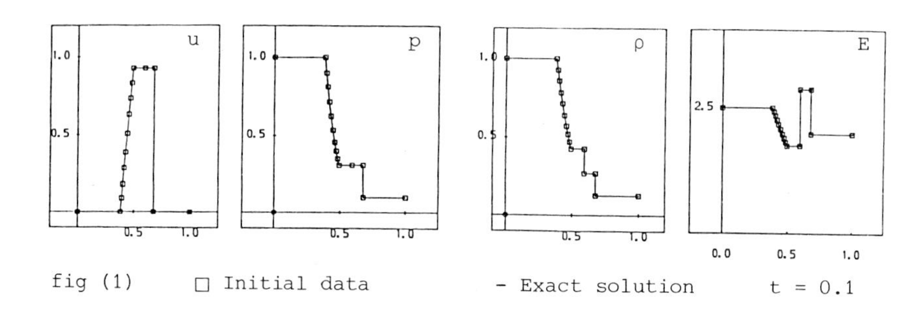

fig (1) □ Initial data - Exact solution t = 0.1

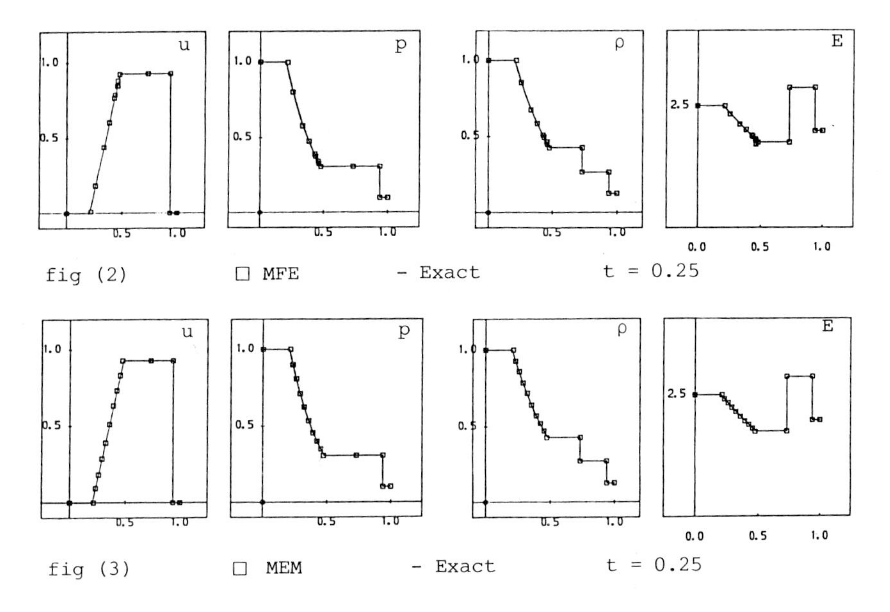

fig (2) □ MFE - Exact t = 0.25

fig (3) □ MEM - Exact t = 0.25

§ 6 MEM AND NUMERICAL STABILITY

In attempting to solve the Riemann problem [7] from initial conditions fig (4), MEM was found to be unstable. We have already seen that the stability of MEM is dependent on the existence of a dominant wave (not present at t=o fig (4)). Also in the results presented here explicit Euler time stepping has been used.

Returning to (3.2), using explicit Euler time stepping and assuming a locally linear flux $\underline{f}$, then under decomposition the discrete approximation (3.2) reduces to

$$\underline{V}_K^{n+1} - \underline{V}_K^{n} = \Delta t \quad b \; (\underline{V}_{K+1}^{n} - V_{K-1}^{n}) \; / \; (\Delta S_K + \Delta S_{K+1}) \tag{6.1}$$

where $b = \dot{x}I - \Lambda$ and $\underline{V}$ are the characteristic variables. The scheme (6.1) is unconditionally unstable. In [4] it is shown that by adding the conservative viscosity

$$(\varepsilon_{K+\frac{1}{2}} \; \Delta u_{K+\frac{1}{2}}^{(p)} - \varepsilon_{K-\frac{1}{2}} \; \Delta u_{K-\frac{1}{2}}^{(p)}) \; / \; \tfrac{1}{2}(\Delta S_K + \Delta S_{K+1}) \tag{6.2}$$

to (3.2) the resulting scheme is first order, stable and total variation diminishing (TVD [5]) under decomposition if

$$\varepsilon = \tfrac{1}{2} \max_p \left| \lambda^{(p)} - \dot{x} \right| \Delta t \qquad \text{and} \qquad \max_p \left| \lambda^{(p)} - \dot{x} \right| \Delta t \leqslant \Delta S_K . \tag{6.3}$$

While the viscosity (6.2) stabilises MEM (run from t=o fig (4)) the results obtained are far too diffusive. To preserve the natural stability of MEM (2.14) the coefficient of viscosity is redefined to be

$$\varepsilon = (\Theta + \gamma\Theta^2) \; \tfrac{1}{2} \max_p \left| \lambda^{(p)} - \dot{x} \right| \Delta t$$

where

$$\gamma \simeq 1/6 \; , \quad \Theta_K = \left\| \frac{D\underline{u}}{Dt}_K \right\|_{\ell_2}^2 \Big/ \max_j \left\| \frac{D\underline{u}}{Dt}_j \right\|_{\ell_2}^2 . \tag{6.4}$$

From (6.4) it follows that Θ and hence ε vanishes in the presence of a dominant wave and the viscosity is removed when natural stability occurs. Using the data of fig (4), the result of applying MEM with (6.2) and (6.4) is shown in fig (5), computed in 70 steps.

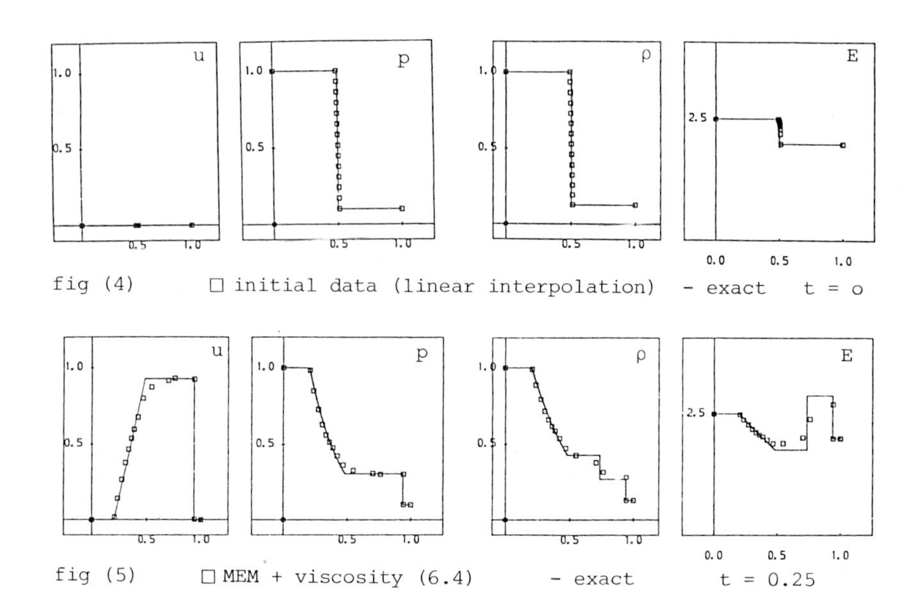

fig (4) □ initial data (linear interpolation) - exact t = o

fig (5) □ MEM + viscosity (6.4) - exact t = 0.25

§ 7 TWO DIMENSIONS (VERY BRIEFLY)

The procedure of (3.1) can be generalised into two dimensions and is fully described together with further observations with analysis and results in [4]. First the system is written in the mobile operator form

$$\frac{D\underline{u}}{Dt} = - \left(\frac{\partial \underline{f}}{\partial x} + \frac{\partial \underline{g}}{\partial y}\right) + \frac{Dx}{Dt}\frac{\partial \underline{u}}{\partial x} + \frac{Dy}{Dt}\frac{\partial \underline{u}}{\partial y} . \tag{7.1}$$

The L_2 integral norm of both sides of (7.1) is taken

$$\left\|\frac{D\underline{u}}{Dt}\right\|^2_{L_2} = \left\|\frac{\partial \underline{f}}{\partial x} + \frac{\partial \underline{g}}{\partial y} - \left(\frac{Dx}{Dt}\frac{\partial \underline{u}}{\partial x} + \frac{Dy}{Dt}\frac{\partial \underline{u}}{\partial y}\right)\right\|^2_{L_2} \tag{7.2}$$

and minimised over Dx/Dt and Dy/Dt to obtain the normal equations (for the grid velocity) which are written as

$$\int_\tau \frac{D\underline{u}}{Dt} \cdot \frac{\partial \underline{u}}{\partial x}\, d\tau = o \quad , \quad \int_\tau \frac{D\underline{u}}{Dt} \cdot \frac{\partial \underline{u}}{\partial y}\, d\tau = o . \tag{7.3}$$

The normal equations are non-singular provided the gradients are non zero and non parallel ($\partial \underline{u}/\partial x \neq \alpha\, \partial \underline{u}/\partial y$). For a non-singular two component system it follows from (7.2) that $D\underline{u}/Dt = o$ on a single characteristic. In particular for the Inviscid Burgers system

$$\begin{aligned} u_t + uu_x + vu_y &= o \\ v_t + uv_x + vv_y &= o \end{aligned} \tag{7.4}$$

it follows from the above procedure (7.3) that $\dot{x} = u$, $\dot{y} = v$, $Du/Dt = o$, $Dv/Dt = o$, further in [4] it is shown that the discrete MEM is exact for (7.4) with a piecewise linear initial data.

Results

The test case considered is to solve (7.4) with an initial data devised by Miller [6] for testing MFE. The initial data is defined by

$$u = -\tau_1 \operatorname{Sin}(\pi x) \ , \ v = 1 + \tau_2 \operatorname{Cos}(\pi x) \quad \text{on} \quad y = 0$$

$$u = \tau_1 \operatorname{Sin}(\pi x) \ , \ v = -1 + \tau_2 \operatorname{Cos}(\pi x) \quad \text{on} \quad y = 1$$

with linear interpolation in between. The boundary conditions (for both components) are Dirichlet at y = 0,1 and reflection at x = 0,1. The initial data for $\tau_1 = 0.6$, $\tau_2 = 0.6$ is shown in fig (6). The result (of MEM) at time t = 0.5 is shown in fig (7) together with the grid. For an initial data with $\tau_1 = -0.6$, $\tau_2 = 0.6$ (not shown) the result (of MEM) at time t = 0.35 is shown in Fig (8).

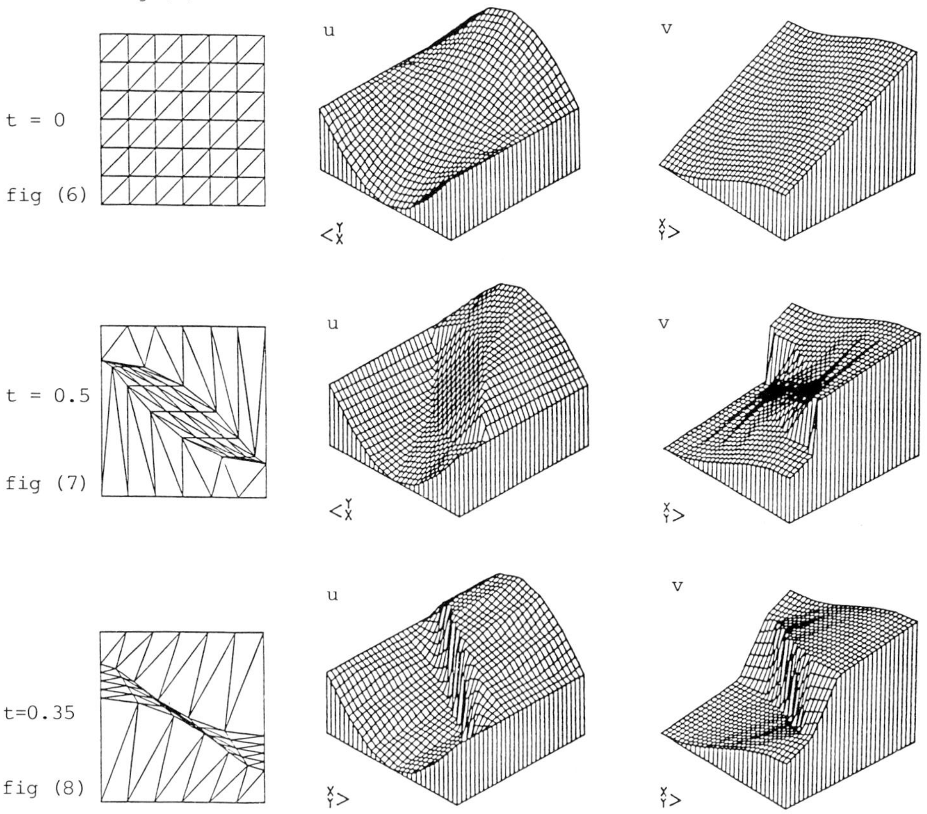

Acknowledgements

I wish to thank my supervisor Dr M.J. Baines for many valuable discussions on MFE and for the encouragement given to me throughout my work. Thanks also to Dr P. Sweby for his exact Rie mann Solver.

I am very grateful to the SERC and the Royal Aircraft Establishment at Farnborough for their financial support.

References

[1] Miller, K. SIAM J. Numer. Anal 18 1019-1057 (1981)

[2[Baines, M.J. Proc. Conf., Reading (1985) (eds Morton & Baines)

[3] Lax, P.D. SIAM Regional Conf. Ser., lectures in Appl. Math Vol II SIAM (1972)

[4] Edwards M.G. Ph.D Thesis, Univ. of Reading (1987)

[5] Harten, A. J. Comp. Phys. 49 357-393 (1983)

[6] Miller, K. Accuracy Estimates and Adaptive Refinements in Finite Element Computations Pub. John Wiley & Sons

[7] Sod, G.A. J. Comp. Phys., 27 1-31 (1978)

[8] Wathen, A.J. and Baines, K.J. I.M.A. Journal of Numer. Anal. 5 (1985)

NEW DEVELOPMENTS IN FAST EULER FINITE ELEMENT SOLVERS

L. FEZOUI, M.H. LALLEMAND, H. STEVE, A. DERVIEUX

INRIA, **Avenue Emile HUGUES, Sophia-Antipolis,**
06560 **VALBONNE, FRANCE**

ABSTRACT

The problem of concern here is the following : given a generally unstructured Finite Element triangular (or tetrahedral) mesh, we investigate the possibilities of applying Multigrid strategies to most efficiently solve the steady Euler equations on this mesh. An automated coarsening algorithm in which Finite Volumes are grouped together is used in combination with upwind schemes. An explicit 2-D algorithm was already presented in [1] ; in this paper, we introduce : firstly, a **3-D extension** of the 2-D explicit algorithm. The reliability of the approach is illustrated by flow computations with a locally refined unstructured tetrahedrization. Secondly a newly derived **implicit 2-D version** is presented : this approach has been already explored in [2,3] for structured grids. In this study, we use the point Jacobi iteration which seems to be a good choice in the case of unstructured meshes.

1. INTRODUCTION

Industrial aerodynamic calculations need non trivial discrete geometries : in FVM / FDM, such as multiblocks for complex geometries and local enrichment by division for complex solutions. Then the discrete geometry generally looses its regular structure. One attractive alternate-approach is to start directly from an a priori **non structured geometry** , as in Finite Element. Now nonstructured discretizations have to be enabled to apply to heavy calculation (many nodes) ; one way to progress in this direction is to adapt the **MG approach** to this context. The **nonstructured MG** subject has been already much studied : since the MG concept initially relied on globally nested grids, and because such a strategy is not efficient in the **unstructured context,** we have to drop several among these conditions : **nested grids** can be generated by **local refinement** [4, 5, 6] ; **conversely, unnested grid** can be employed : we refer to [7], and to [8]. A third way to deal with unstructured MG is for consider abstract coarse levels, as in the **AMG approach** [9] ; this approach is very attractive; however, we think that this method is not enough adapted to compressible CFD since a large amount of informations is handled for coarse level generation, and linearity is much exploited.

The method that we propose in close to AMG but has the following caracteristics : firstly, the degrees of freedom are grouped using a topological criterion, secondly, nonlinear MG schemes can be applied. A first discussion of the method has been presented in [1] with a 2-D explicit algorithm. The purpose of this paper is to present two new developments, namely a 3-D extension and an implicit formulation.

2. COARSENING STRATEGY

We think it is necessary to recall that two basic classes of coarsenings exist, that we can distinguish already in the 1-D context : while in Finite Element coarsening, we get rid of nodes (e.g. the even nodes) : ("elimination"), in Finite Volume coarsening, we replace two neighboring cells by a new one ("grouping"). Because the second way corresponds to a $P0$ interpolation, while the first one corresponds to a $P1$ continuous interpolation, it is more versatile and has been chosen in this study. The first phase is then to introduce a **FVM dual partition** , derived from the original FEM triangulation : this is done by using medians around each node in 2-D as sketched in Fig. 1., median planes in 3-D.

The coarsening algorithm is based on neighboring relations (two cells are neighbors if they contain vertices that are neighbors) ; the algorithm reads as follows : *Consider successively every cell. <1> if the cell C is already included in a group then consider next cell; else: create a new group containing C, and put into this group each cell neighboring C and not already included in a group. <2> if the new group contains only the cell C, then destroy the group and put cell C in an existent group containing a neighbor of C. <3> next cell.*

3. SPATIAL SCHEME

This scheme will be applied to both fine and coarser levels. The time-dependent Euler equations are written in conservative form :

$$W_t + F(W)_x + G(W)_y + H(W)_z = 0, \; with \;\; W = (\rho, \rho u, \rho v, \rho w, E)$$

where ρ is the density, (u,v,w) the velocity and E the total energy per unit volume. An upwind finite volume scheme is derived ; we describe it in the context of the usual explicit time-stepping. Given a cell C_i, the mean value W_i of the dependent variable in this cell is advanced from time level n to time level $n+1$ as follows :

$$vol\ (C_i)[W_i^{n+1} - W_i^n] = -\Delta t \sum_{j\ neighbor\ i} \Phi\ (W_i^n,\ W_j^n,\ \eta^{ij})$$

where η^{ij} is a vector related to the mesh metric, and where Φ is a flux-splitting consistent with $\eta_x^{ij} F + \eta_y^{ij} G + \eta_z^{ij} H$. In the sequel, the following flux-splittings are used : Osher's [10], Steger's-Warming [11], van Leer [12].

A second-order spatial scheme is only applied on the finest grid level which is a true triangulation. The scheme uses some ideas of the MUSCL approach and is described in [13,14] .

4. TIME-EXPLICIT FAS APPROACH

4.1 Basic Iteration Method

Following A. Jameson [15], a Runge-Kutta scheme is applied with four time-steps ; following [16,17] we choose as step lengths : α_1 =.11, α_2 =.2766, α_3 =.5, α_4 =1. The method for evaluating a local time-step is trivially extended to 3-D from the 2-D version [1].

4.2 MG scheme

The basic algorithm uses FAS iterations with sawtooth V-cycles as in [15, 18]. The transfer operators are very simple in the present approach : for fine-to-coarse : values are averaged in a conservative manner, for coarse-to-fine : the trivial injection is applied. Then a full-multigrid algorithm can be applied with the first-order scheme. In order to obtain second-order accurate solutions , the second-order spatial scheme is introduced into the fine-grid solver only for the last phase of the full-multigrid process.

4.3 Numerical experiment

A few 3-D experiments have been performed on a mini-computer. The flux splitting used is van Leer's. Firstly, for comparison with [18], we compute the flow in a channel with 4.2% bump with a tetrahedrization which is essentially a 72x21x3 mesh. The Mach at infinity is .85. The first-order accurate mesh is used. We present in Fig. 2. the history of the convergence when a Full four-grid algorithm is applied ; it can be (favourably) compared to the 3-grid calculation in [18]. Secondly, the flow around a wing was calculated with the second-order version and mesh adaption : Fig. 3 gives an idea of the solutions and shows the history of convergence with multigrid before and after mesh enrichment.

5. IMPLICIT MG SCHEME

The subject of this section is the inclusion of a MG algorithm in the linear phase of an unfactored linearized implicit 2-D scheme. This approach has been already studied by Mulder [2] and Hanel [3] ; it is less ambitious than the fully nonlinear study of Hemker and coworkers [20]. In choosing this first step, we want firstly to upgrade an implicit code which has a satisfactory behavior for a large range of Mach number (from Mach = 10^{-3}, see [21], to Mach = 25, see [22]), and secondly, to get informations related to the linear case before studying the nonlinear one.

5.1 The single grid scheme implicit scheme

The single grid scheme is the result of a series of studies [23, 24, 25, 26]. It is written as a delta-scheme : it consists of an explicit phase with a second-order accurate spatial approximation, denoted by F_2, in which the flux splitting involved is Osher's ; then an implicit phase is applied :

$$\hat{\delta w} = \Delta t \ F_2(w^n)$$

$$[(I-\Delta t A_1(w^n))]\delta w^{n+1}=\hat{\delta w} \tag{1a}$$

$$w^{n+1}=w^n+\delta w^{n+1} \tag{1b}$$

in which the linear operator A_1 is derived from the **Steger-Warming** flux splitting, with first-order approximation, and freezing at value w^n the diagonalized part of the splitting, in short :

$$(A_1(w^n)\delta w^{n+1})_i = \sum_{ij} A^+(w_i^n)\delta w_i^{n+1} + A^-(w_j^n)\ \delta w_j^{n+1}.$$

Furthermore, the linear system (1) defining δw^{n+1} is partially solved by a few collective (ie bloc 4x4 in 2-D) point Jacobi relaxation sweeps, this choice of Jacobi revealed to be advantageous from both point of view of vectorization [25] and memory requirement, since the matrix is not stored but recomputed [26].

5.2 The MG scheme

In the MG version of the above scheme, the linear systeme (1) is solved using a MG algorithm : basic iteration is again the collective Jacobi iteration ; the residual are transfered by using the same transfer operators as for the explicit scheme ; the usual V-cycle is applied.

5.3 Numerical experiments

We present only a few measures and an illustrative experiment : the test case is again a flow in a channel with 4,2% bump (M_∞=.85) ; the mesh contains 2225 points. Two typical convergence histories one-grid, six-grid are presented in Fig. 4.

The cost of the construction of the zones is less than the cost of one single grid iteration ; the cost of one linear cycle is about 1.5 the cost of the corresponding single-grid iteration, as predicted by the theory. The overall gain in efficiency is about 1.8. However, before giving a precise evaluation, the optimization of the MG code has still to be done.

6. CONCLUSION

The MG ideas have been applied to the improvement of two different algorithms for obtaining FEM solution of Euler flows. In both cases a more efficient algorithm is obtained, which applies to the same class of discrete problems as the one-grid algorithm, that is the solution on a given arbitrary unstructured grid ; this purpose is attained by performing an automated coarsening. Further studies are necessary to exploit completely the possibilities of the MG approach.

8. REFERENCES

[1] M-H. LALLEMAND, A. DERVIEUX, A multigrid finite element method for solving the two dimensional Euler equations, Third Copper Mountain Conference on Multigrid Methods, april 6-10, 1987, Copper Mountain, Colorado, U.S.A.

[2] W. A. MULDER, Multigrid Relaxation for the Euler Equations, J. Comp. Phys. 60 (1985), pp. 235-252.

[3] W. SCHRODER, D. HANEL, An unfactored implicit scheme with multigrid acceleration for the solution of the Navier-Stokes equations, Computer & Fluids vol. 15, no 3, pp. 313-336, 1987.

[4] R. BANK, A. SHERMAN, "A multi-level iterative method for solving finite-element equations", Proceedings of the fifth symposium on reservoir simulation, pp. 117-126, Society of petroleum engineers of AIME, (1980).

[5] E. PEREZ, Finite Element and Multigrid solution of the two-dimensional Euler equations on a non-structured mesh, INRIA Report 442 (1985).

[6] F. ANGRAND, P. LEYLAND, Schéma multigrille dynamique pour la simulation d'écoulements de fluides visqueux compressibles, INRIA, Research report no 659.

[7] R. LOHNER, K. MORGAN, Unstructured Multigrid methods, Second European Conference on Multigrid Methods, Koln (RFA), October 1-4, 1985.

[8] D. MAVRIPLIS, A. JAMESON, Multigrid solution of the two-dimensional Euler equations on unstructured triangular meshes, AIAA Paper 87-0353.

[9] A. BRANDT, S.F. MC CORMICK, J. RUGE, Algebraic multigrid (AMG) for sparse matrix equations, in Sparsity and its applications, (D.J. Evans Ed.), Cambridge University Press (1984).

[10] S. OSHER, S. CHAKRAVARTHY, Upwind difference schemes for hyperbolic systems of conservation laws, Math. Computation, April 1982.

[11] J. STEGER, R.F. WARMING, Flux Vector splitting for the inviscid gas dynamic equations with applications to finite difference methods, Journal Comp. Physics, vol. 40, no. 2, pp 263-293, 1981.

[12] B. VAN LEER, Flux-Vector splitting for the Euler equations, Lecture Notes in Physics, Vol. 170, 570-512 (1982).

[13] F. FEZOUI, Résolution des équations d'Euler par un schéma de Van Leer en éléments finis, INRIA Research Report no 358 (1985).

[14] F. FEZOUI, V. SELMIN, H. STEVE, INRIA Report to appear.

[15] A. JAMESON, Numerical solution of the Euler equations for compressible inviscid fluids, Numerical methods for the Euler equations of Fluid Dynamics, F. Angrand et al. Eds., SIAM Philadelphia (1985).

[16] E. TURKEL, B. VAN LEER, Flux vector splitting and Runge-Kutta methods for the Euler equations, ICASE Report 84-27, June 1984.

[17] M-H. LALLEMAND, Etude de schémas Runge-Kutta à 4 pas pour la résolution multigrille des équations d'Euler 2-D. INRIA Research Report (in preparation).

[18] M-H. LALLEMAND, F. FEZOUI, E. PEREZ, Un schéma multigrille en Eléments Finis décentré pour les équations d'Euler, INRIA Research Report 602 (1987).

[19] K. BOHMER, P. HEMKER, H.J. STETTER, The defect correction approach, Computing Suppl., 5, 1-32 (1984).

[20] P. HEMKER, S.P. SPEKREIJSE, Multiple grid and Osher's scheme for the efficient solution of the steady Euler equations, Appl. Num. Math. 2 (1986), pp. 475-493.

[21] F. BENKHALDOUN, A. DERVIEUX, G. FERNANDEZ, H. GUILLARD, B. LARROUTUROU, "Some finite-element investigations of stiff combustion problems : mesh adaption and implicit time-stepping", Proceedings of the NATO advanced research workshop on Mathematical medelling in combustion and related topics, Brauner Schmidt-Lainé eds., to appear, (1987).

[22] B. STOUFFLET, J. PERIAUX, F. FEZOUI, A. DERVIEUX, Numerical simulation of 3-D hypersonic Euler flow around space vehicles using adapted finite elements, AIAA paper 87-0560 (1987).

[23] B. STOUFFLET, Implicit Finite Element Methods for the Euler Equations. Proceedings of the INRIA Workshop on "Numerical Methods for Compressible Inviscid Fluids, 7-9 dec. 1983, Rocquencourt (France), Numerical Methods for the Euler Equations of Fluid Dynamics, F. Angrand et al. Eds., SIAM (1985).

[24] F. FEZOUI, B. STOUFFLET, "A class of implicit upwind schemes for Euler simulation with unstructured meshes", to appear.

[25] F. ANGRAND, J. ERHEL, Vectorized Finite Element codes for compressible Flows, Actes du Colloque "Finite Element in Flow Problem", Antibes (F), 16-20 Juin 1986, à paraître chez Wiley.

[26] H. STEVE, Thesis, in preparation.

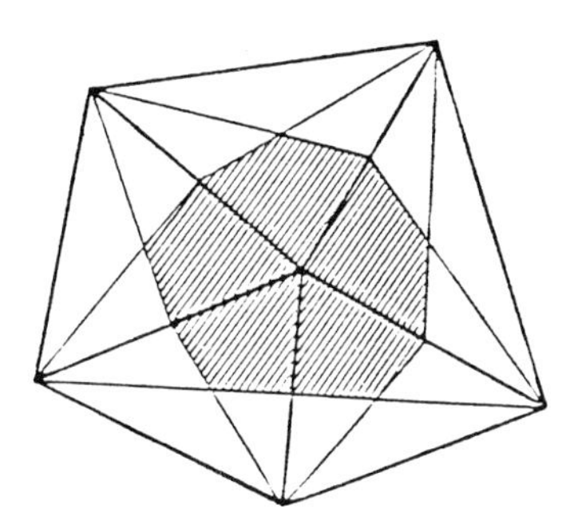

Figure 1 : Control Volume of the dual mesh

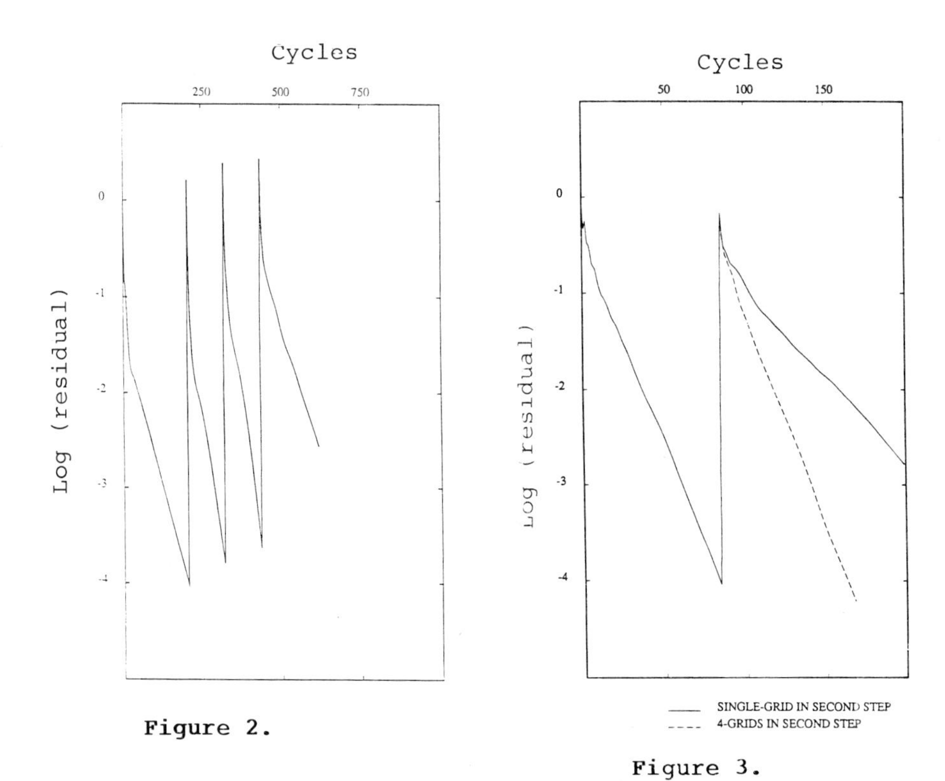

Figure 2.

Figure 3.

Figure 3 : (continued)

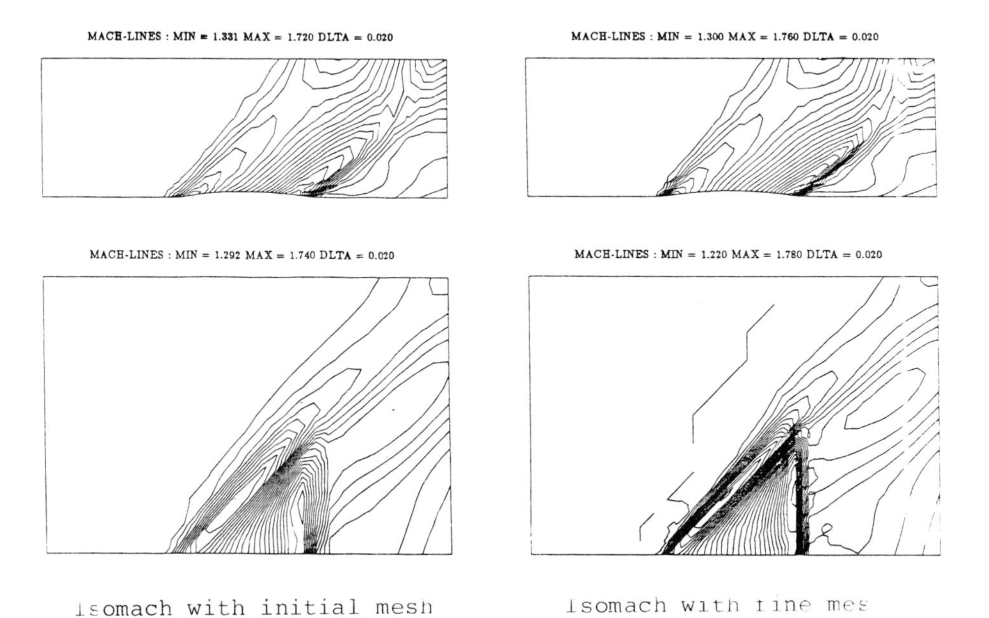

Isomach with initial mesh

Isomach with fine mes

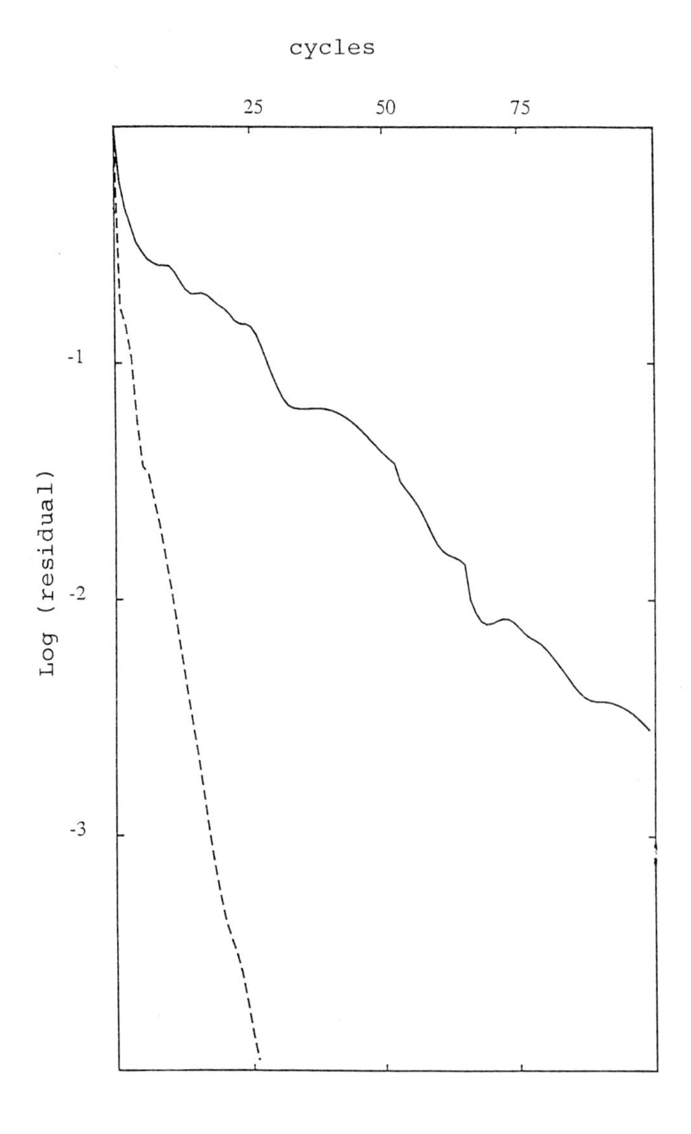

Figure 4 : Comparison of the multigrid implicit Scheme with the single-grid implicit one.

AN ACCURATE EXPLICIT SOLVER FOR 3-D NON-ISENTROPIC INTERNAL FLOWS

B. Fortunato and A. Dadone
Istituto di Macchine ed Energetica, Universita' di Bari,
Via Re David 200, 70125 Bari, Italy

SUMMARY

The present paper provides an accurate explicit Euler solver to compute unsteady three-dimensional non-isentropic internal flows. A Compressible Over INcompressible (COIN) variant of the lambda formulation is presented and a time accurate numerical technique is applied. The merits of the present approach are demonstrated by means of a few applications.

INTRODUCTION

Steady as well as unsteady compressible flow phenomena play a very significant role in most propulsive applications of engineering interest and have been the subject of many theoretical and experimental studies; in the last decades, numerical simulation has become a very useful and widespread tool for predicting compressible flows. Among the several numerical methods developed for computing compressible inviscid flows, the lambda formulation [1, 2] , has several, very desirable features: the time-dependent compressible Euler equations are recast in terms of characteristic type variables and discretized by means of upwind differences, in order to mimic the wave propagation phenomena. Such a methodology has undergone several improvements. In particular Dadone et al. have proposed a perturbative formulation [3] and a COIN (Compressible Over INcompressible) variant [4], which allows to improve the results accuracy by extracting an appropriate incompressible flow solution; indeed, if such a solution is available, it is convenient to reformulate the equations in terms of new variables, which are the differences between those corresponding to the sought compressible solution and those corresponding to the known incompressible one, so that the geometry induced gradients are mostly accounted for by the incompressible flow solution and the smoother perturbative problem can be solved very accurately even on a coarse grid.

Dadone and Napolitano have also developed various implicit integration methods [5, 6] for two-dimensional flows, which allow to improve the efficiency of the methodology by removing the CFL stability restriction of the original explicit schemes. Alternatively, Moretti et al. have devised appropriate relaxation methods for two-dimensional flows [7-9], which allow to reduce a two-dimensional problem to a sequence of two simple quasi one-dimensional problems.

For the case of three dimensional flows, Napolitano et Dadone [10]have

Research supported by C.N.R., under PFE-2 Grant n.86.00884.59.

formulated the 3-D lambda equations, for homentropic flows, and have proposed various numerical schemes to integrate them numerically. Such equations have been generalized to the more relevant case of nonisentropic flows in Ref. 11, where an implicit point-Gauss-Seidel method has been also developed, which however requires too many iterations to converge; the convergence rate has been significantly improved in Ref. 12, where a relaxation method has been proposed, which allows to reduce a three-dimensional flow to a sequence of three simple quasi one-dimensional flows, so that a three-dimensional steady flow can be efficiently computed.

A noteworthy contribution towards the accurate computation of unsteady two-dimensional flows is due to Moretti et al. [13], who have proposed a second-order time accurate numerical technique to integrate the lambda formulation equations; in essence, it is a two-level explicit technique with a very simple treatment of the boundary conditions.

The present paper is concerned with the accurate computation of unsteady three-dimensional internal flows: the COIN variant is extended to this flow case, in order to accurately account for the geometry induced gradients, and the numerical technique suggested in [13] is used, in order to accurately compute unsteady flows. The 3-D COIN lambda equations are derived at first, and the results of some numerical experiments are then presented.

GOVERNING EQUATIONS

The nondimensional continuity, momentum and entropy equations for compressible, inviscid flows are given in vector form [14] as:

$$a_t/\delta + \underline{q}\cdot\underline{\nabla}a/\delta + a\underline{\nabla}\cdot\underline{q} - a\,S_t - a\underline{q}\cdot\underline{\nabla}S = 0 \quad , \tag{1}$$

$$\underline{q}_t + (\underline{q}\cdot\underline{\nabla})\underline{q} + a\underline{\nabla}a/\delta - a^2\,\underline{\nabla}S = 0 \quad , \qquad S_t + \underline{q}\cdot\,\underline{\nabla}S = 0 \quad , \tag{2; 3}$$

where a is the speed of sound, $\underline{q}$ is the velocity vector, $\underline{\nabla}$ is the gradient operator, and subscript t indicates the partial derivative with respect to time. Moreover $\delta = (\gamma-1)/2$, where γ is the specific heat ratio, and S is the entropy, defined as:

$$S = \ln(p/\rho^{\gamma})/(2\,\gamma\,\delta) \tag{4}$$

where p and ρ are the pressure and the density, respectively.

To minimize errors in the regions with large gradients and a consequent decay of total temperature, eqns (1) and (2) can be reformulated, in the spirit of Refs. 3 and 4. Let us split $\underline{q}$ and a into sums of two terms, the ones denoted by ° being the "incompressible" values, computed at the start of the calculations and never changed again, and the ones denoted by a prime the unknowns to be computed:

$$\underline{q} = \underline{q}^{\circ} + \underline{q}' \quad , \qquad a = a^{\circ} + a' . \tag{5}$$

The velocity $\underline{q}^{\circ}$ satisfies the classical incompressible, irrotational flow conditions, while a° is related to q° by the condition of total temperature conservation:

$$\underline{\nabla}\cdot\underline{q}^{\circ} = \underline{\nabla}\times\underline{q}^{\circ} = 0 \quad , \quad \underline{\nabla}[(q^{\circ})^2 + (a^{\circ})^2/\delta] = 0 \quad . \qquad (6;\ 7)$$

By substitution of eqns (6) and (7) into eqns (1) and (2) the following COIN (Compressible Over INcompressible) variant of the continuity and momentum equations in vector form can be obtained:

$$a'_t/\delta + \underline{q}\cdot\underline{\nabla}a'/\delta + a\underline{\nabla}\cdot\underline{q}' - a\,S_t - a\underline{q}\cdot\underline{\nabla}S = -\underline{q}^{\circ}\cdot\underline{\nabla}a^{\circ}/\delta - \underline{q}'\cdot\underline{\nabla}a^{\circ}/\delta \quad , \qquad (8)$$

$$\underline{q}_t + (\underline{q}\cdot\underline{\nabla})\underline{q}' + a\underline{\nabla}a'/\delta - a^2\underline{\nabla}S = -\underline{\nabla}(\underline{q}'\cdot\underline{q}^{\circ}) + (\underline{q}^{\circ}\cdot\underline{\nabla})\underline{q}' + \underline{q}^{\circ}\times(\underline{\nabla}\times\underline{q}') - a'\underline{\nabla}a^{\circ}/\delta \,. \qquad (9)$$

Let ξ, η, and ζ be three orthogonal coordinates, h_1, h_2 and h_3 their metric parameters, u, v and w the three components of the velocity vector, and let us define the following COIN Riemann variables:

$$\begin{aligned} R'^X_1 &= a'/\delta + u' , & R'^Y_1 &= a'/\delta + v' , & R'^Z_1 &= a'/\delta + w' , \\ R'^X_2 &= a'/\delta - u' , & R'^Y_2 &= a'/\delta - v' , & R'^Z_2 &= a'/\delta - w' , \end{aligned} \qquad (10)$$

and the corresponding characteristic slopes:

$$\begin{aligned} \lambda^X_1 &= (u+a)/h_1 , & \lambda^X_2 &= (u-a)/h_1 , & \lambda^X_3 &= u/h_1 , \\ \lambda^Y_1 &= (v+a)/h_2 , & \lambda^Y_2 &= (v-a)/h_2 , & \lambda^Y_3 &= v/h_2 , \\ \lambda^Z_1 &= (w+a)/h_3 , & \lambda^Z_2 &= (w-a)/h_3 , & \lambda^Z_3 &= w/h_3 . \end{aligned} \qquad (11)$$

Adding and subtracting the three components of the COIN momentum equation (9) to the COIN continuity equation (8), six compatibility conditions along six bicharacteristic lines can be obtained (see Ref. 12 for details). The fast solver methodology [8, 9] integrates directly such compatibility conditions as uncoupled equations; here, on the contrary, they are combined to eliminate some non-characteristic terms, so that the following equations can be obtained:

$$\begin{aligned} a'_t &= \delta\,(f^X_1+f^X_2+f^Y_1+f^Y_2+f^Z_1+f^Z_2+f^L_1+f^P_1+2aS_t)/2 , \\ u'_t &= (f^X_1-f^X_2+f^Y_{31}-f^Y_{32}+f^Z_{31}-f^Z_{32}+f^L_2+f^P_2)/2 , \\ v'_t &= (f^Y_1-f^Y_2+f^X_{31}-f^X_{32}+f^Z_{33}-f^Z_{34}+f^L_3+f^P_3)/2 , \\ w'_t &= (f^Z_1-f^Z_2+f^X_{33}-f^X_{34}+f^Y_{33}-f^Y_{34}+f^L_4+f^P_4)/2 , \end{aligned} \qquad (12)$$

while the entropy equation (3) can be written as follows:

$$S_t = f^X_4 + f^Y_4 + f^Z_4 \quad . \qquad (13)$$

Eqns (12) and (13) give the time derivatives of the physical perturbative variables and of the entropy in terms of convective terms (subscripts ξ , η and ζ indicate partial derivatives):

$$
\begin{aligned}
f_1^X &= -\lambda_1^X(R'^X_{1\xi}-aS_\xi), & f_1^Y &= -\lambda_1^Y(R'^Y_{1\eta}-aS_\eta), & f_1^Z &= -\lambda_1^Z(R'^Z_{1\zeta}-aS_\zeta),\\
f_2^X &= -\lambda_2^X(R'^X_{2\xi}-aS_\xi), & f_2^Y &= -\lambda_2^Y(R'^Y_{2\eta}-aS_\eta), & f_2^Z &= -\lambda_2^Z(R'^Z_{2\zeta}-aS_\zeta),\\
f_{31}^X &= -\lambda_3^X(R'^Y_{1\xi}-aS_\xi), & f_{31}^Y &= -\lambda_3^Y(R'^X_{1\eta}-aS_\eta), & f_{31}^Z &= -\lambda_3^Z(R'^X_{1\zeta}-aS_\zeta),\\
f_{32}^X &= -\lambda_3^X(R'^Y_{2\xi}-aS_\xi), & f_{32}^Y &= -\lambda_3^Y(R'^X_{2\eta}-aS_\eta), & f_{32}^Z &= -\lambda_3^Z(R'^X_{2\zeta}-aS_\zeta),\\
f_{33}^X &= -\lambda_3^X(R'^Z_{1\xi}-aS_\xi), & f_{33}^Y &= -\lambda_3^Y(R'^Z_{1\eta}-aS_\eta), & f_{33}^Z &= -\lambda_3^Z(R'^Y_{1\zeta}-aS_\zeta),\\
f_{34}^X &= -\lambda_3^X(R'^Z_{2\xi}-aS_\xi), & f_{34}^Y &= -\lambda_3^Y(R'^Z_{2\eta}-aS_\eta), & f_{34}^Z &= -\lambda_3^Z(R'^Y_{2\zeta}-aS_\zeta),\\
f_4^X &= -\lambda_3^X S_\xi, & f_4^Y &= -\lambda_3^Y S_\eta, & f_4^Z &= -\lambda_3^Z S_\zeta,
\end{aligned} \tag{14}
$$

plus terms carrying local contributions:

$$
\begin{aligned}
f_1^L &= -2\phi'a', & f_2^L &= 2\,(\beta_1'v' - \beta_3'w'),\\
f_3^L &= 2\,(-\beta_1'u' + \beta_2'w'), & f_4^L &= 2\,(\beta_3'u' - \beta_2'w'),
\end{aligned} \tag{15}
$$

and linear corrective terms due to the COIN variant:

$$
\begin{aligned}
f_1^P &= -A_{11}u'-A_{12}v'-A_{13}w'-A_{14}a'-A_{15}, & f_2^P &= -A_{21}u'-A_{22}v'-A_{23}w'-A_{24}a',\\
f_3^P &= -A_{31}u'-A_{32}v'-A_{33}w'-A_{34}a', & f_4^P &= -A_{41}u'-A_{42}v'-A_{43}w'-A_{44}a',
\end{aligned} \tag{16}
$$

with:

$$
\begin{aligned}
\beta_1' &= (h_{2\xi}v' - h_{1\eta}u')/h_1h_2, & \beta_2' &= (h_{3\eta}w' - h_{2\zeta}v')/h_2h_3,\\
\beta_3' &= (h_{1\zeta}u' - h_{3\xi}w')/h_1h_3, & \phi' &= [(h_2h_3)_\xi u'+(h_1h_3)_\eta v'+(h_1h_2)_\zeta w']/h_1h_2h_3,
\end{aligned} \tag{17}
$$

and:

$$
\begin{aligned}
A_{11} &= 2[(h_2h_3)_\xi a^\circ/h_2h_3+a^\circ_\xi/\delta]/h_1, & A_{12} &= 2[(h_1h_3)_\eta a^\circ/h_1h_3+a^\circ_\eta/\delta]/h_2,\\
A_{13} &= 2[(h_1h_2)_\zeta a^\circ/h_1h_2+a^\circ_\zeta/\delta]/h_3, & A_{14} &= 0,\\
A_{15} &= 2(u^\circ a^\circ_\xi/h_1 + v^\circ a^\circ_\eta/h_2+w^\circ a^\circ_\zeta/h_3)/\delta, & &\\
A_{21} &= 2(h_{1\eta}v^\circ/h_2+h_{1\zeta}w^\circ/h_3+u^\circ_\xi)/h_1, & A_{22} &= 2(v^\circ_\xi - h_{2\xi}v^\circ/h_2)/h_1,\\
A_{23} &= 2(w^\circ_\xi - h_{3\xi}w^\circ/h_3)/h_1, & A_{24} &= 2\,a^\circ_\xi/\delta\,h_1,
\end{aligned} \tag{18}
$$

$A_{31} = 2(u^\circ_\eta - h_{1\eta}u^\circ/h_1)/h_2$, $A_{32} = 2(h_{2\xi}u^\circ/h_1 + h_{2\zeta}w^\circ/h_3 + v^\circ_\eta)/h_2$,

$A_{33} = 2\ a^\circ_\eta/\ \delta\ h_2$, $A_{34} = 2(w^\circ_\eta - h_{3\eta}w^\circ/h_3)/h_2$,

$A_{41} = 2(u^\circ_\zeta - h_{1\zeta}u^\circ/h_1)/h_3$, $A_{42} = 2(v^\circ_\zeta - h_{2\zeta}v^\circ/h_2)/h_3$,

$A_{43} = 2(h_{3\xi}u^\circ/h_1 + h_{3\eta}v^\circ/h_2 + w^\circ_\zeta)/h_3$, $A_{44} = 2\ a^\circ_\zeta/\ \delta\ h_3$.

As far as the boundary conditions are concerned, the impermeability condition has been enforced at the channel walls, the pressure has been imposed at the outlet, while the entropy and the total speed of sound have been assigned at the inlet boundary, together with the direction of the velocity vector. As usual for the lambda formulation, the governing equations (12) and (13) are used also at the boundary gridpoints, and the unknown convective terms arriving from outside the computational flow field are evaluated by enforcing the appropriate boundary conditions.

The numerical technique here employed is the natural extension to three-dimensional flows of the one proposed in Ref. 13 for two-dimensional flows; basically it is a two-levels predictor-corrector scheme, which has a second-order accuracy and uses only one node on either side of the node to be computed, at each computational level. Details of the numerical technique and of the boundary conditions treatment are here omitted because of lack of space.

RESULTS

The present approach has been used at first as a steady solver for two simple flow cases, in order to show the merits of the COIN variant. The first considered example is the steady two-dimensional flow around a circular cylinder [15] with an undisturbed Mach number equal to .38; the isoMach lines of the compressible and "incompressible" flows near the leading edge of the cylinder have been plotted in Fig. 1, which shows the two flows to be practically coincident, in this region. This example outlines the COIN problem to be smoother than the full compressible one and accordingly the COIN variant allows to perform accurate computations; using 128x33 gridpoints in the periferal and radial directions respectively, the maximum total temperature error is of the order of .0001 in the present flow case. The quasi one-dimensional free-vortex flow in the 90°-elbow duct shown in Fig. 2 is the second flow case considered: the duct presents square cross sections and constant inlet and outlet radii, equal to 0.9 and 1.1, respectively [11]; the flow is characterized by zero velocity components in the radial (r) and axial (z) directions. Being such a flow the solution of the "incompressible" reference flow too, the COIN variables must result to be zero everywhere in the flow field. Starting from the first iteration, numerical computations have confirmed this theoretical prediction, thus reproducing the exact solution; on the contrary, a code based on the classical lambda formulation without the COIN variant [11] has given only an approximate solution, thus demonstrating once more the better accuracy attainable by means of the COIN variant.

A true three-dimensional unsteady flow has then been considered in order to test the methodology about problems of practical interest. Starting from the previously outlined free-vortex flow, corresponding to a uniform distribution ot total temperature and pressure, the total pressure (p_o) profile has been changed as follows:

$$p_o = 1 - \gamma p M^2 (1 - z)^2.$$

These inlet boundary conditions correspond to prescribe an inlet velocity profile and an entropy distribution which vary in both radial and axial directions, according to the local values of pressure (p) and Mach number (M), and cause a secondary flow to take place in the (r - z) planes. Computations have been performed using 7x25x7 gridpoints in the radial, peripheral and axial directions, respectively, and the time history of the secondary flow development is plotted in Figs. 3 and 4. Cross flow velocity contours in the middle section of the channel ($\theta = \pi/4$) at three different time levels are given in Fig. 3, while the corresponding isentropic lines are plotted in Fig. 4; the first time level corresponds to the early stages of the unsteady flow, while the last time level results represent secondary flows almost completely developed. Figs. 3 and 4 clearly depict the appearance and development of the secondary flows.

The present methodology may also be used as a steady solver. The logarithm (base 10) of the mean square variation of R'^{x}_{1} over the entire flow field is chosen as a representative residual (R). Values of R referring to the computed flow case are plotted in Fig. 5 (continuous line) versus the iteration number k, together with the convergence history pertaining to the same flow case computed by means of the block Point-Gauss-Seidel methodology suggested in Ref. 11 (dotted line). The comparison of the two convergence histories outlines the present methodology to be faster by a factor of about 1.3, while the computational times per iteration step are practically coincident. The present code is very simple; on the contrary, the code developed in accordance with Ref. 11 results to be rather complicate and difficult to be handled because such a technique requires special implicit formulations at each boundary. Finally it must be noticed the present methodology to be slower by a factor of about three with respect to the steady solver proposed in Ref. 12.

CONCLUSIONS

A COIN (Compressible Over INcompressible) lambda formulation for the computation of unsteady internal three-dimensional flows has been provided, which allows to take advantage of the better accuracy of the COIN variant. Moreover a time accurate numerical technique developed for two-dimensional flows has been here extended to three-dimensional flow problems.

The accuracy of the COIN variant has been shown by means of two simple examples, the capability of the suggested methodology to compute unsteady flows has been proved by computing a simple unsteady three-dimensional flow, and finally its moderate ability as a steady solver has been pointed out.

REFERENCES

[1] Moretti, G.: "The λ-scheme", Computers Fluids, 7 (1979) pp. 191-205.

[2] Zannetti, L., Colasurdo, G.: "Unsteady compressible flow: a computational method consistent with the physical phenomena", AIAA J., 19 (1981) pp. 851-856.

[3] Dadone, A., Napolitano, M.: "A perturbative lambda formulation", AIAA J., 24 (1986) pp. 411-417.

[4] Dadone, A.: "A quasi-conservative COIN lambda formulation", Lecture Notes in Physics, 264 (1986) pp. 200-204.

[5] Dadone, A., Napolitano, M.: "An implicit lambda scheme", AIAA J., 21 (1983) pp. 1391-1399.

[6] Dadone, A., Napolitano, M.: "An efficient ADI lambda formulation", Computers Fluids, 13 (1985) pp. 383-395.

[7] Moretti, G.: "A fast Euler solver for steady flows", AIAA 6th Computational Fluid Dynamics Conf. (1983) pp. 357-362.

[8] Dadone, A., Moretti, G.: "Fast Euler solver for transonic airfoils", NASA CR, to be published.

[9] Dadone, A., Fortunato, B., Lippolis, A.: "A fast Euler solver for two and three dimensional internal flows", Symposium Physical Aspects Numerical Gas Dynamics, Farmingdale, NY (August 1987).

[10] Napolitano, M., Dadone, A.: "Three-dimensional implicit lambda methods", AIAA J., 23 (1985) pp. 1343-1347.

[11] Fortunato, B., Napolitano, M.: "Numerical solution of 3-D compressible internal flows using a non isentropic implicit lambda method", Notes Numerical Fluid Mechanics, 13 (1986) pp. 105-112.

[12] Dadone, A., Fortunato, B.: "Fast Euler solver for 3-D internal flows", I.S.C.F.D., Sydney (August 1987).

[13] Moretti, G., Zannetti, L.: "A new improved computational technique for two-dimensional, unsteady, compressible flows", AIAA J., 22 (1984) pp. 758-765.

[14] G. Moretti, "A technique for integrating two-dimensional Euler equations", Computers Fluids, to be published.

[15] Dadone, A.: "Computation of transonic steady flows using a modified lambda formulation", GAMM Workshop Numerical Simulation Compressible Euler Flows, Rocquencourt, France (June 1986). Also Notes Numerical Fluid Mechanics, Vieweg-Verlag, to be published.

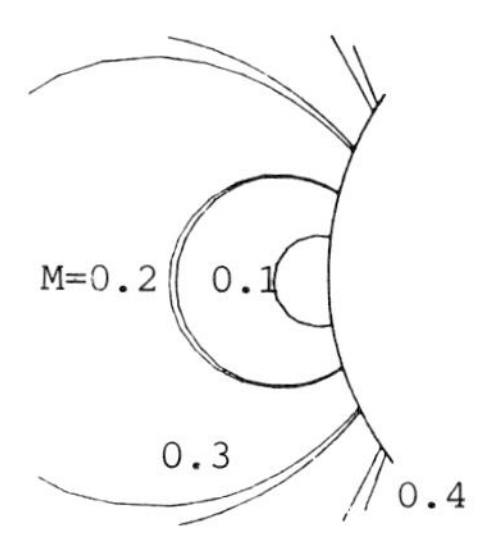

Fig.1 - IsoMach lines in the circular cylinder leading edge region.

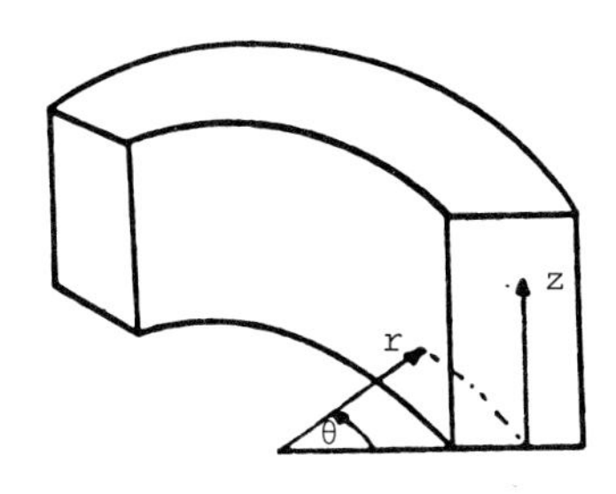

Fig.2 - Sketch of the 3-D elbow geometry.

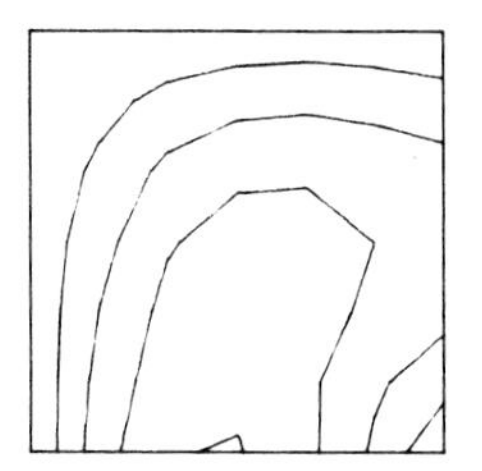

Fig.3 - Cross flow velocity contours in the middle section of the channel at three different time levels.

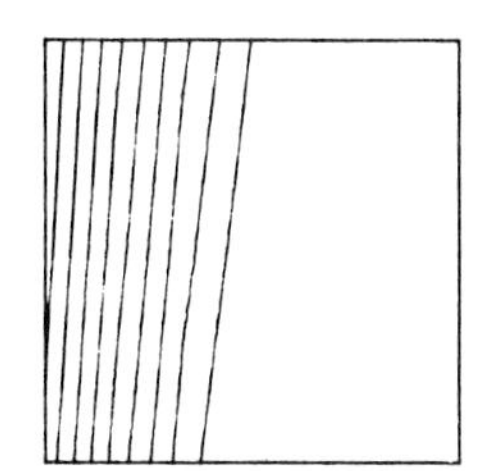

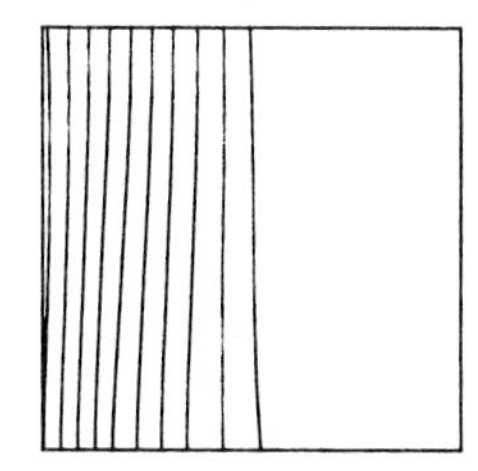

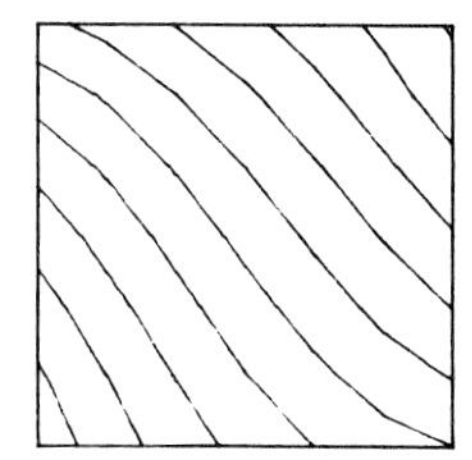

Fig.4 - Isentropic lines in the middle section of the channel at three different time levels.

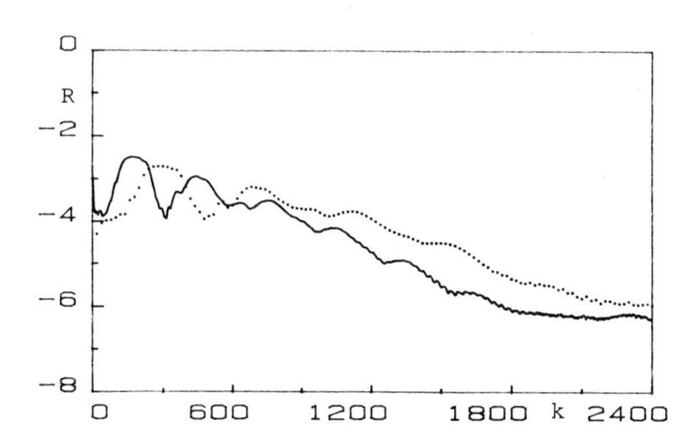

Fig.5 - Convergence history.

CALCULATION OF INCOMPRESSIBLE FLOWS IN MULTIPLY CONNECTED DOMAINS

Laszlo Fuchs
Department of Gasdynamics,
The Royal Institute of Technology,
S-100 44 Stockholm, Sweden.

SUMMARY

The numerical computation of the flow of a viscous incompressible fluid, in a confined domain that contains one or more rigid bodies, is considered. The computational domain may be time-dependent. The streamfunction and vorticity formulation of the Navier-Stokes equation is employed. A system of overlapping grid is used to descretized the equations, and the solution is computed by a Multi-Grid method. The purpose of this paper is to demonstrate the application of the scheme for cases when the value of the streamfunction on the surfaces of the internal bodies is unknown and also for cases when one (or several) solid bodies inside the computational domain accelerate relative to a fixed frame.

INTRODUCTION

The flow of a viscous incompressible liquid in a confined domain is considered. The domain may contain several solid bodies, and these may not be stationary relative to a given frame. The basic numerical approach for such problems is to generate a body-fitted mesh so that solid surfaces become coordinate lines. Such typical grid generations techniques are described in [1]. Finite-Elements gives a better flexibility in describing complex geometries compared to usual finite-difference grids. However, by employing systems of overlapping grids [2-5], one can generate more easily finite-difference grids around complex geometries. Local meshes are generated in each zone independently of each other. The meshes in the different zones may overlap one or more meshes that belong to other zones. The problem can be solved by computing alternatingly a smaller problem in each zone and then updating the 'internal boundary' values. This scheme is known as Schwarz algorithm (see e.g. [2]). An extension to the zonal grid system includes locally refined grids in certain sub-zones. This type of local mesh refinement has been applied by us to different types of problems [3-4]. Successful application of zonal grid methods requires: A. Simple and easy data structure. B. Solution procedures that are fast inspite of the required inter-zonal information exchange. Our zonal-Multi-Grid method, that is described in [6] and the references therein, is capable of handling efficiently these difficulties. Threfore, we found it natural to extend the algorithm to cases where: i). The location of one or several of the internal solid bodies is time-dependent; and ii). The value of the streamfunction on the surface of each solid surface is not known a priori.

In the following we describe the application of the zonal Multi-Grid (MG) algorithm for the solution of the steady and the time-dependent Navier-Stokes equations (in terms of the streamfunction and the vorticity). Typical examples of applications (both steady and unsteady flows) are given in the following. In all the computed cases, the efficiency and good convergence of the scheme has been exhibited.

GOVERNING EQUATIONS AND THEIR APPROXIMATION

The Navier-Stokes equations, for incompressible viscous fluids, in terms of the vorticity, ω, and the streamfunction, ψ, are given by:

$$\omega_t + \nabla^2\omega - Re\, \mathbf{u}\cdot\nabla\omega = 0 \tag{1}$$

$$\omega + \nabla^2\psi = 0 \tag{2}$$

where the velocity vector $\mathbf{u}$ is given by: $\mathbf{u} = \nabla x(0,0,\psi)$ and the vorticity is defined as the z-component of the vorticity vector $(0,0,\omega) = \nabla x\mathbf{u}$. The normalising time scale is taken to be the ratio of the reference length to reference velocity.

The system of equations above, can be solved if two conditions are given on each portion of the boundary. The no-slip condition on solid surfaces, provides such two conditions. If there are solid bodies in the flow field the (constant) value of the streamfunction on such surfaces must be determined as part of the solution. To determine this value uniquely, an additional condition must be satisfied. Some details of the numerical scheme for updating the value of the streamfunction at surfaces of solid bodies are given below. At inflow and outflow boundaries one may specify the streamfunction and the vorticity themselves (instead of the velocity vector). On a partitioned (zonal) domain, one must add 'inter-zonal-boundary conditions'. Such conditions must be defined so that the problem is well posed in each zone. Here, we transfer the values of the streamfunction and the vorticity among the zones (for details see [6]).

A zonal grid system is constructed by generating a local mesh in each zone. Uniform and cartesian grids are used as much as possible with the possible exception near solid bodies, where a local (body-fitted) grid is generated. The local grid is attached to the surface of the solid body even when it undergoes acceleration. In such cases, not only the grids are local, but also the equations are solved in terms of the local frame. The inter-zonal information exchange (and also the equations themselves) must take into account the acceleration and the velocity of the local coordinate system relative to the neighbouring zones. In the case of linear movement of a local coordinate system, only the values of the streamfunction differ. If ψ_f and ψ_m denote the values of the streamfuntion (at the same physical location) as seen in terms of a fixed and a moving coordinate (moving with a velocity V_c along the x-axis), then

$$\psi_f = \psi_m + V_c\, y \tag{3}$$

where y is the distance from the x-axis.

Equations (1) and (2) are approximated by finite-differences on each grid, using central differences for all the terms except the convective terms in (1) which are approximated by upwind differences. It should be noted that in other calculations higher order scheme have been succesfully used. The boundary vorticity (on surfaces of solid bodies) is computed from (2) using the no-slip condition.

SOLUTION PROCEDURE

Here, we distinguish between two basic cases. Steady-state flows are treated by using the basic zonal-MG method, whereas time-dependent flows are computed by a time-marching algorithm and a simplified version of the former scheme. In the following we discuss the two cases separately.

Steady flows

The discrete approximations to equations (1) and (2) are solved by a straightforward MG method: Both equations are 'smoothed' by using a Gauss-Seidel procedure. The non-linear problem is transferred to coarser grids using the so called Full Approximation Storage scheme. The coarse grid corrections are interpolated linearly to the fine grid. Details of the basic scheme are given in [7]. The extensions that are added to the basic MG scheme include the treatment of zonal grids, possibilities for local mesh refinements and a procedure for updating the streamfunction value on the surface of inner bodies. These extensions are discussed in more details in [3] and [6]. Here, we summarize shortly the basic principles.

The basic Schwarz algorithm implies the solution of the discrete equations, in each zone separately and then allowing inter-zonal exchange. This updating procedure is usually very slow. The scheme that we employ transfers data (ω and ψ) transfers data among the zones on the finest grid and on the coarse grids. After an inter-zonal exchange, the approximation is smoothed out by some relaxation sweeps. On the coarsest grid only the corrections are interpolated among the zones. In our code, it is possible to chose either a bi-cubic or a bi-quintic interpolation scheme. In fact, the lower order scheme is adequate on the coarse grids (where the corrections are interpolated), even when the bi-quintic interpolation is used on the finest mesh.

The surface of solid bodies is a streamline along which the value of the streamfunction is constant. The numerical value of streamfunction on the body surface can be determined uniquely by requiring that the pressure shall be integration path independent: The basic principles can be demonstrated for the flow past a cylinder. Using a local (cylindrical coordinates r and θ), the momentum equation in the azimuthal direction, together with the no-slip boundary conditions give that the pressure $p(\theta)$ can be computed from

$$p(\theta)=\int_0^\theta \omega_r\big|_b d\xi . \qquad (4)$$

It is required that the pressure is single valued. Thus

$$p(2\pi)=\int_0^{2\pi} \omega_r\big|_b \, d\xi = 0 . \qquad (5)$$

Numerically, the value of the streamfunction on the surface of the cylinder is adjusted until the integral in (5) vanishes. If this updating procedure is done on the finest grid, the convergence rate is vary slow (since the procedure introduces high frequency errors in the fine grid approximation). Instead, we do not update the surface value of the streamfunction on any of the grids except the coarsest one [6].

Time-dependent flows

At this stage we have studied only explicit methods for solving the vorticity transport equation(1). This equation was advanced in time by using either an explicit Euler, Du-Fort Frankel or a Runge-Kutta scheme. In all these cases, the streamfunction is updated by using the zonal-MG scheme. The surface straemfunction can be calculated in one step since the vorticity is computed by an explicit method. It has been experienced that the streamfunction could be updated after several explicit time steps and thus reducing the total amount of computational work. In the cases that we have computed, however, implicit methods may require less computational effort than the explicit scheme that are mentioned above.

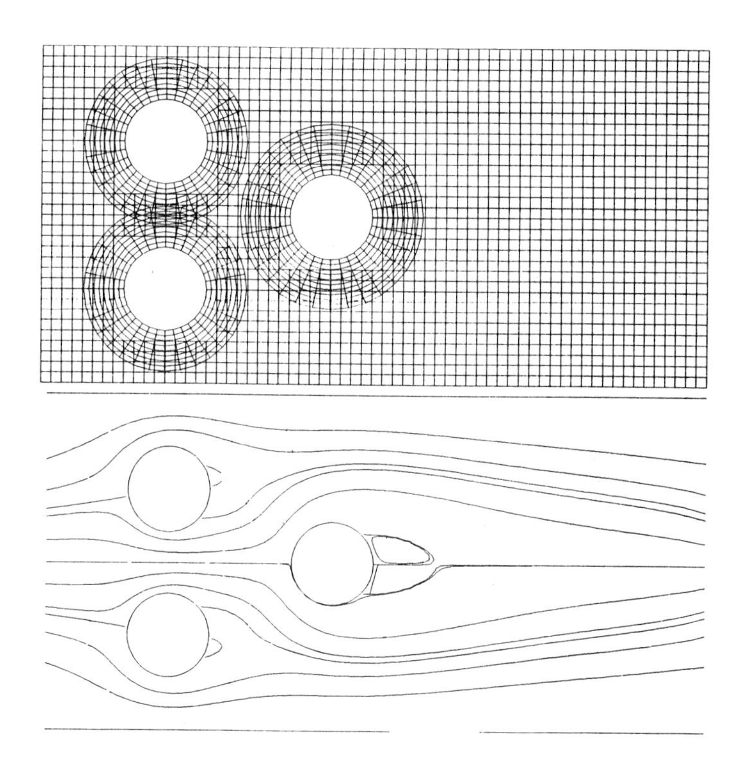

Figure 1: (a. above) The zonal grid system with multiple overlap.
(b. below) Streamline pattern for Re=100.

COMPUTATIONAL EXAMPLES

We consider the flow in a channel that contains three cylindrical objects. These objects could be placed in the channel in an arbitrary manner. The zonal grid system allows the usage of cartesian grids (with possible refinement near the objects) in the channel and a polar grid around each cylinder. In the cartesian grid we 'cut' out a rectangle around each cylinder. Our zonal grid system requires that the polar grids overlap the cartesian grid and that the 'cuts' in the later grid are fully coverd by the former mesh. The polar grids themselves may or may not overlap each other (both situations are shown in Fig. 1a).

Two possible setups are shown in Figures 1.a and 2.a. In the former case the geometry has a plane symmetry and for low Re the flow is also symmetric. The value of the surface streamfunction on the cylindres is known only on one of the cylindres. Thus, the procedure for computing the surface value of the streamfunction is of great practicle importance. Fig. 1.b shows the strealine pattern for Re=100 (based on the radius of the cylindres and the mean inflow velocity) in the symmetrical case. A different arrangement of cylinders is shown in Figures 2. In both cases, the presence of an object downstream causes a substantial decrease in the length of the separation region behind the cylindres that are located further upstream.

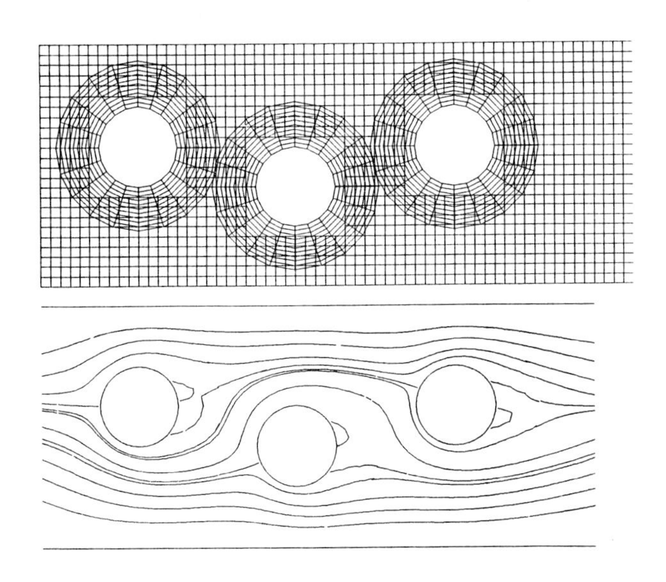

Figure 2: (a. above) The zonal grid system in geometry II.
(b. below) The streamline pattern for Re=100.

Next, consider a valve model in which a cylinrical occluder is moved periodically inside a channel with an inflow orifice. The movement of the occluder is presently flow independent, and the mass flow into the channel is also independent of the position of the occluder. Figures 3 show the central and the two extreme positions of the occluder and the mesh that is used. The streamline pattern is shown in Figures 4 in increments of 1/8 parts of a period. The figures belong to the 4-th period after startup from zero velocity when all the non-periodic transients become small (about one percent). The mesh that is used has the same basic structure for all time steps. The region that is 'cut' out from the cartesian mesh is moved together with the cylinder. At the inter-zonal interfaces, the values of the streamfunctions are transfered according to relation (3).

CONCLUDING REMARKS

The application of a zonal-MG scheme for the computation of steady and time-dependent flows in multiply connected domains was demonstrated. The numerical scheme can be applied to treat non-fixed geometries without the need to generate complex grids in each time step. In all, we have found the method as very flexible and efficient.

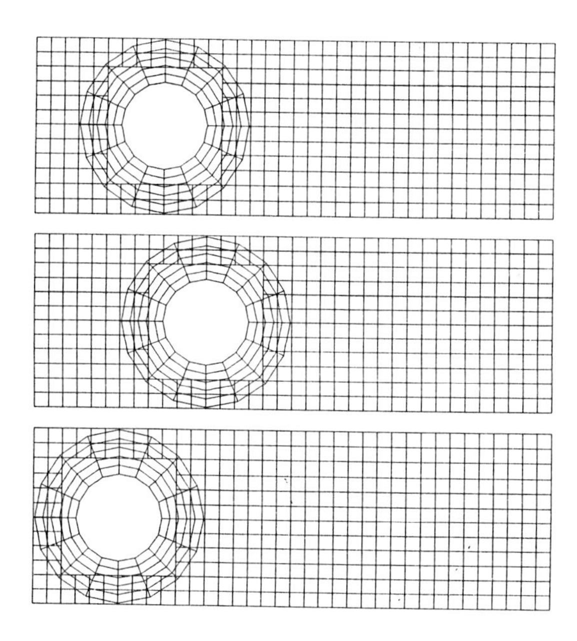

Figure 3: The mesh system with the moving cylinder located at central position (upper) right (middle) and left (lower) positions.

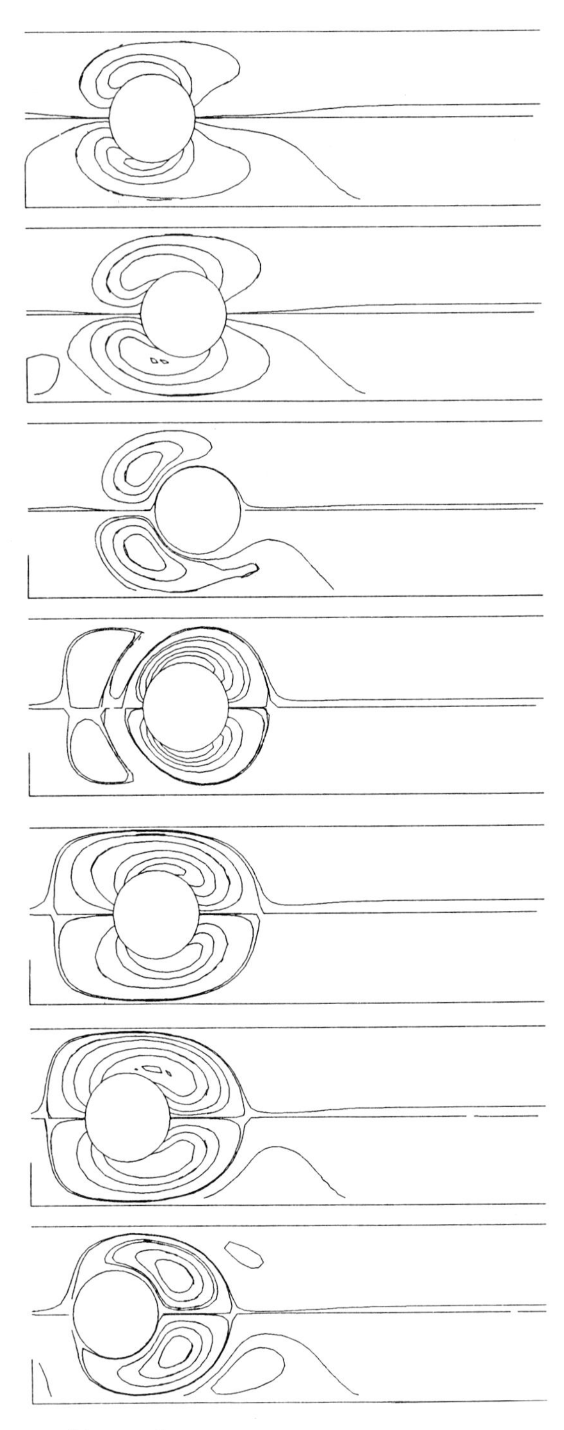

Figure 4: cont'd on next page

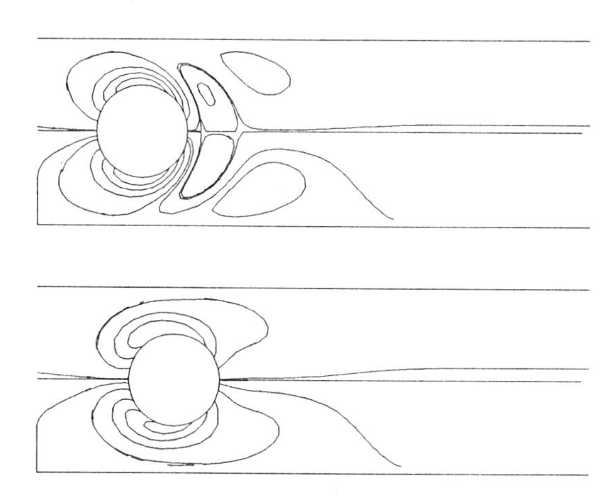

Figure 4: The streamline pattern in a channel with an inflow orifice that blocks half of the channel. Re=100. Full period, starting with the cylinder at central position moving at maximal speed to the right. The time increment between each two frames is 1/8 of the period.

REFERENCES

[1]. J. F. Thompson, Z.U.A. Warsi, and C.W. Mastin; Numerical Grid Generation Foundations and Applications. North-Holland (1985).

[2]. Q.V. Dinh, R. Glowinski, B. Mantel, J. Periaux and P. Perrier; in Computing Methods in Applied Sciences and Engineering, V. Edited by R. Glowinski and J.L. Lions. (North-Holland, 1982), p. 123.

[3]. L. Fuchs; Computers & Fluids, **15** 69 (1986).

[4]. C-Y. Gu and L. Fuchs; in Numerical Methods Laminar and Turbulent Flow-IV, edited by C. Taylor et. al. (Pineridge Press, 1985), p. 1501.

[5]. M.J. Berger; SIAM J. Sci. Stat. Comput. **7** 905 (1986).

[6]. L. Fuchs; - "Computation of viscous flows using a Zonal-Multi-Grid Method". To be published.

[7]. T. Thunell and L. Fuchs; in Numerical Methods Laminar and Turbulent Flow-II. Edited by C. Taylor and B.A. Schrefler. Pineridge Press, p. 141 (1981).

NUMERICAL SIMULATION OF THREE-DIMENSIONAL UNSTEADY VORTEX FLOW USING A COMPACT VORTICITY-VELOCITY ALGORITHM

T. B. Gatski
NASA Langley Research Center
Hampton, VA 23665-5225

C. E. Grosch
Old Dominion University
Norfolk, VA 23508-8508

M. E. Rose
MER Associates
Greensboro, NC 27410
and
R. E. Spall
High Technology Corporation
P. O. Box 9262
Hampton, VA 23666

SUMMARY

A numerical algorithm is presented which is used to solve the unsteady, fully three-dimensional, incompressible Navier-Stokes equations in vorticity-velocity variables. A discussion of the discrete approximation scheme is presented as well as the solution method used to solve the resulting algebraic set of difference equations. Second order spatial and temporal accuracy is verified through solution comparisons with exact results obtained for steady three-dimensional stagnation point flow and unsteady axisymmetric vortex spin-up. In addition, results are presented for the problem of unsteady bubble-type vortex breakdown with emphasis on internal bubble dynamics and structure.

INTRODUCTION

Discrete approximation schemes to the Navier-Stokes (N-S) equations can be extracted using techniques based on finite-difference methods, finite element methods, boundary element methods, spectral methods, or combinations of these. This paper deals with the vorticity-velocity formulation of the N-S equations for unsteady, three-dimensional incompressible flows.

The numerical scheme is compact and applicable to irregularly sized Cartesian grids; although extension to irregularly shaped grids is straightforward. The resulting algebraic equations are solved using iterative techniques. In a previous paper by Gatski, Grosch and Rose [1], an algorithm was presented which solves the corresponding two-dimensional problem using a compact scheme. There, the transport of vorticity was governed by a simple advection-diffusion equation, and the only non-zero component of vorticity was normal to the plane of motion. The governing equation set was closed by imposing the solenoidal condition on the velocity field and using the kinematic definition of the vorticity. The same solution strategy employed in two dimensions is now extended to the three-dimensional case, with, of course, the added complexity of a multi-component vorticity vector and the appearance of a vortex stretching term in the vorticity transport equations.

The idea of a vorticity-velocity formulation itself is not new; however, the method of implementation presented here and in [1] is unique. In previous studies [2,3,4], the velocity field was obtained from Poisson equations. These were derived by taking the curl of the curl of the velocity and using the condition that the velocity is divergence free to eliminate the mixed derivatives. These results were for steady [2] and unsteady [3,4] flows in both two [3,4] and three dimensions [2]. In a recent unsteady, two-dimensional study, Orlandi [5] used a time-splitting method which was analogous to an ADI-type approach. Some of the cross derivatives in the second-order velocity equations were retained in the appropriate sweep directions.

As in the earlier two-dimensional study [1], the velocity field, in the method presented here is obtained from the solution of the Cauchy-Riemann equations resulting from the requirement of a solenoidal velocity field and the defining equations for the vorticity field. However, the discrete approximation scheme to the vorticity transport equation has been restructured in order to use an interative Kaczmarz solver (cf. [1]) on the resulting algebraic set.

The second-order spatial and temporal accuracy of the three-dimensional algorithm is verified by examining both three-dimensional steady stagnation point flow [6] and axisymmetric unsteady vortex spin-up [7].

Finally, results are presented for the unsteady, three-dimensional case of bubble-type vortex breakdown. Plots of the velocity and vorticity variables which illustrate the robust unsteady, asymmetric dynamics occurring within the breakdown bubble are shown.

VORTICITY-VELOCITY ALGORITHM

The equation governing the flow of a Newtonian incompressible fluid can be written as:

$$\nabla \cdot \underline{u} = 0 \tag{1a}$$

$$\nabla \times \underline{u} = \underline{\zeta} \tag{1b}$$

$$\underline{\zeta}_t + (\underline{u} \cdot \nabla)\, \underline{\zeta} - (\underline{\zeta} \cdot \nabla)\underline{u} = \nu \nabla^2 \underline{\zeta}, \tag{1c}$$

where $\underline{u} = \underline{u}\,(\underline{x},t)$ and $\underline{\zeta} = \underline{\zeta}\,(\underline{x},t)$ are the velocity and vorticity fields, respectively, $\underline{x} = (x^1,x^2,x^3)$ is a point in $\mathbf{R}^3$, t is the time, and ν is the kinemetic viscosity. As a consequence of equation (1b), the vorticity field must be solenoidal, so that

$$\nabla \cdot \underline{\zeta} = 0\,. \tag{1d}$$

For a given set of initial conditions within a solution domain D,

$$\underline{u} = \underline{u}_0, \quad \underline{\zeta} = \nabla \times \underline{u}_0, \text{ at } t = 0, \tag{2a,b}$$

and a set of boundary conditions (on the boundary Γ of D)

$$\underline{u} = \underline{u}_\Gamma, \quad \underline{\zeta} = (\nabla \times \underline{u})|_\Gamma \tag{3a,b}$$

a solution is assumed to exist and be unique on a given time interval $0 < t \leqslant T$. Since the equations have the semigroup property, it is only necessary to outline a construction for a time interval within $0 < t \leq T$. Subsequent time intervals are then similarly constructed with the results from the previous time interval used as initial conditions.

As part of the solution strategy, equations (1a) and (1b) are solved simultaneously as a Cauchy-Riemann set. We will refer to this set as the velocity equations, since they determine the velocity field given a divergence-free vorticity field. The vorticity variable $\underline{\zeta}$ is determined next although equation (1c) is not used itself in the solution algorithm. A modified variable, $\underline{\omega}$, is introduced whose governing equation is a simple advection-diffusion equation. The variables $\underline{\zeta}$ and $\underline{\omega}$ are related by a simple exponential transformation. Finally, since there is no a priori requirement that the discrete solution set will satisfy equation (1d), a projection operator is constructed which removes any non-solenoidal motion from the computed vorticity field.

A. Velocity Equation

A discrete approximation scheme for the velocity equations can be derived by integrating equation (1a) over the volume of a Cartesian element and equation (1b) over the faces of the element. The resulting discrete set using box-variables defined at the vertices of the element are;

$$\delta_x(\mu_y\mu_z u_x) + \delta_y(\mu_x\mu_z u_y) + \delta_z(\mu_x\mu_y u_z) = 0 \tag{4}$$

$$\delta_y(\mu_z u_z) - \delta_z(\mu_y u_y) = \hat{\zeta}_x \tag{5a}$$

$$\delta_z(\mu_x u_x) - \delta_x(\mu_z u_z) = \hat{\zeta}_y \tag{5b}$$

$$\delta_x(\mu_y u_y) - \delta_y(\mu_x u_x) = \hat{\zeta}_z \tag{5c}$$

where μ and δ are the respective centered average and difference operations.

B. Transport Equations

Equation (1c) is a coupled set of three scalar partial differential equations. The coupling occurs through the vortex stretching term which redistributes vorticity among components. In two-dimensional flows, there is only one non-zero vorticity component and its transport equation reduces to a simple advection-diffusion equation. In light of the simplification due to this degeneracy, it will be desirable to construct a numerical scheme which also easily degenerates in the case of two-dimensional flow. Since the velocity and vorticity gradients are assumed known at the requisite time levels, a simple transformation of the vorticity variable, of the form;

$$\underline{\zeta} = \exp[\underline{\underline{b}}\,(t - t_m)]\,\underline{\omega} \tag{6a}$$

where
$$\underline{\underline{b}} = \nabla(\underline{u} + \underline{u}^T)^{\ell}/2 \tag{6b}$$

leads to a simple advection-diffusion transport equation. The governing equation for $\underline{\omega}$ is

$$\underline{\omega}_t + (\underline{u}^{\ell} \cdot \nabla)\,\underline{\omega} = \nu\nabla^2\underline{\omega} \tag{7}$$

where the superscript ℓ is simply a converged iteration index which resulted in the solution of the velocity equations. The difference scheme to solve equation (7) is most simply understood by writing it as a first order system. Equation (7) becomes

$$\underline{\omega}_t = \frac{\partial \underline{\phi}_x}{\partial x} + \frac{\partial \underline{\phi}_y}{\partial y} + \frac{\partial \underline{\phi}_z}{\partial z} \tag{8a}$$

$$\underline{\underline{\phi}} = (\underline{\phi}_x, \underline{\phi}_y, \underline{\phi}_z) = \nu \text{ grad } \underline{\omega} - \underline{u}^{\ell} : \underline{\omega} \ . \tag{8b}$$

A discrete approximation scheme to equations (8) can be obtained by imposing discrete boundary integral relationships on the dependent variables on each element within the computational domain. The process is a domain decomposition extension of boundary element methods and results in a compact numerical approximation scheme of the governing differential equations. Based on the weak-element method [8], a suitable approximation basis is chosen (cf. [1]) which results in the following three-dimensional numerical scheme;

$$\delta_t \underline{\omega} = \delta_x \underline{\phi}_x + \delta_y \underline{\phi}_y + \delta_z \underline{\phi}_z \tag{9a}$$

$$\mu_t \underline{\omega} = c_\alpha \mu_\alpha \underline{\omega} - \rho_\alpha \Delta_\alpha \underline{\omega} - \sigma_\alpha \Delta_\alpha \underline{\phi}_\alpha \tag{9b}$$

$$\alpha = x,y,z$$

$$\mu_\alpha \underline{\phi}_\alpha - \rho_\alpha \Delta_\alpha \underline{\phi}_\alpha = \kappa_\alpha \lambda_\alpha^{-1} \Delta_\alpha \underline{\omega} - \mu_\alpha \underline{u}_\alpha^{\ell} (c_\alpha \mu_\alpha \underline{\omega} - \rho_\alpha \Delta_\alpha \underline{\omega}) \tag{9c}$$

where

$$\rho_\alpha = \coth \theta_\alpha^{-1} - \theta_\alpha^{-1}, \quad \sigma_\alpha = \lambda_\alpha \kappa_\alpha^{-1} (\theta_\alpha^{-1} \rho_\alpha) \tag{10}$$

$$c_\alpha = 1 - \Delta_\alpha u_\alpha^{\ell} (\mu_\alpha \mu_\alpha^{\ell})^{-1} \rho_\alpha \tag{11}$$

$$\theta_\alpha = (\mu_\alpha \mu_\alpha^{\ell}) h_\alpha \nu^{-1} \tag{12}$$

$$\lambda_\alpha = \tau h_\alpha^{-1} \tag{13}$$

$$\kappa_\alpha = \nu \tau h_\alpha^{-2}, \tag{14}$$

$h_x = \Delta x/2$, etc., $\Delta_\alpha = h_\alpha \delta_\alpha$ and $\tau = \Delta t/2$. For this set of equations, it is desirable to seek an implicit single stage solution process within a time interval. This can easily be accomplished by requiring that the vorticity, $\underline{\zeta}$, be continuous across consecutive time intervals. This continuity condition can be obtained by simply combining equations (9a) and (9b) and rewriting as

$$\exp\,[\underline{\underline{b}}^{m+1}(t - t_{m+1})]\, \underline{M}_-^{m+1} \underline{\omega}^{m+1} = \exp\,[\underline{\underline{b}}^{m}(t - t_{m})]\, \underline{M}_+^{m} \underline{\omega}^{m} \tag{15}$$

where

$$M_\pm^m \equiv (\mu_t \pm \tau\ \delta_t)^m. \tag{16}$$

Equation (15) coupled with equation (9c) is the set used to solve for the vorticity field within the computational domain.

As alluded to earlier, there is no explicit requirement or guarentee that the resultant vorticity field $\underline{\zeta}$ be solenidal. Thus, utilization of this vorticity field in the solution of the velocity field may be inappropriate. It is therefore necessary to check the divergence of the vorticity field and, if required, project the computed vorticity field onto a space of divergence free vorticity.

C. Helmholtz Projection

The projection of the vorticity field is initiated by using a Helmholtz decomposition to split the vorticity field into a divergence free part and an irrotational part, i.e.,

$$\underline{\zeta} = \hat{\underline{\zeta}} + \nabla\chi \tag{17}$$

where $\hat{\underline{\zeta}}$ is the divergence free part. A Poisson equation for χ can be obtained by taking the divergence of equation (17),

$$\nabla^2\chi = \nabla \cdot \underline{\zeta}, \tag{18a}$$

subject to the boundary conditions,

$$\underline{n} \cdot \nabla\chi = 0. \tag{18b}$$

The boundary conditions, equation (18b), are chosen so that the normal components of vorticity on the boundaries are unaltered by the projection operation.

Once again a compact approximation scheme can be derived for equation (18a). The scheme is given by (cf. [9])

$$\delta_x p_x + \delta_y p_y + \delta_z p_z = \mathrm{div}_e \underline{\zeta} \tag{19a}$$

$$\mu_x p_x = \delta_x\chi, \quad \mu_y p_y = \delta_y\chi, \quad \mu_z p_z = \delta_z\chi \tag{19b,c,d}$$

$$\mu_x\chi - 0.5h_x^2\delta_x p_x = \mu_y\chi - 0.5h_y^2\delta_y p_y = \mu_z\chi - 0.5h_z^2\delta_z p_z . \tag{19e,f}$$

Note that these six equations relate the six discrete values of χ on the faces of each element with the six values of the normal components of $\underline{p}$ on these same faces. In order to obtain the divergence-free vorticity field, it is just a simple matter of subtracting the appropriate component of $\underline{p}$ from the corresponding normal vorticity component on each face of the element.

With the construction of the numerical approximation schemes for the governing differential equations, the issue that remains is the choice of the numerical solver. An iterative technique is used to solve all the discrete equations. A Kaczmarz method is used which has convergence properties similar to SOR techniques. Implementation of this method is described in a previous paper by Gatski, Grosch, and Rose [1].

It is, of course, necessary to comment on the spatial and temporal accuracy of the algorithm. Two test problems were examined. The first was the three-dimensional steady stagnation point flow [6] and the second was the axisymmetric unsteady vortex spin-up problem [7]. The stagnation point flow problem was used to verify the second-order spatial accuracy of the alogrithm. The vortex spin-up problem has an exact time-dependent solution and was used to verify the temporal accuracy of the algorithm. In both cases, L_2 norms of both the velocity and vorticity were used as points of comparison and in both cases second-order spatial and temporal accuracy was achieved.

VORTEX BREAKDOWN SIMULATION

A relevant problem of both academic and practical interest is that of vortex bursting. An analysis of the structure of bubble-type vortex breakdown has been performed using the algorithm described in the previous section. To simulate this phenomenon, an initially axisymmeteric unconfined longitudinal vortex, embedded in a stream of uniform axial velocity, was allowed to both spatially and temporally evolve. The initial vortex consisted of a exponential swirl velocity distribution and zero radial velocity (Burger's vortex). This initial distribution was also used as the inflow boundary condition throughout the computation. A combination of flux and undisturbed free-stream conditions were used as boundary conditions on the remaining computational boundaries for the velocity and vorticity variables. The computations were run until a fully developed time-dependent internal bubble structure was obtained.

Figure 1 is a plot of particle paths with the Cartesian computational grid superimposed. The view is from downstream and above. The upstream stagnation point is located approximately four vortex radii downstream of the inflow plane. Note that there are no axisymmetry requirements imposed on the computations through either the boundary conditions or structure of the numerical algorithm.

A more revealing dynamic picture of the internal bubble structure can be obtained by examining the velocity vectors within the breakdown bubble. Figure 2 shows three-dimensional velocity vectors which are projected onto an x-y plane through the vortex centerline. The flow of figure 2 is an instantaneous representation taken at approximately 81 time units after the start of the computation. The time variable on the problem has been scaled by the inflow axial velocity and inflow vortex core radius. As the figure shows an aft toroidal vortex acts to both inject fluid into the bubble as well as eject fluid -- both of which occur simultaneously but at different azimuthal locations. The net effect, however, is to keep the overall size of the bubble, after becoming fully developed, relatively constant with time. An analysis of the transient results indicates that the internal bubble structure is dominated by unsteady vortical dynamics and, in addition, strongly suggests a cyclic rotational motion of the internal toroidal vortex about the bubble centerline.

Further information about the breakdown bubble can be extracted from reviewing the component vorticity contours. Figure 3 shows the ζ_x and ζ_z component vorticities in the x-y plane at approximately 81 time units after the start of the computation. Even though there are no axisymmetry requirements in the computation, the ζ_y component vorticity was found to be qualitatively similar to the ζ_z component but approximately 90° out of phase.

Figure 3a shows that the inflow vorticity distribution is indeed symmetric, as specified, and that the outflow vorticity distribution is nearly so, although much more intense. This means that the breakdown bubble acts to intensify axial vorticity; that is, the magnitude increases and characteristic size decreases. Within the breakdown bubble, the vorticity has been redistributed to the other vorticity components due to vortex stretching.

Figure 3b shows that near the boundary of the breakdown bubble significant amounts of ζ_z vorticity exists, as well as in the aft portion of the bubble, but off the centerline. In fact, when figure 3b is compared with the velocity vectors of figure 2, the intense vortex structure of the aft vortex is clearly shown. Finally, note that the downstream "tails" of the ζ_z (as well as ζ_y) component vorticity remain "attached" to the main bubble structure.

REFERENCES

[1] Gatski, T. B., Grosch, C. E., and Rose, M. E.: "A Numerical Study of the Two-Dimensional Navier-Stokes Equations in Vorticity-Velocity Variables," J. Comp. Phys., 48 (1982) pp. 1-22.

[2] Dennis, S. C. R., Ingham, D. B., and Cooke, R. N.: "Finite-Difference Methods for Calculating Steady Incompressible Flows in Three-Dimensions," J. Comp. Phys., 33 (1979) pp. 325-339.

[3] Fasel, H. F.: "Numerical Solution of the Complete Navier-Stokes Equations for the Simulation of Unsteady Flows," Lecture Notes in Mathemetics No. 771, Springer-Varlag, Berlin, 1980.

[4] Farouke, B., Fusagi, T.: "A Coupled Solution of the Vorticity-Velocity Formulation of the Incompressible Navier-Stokes Equations," Inst. J. for Num. Methods in Fluids, 5 (1985) pp. 1017-1034.

[5] Orlandi, P.: "Vorticity-Velocity Formulation for High Re Flows," Computers and Fluids, 15 (1987) pp. 137-149.

[6] Howarth, L.: "The Boundary Layer in Three-Dimensional Flow-Part II. The Flow near a Stagnation Point," Phil. Mag. (7) 42 (1951) pp. 1433-1440.

[7] Rott, N.: "On the Viscous Core of a Line Vortex," ZAMP 1XB (1958) pp. 543-553.

[8] Rose, M. E.: "Weak-element Approximations to Elliptic Differential Equations," 24 (1975) pp. 185-204.

[9] Rose, M. E.: "Pressure Calculations for Incompressible Flows," Proceedings of a Workshop in CFD, University of Calfornia at Davis, 1986.

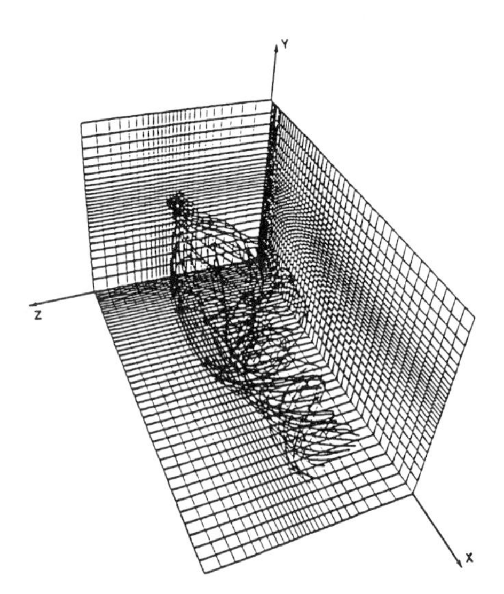

Fig. 1 Particle path trajectories of the vortex breakdown bubble within the computational domain.

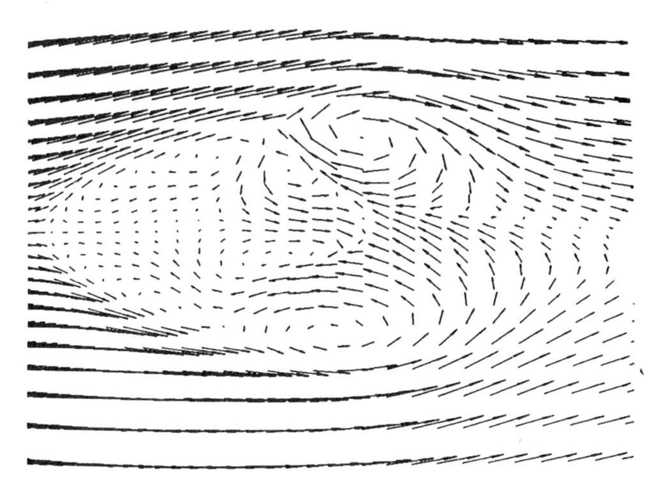

Fig. 2 Planar projected velocity vectors over the interior of the breakdown region at t = 81.28.

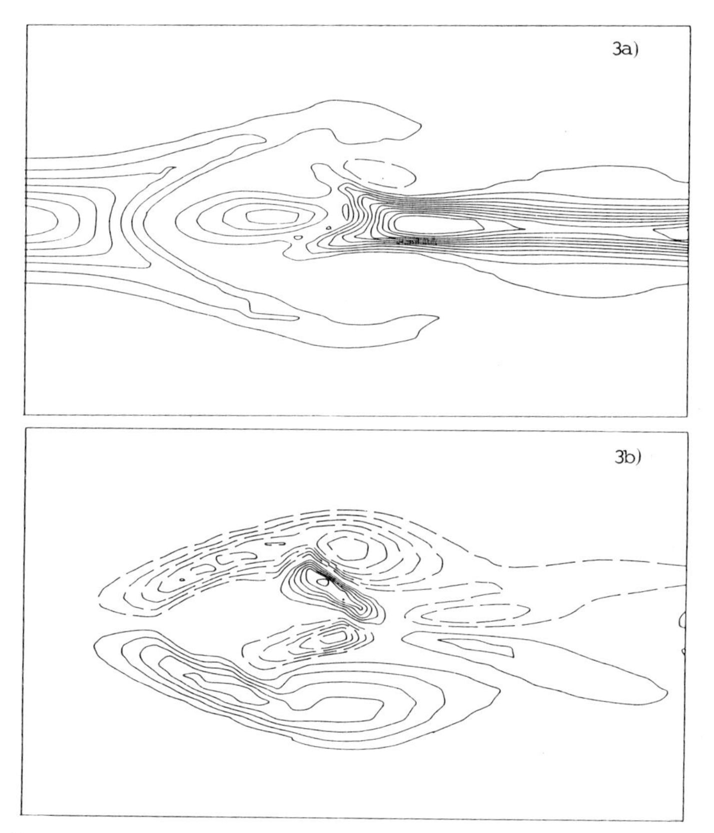

Fig. 3 Contours of constant vorticity: a) ζ_x contours ranging from -0.25 to 2.5; b) ζ_z contours ranging from -1.5 to 1.5. All contour levels in intervals of 0.25.

COMPUTATION OF SEPARATED FLOWS USING A VISCOUS VORTEX METHOD

J.M.R. Graham, P.D. Cozens and B. Djahansouzi
Department of Aeronautics, Imperial College,
London, SW7 2BY

SUMMARY

Results are presented for incompressible unsteady two-dimensional separated flow using a computational method based on a viscous extension of the Cloud-in-cell discrete vortex method. The flows studied are impulsive and oscillatory flow past a 90° corner with varying amounts of rounding and impulsively started flow past a circular cylinder.

INTRODUCTION

The discrete vortex method is a Lagrangian type of method in which vorticity transport in incompressible flow is modelled by tracking discrete fluid elements on which circulation is conserved and which move with the fluid, [1], [2]. Usually the velocity field associated with the vorticity is computed by the Biot-Savart integral. This involves $O(N_V^2)$ operations per time step where N_V is the number of vortex elements and is computationally expensive unless some form of neighbourhood summation is used as in [3]. An alternative, the Cloud-in-cell method [4], projects the moving vortex element onto a fixed mesh and computes the velocity field by finite differences on this mesh. This part of the calculation for two-dimensional flows is usually carried out by a fast solution method for the Poisson equation for the stream function ψ in terms of the vorticity ω on the grid:

$$\nabla^2\psi = -\omega. \tag{1}$$

The velocity field is calculate by central differences of ψ and is interpolated to obtain the velocity at the moving vortex elements. Point vortices and bilinear projection/ interpolation are usually used for two-dimensional flows. The method involves $O(N_g \ell n N_g, N_V)$ operations per time step where N_g is the number of grid points. It is quite feasible to carry out computations with large values of N_g and N_V ($\sim 10^4$). The velocity field is however smoothed by the use of the grid with wavelengths smaller than double the grid spacing removed. This has an advantage in reducing the small wavelength Kelvin-Helmholz instability inherent in the use of discrete vortex methods to represent free vortex sheets, but suggests that the method is more appropriate to flows with diffused rather than sheet like vorticity. The method has however been considered useful as a means of representing high Reynolds number vortex flows because of its apparently low numerical diffusion.

The discrete vortex method has been generalised to include viscous diffusion either by the use of growing viscous cores [5] or by splitting the time step into a convection substep and a diffusion substep modelled statistically by a 'random walk' [6]. The random walk method has also been used in conjunction with the mesh method [7].

In the present paper the split time step is used to separate diffusion and convection but, following a similar approach to [8] for high Peclet number calculations, the diffusion is carried out deterministically by finite differences on the fixed mesh. Since circulation is no longer conserved on each moving element, projection of the diffused vorticity field back onto the moving elements is less straightforward than in the inviscid case.

NUMERICAL METHOD

The flow field is transformed into a rectangular region with the body surface along the (real) x-axis. In the present case conformal transformation was used with 'unwrapping' where appropriate. The 2-D Navier Stokes equations in vorticity streamfunction form in the transformed x,y plane are therefore

$$J^{-1}.\ \omega_t + \psi_y\omega_x - \psi_x\omega_y = \nu(\omega_{xx} + \omega_{yy}) \tag{2}$$

and $$\psi_{xx} + \psi_{yy} = -J^{-1}\omega \tag{3}$$

where J is the Jacobian of the transformation. The vorticity is represented by a distribution of points $(x_k(t),\ y_k(t))$ each carrying circulation $\Gamma_k(t)$, $k=1 \ldots N_v$. A fixed rectangular mesh was used for the computation with closer spacing in the normal direction near the surface. Figure 1 shows an example of the mesh in the physical plane.

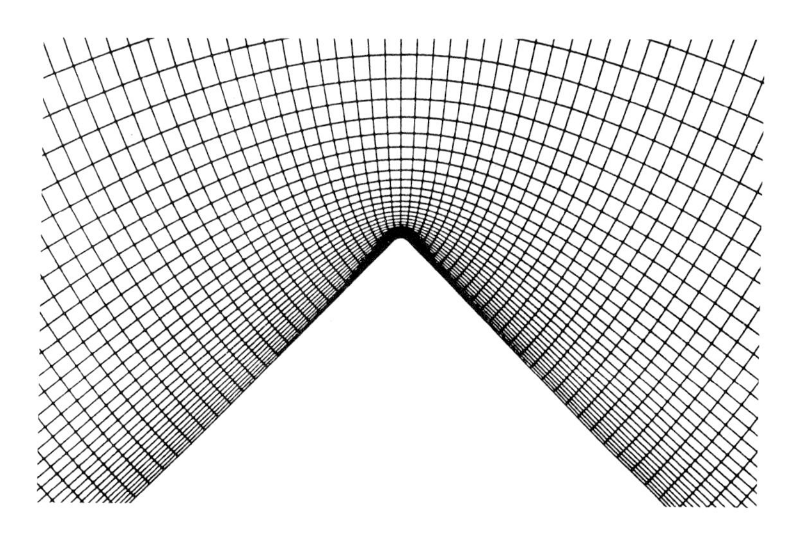

Fig.1 Computation mesh

The circulation from each moving point was projected onto the four nodes of the cell within which it lay using a bilinear function. This preserves the first and second moments of the distribution: $\sum\Gamma_k, \sum\Gamma_k x_k$ and $\sum\Gamma_k y_k$. Equation (3) is solved for ψ by taking a Fourier transform with respect to y and solving the resulting set of ordinary differential equations by finite differences and Gaussian elimination of the tridiagonal matrix. Equation (2) is split into a Lagrangian convection stage:

$$\frac{\partial x_k}{\partial t} = J_k \psi_y^k , \qquad \frac{\partial y_k}{\partial t} = -J_k . \psi_x^k \tag{4}$$

and

$$\frac{\partial \omega}{\partial t} = \nu J \left(\omega_{xx} + \omega_{yy} \right) . \tag{5}$$

J_k , ψ_x^k and ψ_y^k in equation (4) are obtained at the points x_k and y_k by bilinear interpolation from the grid. Forward time differences were used to integrate equation (4) and most of the computations were carried out with first order integration, 3rd order Runge Kutta being used in a few cases only.

In the inviscid Cloud-in-cell method Γ_k remains constant on each moving element, but must vary because of diffusion, eqn (5), for viscous flow. Equation (5) was solved for the grid vorticity using the implicit formulation:

$$\begin{aligned} \Delta\omega = \omega^{(n+1)} - \omega^{(n)} = \nu J \delta t \Big\{ \beta \left(\Delta_x^2 \omega^{(n+1)} + \Delta_y^2 \omega^{(n+1)} \right) \\ + (1-\beta)(\Delta_x^2 \omega^{(n)} + \Delta_y^2 \omega^{(n)}) \Big\} \end{aligned} \tag{6}$$

where δt is the time step, n its number, Δ_x and Δ_y central differences. β was taken in the range $0.5<\beta<1.0$. In order to complete the calculation at each time step it was necessary to project the diffused grid vorticity back onto the moving points. Since the calculation method was intended for high Reynolds number flows $\Delta\omega$ on the grid was regarded as a small change to the inviscid solution which conserved the moving point circulation over the time step. For each grid point (x_g, y_g) $\Delta\omega$, converted to a change in circulation $\Delta\Gamma$, was amalgamated with the moving vortex elements in the four adjacent cells on a weighted basis which conserved $\Delta\Gamma + \sum\Gamma_k$, $\Delta\Gamma\, x_g + \sum\Gamma_k x_k$ and $\sum\Gamma y_g + \sum\Gamma_k y_k$. For those grid points having non-zero $\Delta\omega$ and no adjacent moving elements a new moving vortex was created. This method added only the order of 10 new vortices to the flow at each time step on a 64 x 50 grid, thus permitting calculations for a large number of time steps without an excessive number of moving points. For flow past a solid surface new vorticity enters the flow to maintain the no slip condition. The boundary conditions $\psi = 0$, $\psi_y = 0$ and $\psi_{yy} = -J^{-1}\omega$. on the surface satisfy this and the condition of zero velocity normal to the surface. The overall method is similar to finite difference Navier Stokes methods but with Lagrangian convection.

RESULTS

1) Representation of an oscillatory laminar boundary layer.

The Stokes oscillatory flat plate boundary layer was computed as a test case of the diffusion part of the method. Figure 2 shows good agreement between the computed and analytical velocity profiles for this boundary layer.

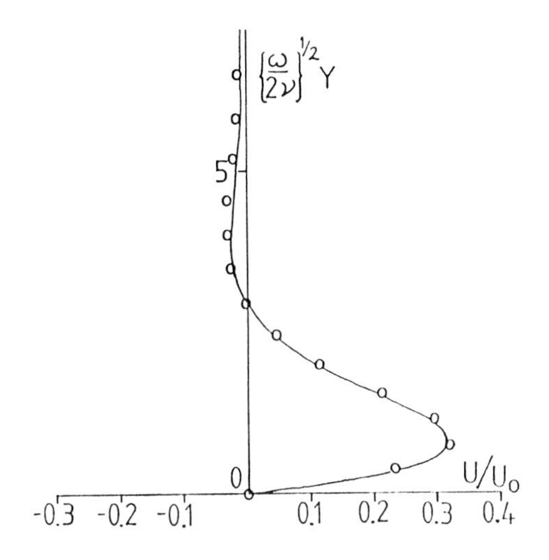

Fig.2 Stokes layer ——— Analytical, o Computed.

2) Impulsively started flow past an edge.

Figure 3 shows a comparison for the vortex distributions and streamlines between results computed for inviscid flow using a self-similar inviscid vortex sheet analysis [9] and the present method with very small values of viscosity and radius of curvature at the edge.

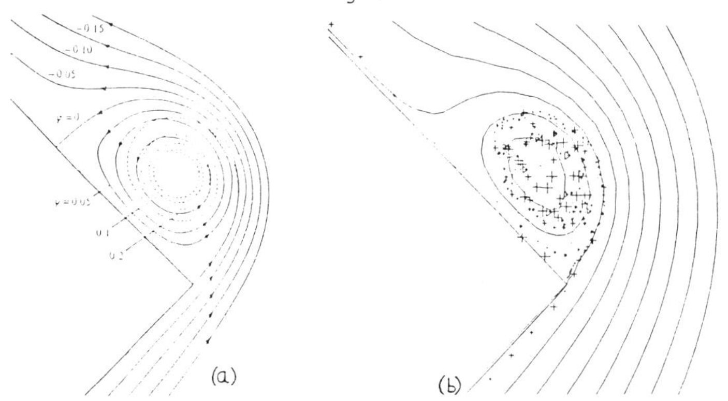

Fig.3 Impulsively started flow past a sharp edge.
(a) Reference [9], (b) Present Computation.

If the same calculation is repeated with a nominally sharp edge and a larger value of the viscosity the result shown in Figure 4 is obtained.

The shed circulation is plotted in terms of similarity variables since in these terms the circulation depends on the Reynolds Number $V^{3/2}\ell^{1/2}t^{1/2}/\nu$ only, where V and ℓ are velocity and length scales of the flow.

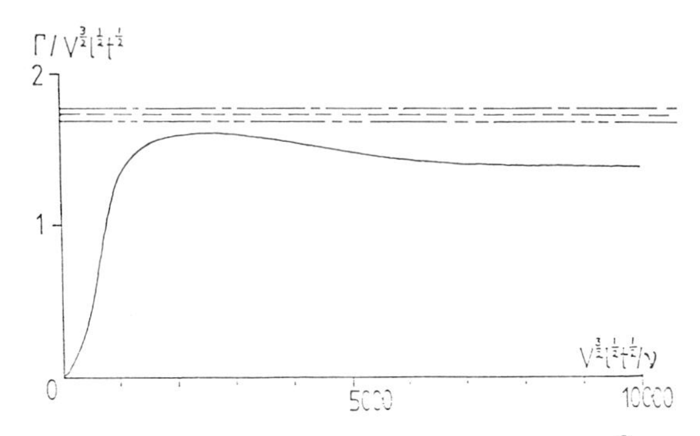

Fig.4 Impulsively started flow past a 90° edge. Circulation growth for inviscid and viscous flow. Inviscid ---[9], — — Present Method; viscous —— present method.

3) Oscillatory flow past edges

Oscillatory flow, $V = V_0 \sin 2\pi t/T$, past a 90° edge with differing degrees of rounding has also been computed using this method. This flow arises at the bilges of barges and other types of ship undergoing roll motion. The vortex shedding from the bilge largely determines the damping and hence the roll response in waves. Figure 5 shows an example of the vortex pairing which occurs for this type of flow. Vortices of alternate sign are formed on each half cycle and form pairs which convect away from the edge under their mutually induced velocity field.

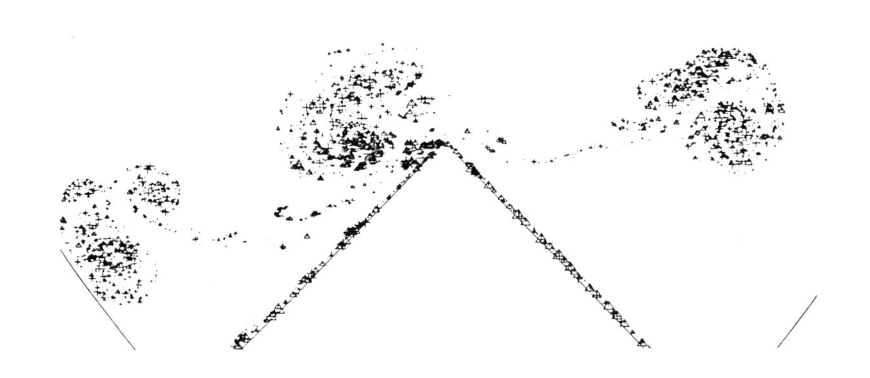

Fig.5 Vortex pairing at an edge in oscillatory flow; vorticity field.

Figure 6 shows the velocity vectors of the vortex field.

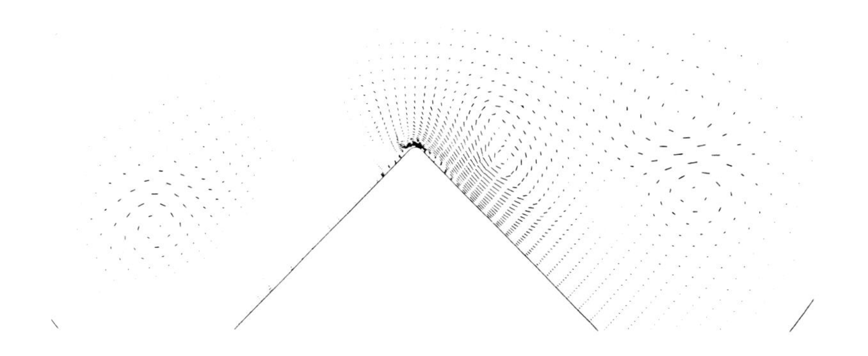

Fig.6 Vortex pairing at an edge in oscillatory flow; velocity vectors.

The associated oscillatory forces induced on the edge by the vortex shedding are shown in Figure 7.

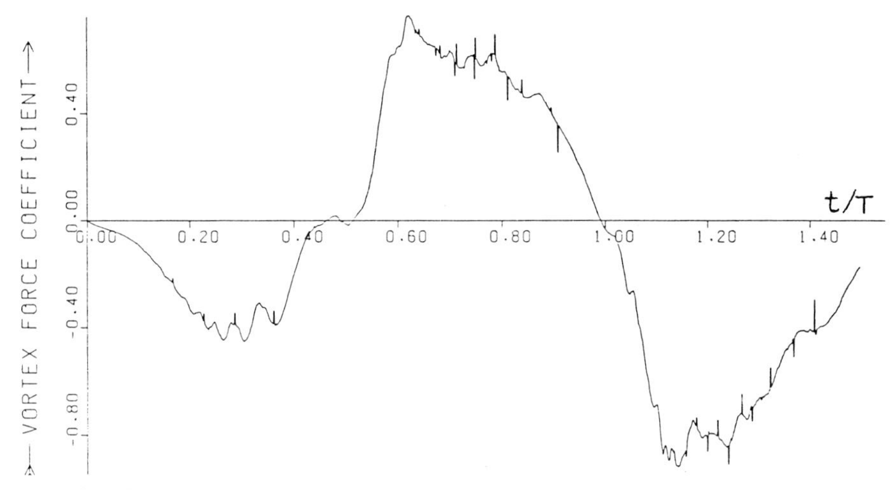

Fig.7 Force induced at an edge by oscillatory flow.

In the above examples the characteristic Reynolds number $V_o^{3/2} t^{1/2} T^{1/2} / \nu = 0(10^4)$.

4) Impulsively started flow past a circular cylinder.

The same method has been used to compute the Von-Karman vortex street which forms behind a circular cylinder of diameter d started impulsively from rest. Figure 8 show the vortex distribution at a nondimensional time $Vt/d = 18.5$ and Reynolds number of 10^4.

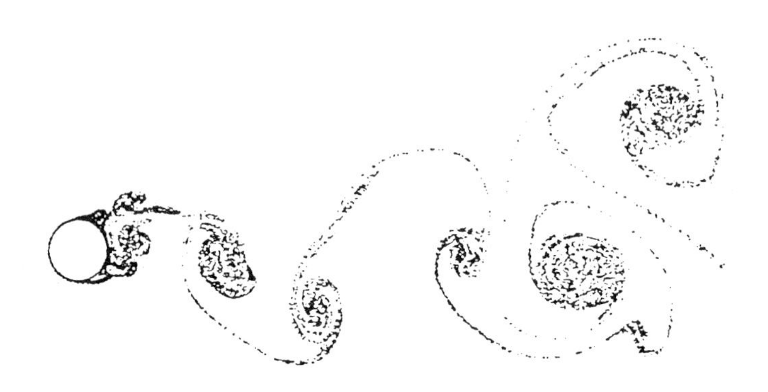

Fig. 8 Impulsively started flow past a circular cylinder.

The wake is purely laminar and shows signs of instability. In reality the wake would be turbulent at this Reynolds number and appear much more diffused. The associated lift and drag force coefficients are shown in Figure 9. They are higher than in reality because of the absence of turbulent diffusion and other three-dimensional effects.

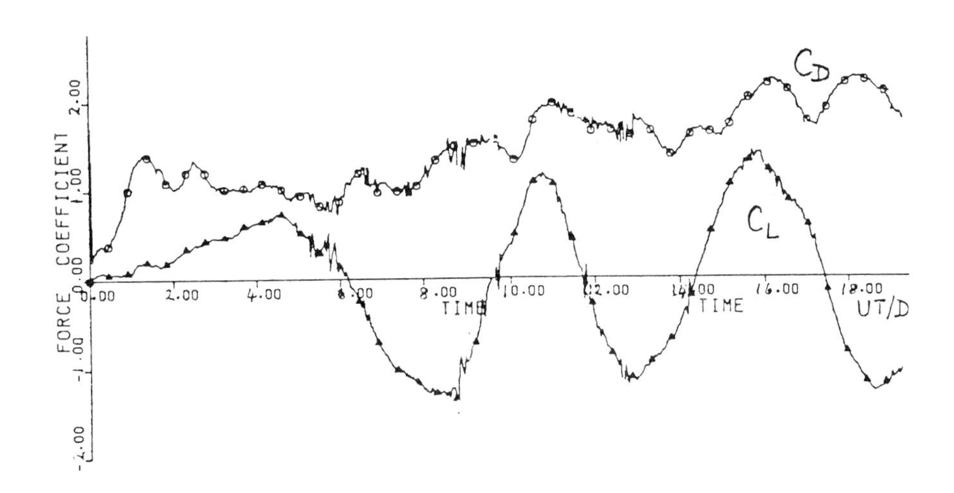

Fig.9 Lift and drag forces on a circular cylinder.

ACKNOWLEDGEMENTS

This work was partly sponsored by BMT Ltd. and partly by the MTD's Fluid Loading Programme, a programme of research jointly funded by SERC, the Department of Energy and the Offshore Industry.

REFERENCES

[1] CLEMENTS, R.R. and MAULL, D.J. The representation of sheets of vorticity by discrete vortices. Prog. Aerosp. Sci. 16, (1975), p.129.

[2] LEONARD, A. Vortex methods for flow simulation. J. Comp. Phys. 37, (1980), p.289.

[3] VAN DOMMELEN, L. Lagrangian techniques for unsteady flow separation. Forum on Unsteady Flow Separation, A.S.M.E. Appl. Mechs. Bioengg., and Fluids Engg. Conf., Cincinnatti, OH., 1987.

[4] CHRISTIANSEN, J.P. Numerical simulation of hydro-dynamics by a method of point vortices. J. Comp. Phys. 13, (1973), p.363.

[5] CHORIN, A.J. Numerical study of slightly viscous flow. J. Fluid Mech. 57, (1973), p.785.

[6] CHORIN, A.J. Vortex sheet approximation of boundary layers. J. Comp. Phys. 27, (1978), p.428.

[7] STANSBY, P.K. and DIXON, A.G. Simulation of flows around cylinders by a Lagrangian vortex scheme. Appl. Ocean Res. 5, (1983), p.167.

[8] FARMER, C.L. and NORMAN, R.A. The implementation of moving point methods for convection diffusion equations. Num Methods for Fluid Dynamics II, IMA. Conf. Series (1986), p.635.

[9] PULLIN, D.I. The large scale structure of unsteady self-similar rolled up vortex sheets. J. Fluid Mech. 88, (1978), p.401.

Application of a 3D Time-Marching EULER Code to Transonic Turbomachinery Flow

H.-W. Happel / B. Stubert

MTU Motoren- und Turbinen-Union München GmbH.
Dachauer Straße 665, 8000 München 50

SUMMARY

This paper describes a time-marching finite volume method to obtain the steady 3D EULER solution within a stator or rotor blade row. The conservation laws are solved in a rotating frame. The numerical scheme is explicit and first order accurate in time and space. To achieve stability, explicit numerical viscosity is added. In order to increase the convergence speed, local time stepping and a sequential grid-refining procedure is used in the program. The accuracy and computational efficiency of the computer code are demonstrated for realistic blade geometries. The comparison with measurements shows good agreement for turbine and compressor blades.

INTRODUCTION

The development of modern gas turbine engines is characterized by the attempt to reduce the number of stages in the turbomachinery components. The resulting increase of stage loading leads to transonic flow fields within the blade rows. Since the unsteady EULER equations are hyperbolic for subsonic as well as for supersonic Mach numbers, time-marching EULER codes are widely used for the prediction and analysis of transonic turbomachinery flow.

The fully 3D time-marching EULER codes, which are now becoming available for industrial use, can at present only be applied to single blade rows or at maximum to one stage, [1]. Therefore multistage turbomachines are designed by using a quasi-3D procedure, in which the fully 3D unsteady turbomachinery flow is approximated by calculating several 2D solutions along blade-to-blade surfaces (S1) as well as along hub-to-tip surfaces (S2) and iterating between the two kinds of 2D solutions, [2]. In practice it is normally satisfactory to perform only one hub-to-tip solution along a mean S2 stream surface and several blade-to-blade solutions along axisymmetric S1 stream surfaces. In a final review the flow through a single blade row is calculated with the 3D EULER code to study the fully 3D influence of the blade geometry on the flow.

GOVERNING EQUATIONS

The unsteady 3D EULER equations are solved in integral form. Suitably written in vector notation the conservation laws are

$$\int_V \frac{\partial \vec{U}}{\partial t}\, dV = - \int_S \left[\vec{F}_r\, dS_r + \vec{F}_\varphi\, dS_\varphi + \vec{F}_z\, dS_z \right] + \int_V \vec{H}\, dV \tag{1}$$

$$\vec{U} := \begin{bmatrix} \rho \\ \rho w_{rc} \\ \rho r(w_\varphi + \omega r) \\ \rho w_z \\ \rho e_{rot} \end{bmatrix} \qquad \vec{F}_r := \begin{bmatrix} \rho w_r \\ (\rho w_r^2 + p)\cos\Delta\varphi - \rho w_\varphi w_r \sin\Delta\varphi \\ \rho r(w_\varphi + \omega r)\, w_r \\ \rho w_z w_r \\ (\rho e_{rot} + p)\, w_r \end{bmatrix}$$

$$\vec{F}_\varphi := \begin{bmatrix} \rho w_\varphi \\ \rho w_r w_\varphi \cos\Delta\varphi - (\rho w_\varphi^2 + p)\sin\Delta\varphi \\ \rho r(w_\varphi + \omega r)\, w_\varphi + p\cdot r \\ \rho w_z w_\varphi \\ (\rho e_{rot} + p)\, w_\varphi \end{bmatrix}$$

$$\vec{F}_z := \begin{bmatrix} \rho w_z \\ \rho(w_r \cos\Delta\varphi - w_\varphi \sin\Delta\varphi)\, w_z \\ \rho r(w_\varphi + \omega r)\, w_z \\ \rho w_z^2 + p) \\ (\rho e_{rot} + p)\, w_z \end{bmatrix} \qquad \vec{H} := \begin{bmatrix} 0 \\ \omega\rho\left[(2w_\varphi + \omega r)\cos\Delta\varphi + 2w_r \sin\Delta\varphi\right] \\ 0 \\ 0 \\ 0 \end{bmatrix},$$

where ρ is the density, and w_r, w_φ and w_z are the components of the relative velocity in radial (r), circumferential (φ) and axial (z) direction, respectively. The total internal energy e_{rot} is specified for a rotating frame with uniform angular velocity ω. The definition of e_{rot} for a calorically perfect ideal gas with constant ratio γ of the specific heats

$$e_{rot} := \frac{\gamma}{\gamma - 1}\frac{p}{\rho} + \frac{1}{2}\left(w_r^2 + w_\varphi^2 + w_z^2 - (\omega r)^2 \right) \tag{2}$$

gives the relation for the static pressure p in the flux terms.

Using cylindrical coordinates, usually source terms must be taken into account in the momentum equations in radial and circumferential direction, because the unit vectors $\vec{e}_r$ and $\vec{e}_\varphi$ vary tangentially. In order to circumvent these difficulties, the momentum equation in the second row of equation (1) is formulated for a mean radial direction $\bar{\vec{e}}_r$, which is constant for all the control volumes. Due to this cartesian formulation the source term results only from the centrifugal and CORIOLIS forces in the rotating relative frame. Furthermore, the momentum equation in circumferential direction is replaced by the corresponding moment of momentum equation in axial direction without any source term in the third row of equation (1).

FINITE VOLUME DISCRETIZATION

Equation (1) is discretized in the physical domain consisting of one blade passage with up- and downstream periodic boundaries approximately along the flow direction, Fig. 1. The domain is divided into basic volumes using H-type grids along hub-to-tip and H- or C-type grids along blade-to-blade mesh surfaces, Fig. 4. The conservation laws are applied to overlapping control volumes. These are composed of eight basic volumes taking the common nodal point close to the center of the control volume as reference point.

For calculating the basic volumes, the hexahedra are composed of five tetrahedra, [3]. The volume of one tetrahedron is determined by

$$V_{T_{abcd}} = \frac{1}{6} \begin{vmatrix} r_a & (r\varphi)_a & z_a & 1 \\ r_b & (r\varphi)_b & z_b & 1 \\ r_c & (r\varphi)_c & z_c & 1 \\ r_d & (r\varphi)_d & z_d & 1 \end{vmatrix} , \tag{3}$$

where the subscripts (a b c d) refer to the four vertices of the tetrahedron, Fig. 2. Then the basic volume is given by

$$V = V_{T_{1452}} + V_{T_{4857}} + V_{T_{5276}} + V_{T_{4723}} + V_{T_{5247}} . \tag{4}$$

The convection term in equation (1) is approximated by

$$\int (\vec{F}_r \, dS_r + \vec{F}_\varphi \, dS + \vec{F}_z \, dS_z) \approx \bar{\vec{F}}_r \iint r \, d\varphi dz + \bar{\vec{F}}_\varphi \iint drdz + \bar{\vec{F}}_z \iint r \, drd\varphi . \tag{5}$$

The surface integrals are determined by a numerical mapping (Fig. 3)

$$S_r = \int_0^1\!\!\int_0^1 r(\xi,\eta) \frac{\partial(\varphi,z)}{\partial(\xi,\eta)} \, d\xi d\eta \qquad S_\varphi = \int_0^1\!\!\int_0^1 \frac{\partial(r,z)}{\partial(\xi,\eta)} \, d\xi d\eta \qquad S_z = \int_0^1\!\!\int_0^1 r(\xi,\eta) \frac{\partial(r,\varphi)}{\partial(\xi,\eta)} \, d\xi d\eta \tag{6}$$

to a unit square using bilinear shape functions for the radius

$$r(\xi,\eta) = r_a(1-\xi)(1-\eta) + r_b\xi(1-\eta) + r_c\xi\eta + r_d(1-\xi)\eta \tag{7}$$

and for the circumferential and axial coordinates, respectively. The fluxes $\vec{F}_r$, $\vec{F}_\varphi$ and $\vec{F}_z$ are averaged by

$$\bar{\vec{F}} = \frac{1}{4}(\vec{F}_a + \vec{F}_b + \vec{F}_c + \vec{F}_d). \tag{8}$$

BOUNDARY CONDITIONS

For subsonic axial velocities the theory of charcteristics provides four boundary conditions in the inlet and only one in the exit plane.

In the inlet plane the radial distributions of

- the relative stagnation temperature,
- the relative stagnation pressure,
- the relative flow angle $\beta_{z\varphi}$ = arc ctg (w_φ/w_z) (for subsonic inflow), and
- the slope of the axisymmetric blade-to-blade stream surfaces ε = arc tg (w_r/w_z)

are imposed assuming uniform values in circumferential direction. If the relative inflow Mach number is supersonic (but w_z is subsonic) the radial distribution of the circumferential component w_φ of the relative velocity is held constant instead of the relative flow angle $\beta_{z\varphi}$. Then the unique incidence condition is automatically satisfied as a part of the solution.

In the exit plane the static pressure is prescribed only at midspan position, because the corresponding radial distribution is a result of the EULER solution. In order to get an uniform static pressure distribution in circumferential direction the radial equilibrium is fulfilled after each time step using volume-averaged values

$$\overline{\left(\frac{\partial p}{\partial r}\right)} = -\frac{1}{V}\left[\int_S \rho\, w_r \vec{w}\, d\vec{S} - \int_V \rho \frac{(w_\varphi + \omega r)^2}{r}\, dV\right] \tag{9}$$

of the static pressure gradient in radial direction. The integration of the right hand side of equation (9) is performed over all basic elements at constant radius in pitchwise direction.

As a 3D flow problem is determined at any point in space by a total of five state variables, the axial momentum equation is additionally used in the inlet plane, and all conservation laws with the exception of the energy equation are discretized in the exit plane.

The periodic boundaries (up- and downstream of the blade) are treated in the same way as interior nodal points, combining informations from the left and right boundary to construct a regular control volume, Fig. 1.

Along the solid walls no mass and energy flux cross the surface, because the normal componet of the velocity is zero everywhere. The transport of momentum and moment of momentum only consists of the pressure forces perpendicular to the surface. The conservation principles are applied to half or quarter elements and the transient changes are attached to the nodal point of reference lying on the solid wall.

NUMERICAL SCHEME

The numerical scheme is explicit and first order accurate in time and space. The first order accuracy in space is caused by the explicit numerical viscosity to achieve stability as well as by using half and quarter elements along the solid walls.

Defining a volume averaged state vector of the control volume $\Delta V_{i,j,k}$

$$\vec{\tilde{U}}_{i,j,k} := \frac{1}{\Delta V_{i,j,k}} \sum_{l=1}^{8} \left(\int \vec{U}\, dV\right)_l \tag{10}$$

with the approximation

$$\int \vec{U}\, dV = \frac{V}{8} \sum_{m=1}^{8} \vec{U}_m \tag{11}$$

of the volume integrals of the eight surrounding basic volumes V and using a lagging correction vector

$$\vec{C}_{i,j,k}^{\,n_0} := (1 - \alpha) \left[\vec{U}_{i,j,k}^{\,n_0} - \vec{\tilde{U}}_{i,j,k}^{\,n_0} \right] \tag{12}$$

the unsteady 3D EULER equations are discretized as

$$\vec{U}_{i,j,k}^{\,n+1} = \vec{\tilde{U}}_{i,j,k}^{\,n} + \vec{C}_{i,j,k}^{\,n_0} - \frac{\Delta t}{\Delta V_{i,j,k}} \left[\int_S (\vec{F}_r dS_r + \vec{F}_\varphi dS_\varphi + \vec{F}_z dS_z)^n - \int_V \vec{H}^n dV \right] \tag{13}$$

for the nodal point of reference (i,j,k) close to the center of the control volume. The transport terms are positive for fluxes leaving the control surface.

In the so-called damping surface technique [4] the considerable numerical viscosity, which is introduced by the volume averaged state vector at time level n, is reduced by the lagging correction vector. This remains constant during N time steps and is then updated for the next period of iterations. With regard to convergence speed the number of iterations between two updates of the correction vector should be chosen as small as possible. In combination with a damping coefficient α in the range of about two percent the updating rate N should not be less than five iterations, because otherwise the calculation becomes unstable.

ACCELERATION OF CONVERGENCE

As explicit time-marching procedures are time consuming with respect to the COURANT-FRIEDRICHS-LEWY condition, the convergence is accelerated using local time stepping, because only the steady state solution is required. This is very useful in connection with grid spacing, in order to achieve a COURANT number close to unity everywhere. Furthermore, a two or three level grid-refining procedure reduces computer time.

COMPUTATIONAL RESULTS

Two calculations were made as examples of transonic 3D turbomachinery flow. Fig. 4 shows the geometry of a turbine stator in the meridional projection and in a blade-to-blade section. The three columns in Fig. 5 show the pressure distribution $p/p_{t\ inlet}$, the isopressure contours and the sonic line at hub, mean, and tip section. The pressure distributions are predicted very well along the pressure side and satisfactory along the suction side using an inviscid code.

The calculation of the 3D flow within a transonic compressor rotor seems to be interesting, although at present no experimental data are available. Fig. 6 contains the mesh and the isomach contours on the suction and the pressure side of the rotor blade. The blade surface distribution and the plots of the isentropic isomach contours are given at hub, mean, and tip section in Fig. 7. The 3D shock region can be very clearly localized.

CONCLUSIONS

The time-marching 3D EULER code is a good tool to review the fully 3D aerodynamic properties of isolated stator or rotor blades after a quasi-3D turbomachinery design. Furthermore, the 3D influence of differently staggered blades can be studied.

ACKNOWLEDGEMENTS

The reported work was performed at MTU Motoren- und Turbinen-Union München GmbH and was supported within research programs of the German Bundesministerium für Forschung und Technologie. The permission to publish the results is gratefully acknowledged.

REFERENCES

[1] DENTON, J. D.: "An improved time-marching method for turbomachinery flow calculation", ASME paper 82-GT-239 (1982)

[2] WU, C. H.: "A general theory of three-dimensional flow in subsonic and supersonic turbomachines of axial-, radial- and mixed-flow types", NACA TN 2604 (1952)

[3] RIZZI, A., ERIKSSON, L.-E.: "Computation of flow around wings based on the EULER equations, J. Fluid Mech., Vol. 148, pp. 45-71 (1984)

[4] COUSTON, M., McDONALD, P.W., SMOLDEREN; J. J.: "The damping surface technique for time dependent solutions to fluid dynamic problems", VKI TN 109 (1975)

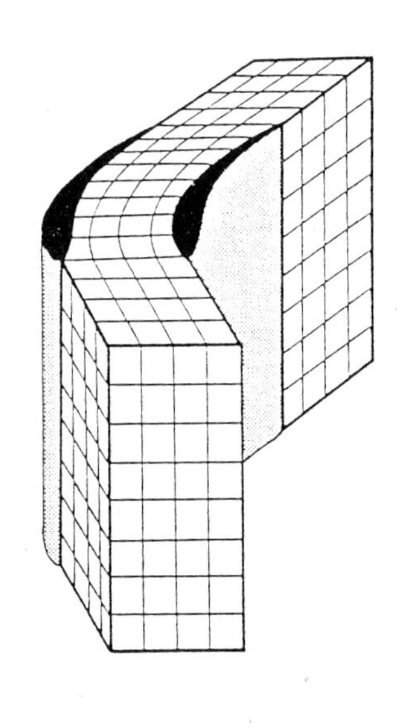

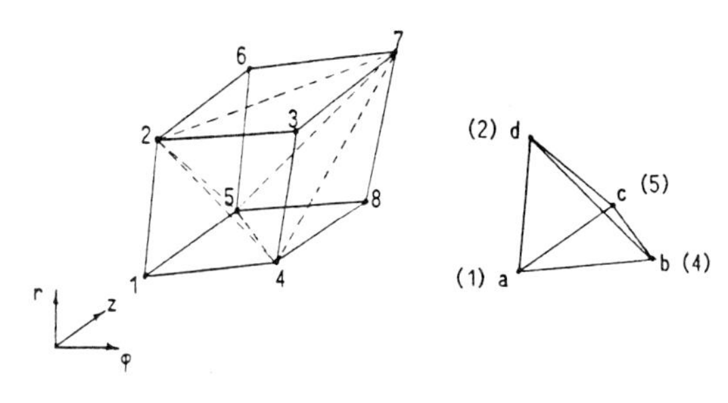

Fig. 2: Basic Volume Composed of 5 Tetrahedra

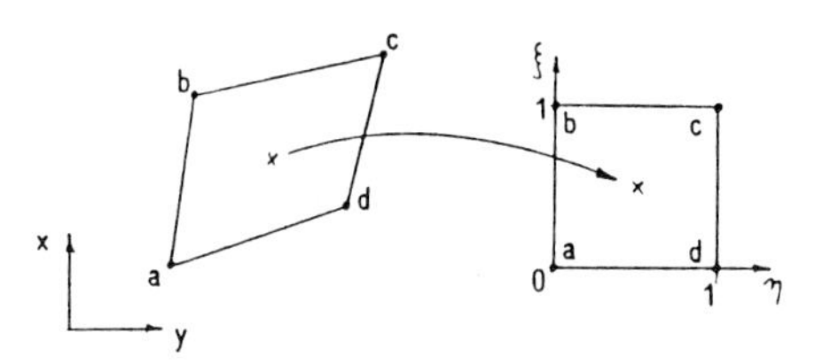

Fig. 1: Discretized Physical Domain

Fig. 3: Mapping of a Surface Integral

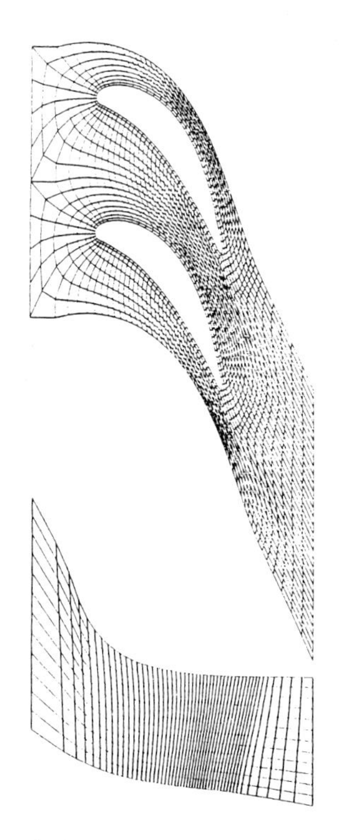

Fig. 4: Geometry of a Turbine Stator

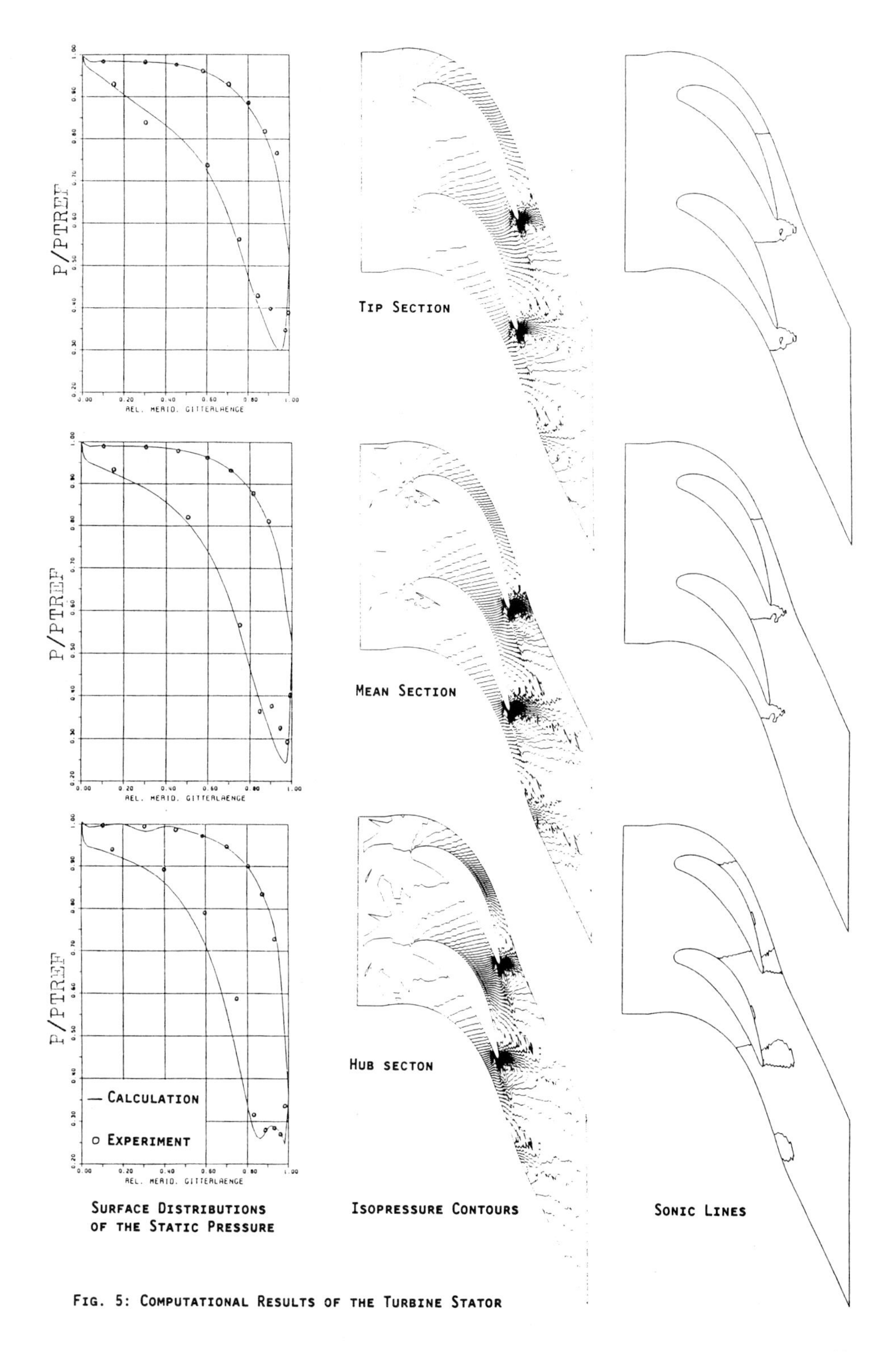

FIG. 5: COMPUTATIONAL RESULTS OF THE TURBINE STATOR

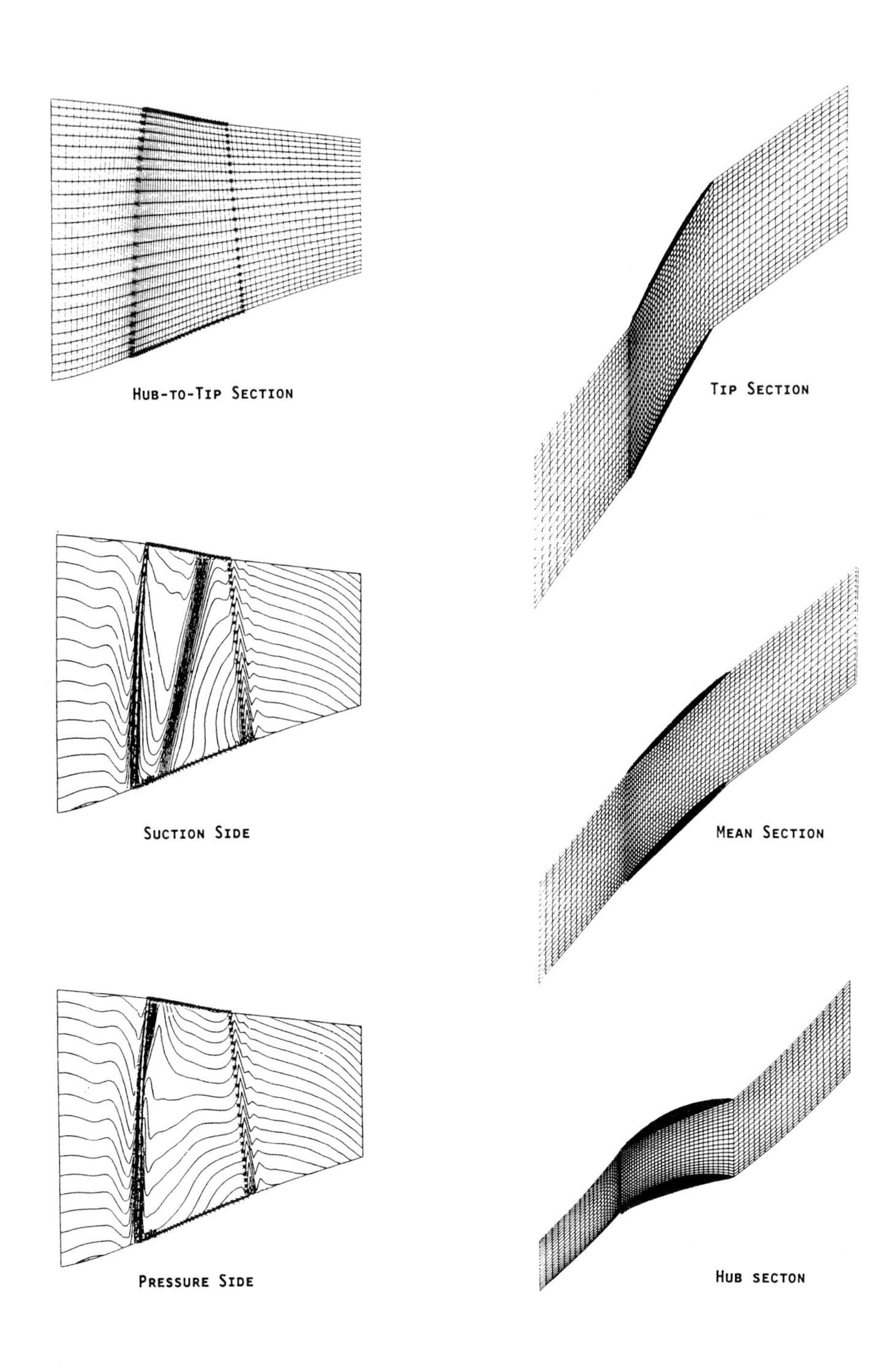

Fig. 6: Geometry of the Compressor Rotor and Isomach Contours on the Blade Surface

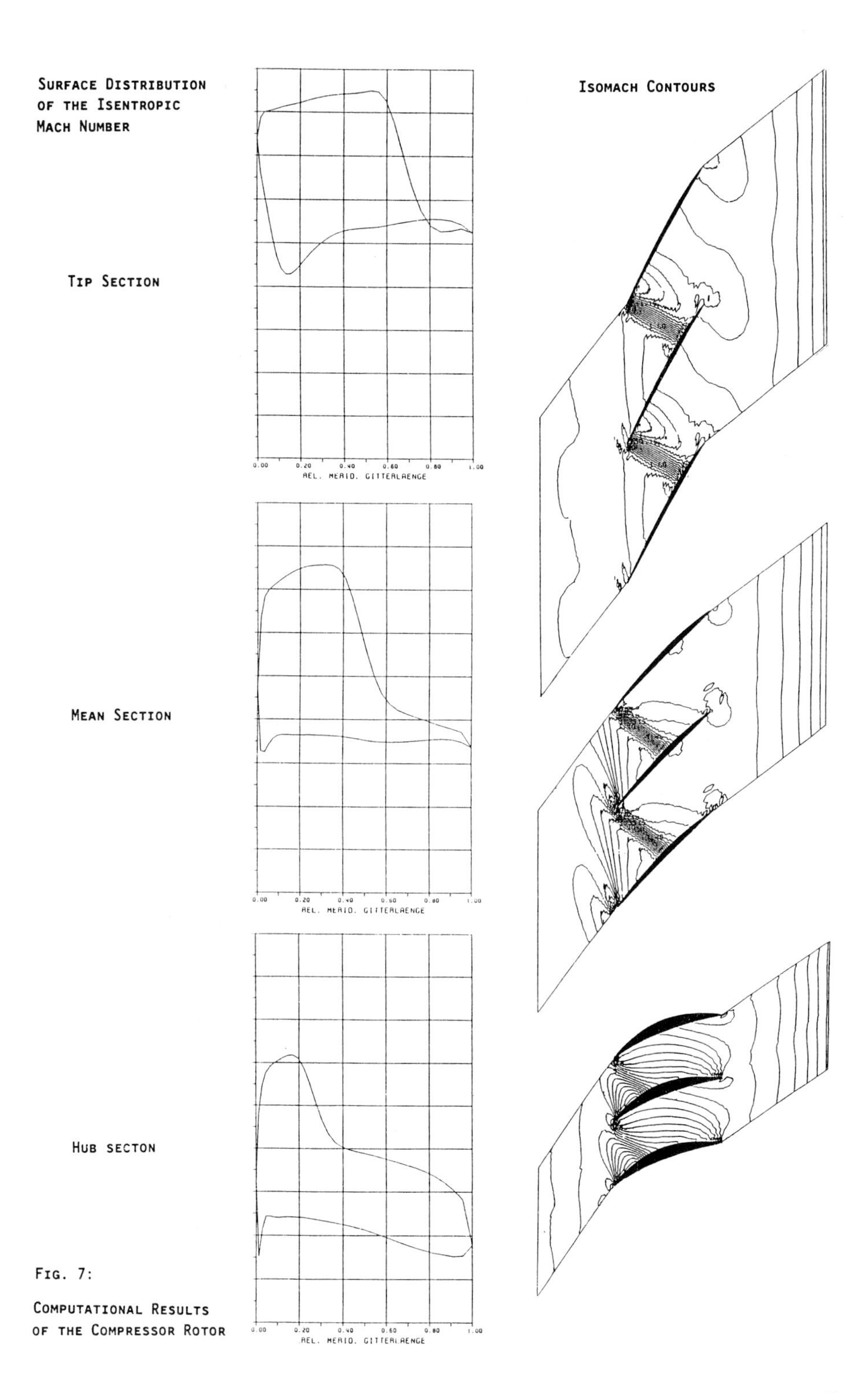

FIG. 7:

COMPUTATIONAL RESULTS OF THE COMPRESSOR ROTOR

A TREATMENT OF MULTIVALUE SINGULARITY OF SHARP CORNER IN INVISCID HYPERSONIC FLOW

Wei Ji
Institute of Mechanics, Chinese Academy of Science,
Beijing, China

Summary

Based on the steady Prandtl-Meyer solution, an approximate method is presented for dealing with the multivalue singularity of sharp corner in an inviscid hypersonic flow. The method, combined with Harten's TVD scheme, is used to treat the corner point of the step in the test problem of a Mach 3 wind tunnel with a step. The excessive numerical error around the corner of the step caused by numerical method is successfully reduced. The occurrence of the spurious "entropy layer" above the step produced by the false entropy increasing around the corner is avoided. More physically reasonable results are obtained.

Introduction

Since 1970's, numerical simulation of gasdynamics by finite difference method has been widely used. Recently, many new high resolution schemes, satisfying entropy condition, have been proposed. But, almost all these schemes were constructed based on one dimensional initial value problems. Their applications to two dimensional problems need to be investigated.

As well known, boundary condition is the source of many important phenomena in fluid dynamics. If numerical treatment of boundary condition is not proper, non-physical solution of initial boundary problem will be obtained by numerical method in initial value problem which has been proved to satisfy entropy condition. A typical problem is the one of "A Mach 3 wind tunnel with a step", which was first proposed by Emery in 1968. Up to now, many schemes have been tested to this problem. The wave structures at time t=4.0 (still unsteady globally) in Fig.2 show the main characteristics of the flow field in Refs. [1-7]. A Mach stem appears above the step and the computed heights of the stem are different for different schemes or different CFL numbers.

In this paper, an approximate method is presented for dealing with the numerical error caused by the multivalue singularity of expansion corner. The results obtained here show that numerical error can be successfully reduced. The shock reflection above the step is normal, which is reasonable physically.

Discription of Problem

For the test problem, one usually chooses computational region shown in Fig.1. Euler equations are used as governing equations. They are:

$$\frac{\partial w}{\partial t} + \frac{\partial F(w)}{\partial x} + \frac{\partial G(w)}{\partial y} = 0 \qquad (1)$$

where $W=[\rho , \rho U, \rho V, E]^T$

$$F(W)=[\rho U, \rho U^2+P, \rho UV, U(E+P)]^T ,$$

$$G(W)=[\rho V, \rho UV, \rho V^2+P, V(E+P)]^T ,$$

$$E=\rho e+\rho(U^2+V^2)/2, \quad e=\frac{1}{\gamma-1}\frac{p}{\rho} .$$

ρ represents density, P pressure, (U,V) velocity and γ gas constant.

Boundary conditions are:

Inflow: P=1.0, ρ =1.4, U=3.0, V=0.0, (2)

Solid wall boundary: Vn=0. (3)

Vn is normal of velocity to the wall.

Initial conditions are:

P = 1.0, ρ =1.4, U=3.0, V=0.0. (4)

For the above initial boundary problem (1-4), the flow quantities will be multivalue at the corner. With numerical experimental results, we find that excessive numerical error will be produced around the corner by using finite discrete point. As matter of fact, this error is spurious entropy increasing which is transported downstream to form a spurious "entropy layer" above the step. When a shock hit on this layer, interaction between the shock and the entropy layer will take place instead of that between the shock and the solid wall.

From the time developing graphs of the computation shown in [7] and the numerical results here (see Table 1), the flow process can be discribed as follows: from some time t* ($t^* \approx 1.0$) on, the flow around the sharp corner is nearly steady, although the global flow field is still unsteady. Under this circumstances, we make an following hypothesis:

Around the corner point, the wave structure of steady state (center-expansion) has been formed when $t>t^*$, although the varieties are still changing very slowly.

According to the hypothesis, we use the steady state relations and some unsteady corrections (even though they are very small) to improve the computation results around the corner.

For the test problem, the following 5 flow properties can be used (some are obvious and other are easy to be proved):

a) Around the corner point, the flow changes from subsonic to supersonic.
b) Around the corner point, the flow is isentropic locally.
c) The sonic line must originate from the corner point.
d) The sonic line is tilted upwind.
e) Behind the sonic line, the flow is the Prandtl-Meyer flow locally.

Two types of mesh distribution (Fig.3.a & Fig.3.b) are employed for computation, mesh A usually for finite difference and mesh B for finite element. Following procedure can be used to solve the problem when using Strang's splitting method[9] to generalize finite difference scheme. Firstly, obtain flow field at a new time level by numerical method, secondly use the physically-based correction given below to treat the corner, finally go on to the next new time level by numerical method in turn. For 2-D finite difference method or 2-D finite element method, the correction step can be implicitly included in the method.

When comes to the treatment of the corner, the direction of velocity and pressure are fixed. The relation (5) is applied to the nearest discrete point at the corner (point B in Fig.3.a or point C in Fig.3.b) which is located in the center-expansion sector, to determine the slope of sonic line at the instant.

$$\alpha = \frac{\gamma+1}{\gamma-1} tg^{-1}\left(\frac{\gamma-1}{\gamma+1}\frac{u_r}{u_\theta}\right). \quad (5)$$

where α--angle between sonic line and position vector of point chosen.

u_r, u_θ --components of velocity in polar coordinate.

Then, the Mach number Mp for every point to be corrected in the sector is obtained from the Prandtl-Meyer relations (6) & (7).

$$\nu(Mp) = \Delta\theta , \quad (6)$$

$$\nu(Mp) = \frac{\gamma+1}{\gamma-1} tg^{-1}\left(\frac{\gamma-1}{\gamma+1}\cdot\sqrt{M_p^2-1}\right) - tg^{-1}(\sqrt{M_p^2-1}). \quad (7)$$

where $\Delta\theta$ is the angle between the velocity direction of point to be corrected and point on sonic line.

The algebraic equations(6) and (7) can be solved by Newton's iterative method. It is easy to show that Eqs.(8) always converge quickly.

$$X_0 = \sqrt{\lambda} ,$$
$$X_n = X_{n-1} - g(X_{n-1})/g'(X_{n-1}), \quad (8)$$
$$g(x) = \lambda tg^{-1}\left(\frac{x}{\lambda}\right) - tg^{-1} x - \Delta\theta .$$

with $\lambda = \sqrt{\frac{\gamma+1}{\gamma-1}}$, $X = \sqrt{Mp^2 - 1}$.

The corrected density can be obtained from the relation of isentropic around the corner and the corrected velocity from equations (9).

$$\begin{cases} up = ap*Mp*cos(rp), \\ vp = ap*Mp*sin(rp). \end{cases} \quad (9)$$

where ap is velocity of sound and rp represents the angle of velocity with horizon.

Remarks:

(1) For mesh B, two points A1 and A2 should be put on the corner point in order to represent the multivalue property of the corner. For each time level, let A1 and A2 be the limit points of the sonic line and the tail of the expansion at the corner, respectively. The pressure, density and velocity of A1 can determined by (10), (11), and (12).

$$P_{A1} = \left(\frac{2+(\gamma-1)M_D^2}{\gamma+1}\right)^{\frac{\gamma}{\gamma-1}}\cdot p_D \quad (10)$$

$$\rho_{A1} = \left(\frac{2+(\gamma-1)M_D^2}{\gamma+1}\right)^{\frac{1}{\gamma-1}}\cdot \rho_D \quad (11)$$

$$\begin{cases} uA1 = aA1*cos(\beta) \\ vA1 = aA1*sin(\beta) \end{cases} \quad (12)$$

where aA1--velocity of sound at point A1, β--the tilted angle upwind of sonic line and D--the nearest point in subsonic region aroung corner (see Fig.3.b).

The quantities of A2 can be obtained by (13), (14), (15), and (16).

$$\nu(MA2) = \alpha \quad (13)$$

$$pA2 = \left(\frac{\gamma+1}{2+(\gamma-1)M_{A2}^2}\right)^{\frac{\gamma}{\gamma-1}} \cdot p_{A1} \tag{14}$$

$$\rho A2 = \left(\frac{\gamma+1}{2+(\gamma-1)M_{A2}^2}\right)^{\frac{1}{\gamma-1}} \cdot \rho A1 \tag{15}$$

$$\begin{cases} uA2 = Ma2 * aA2, \\ vA2 = 0. \end{cases} \tag{16}$$

where aA2 is the velocity of sound at point A2.

(2) Because Harten's TVD scheme is an explicit 5 points one, the point B1 or C1 in Fig.3 should be treated specially in order to diminish the effect of error produced by the corrner on downstream points.

For mesh A, the entropy and enthalpy of point B1 is approximately equal to the ones of point B, which is the same as the method given in [7].

For mesh B, the flow quantities except entropy change larger from time step n to n+1 but change smaller from n to n+2 and the entropy changes very small in both time steps (see table 2). So, the numerical procedure from n to n+2 can beconsidered as an unsteady isentropic oscillating one. Thus, The equation of total enthalpy under the condition of constant entropy can be used here to correct velocity.

$$\frac{\partial H}{\partial t} + \vec{V}.\nabla H = \frac{1}{\rho}\frac{\partial p}{\partial t} \tag{17a}$$

$$H = \frac{\gamma}{\gamma-1}\frac{p}{\rho} + \tfrac{1}{2}|\vec{V}|^2 \tag{17b}$$

The numerical scheme for (17) is as follows:

$$\frac{H_{ij}^{n+1} - H_{ij}^{n}}{\Delta t} = \frac{2}{\rho_{ij}^{n}+\rho_{ij}^{n+1}} \cdot \frac{p_{ij}^{n+1}-p_{ij}^{n}}{\Delta t} - \frac{\tilde{u}_{ij}^{n+1}+u_{ij}^{n}}{2} \cdot \frac{\tilde{H}_{ij}^{n+1}+H_{ij}^{n}-H_{i-1,j}^{n+1}-H_{i-1,j}^{n}}{2\Delta x}$$
$$- \frac{\tilde{V}_{ij}^{n+1}+V_{ij}^{n}}{2} \cdot \frac{\tilde{H}_{ij}^{n+1}+H_{ij}^{n}-H_{i,j-1}^{n+1}-H_{i,j-1}^{n}}{2\Delta y} \tag{18}$$

where $\tilde{f}_{ij}^{n+1}$ and p_{ij}^{n+1} are the numerical results, ρ_{ij}^{n+1} obtained by isentropic.

Harten' TVD scheme

For one dimensional initial value problem (19), monotonic schemes are first order and approximates the modified equation (20) to second order.

$$\partial u/\partial t + \partial f(u)/\partial x = 0. \tag{19}$$

$$\partial u/\partial t + \partial f(u)/\partial x = \Delta t \frac{\partial}{\partial x}[\beta(u,\lambda)\frac{\partial u}{\partial x}] \tag{20}$$

where $\beta(u,\lambda) = \tfrac{1}{2}[\sum_{l=-k}^{k} l^2 \partial H(u,\ldots\ldots, u)/\partial u_l - \lambda^2 a^2(u)]$

$$a(u) = \partial f/\partial u, \qquad \lambda = \frac{\Delta t}{\Delta x}$$

$$H = u^n_{ij} - \lambda(\bar{f}^n_{j+\frac{1}{2}} - \bar{f}^n_{j-\frac{1}{2}})$$

$$\bar{f}^n_{j+\frac{1}{2}} = \bar{f}(u^n_{j-k+1}, \dots, u^n_{j+k}) .$$

Harten [4], by approximating equation (21) with a monotonic scheme, obtained a new high resolution TVD scheme (22).

$$\frac{\partial u}{\partial t} + \frac{\partial}{\partial x}(f(u) + g(u, \lambda, \Delta x)/\lambda) = 0. \tag{21}$$

$$u^{n+1}_j = u^n_j - (\bar{f}^n_{j+\frac{1}{2}} - \bar{f}^n_{j-\frac{1}{2}}) \tag{22a}$$

where

$$\bar{f}^n_{j+\frac{1}{2}} = \tfrac{1}{2}(f^n_j + f^n_{j+1}) - Q(\nu^n_{j+\frac{1}{2}})\Delta_{j+\frac{1}{2}} u^n \tag{22b}$$

$$\nu^n_{j+\frac{1}{2}} = \begin{cases} (f^n_{j+1} - f^n_j)/\Delta_{j+\frac{1}{2}} u^n & \Delta_{j+\frac{1}{2}} u^n \neq 0 \\ 0 & \Delta_{j+\frac{1}{2}} u^n = 0 . \end{cases} \tag{22c}$$

In this paper, we use Strang's spliting method with eqn. (23) to generalize Harten's scheme for two-dimensional Euler equations.

$$W^{n+2}_{ij} = LxLyLyLx\ W^n_{ij} . \tag{23}$$

The formula of Lx and Ly for 2-D Euler equations are given in Ref.[8].

Results and Conclusions

For the space is limited, here we only present the pictures of isopicnic and isentropic of flow field at t=4.0, and of isopicnic at t=16.0 which is in steady state globally.

Fig.4.a gives the isopicnic of mesh A at t=4.0 and Fig.4.b the isentropic of mesh A at t=4.0.

Fig.5.a shows the isopicnic of mesh B at t=4.0 and Fig.5.b the isentropic of mesh B at t=4.0.

From Fig.4 and Fig.5, we can find that neither the entropy layer nor the Mach stem appears above the step, even though meshes are coarse. So, the wave structures at t=4.0 obtained here can be drawn as in Fig.7. The flow field at t=16.0 is given in Fig.6, and is very similar to the one of t=4.0 but not to the one given in Ref.[2].

For the conclusions of the paper, one can find that the treatment of expansion sharp corner presented in this paper really reduces the numerical error caused by multivalue singularity of corner, especially the downstream effect of error. One can also extend the method used here to the supersonic flow over a blunt body with a sharp corner problem and get physically reasonable result.

Acknowlegement-At first, the author would like to thanks his advisor, Prof. D.Huang, very much for his advice and helps. The author also appreciates the useful discussion with Prof. M.Pandolfi from Italy during his visit in P.R.China.

References

[1] A.F. Emery, J.C.P. 2(1968), 306.
[2] B.Van Leer, J.C.P. 32(1979), 101.
[3] P.Woodward & P.Colella, In Proc. 7th Intl. Conf. on Numerical Method in Fluid Dynamics, Lecture Notes in Physics, Vol.141, 434, Springer.
[4] A.Harten, J.C.P. 49(1983), 357.
[5] Y.F. Li, In Proc. 10th Intl. Conf. on Numerical Method in Fluid Dynamics (ed. F.G. Zhuang), 1986.
[6] R.Löhner, K.Morgan & O.C.Zinkiewicz, Computer Method in Applied Mechanics and Engineering, 51(1985), 441.
[7] P.Woodward & P.Colella, J.C.P. 54(1984), 174.
[8] W.Ji, Master thesis, Part I, Math. Dept., Peking University, P.R.China, 1986.
[9] G.Strang, SIAM J. Numer, Anal. 5(1968), 506.

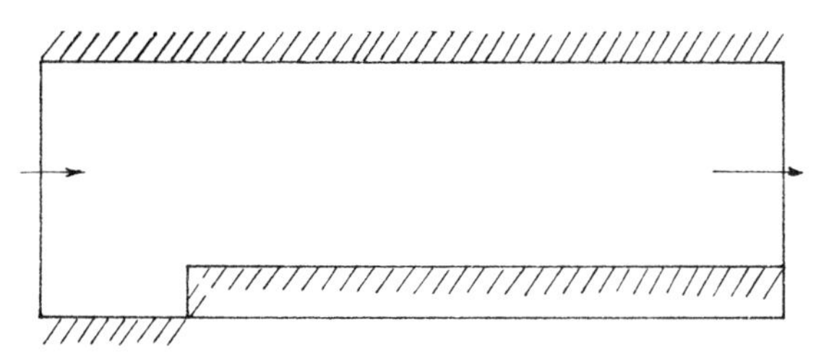

Fig.1 Region of computation

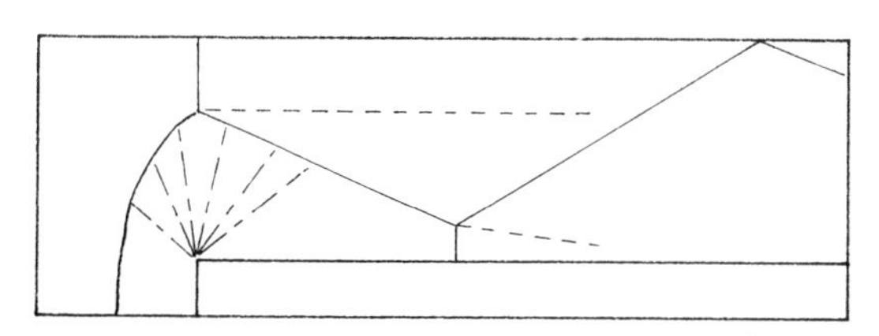

Fig.2 Wave structures at t=4.0 given in Reference[1-6].
—— shock wave
--- slip line
—·— expansion

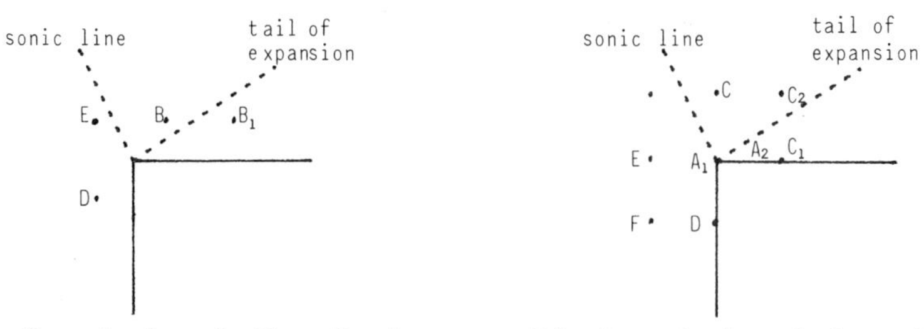

(a) Type A of mesh distribution (b) Type B of mesh distribution

Fig.3

Table 1

Time(t)	Density $(\rho^{n+2} - \rho^{n})/\rho^{n}$	Pressure $(p^{n+2} - p^{n})/p^{n}$	Entropy $(s^{n+2} - s^{n})/s^{n}$
0.998	0.89%	-0.11%	-1.34%
1.020	0.07%	-1.27%	-1.40%
1.043	0.31%	-0.78%	-1.20%
1.066	0.07%	-0.86%	-0.95%
1.089	0.13%	-0.65%	-0.83%
1.112	0.53%	0.10%	-0.63%

Table 2

Time(t)	Density $(\rho^{n+1} - \rho^{n})/\rho^{n}$	Pressure $(p^{n+1} - p^{n})/p^{n}$	Entropy $(s^{n+1} - s^{n})/s^{n}$
0.998	-4.56%	-7.58%	-1.33%
1.020	4.86%	6.80%	0.06%
1.043	-4.35%	-7.10%	-1.14%
1.066	4.61%	6.72%	0.19%
1.089	-4.28%	-6.90%	-1.02%
1.112	5.03%	7.52%	0.38%

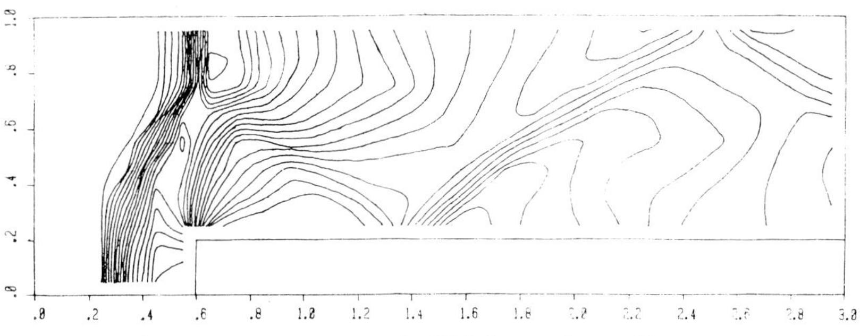

Fig.4a. Isopicnic at t=4.0193 by mesh A. 30 contours.
30 x 10, CFL = 0.995, ε = 0.2.

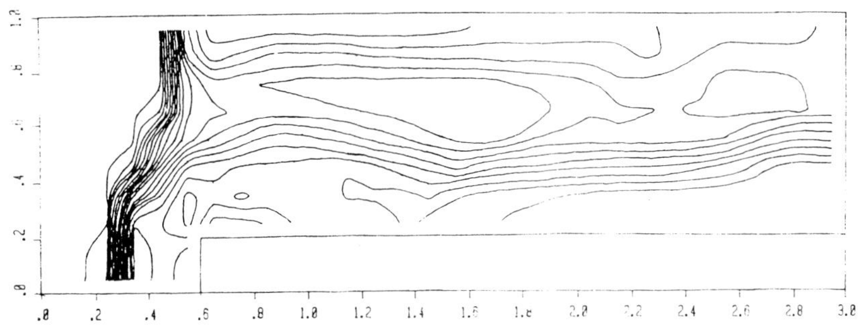

Fig.4b. Isentropic at t=4.0193 by mesh A. 30 contours.
30 x 10, CFL = 0.995, ε = 0.2.

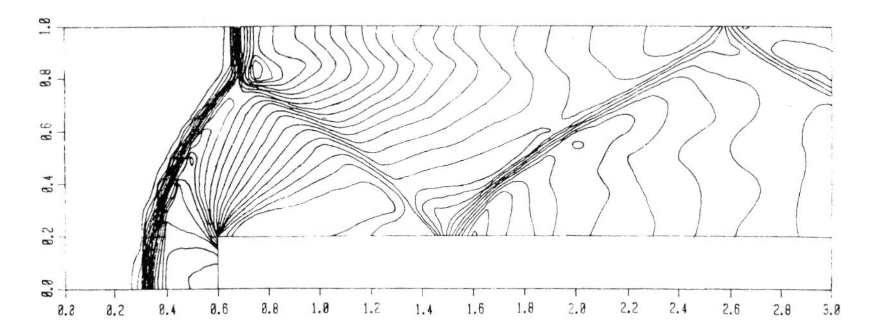

Fig.4a(1). Isopicnic at t=4.0069 by mesh B. 30 contours.
60 x 20, CFL=0.995, $\varepsilon = 0.2$

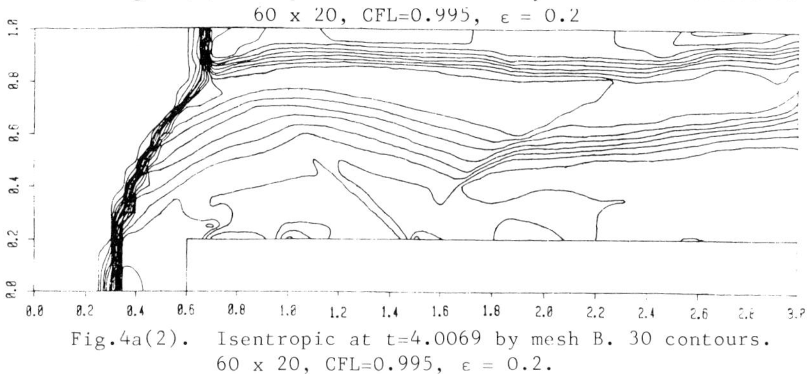

Fig.4a(2). Isentropic at t=4.0069 by mesh B. 30 contours.
60 x 20, CFL=0.995, $\varepsilon = 0.2$.

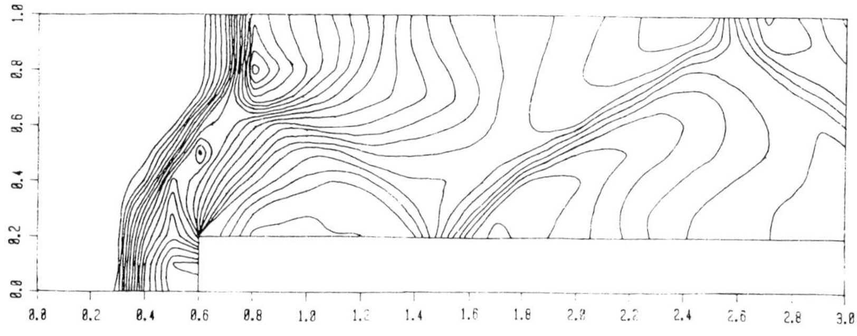

Fig.4b. Isopicnic at t=16.0229 by mesh B. 30 contours.
30 x 10, CFL=0.995, $\varepsilon = 0.2$.

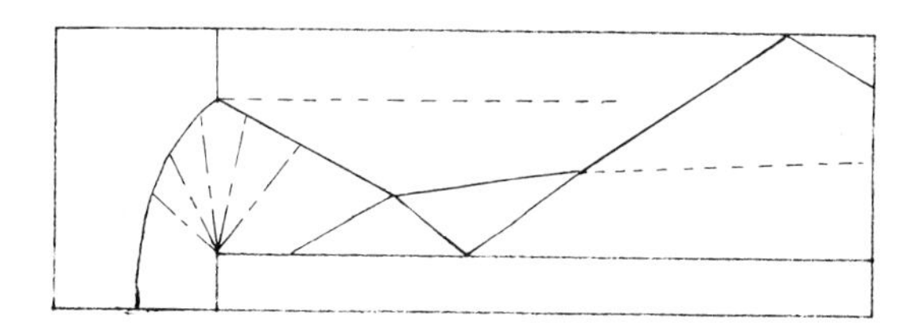

Fig.5 Wave structures at time t=4.0, drawn according to our results.

SHOCK INITIATED IGNITION IN A L-SHAPED DUCT
TWO ASPECTS OF ITS NUMERICAL SIMULATION

R. Klein

Institut für Allgemeine Mechanik, RWTH Aachen
Templergraben 55, D-5100 Aachen, West-Germany

ABSTRACT

The simulation of chemically reacting gas flows requires numerical integrations of inhomogeneous balance equations for the mass, momentum, energy and fuel mass densities. For this purpose a higher order upwind difference scheme is extended for applications to Euler's equations with source terms. - The boundary of the L-shaped duct (cf. Fig. 1) includes a sharp convex corner. A straight forward asymptotic consideration of the vicinity of the edge reveals the quasisteady nature of the corner flow and yields the basis of special numerical corner flow boundary conditions for sub- and supersonic incident flow.

INTRODUCTION

An important aspect of the onset of ignition of explosive gases in confined vessels due to gasdynamic-chemical interactions is the influence of the geometry on these processes. Due to the generally high temperature sensitivity of the reaction rates, changes in the temperature induced by diffractions, reflections and focussing of pressure waves, can significantly affect the evolution of the chemical reactions. As a model system we intend to investigate the L-shaped duct, sketched in Fig. 1. All the three types of pressure wave-boundary interactions are present in this system, if a plane shock wave enters perpendicular to the open side, is diffracted at the convex corner, reflected at the opposite wall and finally focussed in the closed side of the duct.

For a study of this problem by means of numerical simulations, the chemically reacting flow is modelled by Euler's equations for mass, momentum and kinetic and thermal energy densities $\underline{u} = (\rho, m, n, e)$, supplied by a source-term $\underline{q} = (0,0,0,Q\rho r)$, which accounts for the chemical heat release

$$\underline{u}_t + \underline{f}^1(\underline{u})_x + \underline{f}^2(\underline{u})_y = \underline{q} \quad . \tag{1}$$

Here $\underline{f}^1 = (m, m^2/\rho + p, mn/\rho, m(e+p)/\rho)$ and $\underline{f}^2 = (n, mn/\rho, n^2/\rho + p, n(e+p)/\rho)$ denote the fluxes of the quantities $\underline{u}$ in x- and y-direction. Assuming an irreversible one-step reaction, the reaction progress is described by the fuel consumption law

$$\xi_t + (m\,\xi/\rho)_x + (n\,\xi/\rho)_y = -\rho r \quad , \tag{2}$$

where ξ is the fuel mass density and r the reaction rate expression, for which an Arrhenius type temperature dependence

$$r = B(\xi/\rho)\exp(-E/T) \tag{3}$$

is assumed. The constants B and E are the frequency factor and the activation energy respectively. The system is closed by means of the ideal gas equations of state

$$p = (\gamma - 1)(e - (m^2 + n^2)/2\rho) , \qquad T = p/\rho \quad . \tag{4}$$

The source terms in eqs. (1), (2) and the convex corner within the boundary of the duct impose particular complications onto the numerical simulations. The present paper proposes some modifications, which can be supplied to difference schemes in conservation form in order to account for these special difficulties. The discussions refer to a cartesian grid, but the ideas can be transfered to general grid geometries as well.

EXTENDED GODUNOV-TYPE-SCHEME FOR EULER'S EQUATIONS WITH SOURCE TERMS

As a basis for the numerical simulations we employ a second order explicite two-step Godunov type scheme for the homogeneous 1-D-Euler equations. For a review of such algorithms see Munz /1/. The time-update of the conserved quantities in the i-th cell of the uniformly spaced grid reads as

$$\underline{u}_i^{n+1} = \underline{u}_i^n - \frac{\Delta t}{\Delta x}(\underline{g}_{i+1/2}^{n+1/2} - \underline{g}_{i-1/2}^{n+1/2}) , \qquad \underline{g}_{i+1/2}^{n+1/2} = \underline{g}(\underline{u}_{i+}^{n+1/2}, \underline{u}_{(i+1)-}^{n+1/2}) . \tag{5}$$

Here $g(\cdot,\cdot)$ is the numerical flux function of Einfeldt's approximate Riemann-solver /2/ and its arguments are the two states at the cell interface $x_{i+1/2}$, which are obtained by introduction of piecewise linear distributions at time t^n and a preliminary advance in time by $\Delta t/2$. In the present version and for ideal gases this solver is equivalent to Roe's flux difference splitting method /3/, although it requires less computational efforts. The scheme is extended to two dimensions by means of the directional operator splitting technique and is applied on a cartesian mesh with constant spacing in each direction.

The following considerations are based on the 'source wave splitting' ideas of Roe /4/ and Toro /5/. In general, sources of mass, momentum and energy density in one-dimensional compressible flows lead to the generation of pressure waves, traveling in upstream and downstream direction and to changes of the entropy being convected along particle paths. We will now describe an attempt to include these 'source waves' in the basic scheme outlined above, thereby retaining second order accuracy in space and time. The approach will first be discussed for the scalar equation

$$u_t + f(u)_x = u_t + a(u)\, u_x = q(u) , \qquad a = df/du \; . \tag{6}$$

Integration over a space time cell ▨ $= [x_{i-1/2}, x_{i+1/2}] \times [t^n, t^{n+1}]$ yields

$$u_i^{n+1} = u_i^n - \frac{\Delta t}{\Delta x}(f_{i+1/2}^{n+1/2} - f_{i-1/2}^{n+1/2}) + \frac{1}{\Delta x} \int_{▨} q \, dx\, dt \tag{7}$$

where u_i^n, $f_{i+1/2}^{n+1/2}$ are the averages of the conserved quantity and of the flux on the boundaries of the cell.

In the present form eq. (7) does not exhibit any directional biasing of the source term with the characteristic direction. However, this bias can be revealed by inspection of the average fluxes

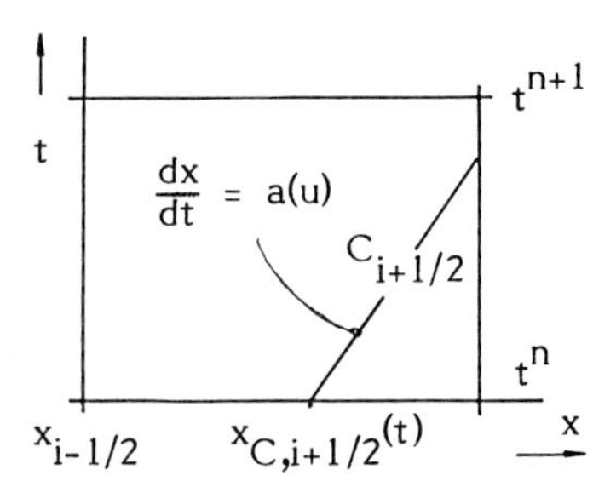

Fig. 2 : Characteristic path of integration in eq. (9)

$$f^{n+1/2}_{i+1/2} = \frac{1}{\Delta t} \int_{t^n}^{t^{n+1}} f(u(x_{i+1/2}, t))\, dt \quad . \tag{8}$$

Taking the distribution $u(x,t^n)$ as initial condition, the exact solution to eq. (6_1) obeys the implicite relation

$$u(x_{i+1/2},t) = u(x_{C,i+1/2}(t),t^n) + \int_{C_{i+1/2}(t)} q(u)\, dt' =: u_h + \Delta u_s \,, \tag{9}$$

where the characteristic path $C_{i+1/2}(t)$ and the position $x_{C,i+1/2}(t)$ are shown in Fig. 2. Thus $u(x_{i+1/2},t)$ consists of a 'hydrodynamic' part u_h, which is due to the advection properties of the l.h.s. of eq. (6_1), and of a source correction Δu_s of order $O(\Delta t)$. A difference approximation to this exact solution should bias the evaluations of the source term with respect to the characteristic direction. By introducing eq. (9) into eq. (8), expanding the flux-function in the integrand as $f = f(u_h) + a(u_h)\Delta u_s + O(\Delta t^2)$, and by integrating with respect to time, one obtains

$$f^{n+1/2}_{i+1/2} = f^{n+1/2}_{i+1/2}\Big|_h + \frac{1}{\Delta t} \int_{◪} a\, q\, dx\, dt \quad , \tag{10}$$

where the nearly triangular domain of integration ◪ covers all the characteristic curve sections $C_{i+1/2}(t)$ for $t \in [t^n,t^{n+1}]$. With respect to eq. (10) we now introduce the numerical flux $\tilde{g}$ of the extended scheme as

$$\tilde{g}^{n+1/2}_{i+1/2} = g^{n+1/2}_{i+1/2} + \frac{\Delta t}{2} (a\, q)^{n+1/2}_{i+1/2} \quad . \tag{11}$$

Here $g^{n+1/2}_{i+1/2}$ denotes the numerical flux of the basic scheme for the homogeneous conservation law (cf. eq. (5)) and (aq) is to be evaluated with an appropriate averaged state at $x_{i+1/2}$ (see below). The extension of this scheme to a system of equations is now straight forward. Let $\underline{u} = (u^1...u^m)$, $\underline{f} = (f^1...f^m)$, $\underline{q} = (q^1...q^m)$ obey the equations

$$\underline{u}_t + \underline{f}(\underline{u})_x = \underline{u}_t + \underline{\underline{A}}(\underline{u})\,\underline{u}_x = \underline{q} \quad , \tag{12}$$

where $\underline{\underline{A}} = \partial \underline{f} / \partial \underline{u}$ has a complete set of left and right eigenvectors $\underline{L}^{(\nu)}$, $\underline{R}^{(\nu)}$ and corresponding real eigenvalues $a^{(\nu)}$ $(\nu = 1,...,m)$.

In order to transfer the scalar result eq. (11) to this system, we consider a Roe-linearized version (cf. /3/) of eq. (12) in the vicinity of each cell interface. The source vector $\underline{q}$ is decomposed with respect to the right eigenvectors in the form

$$\underline{q} = \sum_{\nu=1}^{m} \underline{q}^{(\nu)} \quad , \quad \underline{q}^{(\nu)} = (\underline{L}^{(\nu)} \cdot \underline{q})\, \underline{R}^{(\nu)} \quad . \tag{13}$$

Inspection of the ν-th component of eq. (12), namely

$$\underline{L}^{(\nu)} \cdot (\underline{u}_t + a^{(\nu)} \underline{u}_x)\, \underline{R}^{(\nu)} = \underline{q}^{(\nu)} \tag{14}$$

shows, that in analogy to the scalar case of eq. (6), the portion $\underline{q}^{(\nu)}$ of $\underline{q}$ introduces a disturbance, which propagates along the ν-th characteristic with slope $\partial x / \partial t\big|^{(\nu)} = a^{(\nu)}$. Treating all the portions $\underline{q}^{(\nu)}$ in the way shown for the scalar equation, leads to the source corrected numerical flux

$$\tilde{\underline{g}}_{i+1/2}^{n+1/2} = \underline{g}_{i+1/2}^{n+1/2} + \frac{\Delta t}{2} \sum_{\nu=1}^{n} \left(a^{(\nu)} \underline{q}^{(\nu)} \right)_{i+1/2}^{n+1/2} \quad . \tag{15}$$

The final scheme for the update $t^n \rightarrow t^{n+1}$ in the i-th cell reads as

$$\begin{aligned} \underline{u}_i^{n+1} = \underline{u}_i^n - \frac{\Delta t}{\Delta x} \Big\{ & \Big(\underline{g} + \frac{\Delta t}{2} \sum_{\nu=1}^{m} \left(a^{(\nu)} \underline{q}^{(\nu)} \right) \Big)_{i+1/2}^{n+1/2} \\ & - \Big(\underline{g} + \frac{\Delta t}{2} \sum_{\nu=1}^{m} \left(a^{(\nu)} \underline{q}^{(\nu)} \right) \Big)_{i-1/2}^{n+1/2} \Big\} + \Delta t \, \underline{q}(u_i^{n+1/2}) \quad . \end{aligned} \tag{16}$$

In the calculations presented below we used the Roe-averaged states at the interfaces to evaluate $\underline{q}_{i+1/2}^{n+1/2}$, but one might also think about using both the states, which enter the numerical flux function $g(\cdot,\cdot)$ (eq. (5)) in an upwind fashion. In eq. (16) it is seen that in regions of smooth solutions the source wave contributions nearly cancel to second order corrections, whereas next to steep gradients this cancellation fails and they become as important as the cell center terms.

In order to test the source extended scheme we consider e.g. a circular symetrical problem, similar to Sod's problem /6/, with initial conditions ($\rho = p = 1$ for radii $0 < r \leq 1/2$, $\rho = p = \alpha$ for $1/2 < r \leq 1$ and $\vec{v} \equiv 0$, $\xi/\rho \equiv \beta$; at time $t = 0$). The inert gas case was investigated as well as that of a reactive gas, whose properties were fixed by the parameters E, B, Q given in eqs. (1), (3). For each set of (α, β, E, B, Q) we compared two calculations. The first solved the related one-dimensional problem in terms of polar coordinates, whereas the second one was performed on a two-dimensional cartesian grid. For an inert gas we were thus able to check the performance of the extended 1-D-scheme, when dealing with the geometry-induced source terms in polar coordinates, against the operator splitting device for multidimensional calculations. In order to obtain comparable results even for small radii we prescribed a rigid wall-boundary condition at $r = \delta > 0$ in the 1-D-calculations, thereby choosing δ such that the first pressure peaks at the center, related to the collapse of the incident shock, arrived at the same maxima in both calculations. Now Fig. 4a shows the pressure-time history at the center for a reactive gas case with ($\alpha = 1.245$, $\beta = 0.116$, $E = 10$, $B = 867$, $Q = 109.5$) , and for calculations on a mesh with stepwidth 1/50. In view of the difference in modelling the center region in both calculations, the agreement is acceptable. Fig. 4b gives the radial pressure distribution at time $t = 0.55$, which shows that away from the center both results effectively coincide.

SPECIAL CONVEX CORNER BOUNDARY CONDITIONS

In conservative finite difference schemes for the Euler equations rigid walls are often represented by the 'reflecting wall'-boundary condition. It is based on the assumption, that the wall may, at least locally, be considered as a line of symmetry in a suitably supplemented system. Wherever it is valid, the assumption allows to define artificial reflected states in so-called 'dummy cells' behind the wall, which differ from those before the wall only by a change of sign of the normal velocity. Using these states a higher order difference scheme with symmetric support can be used even for cells next to the wall, and the 'no flux'-condition is satisfied automatically.

For the L-configuration, shown in Fig. 1 this assumption is exactly satisfied by the walls B, C, D, whereas there is no such physically senseful symmetric complement with respect to the walls E and F. At least in the immediate vicinity of the corner, where dummy cells are occupied by reflected states from both adjacent walls, the reflecting wall procedure is unacceptable from a physical point of view. Calculations of Woodward and Colella /7/ indicate, that with the reflecting wall condition used even at the edge, the numerical results undesirably depend on the details of the numerical scheme. Thus there seems to be a need for special corner boundary conditions, which bound these uncertainties by introducing some analytical knowledge into the algorithm. An insight into the structure of the flow near the corner can be obtained by considering the Euler equations in terms of polar coordinates (r,ϕ,t), centered at the edge, i.e.

$$\begin{aligned} r\,(U_t + U\,U_r + p_r/\rho\,) + V\,U_\phi &= V^2 \\ r\,(V_t + U\,V_r\,) + V\,V_\phi + p_\phi/\rho &= -\,V\,U \\ r\,(\rho_t + U\,\rho_r + \rho\,U_r\,) + V\,\rho_\phi + \rho\,V_\phi &= -\,\rho\,U \\ r\,(e_t + U\,(e+p)_r + (e+p)\,U_r\,) + V\,(e+p)_\phi + (e+p)\,V_\phi &= -\,(e+p)\,U \end{aligned} \tag{17}$$

in the limit of vanishing radii. Here U, V represent the radial and azimuthal velocities respectively. We assume, that the quantities ρ, U, V, p are piecewise smooth functions of (r,ϕ,t) and expand them with respect to r in the form

$$\underline{F} = \underline{F}^{(0)}(\phi,t) + r\,F^{(1)}(\phi,t) + \dots \tag{18}$$

$\underline{F} = (\rho,U,V,p)$. Introducing these expansions into eqs. (17) we obtain to leading order

$$\begin{aligned} &V\,(U_\phi - V) = 0\;, && V\,(V_\phi + U) + p_\phi/\rho = 0\;, \\ &\rho\,(V_\phi + U) + V\,\rho_\phi = 0\;, && (e+p)\,(V_\phi + U) + V\,(e+p)_\phi = 0\;. \end{aligned} \tag{19}$$

(The superscript's "(0)" are suppressed for convenience.) Since all time derivatives drop out of these equations, the time enters into their solutions only via the boundary conditions with respect to ϕ, which have to be determined from the instantaneous outer flow. Suitable linear combinations of eqs. $(19_3, 19_4)$ and of eqs. $(19_1, 19_2, 19_3)$ lead to

$$2\,(a^2)_\phi = -\,(\gamma - 1)(U^2 + V^2)_\phi \quad , \quad 2\,(a^2)_\phi = -\,(\gamma - a^2/V^2)(U^2 + V^2)_\phi \tag{20}$$

respectively. These relations are compatible, only if the Machnumber M_ϕ of the transversal velocity obeys $M_\phi = |V/a| \equiv 1$, which characterizes the well known Prandtl-Meyer expansion of stationary gasdynamics. But, since we consider transient processes, the angle of deflection does not neccessarily coincide with the angle of the corner. Instead the extension of the expansion fan depends on the state of the oncoming flow and on the downstream pressure. Therefore one may expect a shear discontinuity, which emerges from the edge and separates the high speed flow behind the fan from the gas in the wake of the step (cf. Fig. 3b).

The constraint $M_\phi \equiv 1$ can be satisfied only in a supersonic flow, such that for subsonic oncoming flow with a finite velocity at the corner to leading order there is only the trivial parallel flow solution, sketched in Fig. 3a. Several other steady flows in the vicinity of a convex corner can be considered, but so far we have restricted to the two cases discussed above, where the pressure in the wake does not exceed the pressure of the oncoming flow. The existence of flow separations in real, though inevitably viscous flows was shown experimentally by Skews /8/ for supersonic, and by Hassenpflug /9/ for subsonic oncoming flow.

Implementation : In a difference scheme in conservation form the change of the cell averages of the conserved quantities during a time step is determined by means of numerical approximations to the fluxes across the cell interfaces. Thus a self suggesting way to implement the leading order corner flow results into a numerical scheme is to modify the fluxes in the immediate vicinity of the corner.

For subsonic oncoming flow the parallel flow solution predicts a tangential departure of a shear discontinuity from the edge. In the present cartesian grid it is in line with a cell interface as is seen in Fig. 3a. In order to simulate the 'no-flux-condition' across this contact surface, any convective transport from the oncoming parallel flow into the wake is inhibited by setting

$$\underline{g}_{splitt} = (0,0,p_{splitt},0)^T \tag{21}$$

for the numerical flux. In the calculations presented below we inserted $p_{splitt} = p_2$, since the O(r)-equations of the perturbation expansions prescribe $p^{(1)} \equiv 0$ in stagnation point flows ($U^{(0)} = V^{(0)} = 0$). Thus no pressure gradient should be allowed in cell 2 within the wake.

To explain the procedure for supersonic incident flow, we use the notations of Fig. 3b. At each time step a Prandtl-Meyer expansion fan is calculated using the state in cell "0" for the oncoming flow and setting

$$p_{wake} = p_3 \quad (v_3 < 0) \, , \qquad p_{wake} = p_3^0 \quad (v_3 \geq 0) \, , \tag{22}$$

where p_3^0 is the stagnation pressure, related to the state in cell "3" and v_3 its velocity component tangential to the wall.

Once the structure of the corner flow is known, the fluxes across all interfaces of the cells "1" and "2" are calculated by means of numerical integrations. These fluxes replace the numerical fluxes of the conservative scheme. Since, in addition, we are using a high order upwind scheme with more than three point support, there is a need to represent the cells "1" and "2" in both coordinate directions, when proceeding with this scheme away from the edge. For example, cell "3" mainly faces the wake, whereas cell "4" is influenced by the high-speed flow behind the fan. We account for this difference by replacing the cell averages of the conserved quantities in cell "2" by their averages over the cell boundaries, $\overline{2\,3}$, $\overline{2\,4}$, which separate cell "2" from "3" and "4" respectively. Analogous considerations hold for cell "1".

To check the influence of the corner flow boundary conditions, we solved the initial-boundary value problem outlined in the introduction, but for an inert gas on a (60 x 96) cell grid. Subsonic incident flow at the corner was obtained with a Mach-number $M_{sh} = 1.6$ of the incident shock wave. Fig. 5 shows two-dimensional distributions of p/ρ^γ, after a vortex has separated from the corner. It is seen, that although unphysical entropy production cannot be completely avoided with the cornerflow modification (Fig. 5b), it is nevertheless substantially reduced compared to the results obtained with the 'reflecting wall'-condition.

For supersonic incident flow the shock Mach-number was increased to $M_{sh} = 6$. In this case we have investigated how a change in the basic one-dimensional scheme influences the structure of the flow separation in the vicinity of the corner. It would be desirable to reduce this influence as far as possible. We have e.g. changed the set of variables, for which piecewise linear distributions are introduced in the preliminary step of the two-step algorithm (see Munz /1/). In particular we either used primitive variables (ρ,u,v,p) or local characteristic variables. Figs. 6 show lines of constant velocity component parallel to the backward facing wall of the corner. Figs. 7a,b were obtained with the 'reflecting wall'-condition and it is seen, that the structure of the flow separation changes substantially with the choice of the slope-variables. In contrary this dependence nearly vanishes, if the Prandtl-Meyer condition is used (cf. Figs. 7c,d). Besides this, the presumably more realistic representation of the shear layer termination in Figs. 7a,c suggests, that even in two-dimensional simulations the slope-calculation in terms of the characteristic fields is favourable.

REFERENCES

[1]. Munz, C.D.: "On the comparison and construction of two-step schemes for the Euler equations", Notes on Numerical Fluid Mechanics, 14, p. 195-217, Vieweg, 1982.

[2]. Einfeldt, B.: "On Godunov type methods for gasdynamics", to appear in SIAM, J. Numer. Anal.

[3]. Roe, P.L.: "Approximate Riemann solvers, parameter vectors and difference schemes", J. Comp. Phys., 43, p. 357-372 (1981).

[4]. Roe, P.L.: "Upwind differencing schemes for hyperbolic conservation laws with source terms", A. Dold, B. Eckmann (Eds.), Lecture Notes in Mathematics, 1270, Springer, 1987.

[5]. Toro, E.F.: "Roe's method in gas dynamical problems associated with the combustion of high energy solids in a closed tube", CoA Rep. No. NFP 86/18, March 1986, College of Aeronautics, Cranfield Inst. of Techn., Cranfield, Bedford MK 43 OAL.

[6]. Sod, A.: "A numerical study of a converging cylindrical shock", J. Fluid Mech., 83, p. 785-794 (1977).

[7]. Woodward, P., Colella, P.: "High resolution schemes for compressible gas-dynamics", Lecture Notes in Physics, 141, Springer, 1981.

[8]. Skews, B.W.: "The perturbed region behind a diffracted shock wave", J. Fluid Mech., 29, p. 705-719 (1967).

[9]. Hassenpflug, H.-U.: "Untersuchungen zur Wirbelbildung durch Beugung einer Stoßwelle", Dissertation RWTH Aachen, 1976.

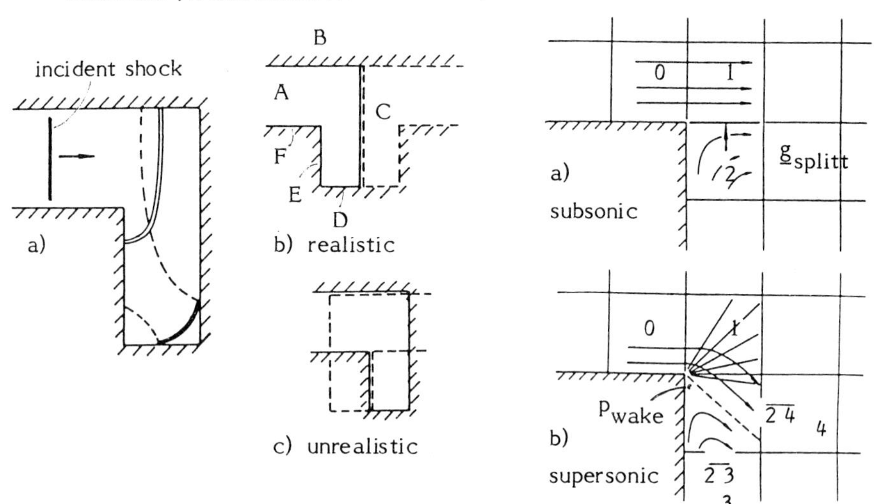

Fig. 1 : The model systems and exemplary symmetric complements

Fig. 3 : The asymptotic structure of the corner flow

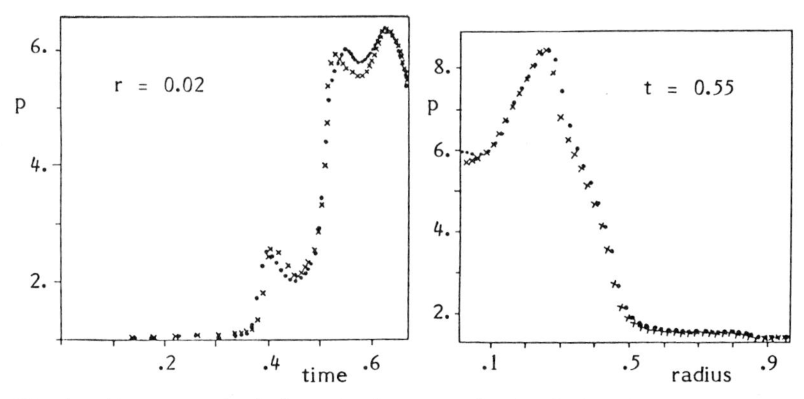

Fig. 4 : Pressure calculations for the generalized cylindrical Sod-problem
x : 1-D-polar coordinates , • : 2-D cartesian coordinates

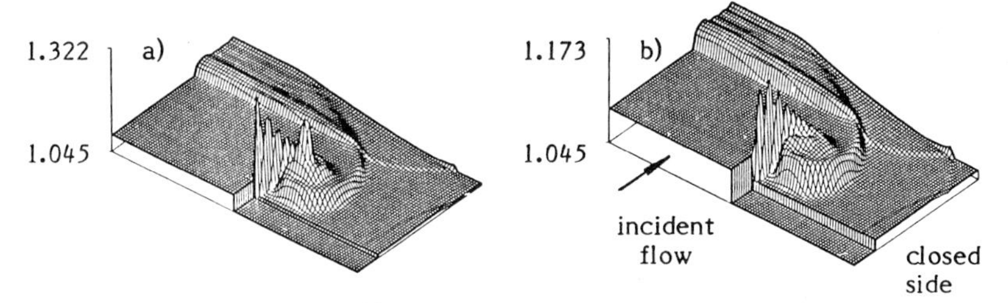

Fig. 5 : Distributions of p/ρ^{γ} for subsonic incident flow
a) without b) with the special corner flow modifications

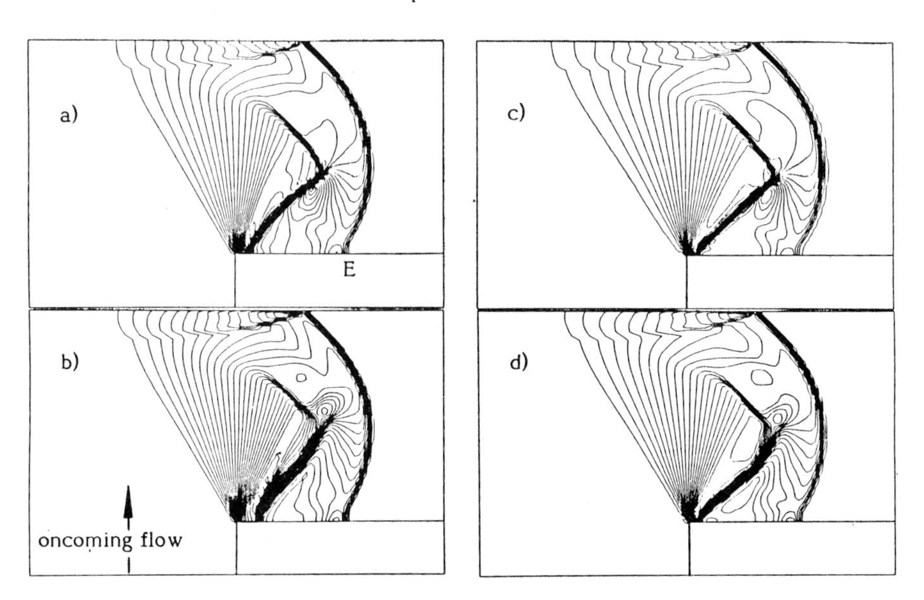

Fig. 6 : Lines of constant velocity parallel to wall E.
a,b) without c,d) with the special corner flow modifications
a,c) slopes in characteristic b,d) in primitive variables

TWO DIMENSIONAL RIEMANN PROBLEMS AND ITS APPLICATIONS: A CONFERENCE REPORT

CHRISTIAN KLINGENBERG
Dept. of Applied Mathematics, University of Heidelberg
Im Neuenheimer Feld 294, 6900 Heidelberg, W. GERMANY

We shall consider nonlinear hyperbolic equations in conservation form

$$(1) \qquad \begin{aligned} u_t + \nabla . f(u) &= 0 \\ u(0,x,y) &= u_o \end{aligned}$$

Even for smooth initial data u_o the solution to (1) in general will lead to jump discontinuities in the solution.

We are interested in the stucture of the solution. The set of jump discontinuities may become quite complicated without any apriori assumptions on the flux function f and initial data.

This is of some importance, when solving (1) numerically. I shall give two examples.

a) One can obtain high resolution by using the method of front tracking. There one marks the jump discontinuities (the front) in the solution by special grid points, which evolve dynamically with the front. From one time step to the next first the front gets moved and the states next to the front get updated. The time step gets finished by solving initial/boundary value problems in the smooth regions between the jump discontinuities using standard finite difference methods. Clearly it is necessary for this approach to qualitatively understand the structure of the jump discontinuities, see [GK].

b) When using finite difference schemes in more than one space dimension, one time step usually consists of sequentially applying a one dimensional scheme in each time direction. This has disadvantages. Roe's [R1] work indicates that the more accurate the one dimensional scheme is, the less accurate the two dimensional split scheme will be. Currently the development of methods which are not relying on the direction of the space axes is under way, e.g. following Roe [R2]. Again a knowledge about the geometry of the jump discontinuities is helpfull for this approach.

In this note we shall consider selfsimilar solutions of (1), this then reduces the number of independent variables by one. In particular we shall consider Riemann problems:

Definition: If (1) together with initial data u_o is invariant under

$$(cx, cy, ct) \longrightarrow (x, y, t) \qquad , \qquad c > 0$$

then it is called a Riemann problem. Thus initial data which is piecewise constant in sectors meeting at the origin together with (1) constituts a Riemann problem.

Definition: A Riemann solution which is invariant under the group action

$$(x , y , t) \longrightarrow (x + \sigma_1\tau, \ y + \sigma_2\tau, \ t + \tau)$$

is an elementary wave with velocity $\vec{\sigma} \in \mathbb{R}^2$.

We believe that the solution to Riemann problems are described qualitatively by the following picture: it consists of elementary waves, moving apart each with their own distinct velocity. These elementary waves are then connected by jump surfaces. We mention in passing that one can give a short list of these generic types of such intersection points for two dimensional gas dynamics, [K], [GK].

For two dimensional Riemann Problems for scalar conservation laws

$$u_t + f(u)_x + g(u)_y = 0$$

we have explicitly constructed the solution [HK] for generic cases. We found that the solution is piecewise smooth. In Fig. 1 we give a list of pieces that this solution is made of. We believe that this constitutes a complete list for polynomial flux functions f and g. In Fig. 2 we give an example of a solution to illustrate how it is composed of the pieces mentioned above.

These constructions lead us to a numerical two dimensional Riemann solver for a certain class of Riemann problems. Suppose we have two shock waves approaching each other, see Fig. 3. At their time of intersection they approximately give rise to a two dimensional Riemann problem, Fig. 4. We think of this as two waves giving rise to a wave fan. Thus in analogy to the time dependent one dimensional case of two waves crossing, there the "timelike" direction corresponds here to a "direction of causality", say ξ . In such a way we introduce new coordintes ξ , η . These transform our system (1) to the problem

$$\frac{\partial P(u)}{\partial \xi} + \frac{\partial Q(u)}{\partial \eta} = 0$$

$$P = P(u-) , \quad \eta < 0$$
$$P = P(u+) , \quad \eta > 0$$

For scalar equations and assuming that $dP/du > 0$, we have implemented a numerical Riemann solver by constructing numerically the Oleinik convex hull, see Fig. 5. The solution then gets transformed back to the x - y plane to give a two dimensional Riemann solution. This constituts a helpful tool for a front tracking algorithm, [KZ].

References:

[GK], Glimm, Klingenberg, McBryan, Plohr, Yaniv, "Front Tracking and Two Dimensional Riemann Problems", Adv. Appl. Math. 6, (1985)

[HK], Hsiao, Klingenberg, "Constructing the solution to a nonconvex two dimensional Riemann problem", Preprint

[K], Klingenberg, "Hyperbolic Conservation Laws in Two Space Dimensions", Report Institut Mittag-Leffler 2, (1986)

[KZ], Klingenberg, Zhu, "A front tracking algorithm including two dimensional Riemann solvers", manuscript

[R1], Roe, "Discontinuous solutions to hyperbolic systems under operator splitting", manuscript

[R2], Roe, "Discrete models for the numerical analysis of time dependent multidimensional gas dynamics", J. Comp. Physics, 63, (1986)

Fig. 1 The solution to $u_t + f(u)_x + g(u)_y = 0$ with initial data constant in the wedges we believe is genericaly made up of the following pieces in the x - y plane:

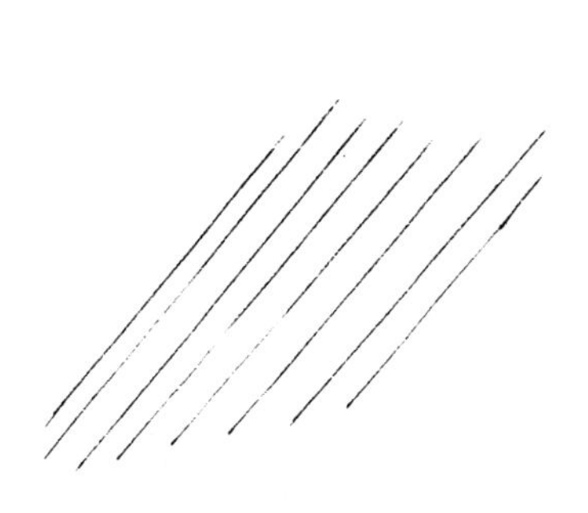

Parallel charactertistics, with the constant state as a special case.

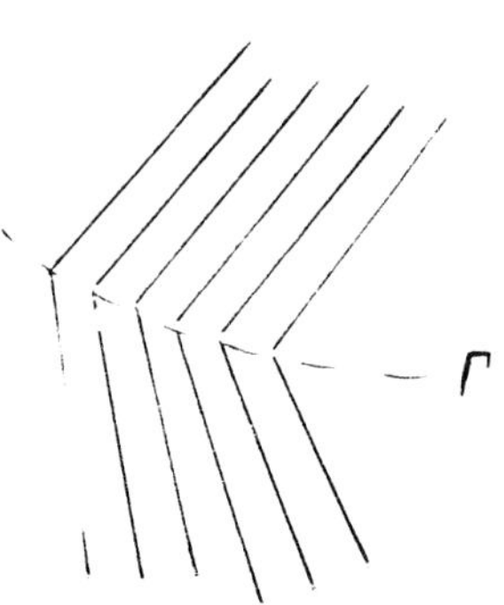

Characteristics meet along $\Gamma = (\ f'(u)\ ,\ g'(u)\ )$.

Contact discontinuities (c.d.), with one sided c.d. as a special case or jumps with constant states on both sides.

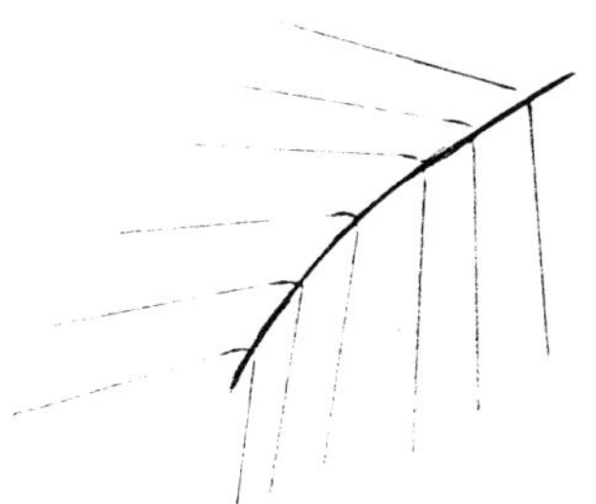

Curved jumps with rarefaction waves impinging on both sides. The characteristics may be tangential on one side or may represent constant states.

Two jumps meet along $\Omega = \left(\frac{[f]}{[u]}, \frac{[g]}{[u]}\right)$

Jump triple points.

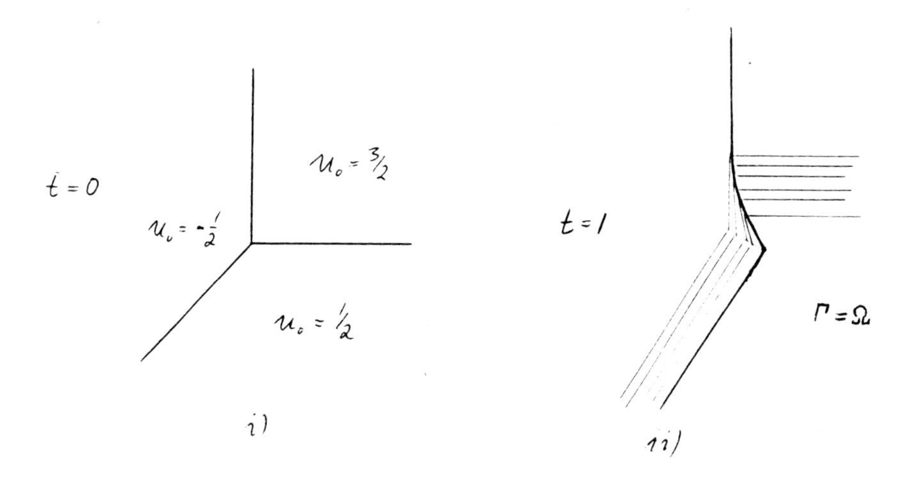

Fig. 2 The solution to $u_t + (u-u^3)_x + (u^3-u)_y = 0$ with initial data given in i) is shown in ii).

Fig. 3 Two shock waves approaching each other.

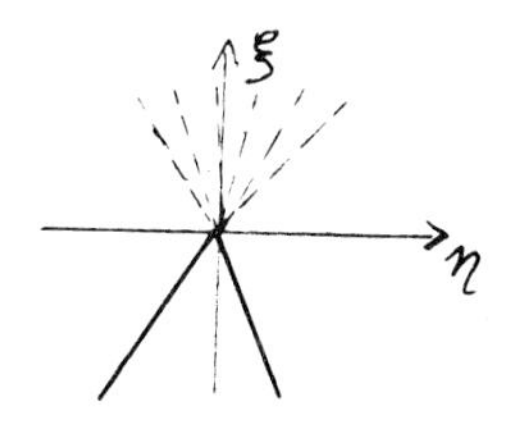

Fig. 4 The two dimensional Riemann problem which arises from Fig. 3 after the intersection of the waves.

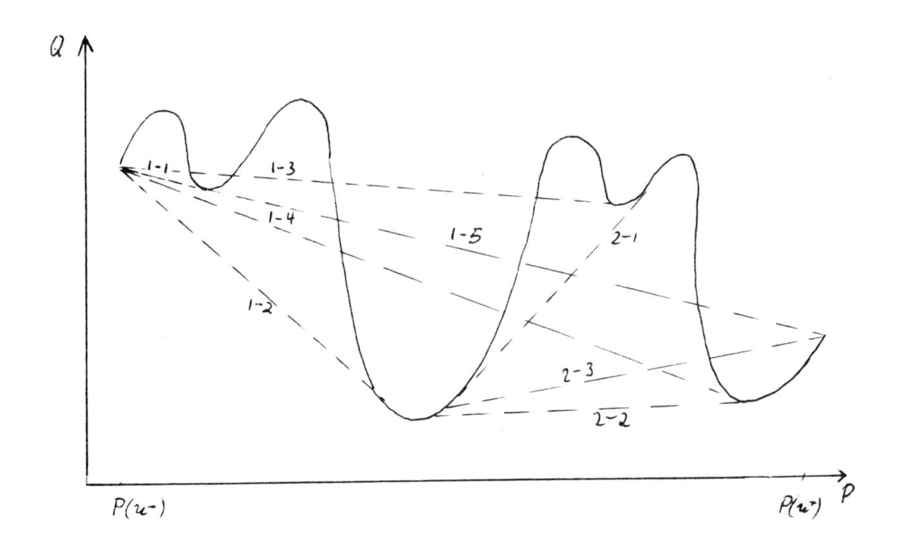

Fig. 5 The numerical construction of the convex hull to the graph of Q versus P, which gives the solution to the Riemann problem. In this case, we first determine the minimum of lines 1-1, 1-2, 1-3, 1-4, 1-5. After having found this to be line 1-2, we continue our search with minimizing slopes of 2-1, 2-2, 2-3, which in this case is 2-2. This then gives the convex hull.

A Numerical Simulation of the Effect of Salinity on a Thermally Driven Flow

Hwar-Ching Ku, Richard S. Hirsh and Thomas D. Taylor

The Johns Hopkins University Applied Physics Laboratory
Laurel, Maryland 20707

ABSTRACT

A numerical simulation of the layered structure of both temperature and salinity in a narrow vertical slot caused when an initially salt stratified fluid is disturbed by a lateral temperature gradient has been performed using the pseudospectral matrix (PSM) method.

To account for the interfacial gradients between each layer a new pseudospectral element method, using the primitive variable formulation of Navier-Stokes equations, was incorporated into the previously developed PSM method. This new element method requires continuity of the field variable itself across the interface between each element (or subdomain), i.e., $C^{(o)}$, while still imposing the zero divergence condition at all points of the domain. Instead of the previously used explicit scheme for the viscous term in the PSM method, an implicit treatment has been applied to enhance both the time-step size and stability of this PSM element method.

Computational results of the layered structures and stepwise temperature profiles indicate that the mechanism of small scale approximation could explain the physical phenomena of the real ocean case.

INTRODUCTION

Layered structures of both temperature and salinity have been found in many parts of the world's ocean. Understanding of the processes involved in the formation of these structures can be obtained experimentally [1,2,3,4] or numerically [3,5] by examining the formation of convective cells in a salinity stratified fluid when a horizontal temperature gradient is imposed between two vertical walls. Without the salinity effect, much information is available about the steady flow patterns which can result, and the simple laminar case is already being used as a standard test case for numerical procedures. If, however, a salinity gradient is present during the heating (or cooling), the number of investigations, especially in the numerical simulation area, do not apppear to be numerous since the steep changes in the field variables make the numerical computations difficult.

Our interest in this problem was initiated by some experiments performed at JHU/APL [4] which modelled the flow near an iceberg. The formation and growth of layers in a density-stratified fluid is an unsteady process which is related to the instability of the side-wall thermal boundary layer as it grows by conduction. Owing to the relative ease of heating diffusion as opposed to salt diffusion, the melted water from the iceberg is colder than the surrounding sea water and so tends to descend but only to a level where the net density is close to that in the interior. Fluid then flows away from the ice, producing a series of layers which slope upward as they extend and heat. These competing density effects result in layers separated by a sharp vertical gradient both in temperature and salinity which may be formed successively. These effects were modelled in the experiment [4] by heating a lateral wall of a tank containing a salt gradient solution to understand the mechanism of double diffusive layer processes. Some aspects of this experiment have been simulated by integrating the Navier-Stokes equations in primitive variables using a pseudospectral element modification of the PSM method which uses the incompressibility constraint at boundaries as the pressure boundary condition.

A $C^{(o)}$ PSEUDOSPECTRAL-ELEMENT MATRIX METHOD

For simplicity the spatial domain is divided into NE elements, each of which has N+1 collocation points, $x_j = 1/2\ [\cos \pi\ (j-1)/N\ (b^e - a^e) + b^e + a^e]$ $(1 \le j \le N+1;\ 1 \le e \le NE)$.

The derivatives at the inter-element points approximated as the weighting of the derivatives from each side are given by the relations

$$f^e_x(x_i) = \frac{1}{L^e} \sum_{m=1}^{N+1} \hat{GX}^{(1)}_{i,m}\ f^e_m \tag{1a}$$

$$f^e_{xx}(x_i) = \frac{1}{(L^e)^2} \sum_{m=1}^{N+1} \hat{GX}^{(2)}_{i,m}\ f^e_m = \frac{1}{L^e} \sum_{m=1}^{N+1} \hat{GX}^{(1)}_{i,m}\ (f^e_x)_m\ , \tag{1b}$$

Here L^e is the length of e-th element defined on the interval $[a^e,\ b^e]$ and $\underline{\hat{G}X}^{(1)}$, $\underline{\hat{G}X}^{(2)}$ are the derivative matrices based on the domain [0,1] (see the detail in [6].

The derivatives at the inter-element points approximated as the weighting of the derivatives from each side, are given by the relations

$$f_x\big|_{interface} = \alpha f^e_x + \beta f^{e+1}_x \tag{2a}$$

$$f_{xx}\big|_{interface} = \alpha f^e_{xx} + \beta f^{e+1}_{xx}; \qquad \alpha + \beta = 1. \tag{2b}$$

With the choice of $\alpha = L^e/(L^e + L^{e+1})$, the fraction of total length of two adjacent elements, Eqs. (2), in view of Eqs. (1), now become

$$FX = \frac{1}{L^e+L^{e+1}} \sum_{m=1}^{N+1} (\hat{GX}^{(1)}_{N+1,m}\ f^e_m + \hat{GX}^{(1)}_{1,m}\ f^{e+1}_m) \tag{3a}$$

$$FXX = \frac{1}{L^e+L^{e+1}} \sum_{m=1}^{N+1} [\hat{GX}^{(1)}_{N+1,m}\ (f^e_x)_m + \hat{GX}^{(1)}_{1,m}\ (f^{e+1}_x)_m]. \tag{3b}$$

By implicitly requiring that both element e and e+1 share the same first derivative f_x at the interface in the expansion of terms of Eq. (3b), it can be reduced to the following form

$$FXX = \frac{1}{L^e+L^{e+1}} [\ \sum_{m=1}^{N+1} (BX^*_{N+1,m}\ f^e_m + BX^{**}_{1,m}\ f^{e+1}_m) + (\hat{GX}^{(1)}_{N+1,N+1} + \hat{GX}^{(1)}_{1,1})\ FX\ ]. \tag{4}$$

It is obvious that the second term in the bracket of Eq. (4) is automatically cancelled out since $\hat{G}X^{(1)}_{1,1} = (2N^2 + 1)/3 = -\hat{G}^{(1)}_{N+1,N+1}$. The elements of modified matrices $\underline{BX}^*$, $\underline{BX}^{**}$ are in Eq. (4).

$$BX^{*}_{N+1,m} = \frac{1}{L^{e}} \sum_{n=1}^{N} \hat{GX}^{(1)}_{N+1,n} \hat{GX}^{(1)}_{n,m} \tag{5a}$$

$$BX^{**}_{1,m} = \frac{1}{L^{e+1}} \sum_{n=2}^{N+1} \hat{GX}^{(1)}_{1,n} \hat{GX}^{(1)}_{n,m} . \tag{5b}$$

The proposed $C^{(o)}$ method has been tested on a standard Burgers' equation

$$\frac{\partial u}{\partial t} + u \frac{\partial u}{\partial x} = \nu \frac{\partial^2 u}{\partial x^2} \qquad |x| \leqq 1 \tag{6}$$

I.C. $u(x,0) = -\sin \pi x$

B.C. $u(\pm 1, t>0) = 0$.

We have chosen six elements (12 points/element) with the length of each element [-1,-0.3], [-0.3,-0.05], [-0.05, 0.], [0.,0.05], [0.05,0.3] and [0.3,1], respectively, to solve this problem. The result $|\partial u/\partial x| =$ 151.535 at the origin computed by $c^{(o)}$ method at time $\pi t = 1.6$ for $\nu =$ $1/100\pi$ compared with the analytical result 152.0051 is in excellent agreement. The profile of u at time $\pi t = 1.6$ is plotted in Fig. 1. Note that even with the same number of points, the $c^{(o)}$ method still can resolve the shock front at the origin when ν is reduced by one order of magnitude.

PRIMITIVE VARIABLE FORMULATION WITH IMPLICIT VISCOUS TERMS

Within each element the method applied to the solution of Navier-Stokes equation is Chorin's [7] splitting technique. According to this scheme, the equations of motion, in tensor form, are

$$\frac{\partial u_i}{\partial t} + \frac{\partial p}{\partial x_i} = \nu \frac{\partial^2 u_i}{\partial x_j^2} + F_i \tag{7}$$

where $F_i = - u_j \, \partial u_i/\partial x_j + S_i$, and S_i is a source term.

The first step is to split the velocity into the sum of a predicted and corrected value. The predicted velocity is determined by Adams-Bashforth time integration of the momentum equation without the pressure and viscous terms

$$\bar{u}_i^{n+1} - u_i^n = \frac{\Delta t}{2} (3F_i^n - F_i^{n-1}) . \tag{8}$$

The second step is developing the pressure and corrected velocity field that satisfies the continuity equation, i.e., the Stokes problem

$$\frac{u_i^{n+1} - \bar{u}_i^{n+1}}{\Delta t} = - \frac{\partial p}{\partial x_i} + \nu \frac{\partial^2 u_i^{n+1}}{\partial x_j^2} \tag{9a}$$

$$\frac{\partial u_i^{n+1}}{\partial x_i} = 0. \tag{9b}$$

We present a fast iterative solution of this Stokes problem without using the influence matrix method [8], which is easily implemented on two or three dimensional problems.

The proposed iteration scheme for the corrector step at each time level is

$$\frac{u_i^{n+1,k} - \bar{u}_i^{n+1}}{\Delta t} = - \frac{\partial p^k}{\partial x_i} + \nu \frac{\partial^2 u_i^{n+1,k}}{\partial x_j^2} \tag{10a}$$

$$u_i^{n+1,k+1} = u_i^{n+1,k} - \Delta t \frac{\partial p'}{\partial x_i} \tag{10b}$$

$$\frac{\partial u_i^{n+1,k+1}}{\partial x_i} = 0 \tag{10c}$$

where k denotes the iteration number and p' the corrected pressure field. The details of the solution for $u_i^{n+1,k+1}$, satisfying the continuity equation in the interior as well as on the boundaries by the PSM method, can be found elsewhere [9,10]. Once $u_i^{n+1,k+1}$ satisfies the divergence-free condition, the updated pressure field is found from $p^{k+1} = p^k + p'$, and then the corrector step is repeated with the current updated pressure until the criterion for the convergence has been met. The novel features of this iteration scheme are summarized as follows: (i) for each iteration continuity is exactly satisfied everywhere; (ii) only a few iterations (usually 2-4) are required; (iii) the continuity equation is used as the boundary condition of pressure, and (iv) with the eigenfunction expansion of the field variables, u_i, p [10] the original multi-dimensional problem can be reduced to a simple one-dimensional matrix operator.

In order to set up a simple and efficient matrix operation for derivatives using the $c^{(o)}$ element method, a global-type operator which combines each local element derivative can be constructed. Therefore, the above-mentioned procedure to obtain the Stokes solution is nothing more than that of the global pseudospectral method [9,10].

DESCRIPTION OF THE PROBLEM

To investigate the layering phenomena, the two-dimensional primitive equations were solved in a narrow vertical slot (see Fig. 2) where a stable stratified fluid was initially at rest. At time t = 0, a horizontal temperature gradient was imposed, and the time-dependent behavior was then followed by integrating the equations.

First define dimensionless variables

$$A = \frac{H}{Z}, \quad u = \frac{u^* Z}{\nu}, \quad v = \frac{v^* Z}{\nu}, \quad t = \frac{t^* Ra \nu}{Z^2}, \quad Ra = \frac{g \beta_T Z^3 (T_W - T_o)}{\nu \kappa}, \quad x = \frac{x^*}{Z}, \quad y = \frac{y^*}{Z},$$

$$T = \frac{T^* - T_o}{T_W - T_o}, \quad p = \frac{p^* Z^2}{Ra \nu^2}, \quad S = \frac{S^* - S_T}{S_B - S_T}, \quad \alpha = \frac{\beta_S (S_B - S_T)}{\beta_T (T_W - T_o)}, \quad Pr = \frac{\nu}{\kappa}, \quad Sc = \frac{\nu}{\mathscr{D}}$$

where g is the acceleration of gravity; β_T, β_S the thermal and salinity expansion coefficients, respectively; ν, the kinematic viscosity; S_T, S_B the initial salinity (parts per thousand) at the top and bottom, respectively, of the vertical slot; κ, $\mathscr{D}$, the thermal and molecular diffusivity; Z, H the width and height of slot; and α the initial ratio of vertical salinity change subject to the horizontal temperature gradient. The differential equations describing the resulting convective flows in a stratified fluid are then

$$\frac{\partial u}{\partial x} + \frac{\partial v}{\partial y} = 0 \tag{11a}$$

$$\frac{\partial u}{\partial t} + \frac{1}{Ra} \left(u \frac{\partial u}{\partial x} + v \frac{\partial u}{\partial y}\right) = - \frac{\partial p'}{\partial x} + \frac{1}{Ra} \left(\frac{\partial^2 u}{\partial x^2} + \frac{\partial^2 u}{\partial y^2}\right) \tag{11b}$$

$$\frac{\partial v}{\partial t} + \frac{1}{Ra} \left(u \frac{\partial v}{\partial x} + v \frac{\partial v}{\partial y}\right) = - \frac{\partial p'}{\partial y} + \frac{1}{Pr} (T - \alpha S') + \frac{1}{Ra} \left(\frac{\partial^2 v}{\partial x^2} + \frac{\partial^2 v}{\partial y^2}\right) \tag{11c}$$

$$\frac{\partial T}{\partial t} + \frac{1}{Ra}\left(u\frac{\partial T}{\partial x} + v\frac{\partial T}{\partial y}\right) = \frac{1}{RaPr}\left(\frac{\partial^2 T}{\partial x^2} + \frac{\partial^2 T}{\partial y^2}\right) \tag{11d}$$

$$\frac{\partial S}{\partial t} + \frac{1}{Ra}\left(u\frac{\partial S}{\partial x} + v\frac{\partial S}{\partial y}\right) = \frac{1}{RaSc}\left(\frac{\partial^2 S}{\partial x^2} + \frac{\partial^2 S}{\partial y^2}\right) \tag{11e}$$

with the perturbation variables $p' = p(x,y) - p_o(y)$, $S'(x,y) = S(x,y) - S_o(y)$ where $\partial p_o(y)/\partial y = -\alpha S_o(y)/Pr$ is the initial static pressure gradient and $S_o(y)$ is the initial salinity profile.

The initial and boundary conditions are

$t = 0$

$$T = u = v = 0, \quad S = S_o(y), \quad 0 \leqq x \leqq 1, \quad 0 \leqq y \leqq A \tag{12a}$$

$t > 0$

$$T = 1, \quad u = v = 0, \quad \partial S/\partial x = 0, \quad x = 1, \quad 0 \leqq y \leqq A \tag{12b}$$

$$T = 0, \quad u = v = 0, \quad \partial S/\partial x = 0, \quad x = 0, \quad 0 \leqq y \leqq A \tag{12c}$$

$$\partial T/\partial y = 0 \quad u = v = 0, \quad S = 0, \quad y = A, \quad 0 \leqq x \leqq 1 \tag{12d}$$

$$\partial T/\partial y = 0 \quad u = v = 0, \quad S = 1, \quad y = 0, \quad 0 \leqq x \leqq 1. \tag{12e}$$

RESULTS AND DISCUSSION

The problem parameters were: aspect ratio $A = 4$, Rayleigh number $Ra = 8.4*10^4$, Prandtl number $Pr = 7$, Schmidt number $Sc = 700$, and the ratio of vertical salinity change subject to the horizontal temperature gradient, $\alpha = 6$. The number of Chebyshev modes used in the horizontal direction was 37, and in the vertical direction 18 elements, each containing 6 points per element, were used. Thus a total of 37 x 109 Chebyshev modes were used to simulate the effect of salinity on the thermally driven flow. The time step Δt is 12 in contrast to $\Delta t = 4$ if the explicit method was used. The solution required 28 seconds per time step to solve two momentum equations with implicit viscous terms plus the pressure Poisson solver on an Alliant Fx-1 computer.

Typical results are shown in Figs. 3 and 4 describing the evolution of the streamlines and iso-salinity lines. It is apparent from Fig. 4 that initially the heated fluid moves upward to an equilibrium density level and two cells immediately form at the top and bottom of slot, which is in agreement with results found by [3,5]. The one at the bottom acts as a solid boundary and gives a faster layer formation compared to the one at the top. The bottom one continuously induces a convective cell above it by the same mechanism until the bottom-generated cells approach the slow-formed top cells in the enclosure.

The salinity profiles of Fig. 3 indicate that the fluid forming each cell only rises up to a potential height above its initial equilibrium density level and then moves away from the heated wall with the inclined downward slope of a cooling process. As the layers become well developed, sharp discontinuities in salinity are exhibited between each layer and this phenomena is observed in Fig. 8, the salinity profiles along the vertical center line, in which results clearly show the profiles reminiscent of the stepwise salinity profiles occurring in the ocean.

As expected at the early state of development, the lateral heat transfer in the middle of vertical slot is mostly by conduction because layers are not formed yet. When layers are well developed, as indicated in Fig. 5, the temperature profiles at the top of each layer move away from the hot wall, while the converse is true near the bottom of each layer. By examining the horizontal velocity profiles along the vertical center line (Fig. 6),one sees that the recirculating convective flow in each layer certainly produces the zigzag profile of temperature along the vertical center line as shown in Fig. 7 (see also Fig. 5).

CONCLUSIONS

A numerical simulation of the effect of salinity on a thermally driven flow has been simulated by a new $c^{(o)}$ pseudospectral-element matrix method. This method solves the Navier-Stokes equations in primitive variable formulation and treats the viscous term implicitly. The predicted results qualitatively give a good explanation of the mechanism of layered structures in temperature as well as salinity profiles.

ACKNOWLEDGEMENT

The authors would like to thank Dr. J. Calman for his helpful discussions and Dr. A. P. Rosenberg for his assistance running a program on the Alliant Fx-1 machine. This work was partially supported by the Office of Naval Research under U.S. Navy (SPAWAR) Contract N00039-87-C-5301.

REFERENCES

[1]. S. A. Thorpe, P. K. Hutt and R. Soulsby, JFM, 38, 375-400, 1969.
[2]. H. E. Huppert and J. S. Turner, JFM, 100, 367-384, 1980.
[3]. R. A. Wirtz, D. G. Briggs and C. F. Chen, GFD, 3, 265-288, 1972.
[4]. J. Calman, "Convection at a Model Ice Edge," APL Tech. Digest, 6 211-215, 1985.
[5]. V. Demay, J. M. Lacroix, R. Peyret and J. M. Vanel, "Numerical Experiment on Stratified Fluids Subject to Heating," Third Intl. Symp. on Density-Stratified Flows, Pasadena, 1987.
[6]. H. C. Ku and D. T. Hatziavramidis, JCP, 56, 495-512, 1984.
[7]. A. J. Chorin, Math. Comp., 22, 745-762, 1968.
[8]. Kleiser and U. Schumann, Notes on Numerical Fluid Mechanics, (E. H. Hirschel, Ed.) Vieweg, Braunschweig, 1980.
[9]. H. C. Ku, T. D. Taylor and R. S. Hirsh, C&F, 15, 195-214, 1987.
[10]. H. C. Ku, R. S. Hirsh and T. D. Taylor, JCP, 70, 439-462, 1987.

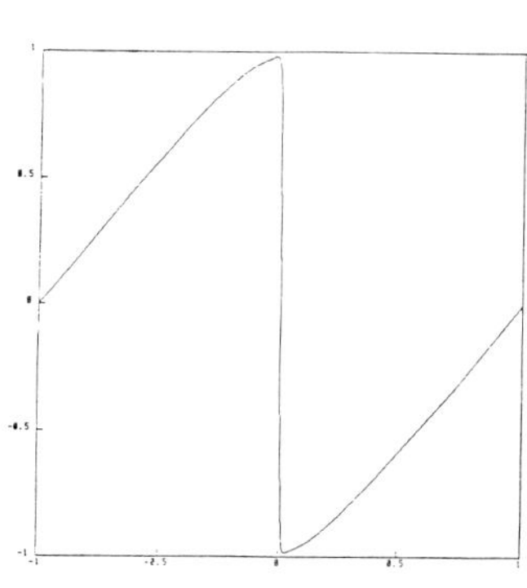

Fig. 1. Solution of the Burgers equation by the pseudospectral-element matrix method at time $\pi t = 1.6$

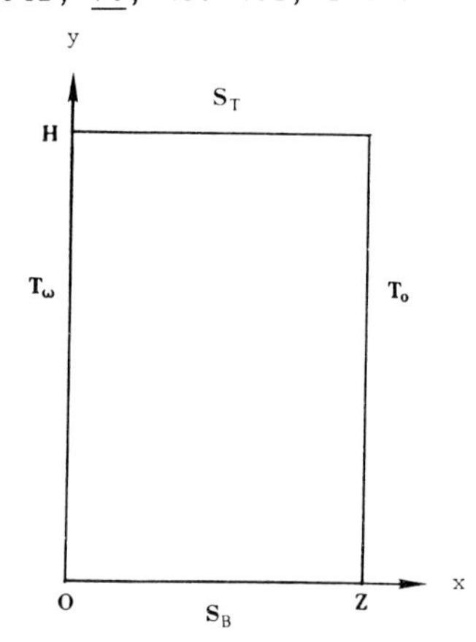

Fig. 2. Geometry and boundary conditions of thermally driven vertical slot with salinity gradient

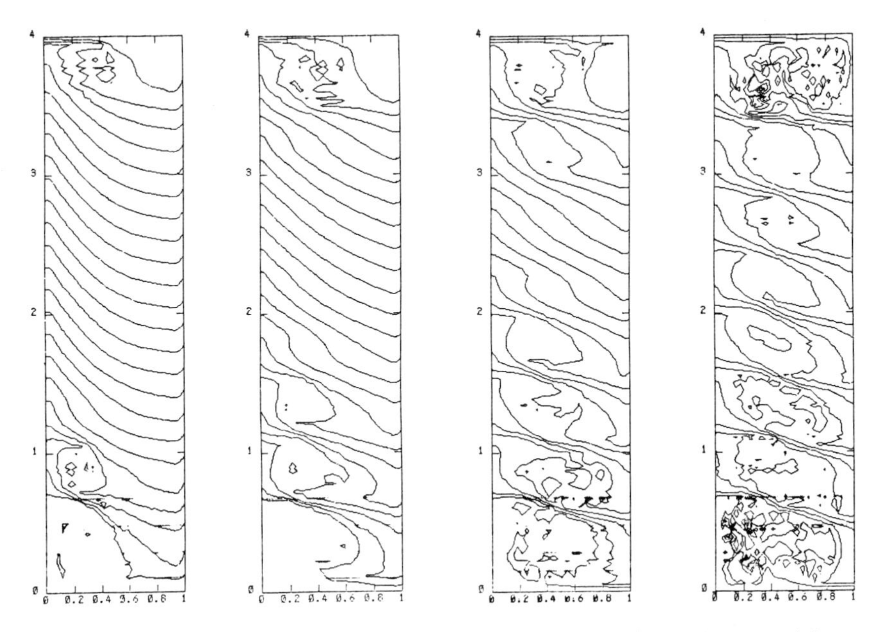

Fig. 3. Evolution of isosalinity lines at time t/Ra (a) 0.5, (b) 1.0, (c) 1.5, (d) 2.0

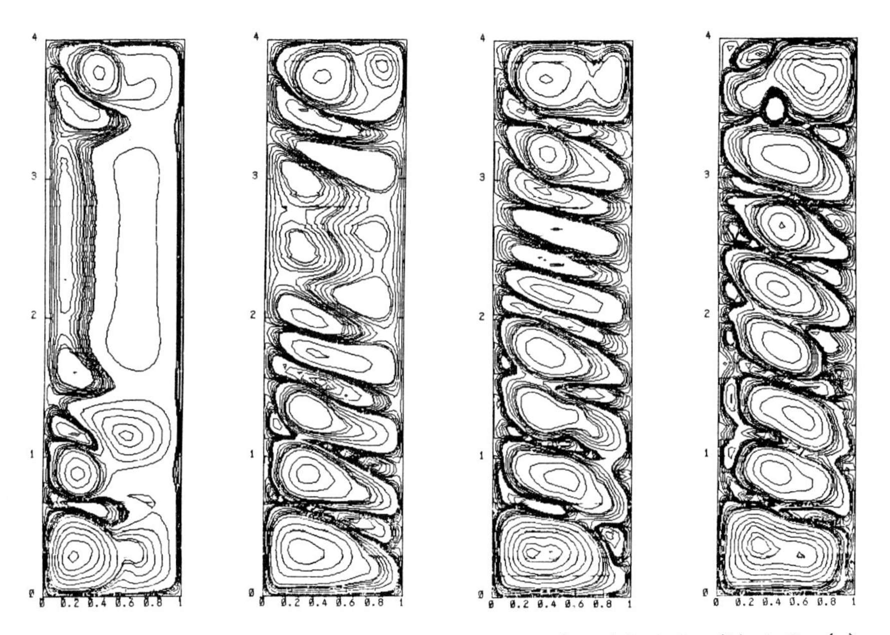

Fig. 4. Evolution of streamlines at time t/Ra (a) 0.5, (b) 1.0, (c) 1.5, (d) 2.0

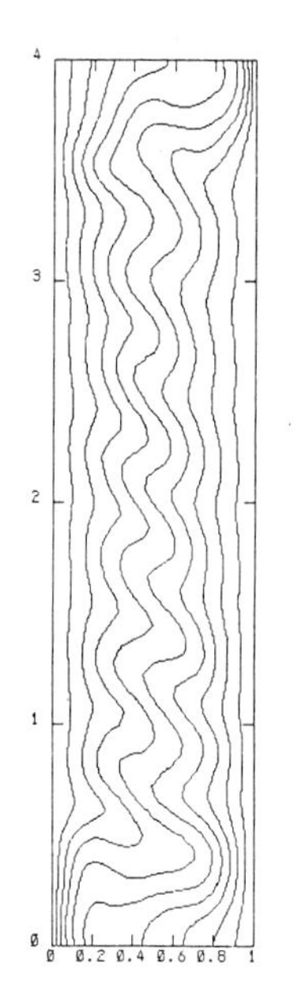

Fig. 5. Temperature contour maps at time t/Ra = 2.0

Fig. 6. Horizontal velocity profile on vertical centerline at time t/Ra = 2.0

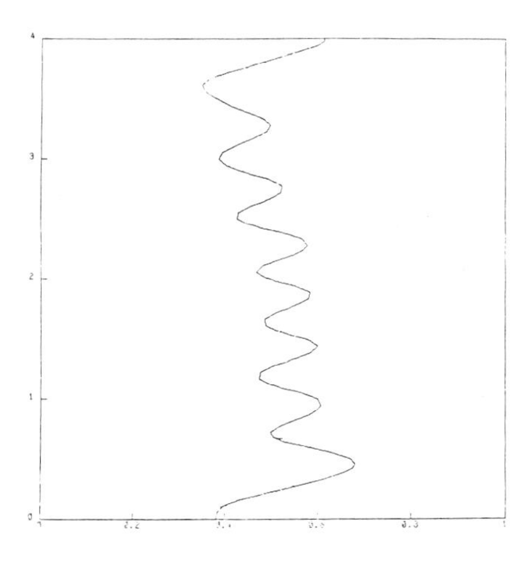

Fig. 7. Temperature profile on vertical centerline at time t/Ra = 2.0

Fig. 8. Salinity profile on vertical centerline at time t/Ra = 2.0

NUMERICAL RESOLUTION OF NAVIER-STOKES EQUATIONS IN VELOCITY-VORTICITY FORMULATION: APPLICATION TO THE CIRCULAR CYLINDER

LABIDI Wathik - TA PHUOC Loc
LIMSI(CNRS) - B.P 30 -91406 ORSAY CEDEX - FRANCE

ABSTRACT

An A.D.I method is applied to the resolution of the unsteady Navier-Stokes equations in velocity-vorticity formulation for incompressible viscous flow around a circular cylinder. The divergence free condition is treated and a high order numerical scheme is used. Numerical results can be obtained with a small number of points even for high Reynolds number.

I- INTRODUCTION

The study of the two-dimensionnal unsteady flow of an incompressible viscous fluid around a circular cylinder has been the subject of several numerical works. Some of them used primitive variables for which the boundary condition on the pressure presents some difficulty, others used a stream function-vorticity formulation.

The purpose of this work is:

1) To present a method of resolution using the vorticity-velocity formulation, proposed in [1-4], who gives directly the velocity and vorticity fields. The advantage of this formulation is the relatively easy treatment of the boundary condition for the vorticity.

2) To use and compare two finite differences schemes respectively second order and combined second and fourth order to solve the Poisson's equations. The use of this high order combined scheme allows us : a) to use, for the same Reynolds number, less points than for the fully second order scheme. b) to use a larger time step than for the second order scheme. c) to better fulfill the continuity equation (analysed in [1-4]) using appropriate conditions on the radial and tangential velocities.

3) To study the impulsively started flow around a circular cylinder, particularly the onset and the evolution of the separation on the body and the recirculation area, the secondary vortices formation in the flow during the first stages of the motion.

II- FUNDAMENTAL EQUATIONS

For unsteady flow of incompressible viscous fluid, the Navier-Stokes equations in polar coordinates are:

Continuity equation:

$$\frac{1}{r}\frac{\partial(ru)}{\partial r} + \frac{1}{r}\frac{\partial v}{\partial \theta} = 0 \quad . \tag{1}$$

Vorticity transport equation:

$$\frac{\partial\omega}{\partial t}+u\frac{\partial\omega}{\partial r}+\frac{v}{r}\frac{\partial\omega}{\partial\theta}=\nu[\frac{1}{r}\frac{\partial(r\frac{\partial\omega}{\partial r})}{\partial r}+\frac{1}{r^2}\frac{\partial^2\omega}{\partial\theta^2}] \quad . \tag{2}$$

Poisson equations:

These equations are derived from the continuity one and can be written as:

$$\frac{1}{r}\frac{\partial(r\frac{\partial u}{\partial r})}{\partial r}+\frac{1}{r^2}\frac{\partial^2 u}{\partial\theta^2}=-\frac{1}{r}\frac{\partial\omega}{\partial\theta} \quad , \tag{3}$$

$$\frac{1}{r}\frac{\partial(r\frac{\partial v}{\partial r})}{\partial r}+\frac{1}{r^2}\frac{\partial^2 v}{\partial\theta^2}=\frac{\partial\omega}{\partial r} \quad . \tag{4}$$

The dimensionless variables are deduced from the physical ones by the relations:

$$r=\frac{\tilde{r}}{a},\theta=\frac{\tilde{\theta}}{\pi},\vec{U}=\frac{1}{U_0}\tilde{\vec{U}}$$

$$\omega=\frac{a}{U_0}\tilde{\omega},t=\frac{U_0}{a}\tilde{t} \ .$$

The Reynolds number of the flow is defined as:

$$Re=\frac{2aU_0}{\nu}$$

a : radius of the cylinder , U_0 : velocity at the infinity

ν : kinematic viscosity.

Using a log-polar transformation, the circular domain is mapped into a rectangle. The equations are simplified and the mesh is regular.

With $r=exp(\pi\cdot R)$, $\theta=\Theta$

and $E=\pi\cdot exp(\Pi\cdot R)$.

The radial and tangential components of the velocity are replaced by the following quantities: $U=E\cdot u$, $V=E\cdot v$.

Hence, the equations are:

- Continuity equation:

$$\frac{1}{E^2}(\frac{\partial U}{\partial R}+\frac{\partial V}{\partial\Theta})=0 \quad . \tag{5}$$

- Vorticity transport equation written in conservative form:

$$E^2\frac{\partial\omega}{\partial t}+\frac{\partial(U\omega)}{\partial r}+\frac{\partial(V\omega)}{\partial\Theta}=\frac{2}{Re}(\frac{\partial^2\omega}{\partial R^2}+\frac{\partial^2\omega}{\partial\Theta^2}) \quad . \tag{6}$$

- Poisson equations:

$$\frac{\partial^2 U}{\partial R^2}+\frac{\partial^2 U}{\partial\Theta^2}=-E^2\frac{\partial\omega}{\partial\Theta} \quad , \tag{7.a}$$

$$\frac{\partial^2 V}{\partial R^2}+\frac{\partial^2 V}{\partial\Theta^2}=\frac{\partial(E^2\omega)}{\partial R} \quad . \tag{7.b}$$

The boundary conditions for the velocity are:

- On the surface of the cylinder: we impose the no-slip condition.

$$u = v = 0 \quad . \tag{8}$$

- At the external boundary, two kinds of boundary conditions were used:
  * At infinity, the flow is assumed to be irrotational and the boundary conditions are:

$$u = -(1 - \frac{\Pi^2}{E^2}) \cdot cos\Theta \quad , \tag{9.a}$$

$$v = (1 + \frac{\Pi^2}{E^2}) \cdot sin\Theta \quad . \tag{9.b}$$

  * To take into account the continuity equation, we use boudary conditions based on the first order derivatives of the velocity components (cf III.2).

III - NUMERICAL METHODS

III.1) The vorticity transport equation is solved by an A.D.I method proposed by K.Aziz and J.D.Hellums [5] for the three-dimensional parabolic equations.

If we consider the following equation:

$$\frac{\partial \omega}{\partial t} + \varphi_w = (\delta_{RR} + \delta_{\Theta\Theta}) \cdot \omega \quad . \tag{10}$$

with $\varphi_\omega = 0$

The calculation is splitted in the two following time steps:

$$(\delta_{\Theta\Theta} - \frac{2}{\Delta t})\omega^{(n+\frac{1}{2})} = -(\delta_{\Theta\Theta} + 2\delta_{RR} + \frac{2}{\Delta t})\omega^{(n)} + 2\varphi_w \quad , \tag{11.a}$$

$$(\delta_{RR} - \frac{2}{\Delta t})\omega^{(n+1)} = \delta_{RR}\omega^{(n)} - \frac{2}{\Delta t}\omega^{(n+\frac{1}{2})} \quad . \tag{11.b}$$

with $\delta_{RR}. = \frac{1}{E^2}\frac{\partial^2 .}{\partial R^2} - \frac{Re}{2E^2}\frac{\partial (U.)}{\partial R}$

and $\delta_{\Theta\Theta}. = \frac{1}{E^2}\frac{\partial^2 .}{\partial \Theta^2} - \frac{Re}{2E^2}\frac{\partial (V.)}{\partial \Theta}$.

To close the linear system of equations, we use the following boundary conditions:

- On the cylinder: The vorticity can be obtained from the the following formula

$$\omega = \frac{1}{E^2}\frac{\partial V}{\partial R}$$

which is discretized by a Taylor expansion of the tangential velocity.

$$V(2,J) - V(1,J) = \Delta R(\frac{\partial V}{\partial R})_{I=1} + \frac{\Delta R^2}{2}(\frac{\partial^2 V}{\partial R^2})_{I=1} + O(\Delta R^3) .$$

Expressing $\frac{\partial^2 V}{\partial R^2}$ in term of the vorticity ω we obtain:

$$(E^2 \cdot \omega)_{I=2} + (E^2 \cdot \omega)_{I=1} = \frac{2}{\Delta R}(V_{I=2} - V_{I=1}) \quad . \tag{12}$$

- at the external boundary, two kinds of boundary conditions can be used:

* The flow is supposed to be irrotational at infinity

$$\omega = 0 \quad . \tag{13}$$

This condition is used for Reynolds number up to 1000.

* We can also use an open boundary condition which is equivalent to a transport without viscosity of the vorticity.

$$E^2 \frac{\partial \omega}{\partial t} + \frac{\partial (U\omega)}{\partial r} + \frac{\partial (V\omega)}{\partial \Theta} = 0 \quad . \tag{14}$$

III.2) The velocity component equations are solved by an A.D.I method derived from the previous one. Both are discretized with the following finite differences schemes:

-For first order derivative:

$$(\frac{\partial S}{\partial R})_{I,J} = \frac{1}{2\Delta R}(S_{I+1,J} - S_{I-1,J}) + O(\Delta R^2)$$

-For the second order derivatives, the two different numerical schemes adopted are:

a. For the second order accurate scheme, which will be called method "a", the second order derivatives are discretized by second order accurate finite differences for the first and the second step of the A.D.I algorithm. The first step corresponds to the tangential direction and the second to the radial one.

b. For the combined second and fourth order accurate scheme, which will be called method "b", the second order derivatives are discretized by second order finite differences for the first step of the A.D.I algorithm and by fourth order accurate finite differences for the second one.

a. $(\frac{\partial^2 S}{\partial R^2})_{I,J} = \frac{1}{\Delta R^2}(S_{I+1,J} - 2S_{I,J} + S_{I-1,J}) + O(\Delta R^2)$

$(\frac{\partial^2 S}{\partial \Theta^2})_{I,J} = \frac{1}{\Delta \Theta^2}(S_{I,J+1} - 2S_{I,J} + S_{I,J-1}) + O(\Delta \Theta^2)$

b. $(\frac{\partial^2 S}{\partial R^2})_{I,J} = \frac{1}{10\Delta R^2}(-S_{I+2,J} - S_{I-2,J} + 14S_{I+1,J} + 14S_{I-1,J} - 26S_{I,J}) + O(\Delta R^4)$

for the second step of the A.D.I algorithm

and

$(\frac{\partial^2 S}{\partial R^2})_{I,J} = \frac{1}{\Delta R^2}(S_{I+1,J} - 2S_{I,J} + S_{I-1,J}) + O(\Delta R^2)$

$(\frac{\partial^2 S}{\partial \Theta^2})_{I,J} = \frac{1}{\Delta \Theta^2}(S_{I,J+1} - 2S_{I,J} + S_{I,J-1}) + O(\Delta \Theta^2)$

for the first step of the A.D.I algorithm

The method "b" applied to the Poisson equations takes implicitely into account the continuity equation on the surface of the cylinder.

At the surface of the cylinder, we can write

for the radial velocity:

$$\frac{\partial U}{\partial R} = 0 \quad . \tag{15}$$

and for the tangential velocity:

$$\frac{\partial V}{\partial R} = E^2 \cdot \omega \quad . \tag{16}$$

The two steps of the resolution of Poisson equation for the radial component are:

$$(\delta_{\Theta\Theta} - \frac{2}{\Delta\Theta})U^{(\nu+\frac{1}{2})} = -(\delta_{\Theta\Theta} + 2\delta_{RR} + \frac{2}{\Delta\Theta})U^{(\nu)} + 2\varphi_u \quad , \tag{17.a}$$

$$(\delta_{RR} - \frac{2}{\Delta\Theta})U^{(\nu+1)} = \delta_R R U^{(\nu)} - \frac{2}{\Delta\Theta}U^{(\nu+\frac{1}{2})} \quad . \tag{17.b}$$

with $\varphi_u = -E^2 \frac{\partial\omega}{\partial\Theta}$.

The same scheme is used for the tangential component and:

with $\varphi_v = \frac{\partial E^2\omega}{\partial\Theta}$.

In order to close the linear system of equations, the following boundary conditions are considered:

Method "a"

On the surface of the cylinder, we use the equation (8).

On the external boundary we use the condition of irrotational flow. It must be noticed that the use of Neumann's conditions improves the results for the divergence free condition.

Method "b"

At the surface of the cylinder, we impose explicitely the no-slip condition and use implicitely the conditions described by the equations (15) and (16) to ensure the continuity equation.

At the external boundary, we use the condition of irrotational fluid.

IV - RESULTS

The numerical calculations have been made on a vector computer VP 200. The evolution with time of the flow structure and of the distribution of vorticity on the cylinder surface are reported for Reynolds number equal to 1000 and 3000 . A comparison with results obtained in [6],[7] shows a good similarity for the structure of the flow but a little delay on the separation development can be seen.

It appears that the divergence free condition is not satisfied correctly for points closed to the cylinder. The method "b" permits to calculate the components of the velocity with more accuracy. The results obtained with method "b" in comparison with method "a" show that the continuity equation is better fulfilled and the separation is better calculated.

IV.1 - Re=1000.

To ensure the convergence of the algorithm, we must use more nodes for the method "a" than the "b" one. A grid system of 81*121 points is adopted for the fully second order accurate scheme with time step equal to 0.02 and the C.P.U time is equal to 0.06 s for each time step. A grid system of 31*61 nodes is chosen for the combined scheme with time step equal to 0.05 and the C.P.U time is equal to 0.02 s for each time step.

The distributions of the vorticity on the cylinder surface are shown in figure 1. The evolution with time of the flow structure is shown in figure 2 for the two methods and we can see for t=3 the appearence of a pair of secondary vortices and this phenomena appears only for Reynolds number greater than 1000.

IV.2 - Re=3000.

For Re=3000., a grid sytem of 161*241 nodes for the second order accurate scheme and 61*121 nodes for the combined one are adopted. The dimensionless time-step is taken equal to 0.01 for the two schemes. The C.P.U time for each time step is equal to 0.22 s for the second accurate scheme and 0.09 s for the combined one.

The evolution of the flow structure with time is shown in figure 3 for the two numerical schemes. The distribution of the vorticity at the cylinder surface given in figure 4 shows that the time of appearence of secondary vortex is approximatively t=3.5 and it confirms the existence of a pair of secondary vortices.

V - CONCLUSION

The comparison between the methods "a" and "b" for Re=1000 and Re=3000 shows that the continuity equation is better fulfilled with the combined second and fourth order accurate scheme. The velocities are calculated with more accuracy and with less points than for the second order accurate scheme.

The perspective of work in this way will consist in the resolution of unsteady three-dimensional Navier-Stokes equations for external flow with the combined finite difference scheme. The problems of flow around finite length body, for which the three-dimensional effects cannot be neglected can be studied with this formulation. A recent study [8] concerning the case of quasi-confined configuration shows that it is possible to study this kind of problems with this formulation.

REFERENCES

[1]. H. Fasel, Investigation of boundary layers by a finite difference model of Navier-Stokes equations. J. Fluid. Mech. 78,353-383(1976).

[2]. S. C. R. Dennis, D. B. Ingham and R. N. Cook, Finite difference methods for calculating steady incompressible flows in three dimensions. J. Comput. Phys.33,325-339(1979).

[3]. J. B. Gatski, C. E. Grosch and M. E. Rose, A numerical study of the two-dimensional Navier-Stokes equations in vorticity-velocity variables. J. Comput. Phys.48 1-22(1982).

[4]. P. Orlandi,Vorticity-velocity formulation for high Re flows. Computers and Fluids. Vol.15, No.2, pp137-149(1987).

[5]. K. Aziz and J. D. Hellums, Numerical solution of three-dimensional equations of motion for laminar convection, The Physics of Fluid, Vol 10, Nb 2.feb 67.

[6]. Ta Phuoc Loc, Numerical analysis of unsteady secondary vorticies generated by impulsively started circular cylinder, J.Fluid. Mech, Vol.100, part 1, pp 111-128.

[7]. Ta Phuoc Loc and R.Bouard, Numerical solution of the early stage of the unsteady viscous flow around a circular cylinder: a comparison with experimental visualisation and measurements, J. Fluid. Mech, Vol.160, pp93-117(1985).

[8]. Ta Phuoc Loc et A.Toumi, Etude numérique de l'écoulement tridimensionnel d'un fluide visqueux incompressible en formulation vecteur vitesse-vecteur tourbillon, C.R.acad.Sc,T340,Série II,n^{0}1.1987.

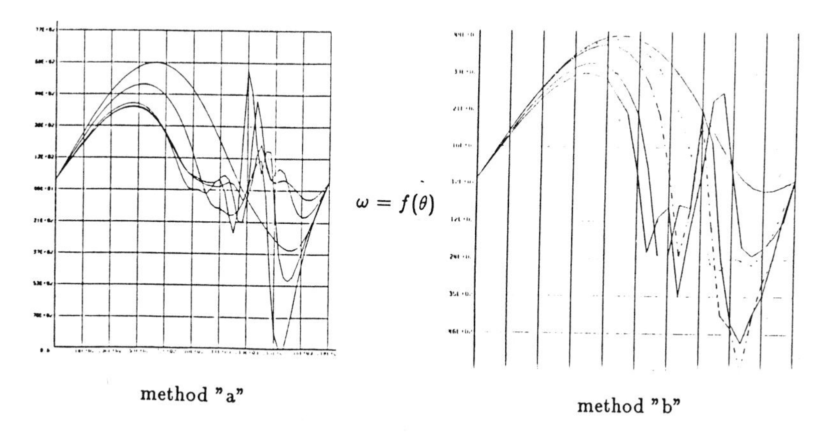

fig 1- Distribution of the vorticity on the cylinder for Re=1000.

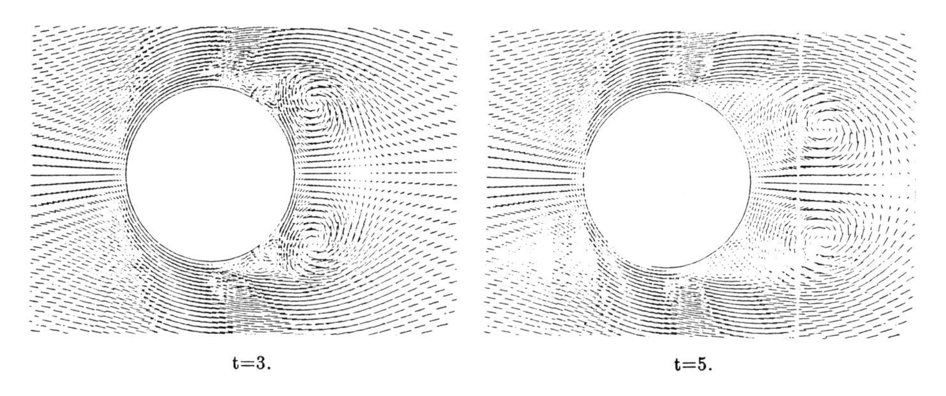

fig 2.a- Structure of the flow for Re=1000. (Method a)

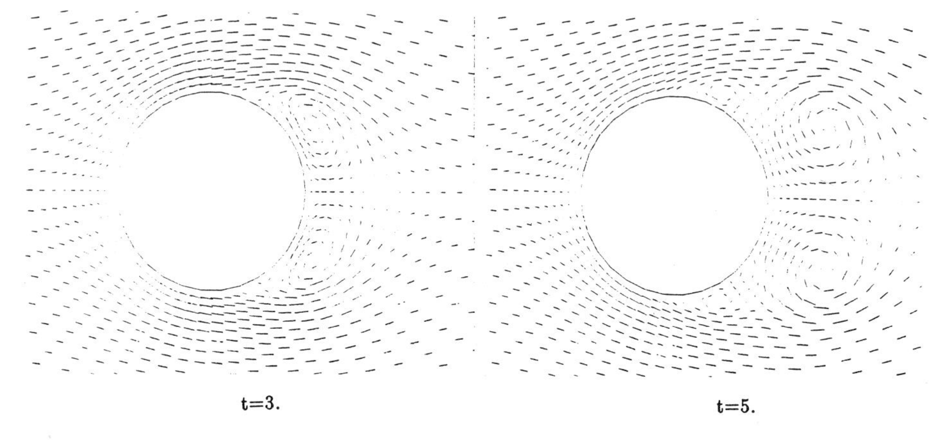

fig 2.b- Structure of the flow for Re=1000. (Method b)

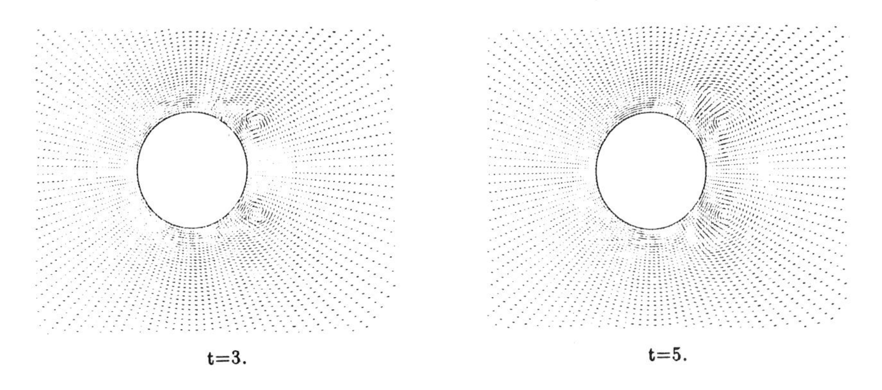

fig 3.a- Structure of the flow for Re=3000. (Method a)

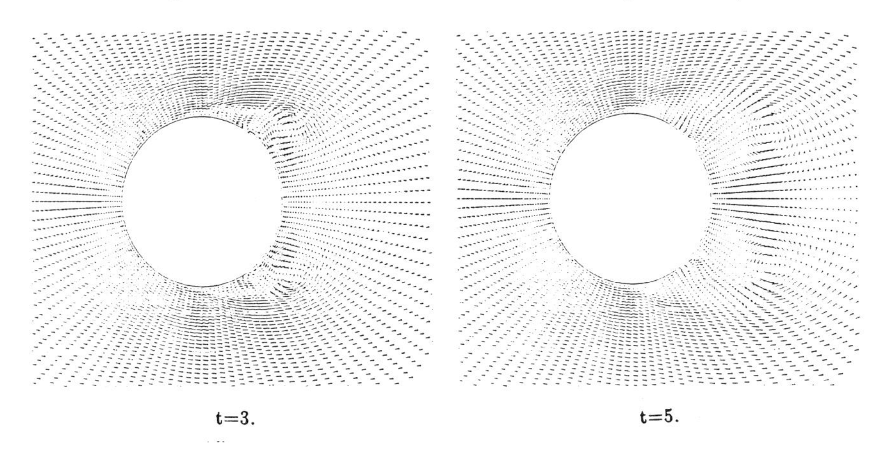

fig 3.b- Structure of the flow for Re=3000. (Method b)

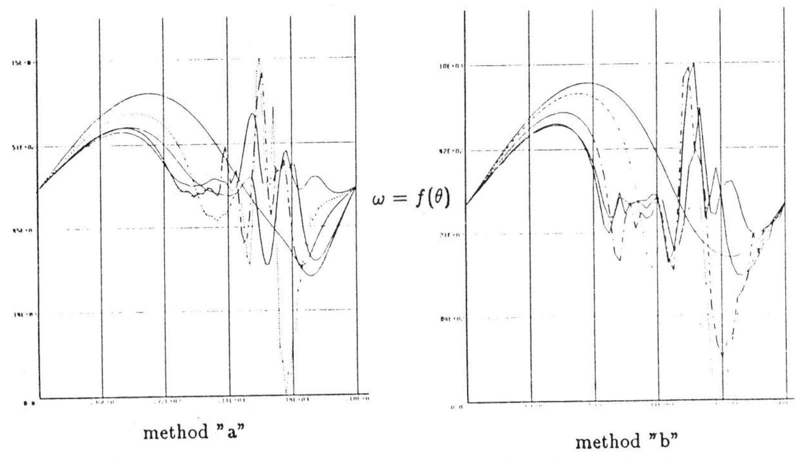

fig 4 - Distribution of the vorticity on the cylinder for Re=3000.

A PSEUDOSPECTRAL MULTI-DOMAIN METHOD FOR THE NAVIER-STOKES EQUATIONS WITH APPLICATION TO DOUBLE-DIFFUSIVE CONVECTION

by

J.M. LACROIX , R. PEYRET and J.P. PULICANI
Laboratoire de Mathématiques
Université de Nice
Parc Valrose
F-06034 NICE cedex

SUMMARY

A Fourier-Chebyshev multi-domain method based on an extensive use of the influence matrix technique is presented and applied to the calculation of the flow induced by the presence of a heat source in a fluid stratified by salinity.

INTRODUCTION

It is known that Chebyshev-spectral methods are well adapted to the computation of boundary layer-type flows. On the other hand, the accurate representation of solutions exhibiting large gradients in regions far from the boundaries of the computational domain necessitates a very large number of Chebyshev polynomials [1]. This difficulty can be removed by means of various techniques : adaptive change of variable [2], multi-domain [3], spectral element [4].

The present paper is devoted to the presentation and application of a Fourier-Chebyshev multi-domain method for solving the Navier-Stokes equations in vorticity-stream function variables. The method is based on an extensive use of the influence matrix technique for handling the boundary conditions as well as the patching conditions (continuity of functions and normal derivatives) at the interfaces of the subdomains. The method presents the advantages to be direct and to lead to the solution of Helmholtz equations with only Dirichlet boundary conditions.

The method is applied to the calculation of the flow induced by the presence of a heat source in a fluid stratified by salinity.

PHYSICAL PROBLEM AND MODELISATION

We consider a fluid initially at rest within an infinite horizontal strip (Fig.1) of height H $(-H/2 \leq y \leq H/2)$. This fluid is at uniform temperature $T=T_0$ and is stably stratified by a constant negative salinity gradient $dS/dy = (S_0-S_1) / H$. Then, the fluid is heated in its lower part by an infinite sequence of heat sources of small extent ℓ spaced one from the other by a constant distance L. With such a configuration, the solution is periodical in the horizontal direction with a period equal to L and it will be computed in the interval $-L/2 \leq x \leq L/2$.

The governing equations are the Navier-Stokes equations within the Boussinesq approximation (dimensionless dependent variables : vorticity ω, stream function ψ, temperature θ) and a transport-diffusion equation for the saline concentration C. These equations are :

$$\theta_t + \mathbf{V}.\nabla\theta = \Delta\theta \quad (1)$$

$$C_t + \mathbf{V}.\nabla C = \tau\,\Delta C \quad (2)$$

$$\omega_t + \mathbf{V}.\nabla\omega = Pr\,\Delta\omega + a\,Pr\,(\,Rt\,\theta_X - Rs\,C_X\,) \quad (3)$$

$$\Delta\psi - \omega = 0\,. \quad (4)$$

They have to be solved in the domain $0 \le X \le 2\pi$, $-1 < Y < 1$, where $X = 2\pi x / L + \pi$, $Y = 2y / H$, the caracteristic time being $H^2/4\kappa_T$; $\theta = (T - T_0) / \delta T$ with $\delta T = H\,q_0 / k_T$, q_0 being the maximum value of the heat flux q(x) prescribed on the lower wall $y = -H/2$, and k_T is the thermal conductivity, $C = (S - S_0) / \delta S$ with $\delta S = (S_1 - S_0)$, $\mathbf{V} = (u,v) = (-\psi_Y\;, a\,\psi_X), \nabla = (a\,\partial_X, \partial_Y)$, $\Delta = a^2\partial_{XX} + \partial_{YY}$. Where $a = \pi / A$, $A = L / H$. The other dimensionless parameters are :

$$Rt = g\alpha H^3\delta T / 8\kappa_T\nu\,,\quad Rs = g\beta H^3\delta S / 8\kappa_T\nu\,,\quad Pr = \nu / \kappa_T\,,\quad \tau = \kappa_S / \kappa_T$$

where g is the gravitational acceleration ; α and β are the thermal and saline expansion coefficients, respectively ; κ_T and κ_S are the thermal and saline diffusivities, respectively ; ν is the kinematic viscosity.

The initial conditions ($t \le 0$) are

$$\theta = 0,\; C = 0.5\,(1-Y),\; u = v = \omega = 0 \quad (5)$$

The boundary conditions ($t > 0$) are

$$\theta_Y = \varphi(X) = -\exp[\,-\lambda\,(X-\pi)^2\,] \text{ on } Y = -1,\; \theta_Y = 0 \text{ on } Y = 1 \quad (6)$$

$$C = 1 \text{ on } Y = -1\;;\; C = 0 \text{ on } Y = 1 \quad (7)$$

$$\psi = \psi_Y = 0 \text{ on } Y = \pm 1 \quad (8)$$

where λ characterizes the width of the heat source ($\ell / L \ll 1$ and $\lambda \gg 1$).

Due to the heating, a vertical plume is created which rises inside the stratified fluid up to some level characterized by the ratio between heating and stratification. The upward motion of the fluid has the effect to bring up salted fluid inducing, at the top of the plume, a thin layer in which the salinity exhibits a large variation mainly in the vertical direction. The accurate representation of this thin layer moving upwards, necessitates the use of a large number of Chebyshev polynomials if a one-domain pseudospectral method is used as in [6]. One of the purposes of the present study is to show this number can be reduced by using a multi-domain approach.

TIME-DISCRETIZATION

The time-discretization scheme is semi-implicit : 2nd-order Backward Euler scheme for the diffusive terms with an Adams-Bashforth-type evaluation of the convective term, [7] i.e. :

$$(\,3\theta^{n+1} - 4\theta^n + \theta^{n-1}) / 2\delta t + 2(\mathbf{V}.\nabla\theta)^n - (\mathbf{V}.\nabla\theta)^{n-1} = \Delta\theta^{n+1} \quad (9)$$

$$(\,3C^{n+1} - 4C^n + C^{n-1}) / 2\delta t + 2(\mathbf{V}.\nabla C)^n - (\mathbf{V}.\nabla C)^{n-1} = \tau\,\Delta C^{n+1} \quad (10)$$

$$(\,3\omega^{n+1} - 4\omega^n + \omega^{n-1}) / 2\delta t + 2(\mathbf{V}.\nabla\omega)^n - (\mathbf{V}.\nabla\omega)^{n-1} =$$
$$Pr\,\Delta\omega^{n+1} + a\,Pr\,(\,Rt\,\theta_X^{\,n+1} - Rs\,C_X^{\,n+1}\,) \quad (11)$$

$$\Delta\psi^{n+1} - \omega^{n+1} = 0 \quad (12)$$

where n refers to the time $t_n = n\,\delta t$.

Hence, the temperature and the salinity are independently computed as solution of Helmholtz equations. Then, their values are used in the buoyant term in the vorticity equation. The vorticity and the stream function are solution of a Stokes-type problem.

SPACE-APPROXIMATION

First, the dependent variables are expanded in truncated Fourier series in the X-direction. The Fourier coefficients at time $(n+1)\delta t$ satisfy in $-1< Y <1$ to (we omit the superscrit n+1) :

$$\theta_k'' - \varepsilon\,\theta_k = F_k \quad , \quad \theta_k'(-1) = \varphi_k \quad , \quad \theta_k'(1) = 0 \tag{13}$$

$$C_k'' - \mu\,C_k = G_k \quad , \quad C_k(-1) = 1 \quad , \quad C_k(1) = 0 \tag{14}$$

$$\omega_k'' - \gamma\,\omega_k = H_k \quad , \quad \psi_k'' - \chi\,\psi_k - \omega_k = 0 \tag{15}$$

$$\psi_k(-1) = \psi_k'(-1) = 0\,, \quad \psi_k(1) = \psi_k'(1) = 0 \tag{16}$$

where k refers to the k-th Fourier mode ($-K+1 \le k \le K$) and $\chi = a^2\,k^2$, $\varepsilon = \chi + 3\,/\,(2\,\delta t\,)$, $\mu = \chi + 3\,/\,(2\,\delta t\,\tau\,)$, $\gamma = \chi + 3\,/\,(2\,\delta t\,Pr)$.

The above problems are solved by a Tau-Chebyshev method using a multi-domain approach. The terms F_k, G_k and H_k are evaluated through the pseudo-spectral technique.

MULTI-DOMAIN METHOD FOR STOKES-TYPE PROBLEM

We describe the method for solving the Stokes-type problem (15)-(16). The same method with obvious modifications applies also to the Helmholtz problems satisfied by the temperature and salinity. More details are given in [3].

By omitting the subscript k, the problem to be solved is of the general form :

$$\omega'' - \gamma\,\omega = H, \qquad \psi'' - \chi\,\psi - \omega = 0 \qquad \text{in } \Omega:\ -1 < Y < 1 \tag{17}$$

$$\psi(-1) = g_-, \qquad \psi'(-1) = h_-, \qquad \psi(1) = g_+, \qquad \psi'(1) = h_+ \tag{18}$$

The domain Ω is decomposed into three subdomains $\Omega_j :]\,Y_{j-1}\,;\,Y_j\,[$, $j = 1,2,3$; $Y_0=-1$, $Y_3=1$. At the interfaces $Y = Y_1$ and $Y = Y_2$, we prescribe continuity of the functions ω, ψ and of their derivatives ω', ψ'. That gives 4 conditions to be enforced at each interface which, added to the 4 boundary conditions at $Y=\pm1$, give a total of 12 conditions. Hence, the solution (ω_j,ψ_j) in each subdomain Ω_j is decomposed according to :

$$\begin{Bmatrix} \omega_j \\ \psi_j \end{Bmatrix} = \begin{Bmatrix} \overline{\omega}_j \\ \overline{\psi}_j \end{Bmatrix} + \sum_{i=1}^{4} \lambda_{ij} \begin{Bmatrix} \omega_{ij} \\ \psi_{ij} \end{Bmatrix} \qquad j = 1,2,3$$

where ($\overline{\omega}_j$, $\overline{\psi}_j$), $j = 1,2,3$ are solutions of :

$$\begin{cases} \overline{\omega}_j'' - \gamma\,\overline{\omega}_j = H \\ \overline{\omega}_j(Y_{j-1}) = \overline{\omega}_j(Y_j) = 0 \end{cases} \qquad \begin{cases} \overline{\psi}_j'' - \chi\,\overline{\psi}_j = \overline{\omega}_j \\ \overline{\psi}_j\,(Y_{j-1}) = \overline{\psi}_j(Y_j) = 0 \end{cases}$$

and (ω_{ij}, ψ_{ij}), $i = 1,\ldots,4$, $j = 1,2,3$ are solutions of :

if j = 1,2,3 and i = 1,2

$$\begin{cases} \omega_{ij}'' - \gamma\, \omega_{ij} = 0 \\ \omega_{ij}(Y_{j-1}) = -i+2 \;\; ; \;\; \omega_{ij}(Y_j) = i-1 \end{cases} \qquad \begin{cases} \psi_{ij}'' - \chi\, \psi_{ij} = \omega_{ij} \\ \psi_{ij}(Y_{j-1}) = \psi_{ij}(Y_j) = 0 \end{cases}$$

If j = 1,2,3 and i = 3,4

$$\begin{cases} \omega_{ij}'' - \gamma\, \omega_{ij} = 0 \\ \omega_{ij}(Y_{j-1}) = \omega_{ij}(Y_j) = 0 \end{cases} \qquad \begin{cases} \psi_{ij}'' - \chi\, \psi_{ij} = \omega_{ij} \\ \psi_{ij}(Y_{j-1}) = -i+4 \;\; ; \;\; \psi_{ij}(Y_j) = i-3 \, . \end{cases}$$

Therefore, the part $(\overline{\omega}_j, \overline{\psi}_j)$ satisfies the equations of the problem while the linear combinaison of the elementary solutions (ω_{ij}, ψ_{ij}) allows us to satisfy the boundary conditions and the patching conditions at interfaces. By prescribing these conditions, we obtain an algebraic system determining the 12 parameters λ_{ij}, i = 1,...,4, j = 1,2,3. As a matter of fact, the conditions $\psi(\pm 1) = g_{\pm}$ simply give $\lambda_{31} = g_-$ and $\lambda_{43} = g_+$. Morever, the continuity of ω and ψ at the interfaces yields 4 equalities between some of the λ_{ij}. Hence, the order of the system is reduced to 6. The elementary solutions are time-independent : they can be computed once for all in a preprocessing stage and stored (note that $\omega_{3j} = \omega_{4j} = 0$, j = 1,2,3). It is the same for the matrix ("influence matrix") of the system determining the λ_{ij} 's : it is calculated and inverted once for all and stored.

Hence, at each time-cycle and for each Fourier mode, the computational effort reduces to the solution of uncoupled Helmholtz equations with homogeneous Dirichlet boundary conditions for $\overline{\omega}_j$ and $\overline{\psi}_j$, j = 1,2,3. The various Helmholtz problems are solved by a Tau-method using Chebyshev polynomial expansions and leading to uncoupled quasi-tridiagonal systems for even and odd Chebyshev modes, respectively [5]. These systems are solved by a factorization procedure.

NUMERICAL RESULTS FOR THE PHYSICAL PROBLEM

The application concerns the heating of a salted water, so that Pr = 7, $\tau = 0.01$. The thermal Rayleigh number is Rt = 2 10^6. The dimensionless distance between two consecutive sources is A = 1 and the extent of each source is defined by $\lambda = 100 / 3$. Due to the small value of τ the layer in which the salinity exhibits a large variation is rather thin.

Firstly, we consider the case of a strong stratification (Rs = 5 10^5) such that the plume remains in the lower part of the domain. Fig.2 compares results obtained, at t = 0.2 , in the single-domain case with 81 polynomials in the vertical direction and in a three-domain case using a total number of 49 polynomials : 31 polynomials in the region Ω_1 : $-1 < Y < -0.5$, 10 in Ω_2 : $-0.5 < Y < 0.25$ and 8 in Ω_3 : $0.25 < Y < 1$, both cases using 100 Fourier modes (i.e. 200 points in $0 \le X \le 2\pi$). In both cases the time-step is $\delta t = 5\ 10^{-5}$. Looking at the salinity profiles at $X = \pi$ and the isosalinity contours, we observe that the one-domain method needs much more polynomials to produce a smooth salinity solution. For these calculations, the computing time per time-cycle is 0.20 sec. for the single-domain solution and 0.122 sec. for the three-domain solution on the CRAY 1S computer. In a more general way, for a same number of polynomials, the three-domain method is no much more time-consuming that the one-domain method.

In the above computations, we have chosen the location of the interfaces and the repartition of the number of polynomials in each subdomain in the most favourable manner. In order to make more evident the versatility of the method, we consider now the situation in which the upward motion of the plume is more important, so that the large gradient layer crosses the interface between two subdomains and reaches the middle-part of the global domain. This is obtained by decreasing the magnitude of the stratification ($Rs = 1.25\ 10^5$). Here, we take 150 Fourier modes for each method and in the three-domain case, we simply use a dichotomous partition, i.e. : $Y_1 = -0.5$, $Y_2 = 0$. Fig.3 compares results obtained at $t = 0.2$ with the different techniques. In the single domain case with 81 Chebyshev modes the salinity profiles at $X = \pi$ and $X = 1.2\ \pi$ exhibit oscillations while the solution in the three-domain case using 72 Chebyshev modes (28 in the region Ω_1, 36 in Ω_2 and 8 in Ω_3) are much more smooth. Further by taking the same total number of polynomials (28 in Ω_1 , 45 in Ω_2 and 8 in Ω_3) than in the one-domain case, we obtain a perfectly smooth solution. For the latter case, Figs.4 and 5 show that any degradation occurs in the solution where the plume moves from Ω_1 to Ω_2.

In conclusion, the multi-domain method can be efficiently used even if the large gradient zone is moving through the interfaces or if its location is not known very precisely, provided the resolution in each subdomain is sufficiently high. Globally, it is less expensive than the one-domain method and its efficiency could yet be increased on a parallel computer.

This work was partly supported by a contract DRET No. 85 / 021. The computations have been supported by " Centre de Calcul Vectoriel pour la Recherche ".

REFERENCES

[1] BASDEVANT, C., DEVILLE, M., HALDENWANG, P., LACROIX, J.M., OUAZZANI, J., PEYRET, R., ORLANDI, P., and PATERA, A.T., Spectral and finite difference solutions of Burgers equation, Computers and fluids, vol. 14, p. 23-41, 1986.

[2] GUILLARD, H., PEYRET, R., On the use of spectral methods for the numerical solution of stiff problems, Comp. Meth. Appl. Mech. Eng. (to be published).

[3] PULICANI, J.P., A spectral multi-domain method for the solution of 1D-Helmholtz and Stokes-type equations, Computers and fluids (to be published).

[4] PATERA, A. T., A Spectral element method for Fluid Dynamics : laminar flow in channel expansion, J. Comp. Phys., vol. 54, p. 466-488, 1984.

[5] GOTTLIEB, D., ORSZAG, S.A., Numerical Analysis of Spectral Methods : Theory and Application, CBMS Regional Conference Series in Appl. Math., SIAM, 1977.

[6] DEMAY, Y., LACROIX, J.M., PEYRET, R., and VANEL, J.M., Numerical experiments on stratified fluids subject to heating, Third International Symposium on Density-Stratified Flows, Pasadena, February 3-5, 1987.

[7] VANEL, J.M., PEYRET, R., and BONTOUX, P., A pseudo-spectral solution of vorticity-stream function equations using the influence matrix technique, in "Numerical methods for fluid dynamics II", (K.W. Morton, M.J. Baines, Eds.), p. 463-475, Clarendon Press, Oxford, 1986.

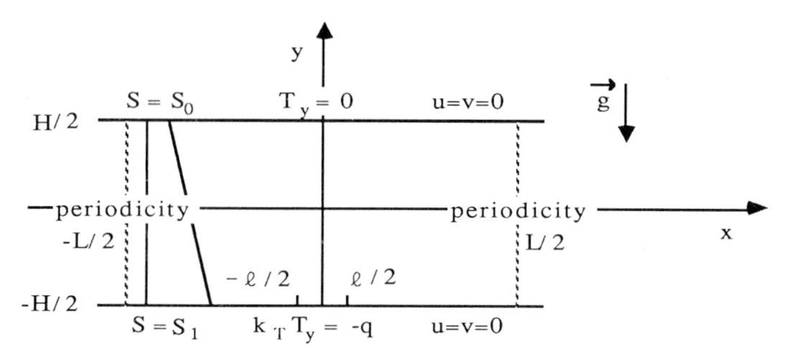

Fig.1 : Geometrical configuration

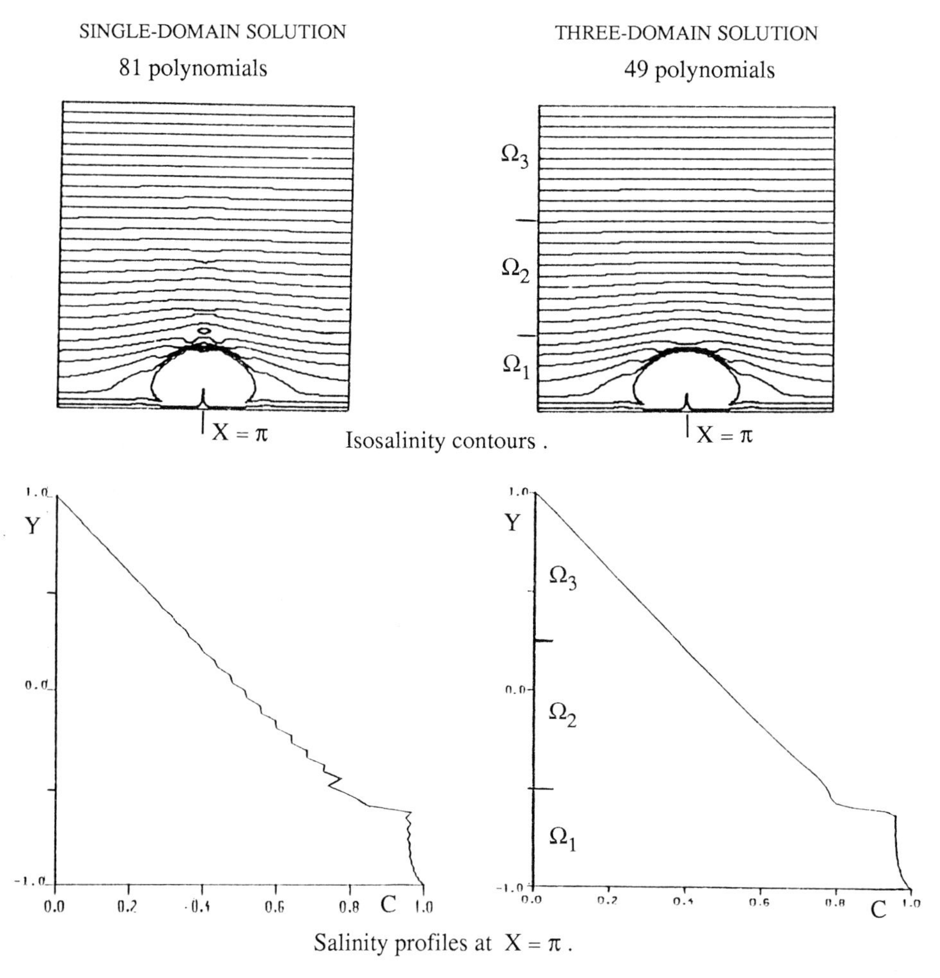

Isosalinity contours .

Salinity profiles at $X = \pi$.

Fig.2 : $Pr = 7$, $\tau = 10^{-2}$, $Rt = 2\ 10^6$, $Rs = 5\ 10^5$.

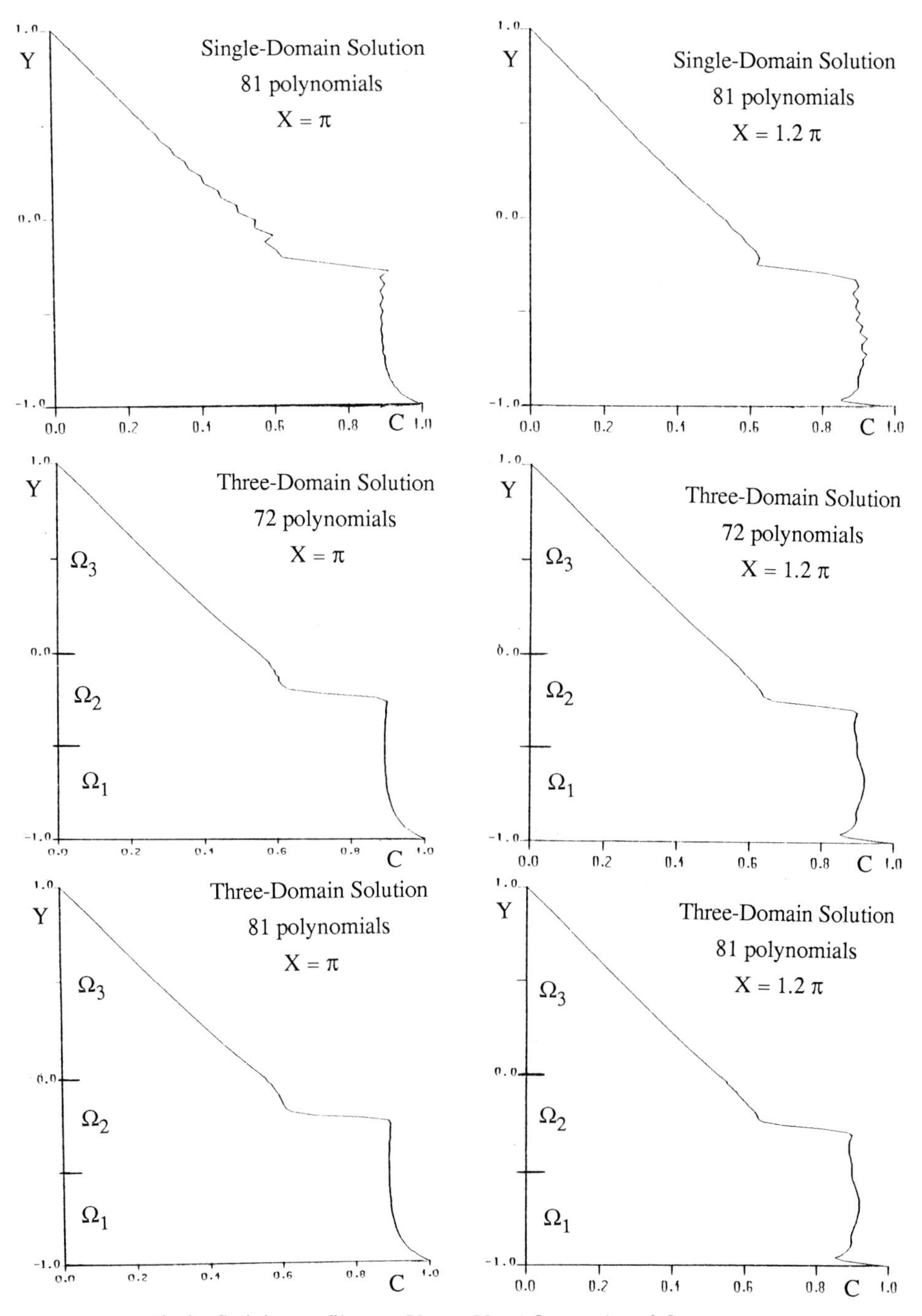

Fig.3 : Salinity profiles at $X = \pi$, $X = 1.2\,\pi$ and $t = 0.2$.

$Pr = 7$, $\tau = 10^{-2}$, $Rt = 2\ 10^{6}$, $Rs = 1.25\ 10^{5}$.

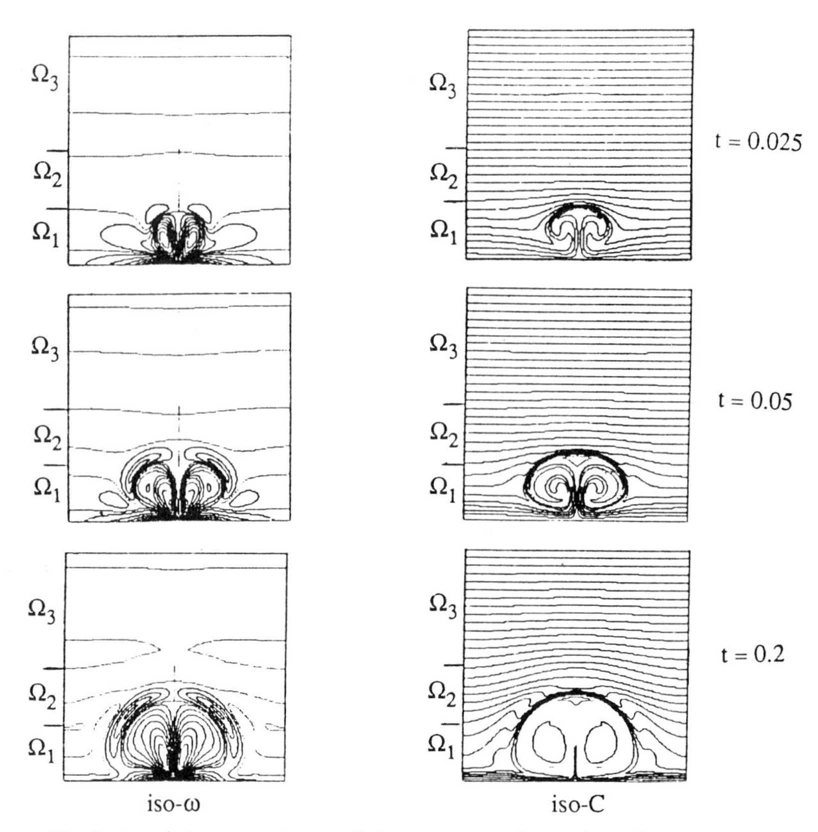

Fig.4 : isovorticity and isosalinity contours for various times.

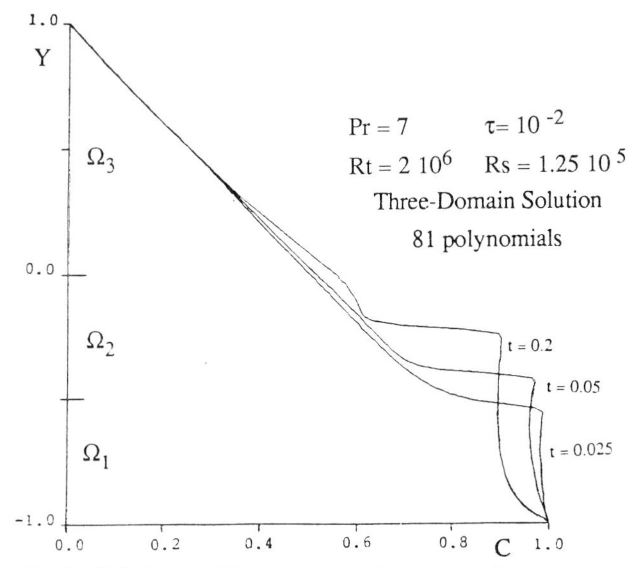

Fig.5 : Salinity profiles at $X = \pi$ for various times.

A SEMI-IMPLICIT SCHEME FOR THE NAVIER-STOKES EQUATIONS

Bernard Loyd
Earll M. Murman
Saul S. Abarbanel

Computational Fluid Dynamics Laboratory
Massachusetts Institute of Technology
Cambridge, Massachusetts 02139, USA

Abstract

This paper presents a novel *Semi-Implicit Navier Stokes Solver (SINSS)*. *SINSS* combines the advantages of implicit with those of explicit schemes: temporal integration is implicit in the direction normal to a body and explicit in the direction(s) tangential to it. Numerical stiffness due to disparate physical scales in the normal direction is eliminated, and the stability of the algorithm depends only on relatively coarse streamwise grid spacing – not on the typically fine normal spacing. Approximate factorization is unnecessary and only one matrix inversion per stage is required.

The semi-implicit solver is applied to a finite volume formulation of the $2-D$ thin layer Navier-Stokes equations. Efficiency of the algorithm is studied by comparison of convergence histories of the semi-implicit algorithm with those of a multigrid explicit scheme and a fully implicit approximately factorized scheme. The effect of residual smoothing is also considered. Computations show that the explicit, semi-implicit, and fully implicit schemes are of comparable efficiency for inviscid calculations. *SINSS* is superior in high Reynolds number flows, where multigrid loses effectiveness and the implicit scheme appears to have convergence problems.

1 Introduction

Fine spatial resolution for Navier-Stokes simulations is often necessary only in the direction normal to a body. At the differential equation level this implies that only the boundary layer like viscous terms need to be retained, leading to the so called thin layer Navier-Stokes equations. At the discrete level this implies mesh cells will have a much smaller dimension in the normal direction compared to the streamwise direction. Thus, the stability restriction for explicit schemes in body fitted grid systems is usually dominated by the small normal spacing and results in numerical "stiffness" in the equations.

The *SINSS* solver eliminates the stability restriction due to the normal spacing by solving the flow equations implicitly in the normal direction. However, the solver is explicit in the tangential (flow) direction, thereby avoiding factorization (AF) schemes, *CFL* limitations associated with the AF error, and a second or second and third block-tridiagonal inversions in two or three dimensions. This scheme was first presented by [Loyd et al. 86]. Details may be found in [Loyd et al. 87].

This paper presents the characteristics and efficiency of a semi-implicit algorithm as applied to a popular multi-stage scheme. First, we present the Navier-Stokes equations, and a simple method of implementing the thin layer approximation that preserves the conceptual simplicity and conservative property of the finite volume approach. The semi-implicit algorithm is then derived by considering time linearization of only the cross flow flux terms. The *CFL* limit on the time step for explicit schemes reduces to a one-dimensional *CFL* restriction based only upon the streamwise terms [Loyd et al. 87].

To evaluate the relative efficiency of the semi-implicit we compare convergence histories obtained with it to those obtained with an explicit and a fully implicit solver [Beam & Warming 76]. Multigrid (MG) and residual smoothing (RS) are applied in the explicit solver to accelerate convergence, and RS is also applied in the streamwise (explicit) direction in the semi-implicit scheme. The paper is concluded with results for inviscid and viscous flow cases.

2 2–D Navier-Stokes Equations

The two-dimensional Navier–Stokes equations integrated in Cartesian coordinates over a control surface Ω with boundary $\partial\Omega$ are:

$$\frac{\partial}{\partial t}\iint_{\Omega} \mathbf{W}\, dS + \oint_{\partial\Omega} (\mathbf{F}\, dy - \mathbf{G}\, dx) = 0. \tag{1}$$

$\mathbf{W} = (\rho \;\rho u \;\rho v \;E)^T$ is the vector of state variables, where ρ, u, v are density, x and y components of velocity, and E is total energy. The flux vectors $\mathbf{F}$ and $\mathbf{G}$ are

$$\mathbf{F} = \begin{pmatrix} \rho u \\ \rho u^2 + P + \tau_{xx} \\ \rho uv + \tau_{xy} \\ \rho uH + u\tau_{xx} + v\tau_{xy} - q_x \end{pmatrix}, \qquad \mathbf{G} = \begin{pmatrix} \rho v \\ \rho uv + \tau_{yx} \\ \rho v^2 + P + \tau_{yy} \\ \rho vH + u\tau_{yx} + v\tau_{yy} - q_y \end{pmatrix}. \tag{2}$$

H is the stagnation enthalpy ($H = E + P/\rho$), and the viscous stresses are, with the Stokes hypothesis:

$$\begin{aligned} \tau_{xx} &= \tfrac{2}{3}\mu\left(2\tfrac{\partial u}{\partial x} - \tfrac{\partial v}{\partial y}\right) & \tau_{yy} &= \tfrac{2}{3}\mu\left(2\tfrac{\partial v}{\partial y} - \tfrac{\partial u}{\partial x}\right) \\ \tau_{xy} &= \tau_{yx} = \mu\left(\tfrac{\partial u}{\partial y} + \tfrac{\partial v}{\partial x}\right) & & \\ q_x &= \tfrac{\mu}{Pr}\left(\tfrac{\partial H}{\partial x} - u\tfrac{\partial u}{\partial x} - v\tfrac{\partial v}{\partial x}\right) & q_y &= \tfrac{\mu}{Pr}\left(\tfrac{\partial H}{\partial y} - u\tfrac{\partial u}{\partial y} - v\tfrac{\partial v}{\partial y}\right). \end{aligned} \tag{3}$$

Pr is the Prandtl number, $Pr = \frac{\mu C_p}{k}$. The equation of state for a perfect gas closes the system:

$$P = (\gamma - 1)\left[E - \frac{1}{2}\rho(u^2 + v^2)\right]. \tag{4}$$

2.1 Thin Layer Approximation

The thin layer approximation is appropriate for high Reynolds number flow cases where shear stresses in the body normal direction are much larger than other viscous stresses. We restrict our attention to flows where this is the case and apply the thin layer approximation.

Because of the assumptions in the thin layer form of the Navier Stokes equations, the computational (ξ, η) grid must contain a family of lines that is body normal or nearly so. Since the normal direction η in general does not correspond to either the x or y coordinate direction, the derivatives in the viscous stresses must be transformed to ξ and η coordinates via the generalized transformation:

$$\frac{\partial}{\partial x} = \frac{\partial \xi}{\partial x}\frac{\partial}{\partial \xi} + \frac{\partial \eta}{\partial x}\frac{\partial}{\partial \eta} = \xi_x \frac{\partial}{\partial \xi} + \eta_x \frac{\partial}{\partial \eta} \qquad \frac{\partial}{\partial y} = \frac{\partial \xi}{\partial y}\frac{\partial}{\partial \xi} + \frac{\partial \eta}{\partial y}\frac{\partial}{\partial \eta} = \xi_y \frac{\partial}{\partial \xi} + \eta_y \frac{\partial}{\partial \eta}. \tag{5}$$

Using the assumption $\frac{\partial}{\partial \eta} >> \frac{\partial}{\partial \xi}$, all ξ derivative terms may be dropped to obtain:

$$\frac{\partial}{\partial x} = \eta_x \frac{\partial}{\partial \eta}, \qquad \frac{\partial}{\partial y} = \eta_y \frac{\partial}{\partial \eta}. \tag{6}$$

The thin layer shear stresses and heat fluxes become:

$$\begin{aligned} \tau_{xx} &= \tfrac{2}{3}\mu\left(2\eta_x\tfrac{\partial u}{\partial \eta} - \eta_y\tfrac{\partial v}{\partial \eta}\right) & \tau_{yy} &= \tfrac{2}{3}\mu\left(2\eta_y\tfrac{\partial v}{\partial \eta} - \eta_x\tfrac{\partial u}{\partial \eta}\right) \\ \tau_{xy} &= \tau_{yx} = \mu\left(\eta_y\tfrac{\partial u}{\partial \eta} + \eta_x\tfrac{\partial v}{\partial \eta}\right) & & \\ q_x &= \tfrac{\mu}{Pr}\left(\eta_x\tfrac{\partial H}{\partial \eta} - u\eta_x\tfrac{\partial u}{\partial \eta} - v\eta_x\tfrac{\partial v}{\partial \eta}\right) & q_y &= \tfrac{\mu}{Pr}\left(\eta_y\tfrac{\partial H}{\partial \eta} - u\eta_y\tfrac{\partial u}{\partial \eta} - v\eta_y\tfrac{\partial v}{\partial \eta}\right). \end{aligned} \tag{7}$$

The viscous terms need only be computed at faces 1 and 3 (Figure 1). The viscous fluxes across faces 2 and 4 are discarded since, by assumption, they are small in comparison.

3 Spatial Discretization

The governing equations must be discretized before numerical solution can be attempted. Discretization may proceed in two steps, spatial and temporal. Although the two steps are not independent, since stability of the temporal integration depends on the form of the spatial operator, it is convenient to consider them separately. In this section we discuss the spatial discretization, as well as the related topics of boundary conditions and artificial viscosity.

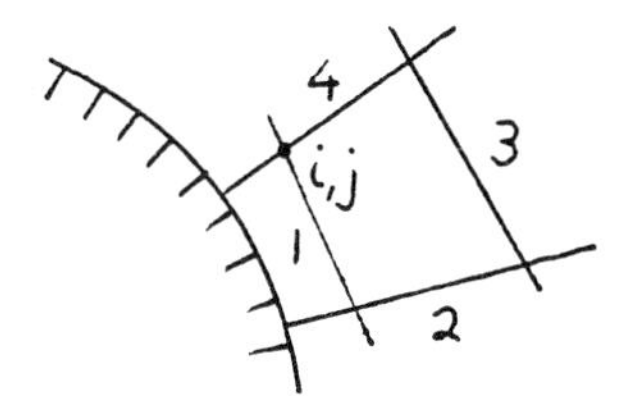

Figure 1: Body Normal Geometry

3.1 Spatial Difference Operators

The line integrals in (1) are replaced by a discrete summation of fluxes across each cell face:

$$\oint_{\partial\Omega} (\mathbf{F}\,dy - \mathbf{G}\,dx) \cong \sum_{l=1}^{4} \mathbf{F}_l \Delta y_l - \mathbf{G}_l \Delta x_l \tag{8}$$

$\mathbf{F}_l$ and $\mathbf{G}_l$ are approximated by simple averages of the flux vectors at the adjacent cells, and, at *faces* 3 and 1, a contribution from the viscous terms:

$$\begin{aligned} \mathbf{F}_2 &= .5 \times (\mathbf{F}_{i,j} + \mathbf{F}_{i+1,j}) \\ \mathbf{F}_3 &= .5 \times (\mathbf{F}_{i,j} + \mathbf{F}_{i,j+1}) + \mathbf{F}_{vis}\ . \end{aligned} \tag{9}$$

Δ's indicate differences taken in the counterclockwise direction, e.g.:

$$\begin{aligned} \Delta y_2 &= y_{i+1,j+1} - y_{i+1,j} \\ \Delta y_3 &= y_{i,j+1} - y_{i+1,j+1}\ . \end{aligned} \tag{10}$$

The viscous operator is written in finite difference form as either

$$\frac{\partial(\,)}{\partial\eta} = \frac{(\,)_{j+1} - (\,)_j}{\Delta\eta} \quad \text{or} \quad \frac{\partial(\,)}{\partial\eta} = \frac{(\,)_j - (\,)_{j-1}}{\Delta\eta}\,, \tag{11}$$

depending on whether the viscous flux at *face 3* or at *face 1* is desired. The appropriate metric η_x or η_y is obtained from the definitions:

$$\eta_x = -\frac{1}{J}\frac{\partial y}{\partial\xi} \qquad \eta_y = \frac{1}{J}\frac{\partial x}{\partial\xi}\,, \tag{12}$$

where J is the Jacobian $J = x_\xi y_\eta - x_\eta y_\xi$. The metrics defining J are calculated with centered differences.

The discretization reduces to a second order accurate centered difference approximation on a Cartesian grid. That order of accuracy is not maintained on arbitrary grids, and it is important that grids vary smoothly to limit degradation in accuracy. It is generally true that for a given order of accuracy a compact stencil such as this will result in a smaller truncation error than a less compact stencil.

Upon forming the viscous terms at each face, the finite volume line integration proceeds as before. The definitions (11) are convenient because they require only a three point stencil for the thin layer Navier Stokes terms. Because it is written in finite volume form the scheme identically conserves mass, momentum, and energy.

3.2 Artificial Viscosity

The discretization described above allows odd-even decoupling of state vector values at adjacent points. Due to aliasing errors, this decoupling may inhibit convergence of the temporal integration. Also, in inviscid regions one finds that discontinuities in the flow field may cause divergence of the algorithm. Artificial viscosity is added to the physical fluxes to damp out the non-physical odd-even oscillations and stabilize the integration in areas of discontinuos flows such as found around a shock wave. We use a pressure weighted blend of second and fourth differences that has been established as particularly effective [Jameson et al. 81]. In regions of viscous flow the artificial viscosity is turned off.

3.3 Boundary Conditions

Numerical boundary conditions must be imposed at the body and at inflow and outflow. For viscous cases we set the velocity at the body equal to zero and maintain only viscous and pressure contributions to the flux integrals (8). Riemann invarient boundary conditions are specified at the inflow, and, for inviscid cases, at the outflow. For viscous flows, all quantities except the pressure, which is set from the freestream, are extrapolated from the interior at the outflow boundary.

4 Semi-Implicit Approach

This section describes the temporal discretization of the governing equations. We begin with a synopsis of an explicit multistage approach, which also serves to introduce the nomenclature. More importantly, the semi-implicit approach presented in the following section is easily derived via consideration of the explicit integration.

Various techniques for accelerating convergence are often used with explicit schemes. To enable a fair comparison we implement residual smoothing and multigrid in the explicit scheme in Sections 4.4 and 4.5. We also apply residual smoothing in the explicit direction in *SINNS*.

4.1 Explicit Multi-Stage Integration

A popular multistage scheme for fluid dynamic calculations [Jameson et al. 81] is given as:

$$\begin{aligned}
\mathbf{W}^0 &= \mathbf{W}^n \\
\mathbf{W}^1 &= \mathbf{W}^0 - \alpha_1 \tfrac{\Delta t}{S}\left[\textstyle\sum_{l=1}^4 \mathbf{F}_l^0 \Delta y_l - \mathbf{G}_l^0 \Delta x_l - \mathbf{D}^0\right] \\
\mathbf{W}^2 &= \mathbf{W}^0 - \alpha_2 \tfrac{\Delta t}{S}\left[\textstyle\sum_{l=1}^4 \mathbf{F}_l^1 \Delta y_l - \mathbf{G}_l^1 \Delta x_l - \mathbf{D}^0\right] \\
\mathbf{W}^3 &= \mathbf{W}^0 - \alpha_3 \tfrac{\Delta t}{S}\left[\textstyle\sum_{l=1}^4 \mathbf{F}_l^2 \Delta y_l - \mathbf{G}_l^2 \Delta x_l - \mathbf{D}^0\right] \\
\mathbf{W}^4 &= \mathbf{W}^0 - \tfrac{\Delta t}{S}\left[\textstyle\sum_{l=1}^4 \mathbf{F}_l^3 \Delta y_l - \mathbf{G}_l^3 \Delta x_l - \mathbf{D}^0\right] \\
\mathbf{W}^{n+1} &= \mathbf{W}^4 \; .
\end{aligned} \tag{13}$$

Superscripts denote temporal stages. The artificial viscosity, operator $\mathbf{D}$ is frozen at the first stage to minimize computational effort. The constants $(\alpha_1, \alpha_2, \alpha_3)$ are equal $(1/4, 1/3, 1/5)$, except in the multigrid scheme. Vectors are in bold print. Matrices will be denoted by [ ].

The time step Δt that may be taken is limited by the *CFL* condition. The four stage scheme above, with

$$\Delta t = \lambda \frac{S}{|\mathbf{V} \cdot d\mathbf{L}|_{max} + C|d\mathbf{L}_{max}|} , \tag{14}$$

where $\mathbf{V} = u\vec{\imath} + v\vec{\jmath}$, C is speed of sound, and $d\mathbf{L} = \Delta x\vec{\imath} + \Delta y\vec{\jmath}$, is stable for $\lambda \leq 2\sqrt{2}$.

4.2 Semi-Implicit Integration

The explicit temporal time stepping (13) can easily be converted to a semi-implicit discretization. Consider the first stage of a multistage scheme with implicit semi-discretization of the normal component of the flux vectors:

$$\begin{aligned}
\mathbf{W}^1 - \mathbf{W}^0 = -\alpha_1 \tfrac{\Delta t}{S} \; &[(\mathbf{F}^1 \Delta y - \mathbf{G}^1 \Delta x)_1 + (\mathbf{F}^0 \Delta y - \mathbf{G}^0 \Delta x)_2 + \\
&(\mathbf{F}^1 \Delta y - \mathbf{G}^1 \Delta x)_3 + (\mathbf{F}^0 \Delta y - \mathbf{G}^0 \Delta x)_4 - \mathbf{D}^0] \; .
\end{aligned} \tag{15}$$

Each term in (15) gives the flux across one of the grid faces. Using the standard Newton type linearization for $\mathbf{F}$ and $\mathbf{G}$, we write,

$$\begin{aligned}
\mathbf{F}^1 &= \mathbf{F}^0 + \frac{\partial \mathbf{F}^0}{\partial t}\Delta t + O(\Delta t^2) & \mathbf{G}^1 &= \mathbf{G}^0 + \frac{\partial \mathbf{G}^0}{\partial t}\Delta t + O(\Delta t^2) \\
&= \mathbf{F}^0 + [A]\Delta \mathbf{W} + O(\Delta t^2) & &= \mathbf{G}^0 + [B]\Delta \mathbf{W} + O(\Delta t^2)
\end{aligned} \tag{16}$$

where $[A]$ and $[B]$ are defined as the 4×4 Jacobian matrices $[\partial \mathbf{F}^0/\partial \mathbf{W}^0]$ and $[\partial \mathbf{G}^0/\partial \mathbf{W}^0]$, respectively, and $\Delta \mathbf{W}^1 \equiv \mathbf{W}^1 - \mathbf{W}^0$. Inserting (16) into (15) and reordering gives

$$\left[[I] + \alpha_1 \frac{\Delta t}{S}\left([A]_1 \Delta y_1 + [A]_3 \Delta y_3 - [B]_1 \Delta x_1 - [B]_3 \Delta x_3\right)\right] \Delta \mathbf{W} = -\alpha_1 \frac{\Delta t}{S}\left[\sum_{l=1}^{4} \mathbf{F}_l \Delta y_l - \mathbf{G}_l \Delta x_l - \mathbf{D}\right]^0 . \tag{17}$$

This is an implicit matrix equation for $\Delta \mathbf{W}$. The *RHS* is the usual semi-discrete form of the residual, while the *LHS* differs from $[I]$ due to introduction of the terms from the time linearization of $\mathbf{F}$ and $\mathbf{G}$. Because the *LHS* contains only dependent variables at $(j-1, j, j+1)$, it is a block tridiagonal system of equations.

Subsequent steps in a multistage scheme have the same form. For a four step scheme:

$$\begin{aligned}
\mathbf{W}^0 &= \mathbf{W}^n \\
[\mathrm{LHS}]^0 \Delta \mathbf{W}^1 &= -\alpha_1 \tfrac{\Delta t}{S}\left[\textstyle\sum_{l=1}^{4} \mathbf{F}_l^0 \Delta y_l - \mathbf{G}_l^0 \Delta x_l - \mathbf{D}^0\right] \\
[\mathrm{LHS}]^1 \Delta \mathbf{W}^2 &= -\alpha_2 \tfrac{\Delta t}{S}\left[\textstyle\sum_{l=1}^{4} \mathbf{F}_l^1 \Delta y_l - \mathbf{G}_l^1 \Delta x_l - \mathbf{D}^0\right] - \Delta \mathbf{W}^1 \\
[\mathrm{LHS}]^2 \Delta \mathbf{W}^3 &= -\alpha_3 \tfrac{\Delta t}{S}\left[\textstyle\sum_{l=1}^{4} \mathbf{F}_l^2 \Delta y_l - \mathbf{G}_l^2 \Delta x_l - \mathbf{D}^0\right] - (\Delta \mathbf{W}^2 + \Delta \mathbf{W}^1) \\
[\mathrm{LHS}]^3 \Delta \mathbf{W}^4 &= - \tfrac{\Delta t}{S}\left[\textstyle\sum_{l=1}^{4} \mathbf{F}_l^3 \Delta y_l - \mathbf{G}_l^3 \Delta x_l - \mathbf{D}^0\right] - (\Delta \mathbf{W}^3 + \Delta \mathbf{W}^2 + \Delta \mathbf{W}^1) \\
\mathbf{W}^{n+1} &= \mathbf{W}^3 + \Delta \mathbf{W}^4
\end{aligned} \tag{18}$$

where $[LHS] = [I] + \alpha \frac{\Delta t}{S}\left([A]_1 \Delta y_1 + [A]_3 \Delta y_3 - [B]_1 \Delta x_1 - [B]_3 \Delta x_3\right)$ and $\Delta \mathbf{W}^S = \mathbf{W}^S - \mathbf{W}^{S-1}$. Although the *RHS* of stages three and four contain more than one vector $\Delta \mathbf{W}$, only one $\Delta \mathbf{W}$ needs to be stored. Subsequent $\Delta \mathbf{W}$'s are simply added to the stored vector to give $\sum_{i=1}^{s-1} \Delta \mathbf{W}$. The system (18) can be efficiently inverted with a block tridiagonal Gauss elimination routine.

The time step in the semi-implicit integration is limited only by the tangential flux terms, which were treated explicitly. Equation 14 reduces to:

$$\Delta t = \lambda \frac{\Delta l_t}{u_t + C} , \tag{19}$$

where l_t is the tangential spacing, and $\lambda \leq 2\sqrt{2}$ [Loyd et al. 87]. The normal spacing is no longer restrictive.

4.3 Matrix Conditioning

For grids with high aspect ratio cells ($\Delta x / \Delta y >> 1$) the matrices [LHS] become increasingly ill conditioned. Consider, for example, a rectangular mesh with $\Delta x = Const.$ and $\Delta y = \Lambda \Delta x$, where $\Lambda << 1$. Then,

$$\begin{aligned}
[\mathrm{LHS}] &= [I] - \alpha \tfrac{\Delta t}{\Lambda \Delta x^2}\left[[B]_1 \Delta x_1 + [B]_3 \Delta x_3\right] \\
&= [I] + \alpha \tfrac{\Delta t}{\Lambda \Delta x}\left[[B]_{j+1} - [B]_{j-1}\right] ,
\end{aligned} \tag{20}$$

and diagonal dominance is lost as the off diagonal terms increase with $1/\Lambda$.

We increase the diagonal dominance of $[LHS]$ by adding implicit smoothing. It is applied by adding to $[LHS]$ the undivided second difference operator:

$$-\mu_{IS}\left[\mathbf{W}_{j+1} - 2\mathbf{W}_j + \mathbf{W}_{j-1}\right] . \tag{21}$$

The implicit smoothing does not affect the steady state solution; however, large μ_{IS} may inhibit convergence to steady state.

4.4 Residual Smoothing

The Courant number limitation both for the explicit and semi-implicit schemes can be relaxed by smoothing the residuals. In effect, this increases the stencil of influence of the difference scheme and

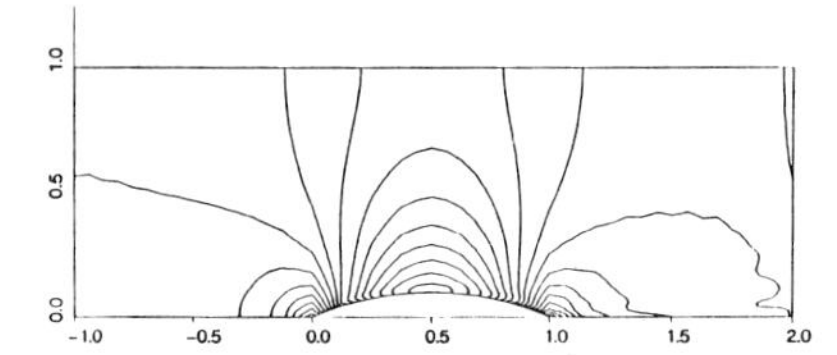

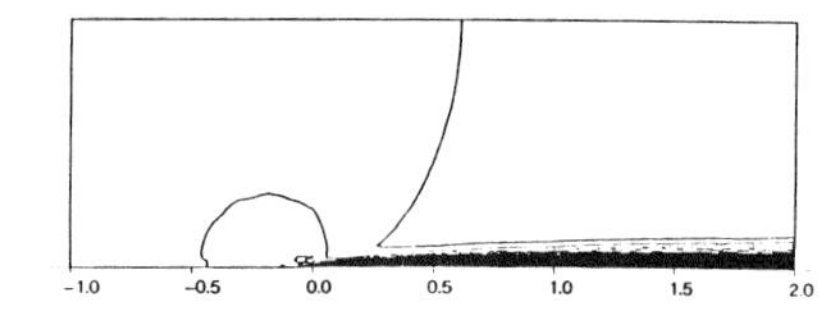

Figure 2: Mach Contours in Channel Flow: Inviscid & Viscous ($Re = 2000$)

thus increases the permissible time step. Convergence acceleration is a result of both the increase in Δt and the damping of the residuals. Residual smoothing is best applied implicitly. In two dimensions:

$$\begin{aligned} (1-\mu_{RS}\delta_{yy})\mathbf{R}' &= \mathbf{R} \\ (1-\mu_{RS}\delta_{xx})\mathbf{R}'' &= \mathbf{R}' \end{aligned} \tag{22}$$

where $\mathbf{R}$ is the vector of residuals and δ is the undivided second difference operator. In the semi-implicit scheme residual smoothing is applied only in the explicit direction, since the implicit discretization and smoothing has a similar effect in the cross stream direction. Optimal values of the smoothing coefficient and the new time step may be found by numerical experimentation.

4.5 Multi-Grid Convergence Acceleration

Multigrid can also be used to accelerate convergence. Multigrid works by accelerating the propagation of information across the grid. The conservation equations are solved on successively coarser grids to achieve, in effect, a larger difference stencil. Coarse corrections are interpolated back up to the fine grid which drives the scheme. On coarse grids, substantial 2nd and 4th difference damping is added to the fluxes to kill short wave length disturbances. The restriction operator, or forcing function, is also smoothed with a second difference operator. We use a simple *V*-type strategy to cycle from fine through coarse grids and back to fine. Details are given in [Loyd et al. 87].

5 Results

Following are results of a set of inviscid and viscous flow cases. Figure 2 gives the geometries and typical Mach number contours of the two sets of cases. The inviscid case is a $M_\infty = .5$ channel flow with a $t/c = 0.1$ circular arc bump. The viscous test case is laminar flow at $M_\infty = 0.5$ in a channel, with a lower wall beginning 1/3 of the way into the channel. Symmetry is assumed at the upper boundaries. Because both flows are subsonic no second difference smoothing is added to the spatial operator. A very small amount of fourth difference smoothing ($\nu_4 = 0.002$) is added.

The cases were run with each of the three schemes. Considerable care was taken to make a fair comparison, and for each method, parameters used were those that gave the most efficient solution. All three codes were written from scratch by the first author. The explicit scheme simply requires setting $[B] = [A] = 0$ in Equation 18. The Beam & Warming scheme uses a three point backward temporal integration, although backward Euler temporal integration was also tried with similar results. The *CFL* number resulting in quickest convergence was chosen. Boundary conditions, smoothing formulation, and flux balance formulations in the codes are identical.

Convergence was taken as

$$\frac{1}{I\times J}\sum_{i,j=1}^{I,J}\left\{\left|\frac{\Delta\rho u}{\Delta t}\right|+\left|\frac{\Delta\rho v}{\Delta t}\right|+\left|\frac{\Delta E/E_\infty}{\Delta t}\right|\right\}_{i,j}\leq 5\times 10^{-4}\,. \tag{23}$$

This criterion allows fair comparison of convergence histories calculated with different methods and time steps. All calculations were made on a *DEC Microvax II* which is approximately equivalent to a *VAX 750.*

5.1 Inviscid Flow Cases

Table 1 gives the iteration histories for the inviscid cases calculated with the explicit scheme. The second and third columns give the *CPU* time (in minutes) and iteration count for the scheme without acceleration devices. The fourth and fifth give those results for the scheme with multigrid (MG) and residual smoothing (RS). Each calculation was made on a grid with 48 streamwise cells and 16, 24, or 32 cells in the cross stream direction.

Table 1: Inviscid Solution with Explicit Scheme

	NO. ACCEL.		MG & RS	
# cells	CPU(m)	ITER	CPU(m)	ITER
16	291	4177	19	155
24	754	7295	37	208
32	1370	9970	71	306

Solutions without acceleration mechanisms are characterized by a very slow convergence rate, due, in part, to slow damping of pressure waves. Multigrid and residual smoothing are are very effective for this case, resulting in an iteration count reduction of up to a factor of 35 which gives a factor of 20 savings in *CPU* time.

Table 2 gives convergence histories for the same flow cases using the semi-implicit and fully implicit schemes. The iteration count is much smaller with both the semi-implicit and the fully implicit algorithm than with the explicit scheme. However, each iteration takes proportionally more *CPU* time, resulting in similar efficiency. Note that *SINSS* iteration counts, especially, appear to be unaffected by the number of normal grid cells.

Table 2: Inviscid Solutions with *SINSS* and Beam & Warming Scheme

	S-I		**B & W**	
# cells	CPU(m)	ITER	CPU(m)	ITER
16	42	118	30	190
24	63	117	62	264
32	83	115	63	203

5.2 Viscous Flow Cases

In viscous flow cases, the accurate prediciton of skin friction is usually of importance. To ensure that this quantity is converged we require, *in addition* to (23), that the percentage change in skin friction coefficient over time is small:

$$\sum_{n=i}^{i-2}\left[\frac{\%\Delta(C_f)}{\Delta t_{ave}}\right]^n \leq 5\cdot 10^{-4} \tag{24}$$

where $\%(\Delta C_f) = (C_f^i - C_f^{i+1})/C_f^i$ and Δt_{ave} is a representative time step. Summing over three iterations helps eliminate spurious small values of $\Delta\%C_f$ due to oscillatory convergence of the skin friction coefficient.

Reynolds numbers (based on channel height) are 2×10^3, 10^4, and 10^5. All cases have 48 cells in the streamwise direction and 24, 32, or 32 cells across the half-channel. The grids were generated with stretchings $A = \Delta y_{j+1}/\Delta y_j$ of 1.12, 1.15, or 1.18, respectively.

Iteration histories and CPU requirements are given in Table 3. The second column gives the acceleration mechanism (RS and/or MG) for the explicit code. The semi-implicit code used only residual smoothing.

SINSS does significantly better than the explicit or the implicit scheme at all Reynolds numbers. However, at $Re = 10^5$ an implicit smoothing coefficient value of $\mu_{IS} = 0.2$ was necessary for convergence. The implicit scheme converged with difficulty at the highest Reynolds number, despite attempts with a

Table 3: Viscous Channel Flow

	EXPLICIT			S-IMPLICIT		B & W	
Re #	Accel.	CPU(m)	ITER	CPU(m)	ITER	CPU(m)	ITER
2×10^3	MG & RS	99	493	66	109	75	298
	no acc.	219	1678	71	118	-	-
1×10^4	MG & RS	366	1385	69	85	142	421
	no acc.	574	3345	98	121	-	-
1×10^5	RS	548	2820	136	168	550	1500
	no acc.	758	4418	232	287	-	-

variety of parameter values. The grid stretching at high Reynolds numbers renders the explicit solver increasingly stiff and decreases the effectiveness of multigrid. The Reynolds number 10^5 case converged only *without* multigrid.

6 Conclusions

A semi-implicit algorithm for solving the thin layer Navier-Stokes equations in finite volume form is presented. The method retains much of the flexibility of explicit schemes while eliminating the numerical stiffness due to the disparate physical scales found in typical viscous calculations. It is applied to viscous and inviscid flows and is compared to a fully implicit scheme and an explicit scheme equipped with multigrid and residual smoothing. While the explicit scheme is slightly more efficient for inviscid solutions on coarse grids, the semi-implicit algorithm is up to 5 times more efficient than the explicit and implicit schemes on the computed viscous cases.

The algorithm is easy to implement on vector and parallel architecture machines, and preliminary calculations with a fully vectorized code have been made on a *Cray XMP*. An attractive possibility is to use an explicit solver in the outer inviscid flow coupled to *SINSS* in the viscous layer. Future computations will focus on turbulent flows on highly stretched grids and extension to three dimensions.

Acknowledgements

We would like to acknowledge stimulating technical discussions with Bob Melnick and Frank Marconi of Grumman Aerospace Corporation, Bernard Mueller of *FFA*, and Michael Giles of MIT. We gratefully acknowledge *NSF* fellowship support to the first author for the duration of this work, and support by Grumman Aerospace Corporation and *AFOSR* Grant # 82-0136, monitored by Jim Wilson.

References

[Beam & Warming 76] Beam, R.W. and Warming, R.F., "An Implicit Factored Scheme for the Compressible Navier-Stokes Equations", *AIAA* Journal Vol. 16, pp. 393-401.

[Jameson et al. 81] Jameson, A., Schmidt, W., and Turkel, E. "Numerical Solutions of the Euler Equations Using Runge-Kutta Time Stepping Schemes", *AIAA* Paper 81-1259, 1981.

[Loyd et al. 86] Loyd, B., Murman, E. M., and Abarbanel, S. S., "A Semi-Implicit Scheme for the Navier Stokes Equations", Presented at the Society of Industrial and Applied Mathematics National Meeting, Boston, 24 July 1986.

[Loyd et al. 87] Loyd, B., Murman, E. M., and Abarbanel, S. S. "Semi-Implicit Solution of the Navier-Stokes Equations (*in preparation*)", CFDL Report #87-7, M. I. T., 1987.

NUMERICAL SIMULATION OF THE STRONG INTERACTION BETWEEN A COMPRESSOR BLADE CLEARANCE JET AND STALLED PASSAGE FLOW

N.M. McDougall & W.N. Dawes

Whittle Laboratory, University of Cambridge, U.K.

SUMMARY

A number of groups are developing three dimensional Navier-Stokes solvers for application to both internal and external flows [1, 2, 3]. Regrettably, the sole objective of this work is often merely to predict aerodynamic performance. Successful a priori prediction of turbine efficiency, for example, lies some years into the future, limited as we are by computer size and speed and, more gravely, by inability to model turbulence. The present authors believe the most valuable use of numerical simulation is to provide physical insight into the fluid mechanics; it is this insight which leads to improved component design.

The current paper describes a 3D Navier-Stokes solver, written for turbomachinery use. The compressible equations of motion, with simple mixing length turbulence closure, are discretised in finite volume form and time marched by a hybrid two step explicit-one step implicit algorithm. Convergence rate to a steady solution is enhanced by a multigrid algorithm based on the Brandt Full Approximate Scheme.

The solver is applied to an annular cascade of compressor stator blades both with and without a clearance gap at the hub. With the hub gap sealed the blade is effectively stalled with reversed flow over much of the suction surface. With the hub gap open, the clearance jet entrains and re-energises much of the stalled passage flow, radically modifying the flowfield.

EQUATIONS OF MOTION

The three-dimensional Reynolds averaged Navier-Stokes equations are written in finite volume form and cast in the blade-relative frame using cylindrical coordinates (r, θ, x) [3]:

$$\frac{\partial}{\partial t} \oint_{VOL} \overline{U} dVOL = \oint \overline{H} \cdot d\overline{AREA} + \oint \rho \overline{S} dVOL \qquad (1)$$

where $\overline{U} = [\rho, \rho W_x, r\rho W_\theta, \rho W_r, \rho E]$, $\overline{H} = [\rho\overline{q}, \rho W_x\overline{q} + \overline{\tau}\hat{\imath}_x, r\rho W_\theta\overline{q} + \overline{\tau}\hat{\imath}_\theta, \rho W_r\overline{q} + \overline{\tau}\hat{\imath}_r, \rho I\overline{q}]$, $\overline{S} = [0, 0, -2\Omega r W_r, W_\theta^2/r + r\,\Omega^2 + 2\,\Omega W_\theta, 0]$

with $\overline{q} = W_x \hat{\imath}_x + W_r \hat{\imath}_r + W_\theta \hat{\imath}_\theta$, the relative velocity; Ω = rotation speed; $\overline{\tau}$ = the stress tensor (containing both the static pressure and the viscous stresses); and I = the rothalpy. The system is closed by an equation of state and a two-layer mixing length turbulence model patterned after Baldwin and Lomax [4].

NUMERICAL SOLUTION PROCEDURE

Finite Volume Formulation

The governing equations (1) are written in integral conservation form and the numerical discretization is designed to mimic this. We divide the computational domain into hexahedral cells and store the variable $\overline{U}$ at cell centres (ijk). The integrals in the equations are replaced by discrete summation around the faces of the computational cell.

$$\frac{\Delta \overline{U}_{ijk}}{\Delta t} \Delta VOL_{ijk} = \left\{ \sum_{CELL\,(ijk)} \overline{H}.\Delta\overline{AREA} + \rho S_{ijk} \Delta VOL_{ijk} + \overline{D}_{ijk} \right\} = \overline{R}_{ijk} . \quad (2)$$

Fluxes through cell faces are found by linear interpolation of density, velocity etc., between cell centres and so the formal spatial accuracy is second order on smoothly varying meshes and global conservation is ensured. Viscous stresses are computed by defining a local curvilinear coordinate system and the chain rule. $\overline{D}_{ijk}$ represents the adaptive artificial diffusion term recommended by Jameson [5] and added to the discretized equations (2) to control odd-even point solution decoupling and to suppress oscillations in regions with strong pressure gradients. This diffusion consists of low-level fourth difference background smoothing which is replaced locally by more vigorous second difference smoothing when a strong pressure gradient is detected.

Boundary Conditions

At inflow, total temperature and pressure are fixed and either flow angle or absolute swirl velocity held constant. At outflow the hub static pressure is fixed and radial variation derived from the simple radial equilibrium equation.

For cells adjacent to solid boundaries, zero fluxes of mass, momentum, and energy are imposed through the cell face aligned with the solid boundary. Wall static pressure is found by setting the derivative of pressure normal to the wall equal to zero. To prescribe the wall shear stress the velocities stored at cell centres adjacent to the wall and the known zero value of velocity on the wall are used to compute the velocity gradients at the wall. These gradients, together with the wall viscosity, are used with a locally defined curvilinear coordinate system to compute the wall shear stresses. If required, the wall shear stress can be computed from a universal logarithmic skin friction law.

Preprocessed Algorithm

The discretized equations are time-marched using an implicit preprocessed algorithm described in references [1] and [6]. In outline, two explicit steps are performed,

$$\Delta\overline{U}^{(1)} = \frac{\Delta t}{\Delta VOL} \cdot \overline{R}_{ijk}^{n} \ ; \ \overline{U}^{(1)} = \overline{U}^{n} + \Delta\overline{U}^{(1)}$$

$$\Delta\overline{U}^{(2)} = \frac{\Delta t}{\Delta VOL} \cdot \overline{R}_{ijk}^{(1)} \qquad (3.1)$$

followed by the solution of an implicit set of equations

$$\left[I - \varepsilon_I \frac{\Delta t^2}{\Delta VOL} \lambda_I^2 \delta_{I\,I}^2\right] * \left[I - \varepsilon_J \frac{\Delta t^2}{\Delta VOL^2} \lambda_J^2 \delta_{J\,J}^2\right]$$

$$* \left[I - \varepsilon_K \frac{\Delta t^2}{\Delta VOL} \lambda_K^2 \delta_{K\,K}^2\right] \Delta\overline{U} = \Delta\overline{U}^{(2)} ; \ \overline{U}^{n+1} = \overline{U}^{n} + \Delta\overline{U} \quad (3.2)$$

where $\varepsilon_I, \varepsilon_J$ and ε_K are free parameters (of order unity), λ_I, λ_J and λ_K are the spectral radii of the Jacobians associated with the convective fluxes in the I, J and K directions, and, for example

$$\delta_{I\,I}^2 \phi = \phi_{I+1} - 2\phi_I + \phi_{I-1}.$$

The left-hand side is factored into three tridiagonal matrices for efficient inversion.

Although in principle the algorithm can be made stable for any size time step by suitable choice of the free parameters, $\varepsilon_I, \varepsilon_J$ and ε_K, in practice it is found that there is an optimum range of time steps which leads to minimum number of steps to convergence. It was found that CFL numbers in the range 1-3 were optimal with associated values of ε equal to unity. For the results presented in this paper, the CFL number was 2 and $\varepsilon_I = \varepsilon_J = \varepsilon_K = 1$.

Multigrid convergence acceleration

The present effort is concerned with steady solutions, so a multigrid acceleration technique [5] [7] was used. The basic principal of multigrid is to take advantage of the fact that the finite volume residue, R_{ijk} (equation 2), has errors over the whole range of wavelengths which can be supported by the mesh. The time marching algorithm, equation (3), is efficient at eliminating the short wavelength components of this error, but much less so for higher wavelengths. So we define a succession of ever coarser meshes, by deleting every other mesh line, derive an appropriate representation of the residue on each mesh and use the basic time marching solver, equation (6), to reduce the level of the error associated with the current mesh. Thus, sweeping through the meshes should allow each of the wavelengths in the residue error to be attacked with optimum efficiency. One level of multigrid only was used in the current effort with a simple sawtooth strategy.

EXPERIMENTAL FACILITY

The experiments were carried out in the Deverson Rig, a low speed axial compressor test facility in the Whittle Laboratory. The compressor is a single stage machine (Table 1) with geometry similar to a typical stage from a modern gas turbine compressor.

TABLE 1 Compressor Geometry and Flow Details

	ROTOR	STATOR
Aerofoil section	C4	C4
Aerofoil chord	0.110m	0.114m
Aspect ratio	1.38	1.33
Tip diameter	1.52m	1.52m
Clearance/chord ratio	0.005	0.00/0.03
Inlet axial velocity	14.36m/s	
Rotational speed	500rpm	
Reynolds No. (blade chord)	3×10^5	

The operating conditions for the test were set so that the mass flow was slightly greater than that at which rotating stall was observed, far from the design point.

Traverses using small pneumatic probes were carried out upstream of the stator row to establish the conditions at inlet to the blades. In order to visualise the flow close to the blade surface, a suspension of diesel oil and fluorescent dye was applied to the stator blade surfaces. As gravity can affect the patterns shown, only horizontal surfaces were used during the tests.

APPLICATION OF THE FLOW SOLVER

The stators of the compressor were discretised using the 17 x 41 x 17 mesh illustrated in Figures 1 and 2. The mesh is too coarse for detailed predictions but does allow a quite satisfactory simulation. Two cases were computed: with and without a clearance gap at the hub. To model the clearance gap the blade thickness was reduced to zero and periodic boundary conditions applied between the blade tip and the hub endwall, as illustrated in Figure 2. Computer memory constraints meant that no attempt was made to refine the mesh near the casing so that here the simulation is viscous only in the blade-blade sense. At inflow the flow angle imposed via the boundary conditions was that measured by probe traverse, Figure 3. The code runs at about 6.8 x 10^{-3} seconds/point/time step on a Perkin Elmer 3230 mini-system and 1500 time steps were used to obtain satisfactory convergence.

DISCUSSION OF RESULTS

Close to the hub the stator is subjected to over ten degrees of positive incidence. The resulting separated flow is shown in Figure 4, together with the predicted flow vectors close to the blade suction surface. The passage flow is dominated by a separation which extends to cover almost the entire chord at the hub. Traverses downstream of the bladerow (not presented here), together with the predictions, confirm that this region extends circumferentially across a large portion of the blade passage. Although the mesh used in the predictions is clearly unable to resolve the features of the flow close to the casing, the extent of the hub corner separation is predicted correctly.

Once the hub clearance is opened, the fluid emerging as a clearance jet counteracts the separation by entrainment and also by displacing the low-momentum fluid towards the opposite (pressure) side of the passage, Figure 5. In the compressor, the hub clearance is a constant gap between the hub and the blade tip. This is modelled in the computations by progressively decreasing the thickness of the blade towards the gap, resulting in the rounded blade tip shown in Figure 2. Although this may change the detail of the leakage jet emerging from beneath the blade, it is close enough to the real geometry to predict the gross effects seen in the compressor. Initially, the clearance was set at 3% chord for the computations, to match the clearance found in the real machine. This had the expected effect of reducing the extent of the predicted separation, but only so that the overall size was reduced by approximately 50%. The results presented (Figure 5) show predictions with a tip clearance of 5% of chord, and the agreement between experiment and prediction can be seen to be good. The corner separation has been completely inhibited by the clearance flow, leaving a small separation towards the trailing edge at mid-span. The discrepancy in the results is probably due to the differing tip geometries between the calculation and experiment. The rounded blade tip used in the calculation will almost certainly produce a more diffuse leakage jet which will be less effective in entraining and displacing the separated flow.

CONCLUSIONS

The ability to predict a strongly separated flow in a compressor blade row has been demonstrated. Although minor details of the geometry required adjustment to obtain complete agreement with the experimental data, the predictions showed the correct qualitative detail, and correctly predicted the improvement in the passage flow when a tip clearance is included. Better overall agreement is anticipated when the geometry is described by a fine mesh and the tip clearance modelling is modified to more nearly represent the real geometry.

REFERENCES

[1] DAWES, W. N.: "Application of full Navier-Stokes solvers to turbomachinery flow problems." VKI Lectures Series 2: Numerical techniques for viscous flow calculations in turbomachinery bladings, Jan. 1986.

[2] MOORE, J.: Performance evaluation of flow in turbomachinery blade rows. AGARD lecture series No. 140 "3D Computation Techniques Applied to Internal Flows in Propulsion Systems", Rome, June 1985.

[3] SARATHY, K. P.: "Computation of Three-Dimensional Flowfields Through Rotating Blade rows and Comparison with Experiment." ASME Journal of Engineering for Power, Vol. 104, (1982).

[4] BALDWIN, B., LOMAX, H.: "Thin Layer Approximation and Algebraic Model for Separated Turbulent Flows." AIAA Paper No.78-257, 1978.

[5] JAMESON, A., BAKER, T. J.: "Multigrid Solution of the Euler Equations for Aircraft Configurations." AIAA Paper No. 84-0093, (1984).

[6] DAWES, W. N.: "A pre-processed implicit algorithm for 3D viscous compressible flow." Notes on Numerical Fluid Mechanics, Vol. 12, Vieweg 1986.

[7] BRANDT, A.: "Multi-level adaptive solutions to boundary-value problems." Math. Comp. Vol. 31, No. 138, (1977).

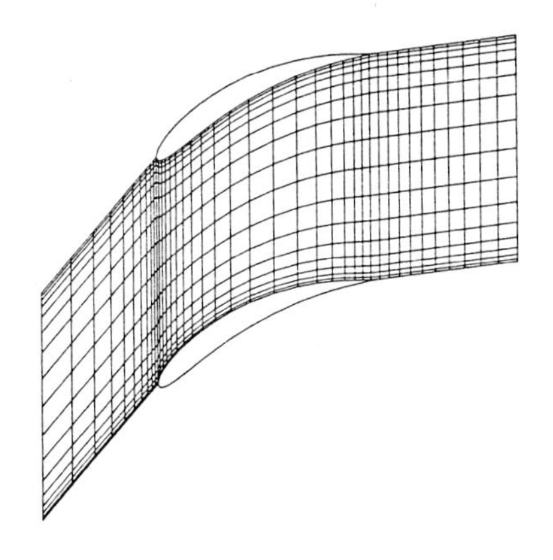

FIGURE 1 THE 17 X 41 BLADE-BLADE MESH

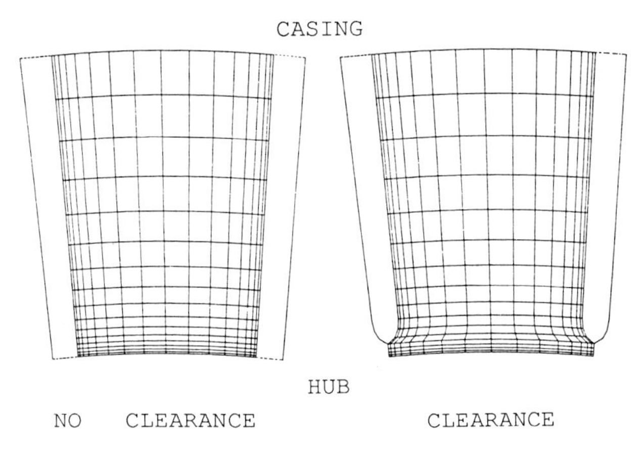

FIGURE 2 THE 17 x 17 CROSS FLOW MESH

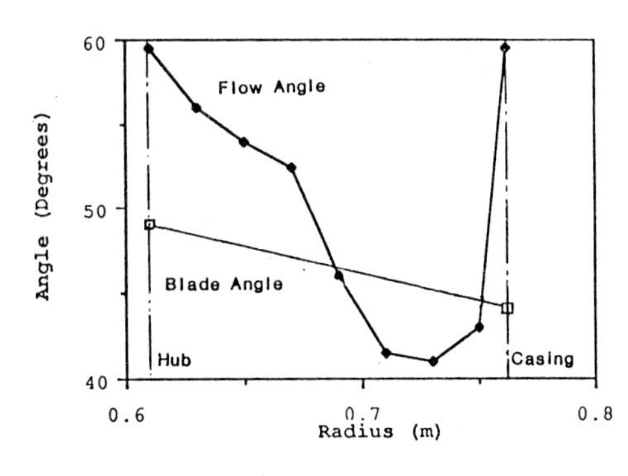

FIGURE 3 THE MEASURED FLOW ANGLE UPSTREAM OF THE STATOR

PREDICTED VELOCITIES, HUB SEALED

LE

REVERSED FLOW

HUB

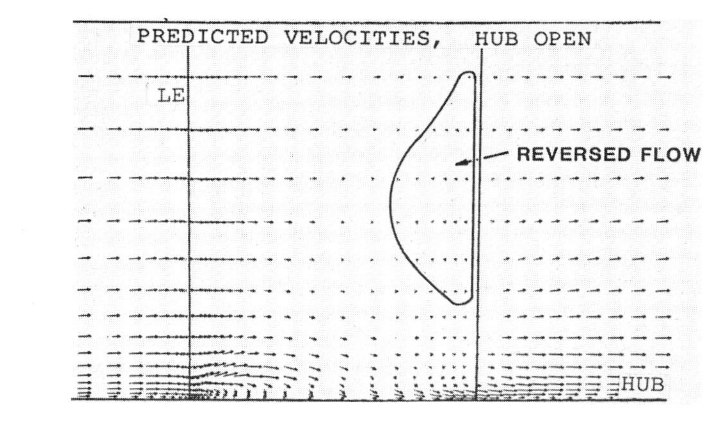

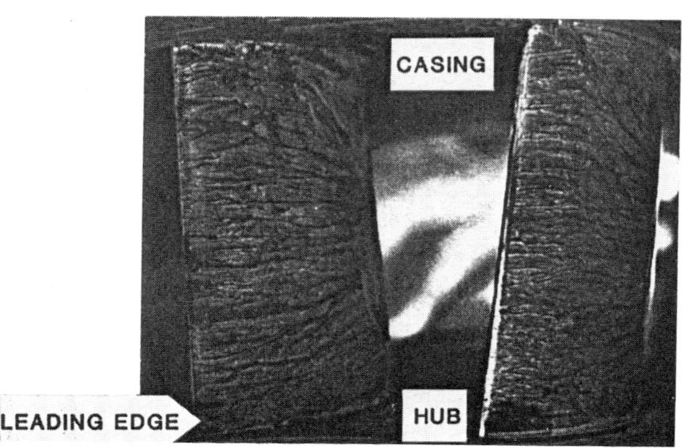

COMPARISON BETWEEN PREDICTED FLOW VECTORS NEAR THE SUCTION SURFACE AND OIL FLOW VISUALISATION

FIGURE 4

FIGURE 5

A FINITE VOLUME METHOD FOR STEADY TWO-DIMENSIONAL INCOMPRESSIBLE FLOWS USING NON-STAGGERED NON-ORTHOGONAL GRIDS

S.Majumdar, B.Schoenung, W.Rodi

Institute for Hydromechanics,
University of Karlsruhe,
Karlsruhe,F.R.Germany

SUMMARY

This paper describes a finite-volume method for the numerical prediction of two-dimensional,steady,incompressible flows with complex irregular boundaries. Non-orthogonal boundary-fitted grid with cell-centred variable arrangement and Cartesian velocity components as dependent variables are used in this method. The accuracy and numerical stability of the procedure are validated against experimental or other numerical results.

INTRODUCTION

One of the main problems encountered in predicting a real-life flow situation is the complex irregular geometry of the flow domain. Conventional finite volume methods employ rectangular numerical grids which are not suitable for irregular geometries. Flows with such geometries can be resolved much better with curvilinear boundary-fitted grids. As orthogonal grids are quite difficult to generate especially for three-dimensional flows, a generally applicable finite-volume method must be able to cope with boundary-fitted non-orthogonal grids which can often be generated by using simple algebraic procedures.

Following the key-idea proposed by Rhie & Chow[1]and further work of Peric[2],a finite-volume method for non-orthogonal,boundary-fitted grids has been developed which employs a non-staggered variable arrangement but does not suffer from the so-called 'chequerboard splitting' of pressure field,usually encountered when such grid is used. The present paper describes in brief the governing equations in two-dimensional plane and/or axisymmetric non-orthogonal cooordinates, the discretisation procedure and in particular the modified treatment of pressure-determining equation for cell-centered variable arrangement employing the principle of SIMPLEC algorithm [3] . Finally the capabilities of the proposed method are demonstrated through calculation results for a few test cases.

MATHEMATICAL MODELLING

Governing Equations . The governing equations in a general coordinate system may be expressed either with Cartesian[1,2] or grid-oriented [4] velocities as dependent variables. The use of Cartesian velocities avoids the calculation of any curvature terms which are extremely grid-sensitive. The equations for steady,two-dimensional flows in non-orthogonal coordinates using Cartesian velocity components may be written in a generalised form as follows:

$$\frac{\partial}{\partial x_1}(C_{1\phi}\cdot\Phi + D_{1\phi}) + \frac{\partial}{\partial x_2}(C_{2\phi}\cdot\Phi + D_{2\phi}) = J\ S_\phi\ r \qquad (1)$$

where, the convective coefficients $C_{1\phi}$, $C_{2\phi}$, the diffusive coefficients, $D_{1\phi}$, $D_{2\phi}$ and the source terms S_ϕ for different variable ϕ and the Jacobian J are shown in Table 1.

Turbulence Model. For turbulent flows, the viscosity μ in the diffusion terms needs to be replaced by an effective viscosity, which is the sum of the laminar and a turbulent viscosity. This turbulent viscosity is calculated in the present work using the standard k-ε turbulence model [5]. The turbulent kinetic energy, k and its dissipation rate, ε are obtained by solving the relevant transport equations which are also of the same general form shown in Table 1.

NUMERICAL ANALYSIS

Grid and Variable Arrangement. The kind of grid and the variable arrangement used in the present procedure is shown in Fig.1(a). The grid nodes are joined by simple linear segments to form control volumes. All the variables are stored at the geometric centre of the control volumes. The use of this single control volume approach reduces the computation of geometrical quantities and internodal interpolation to a minimum; however it needs a special treatment of the continuity equation to avoid the so-called 'chequerboard splitting' of pressure.

Discretisation of the Equations. Formal integration of the differential equations over a control volume leads to the following difference equation:

$$I_e - I_w + I_n - I_s = \int_{\Delta V} S_\phi \, dV \qquad (2)$$

where, I_e, for example, represents the total flux of across the face e. The surface fluxes I_e, I_w etc. consist of three distinct parts, a convective contribution, I^C and a normal diffusive contribution I^{DN} and a cross-derivative diffusive part, I^{DC}. The numerical evaluation of the different terms of Eqn.(2) requires the calculation of a few geometrical factors and suitable interpolation practice for internodal variation of the dependent variable.

Geometrical Factors. The most important geometrical information required in the present formulation are the factors, β^i_j which may numerically be expressed as follows.

$$\beta^1_1 = \frac{\partial y_2}{\partial x_2} \simeq b^1_1 / \Delta x_2 \; ; \quad \beta^1_2 = -\frac{\partial y_1}{\partial x_2} \simeq b^1_2 / \Delta x_2$$
$$\beta^2_1 = -\frac{\partial y_2}{\partial x_1} \simeq b^2_1 / \Delta x_1 \; ; \quad \beta^2_2 = \frac{\partial y_1}{\partial x_1} \simeq b^2_2 / \Delta x_1 \qquad (3)$$

where the lengths b^i_j for the cell around P are shown in Fig.1(b). The cell volume, the normal distance from the wall required in the 'Wall-function' approach, the wall inclination to the Cartesian axes etc., are easily evaluated from the cell-corner coordinates.

Assumptions regarding Internodal Variation.

(i) The value of any variable solved for is assumed to be uniform and equal to ϕ_P over the control volume surrounding the point P.

(ii) Linear interpolation is used to calculate the values of ϕ or pressure at the cell faces for evaluation of normal diffusion flux, I^{DN} and the cross-derivative diffusion flux, I^{DC} or for pressure-gradient terms in the momentum equations.

(iii) For the convective fluxes, I^C at the cell-faces Upwind/Central Hybrid differencing scheme is presently incorporated.

With these assumptions, the difference equations may be expressed in a general form as follows.

$$A_P \Phi_P = A_E \Phi_E + A_W \Phi_W + A_N \Phi_N + A_S \Phi_S + S_\Phi^U \Delta V_P \tag{4}$$

$$A_P = A_E + A_W + A_N + A_S - S_\Phi^P \Delta V_P \; ; \quad S_\Phi = S_\Phi^U + S_\Phi^P \Phi_P. \tag{4a}$$

The detailed derivation is given in Ref.[6]. The continuity equation for incompressible flows, having no pressure terms in it, however needs a special treatment which is now described.

Special treatment of the Continuity equation. The basic concept of the SIMPLE[7] or SIMPLEC [3] algorithm is (i) to assume an artificial linkage between the cell-centered pressure and the cell-face velocities from the discretised momentum equations, (ii) to frame a system of linear equations for pressure correction that satisfy continuity and (iii) finally to correct the pressure and velocity accordingly. A similar strategy is followed in the present procedure too with the only basic difference that unlike the "classical staggered arrangement" the velocities at cell-face locations, not being available there, are to be calculated from the adjacent cell-centered quantities.

The discretised momentum equations for cell-centered velocities using underrelaxation parameters α_u and α_v are as follows:

$$\begin{aligned} v_{1P} &= \alpha_u \left[H_P^u + D_{1P}^u (p_w - p_e) + D_{2P}^u (p_s - p_n) \right] + (1-\alpha_u) v_{1P}^o \\ v_{2P} &= \alpha_v \left[H_P^v + D_{1P}^v (p_w - p_e) + D_{2P}^v (p_s - p_n) \right] + (1-\alpha_v) v_{2P}^o \end{aligned} \tag{5}$$

$$H_P^u = (\Sigma A_{nb}^u v_{1nb} + S_u^U \cdot \Delta V_P) / A_{PP}^u \tag{5a}$$

where, A_{nb} represents the coefficients for neighbouring nodes; v_{1P}^o, v_{2P}^o are their old values at the previous iteration and A_{PP}^u and A_{PP}^v are the values of A_P for v_1 and v_2 at the node P.

$$\begin{aligned} D_{1P}^u &= b_{1P}^1 \, r_P / A_{PP}^u \, , \quad D_{2P}^u = b_{1P}^2 \, r_P / A_{PP}^u \\ D_{1P}^v &= b_{2P}^1 \, r_P / A_{PP}^v \, , \quad D_{2P}^v = b_{2P}^2 \, r_P / A_{PP}^v . \end{aligned} \tag{5b}$$

Now while evaluating the cell-face velocities v_{1w}, v_{2w} etc., linear interpolation between cell-centered components leads to 'chequerboard splitting' which may be avoided using the 'Momentum Interpolation', suggested in [1,2]. According to this interpolation, the face velocities v_{1w} or v_{2w} with proper underrelaxation may be expressed as:

$$\begin{aligned} v_{1w} &= \alpha_u \left[\overline{H_P^u + D_{2P}^u (p_s - p_n)} + \overline{D_{1w}^u} \, (p_W - p_P) \right] + (1-\alpha_u) v_{1w}^o \\ v_{2w} &= \alpha_v \left[\overline{H_P^v + D_{2P}^v (p_s - p_n)} + \overline{D_{1w}^v} \, (p_W - p_P) \right] + (1-\alpha_v) v_{2w}^o \end{aligned} \tag{6}$$

where, the expression with overbars represents the linear average of the same quantities evaluated at the cell-centers P and W adjacent to the face 'w'; v°_{1w} and v°_{2w} are the values at the previous iteration level.

The SIMPLEC algorithm assumes that the change of pressure at any node(P) leads to a change of velocity which is equal at all other four nodes (E,W,N and S in Fig.1(a)) surrounding P. One further assumption in the present non-orthogonal formulation is that the change in face velocities(v_{1w} for example) is pre-dominant due to change in pressure-gradient along the main direction(along W-P for face 'w'),whereas the change due to the cross-direction gradient (along sw-nw for face 'w')is negligible. Using these two assumptions,one may express the pressure-velocity correction relationship for the cell centre as:

$$
\begin{aligned}
v'_{1P} &= \alpha_u \left[D^u_{1P} (p'_w - p'_e) + D^u_{2P} (p'_s - p'_n) \right] / \left[1. - \alpha_u \, \Sigma A^u_{nb} / A^u_{PP} \right] \\
v'_{2P} &= \alpha_v \left[D^v_{1P} (p'_w - p'_e) + D^v_{2P} (p'_s - p'_n) \right] / \left[1. - \alpha_v \, \Sigma A^v_{nb} / A^v_{PP} \right].
\end{aligned} \tag{7}
$$

The corresponding relationships for the pressure-velocity linkage at the cell-face are for v_{1w} for example,

$$
v'_{1w} = \alpha_u \, \overline{D^u_{1w}} \, (p'_W - p'_P) / \left[1. - \alpha_u \, \overline{\Sigma A^u_{nb} / A^u_{PP}} \right] . \tag{8}
$$

The continuity equation is now expressed as following:

$$
C_e - C_w + C_n - C_s = 0 \tag{9}
$$

where, the mass-flux C_w ,for example may be written as

$$
C_w = \rho_w \, r_w \left[b^1_{1w} (v_{1w} + v'_{1w}) + b^1_{2w} (v_{2w} + v'_{2w}) \right]. \tag{9a}
$$

Substituting the face velocities v_{1w} etc., from Eqn.(6) and their corrections v'_{1w} etc., from Eqn.(8) one obtains a system of linear equations for p',expressible in the form of general Eqn.(4). Cell-centre pressures are then corrected by the amount $p'$and corrections are applied to two cell-centre velocities($v_{1P}$and v_{2P}) according to Eqn.(7) and to all the eight cell-face velocities according to Eqn.(8) respectively.

Boundary Conditions . The most frequently encountered boundary conditions are (i) Inlet, (ii)Outlet,(iii)Symmetry planes and (iv)Rigid walls. At the inlet usually the known boundary values are prescribed. For all the other kind of boundaries the links to the adjacent control volumes are disconnected by setting the convective flux,the normal as well as the cross-derivative diffusive flux to zero. At the outlet,symmetry planes or boundaries with specified flux the boundary values need to be updated using the interior field values so that the required gradient condition is fulfilled. For the near-wall control volumes however the Wall function approach of Launder & Spalding [5]is used where the wall shear-stress vector is expressed as a function of the nodal velocity component parallel to the wall. The shear stress acting parallel to the wall is again decomposed in two components along v_1 and v_2 respectively to be used as source terms in the corresponding equations. In the equation for k and ε also,the coefficient linking the wall node is set to zero and the source terms are modified according to the assumption of logarithmic law of wall[5].

Solution Algorithm. The Strongly Implicit Procedure due to Stone[8]is used for solving the system of linear equations for each variable. For velocities and the scalars only one inner sweep is performed in the solver. Since a good mass-conservation is crucial at every outer iteration sweep for efficient and reliable solution using a seggragated or sequential strategy , a low value of residual reduction factor is prescribed as termination of the pressure correction sweep in each iteration cycle. A combination of underrelaxation factor of 0.9 for momentum equations with a residual reduction factor of 0.1 for continuity and the partial cancellation parameter of 0.9 for Stone's method have been found to work reasonably well for all the test problems.

RESULTS AND DISCUSSION

In this section,applications of the calculation procedure described above to three different flow calculations are presented and discussed.

Laminar flow through a diffuser with curved wall. The flow geometry,the grid and the boundary conditions used are shown in Fig.2(a). Fig.2(b) compares the wall shear stress predicted by the present procedure to the prediction of Demirdzic[4] who used a staggered variable arrangement with grid-oriented velocities as dependent variables. Reasonable agreement validates the correct treatment of inertia,viscous and pressure terms in the present procedure.

Laminar flow around a circular cylinder. Fig.3(a) shows the non orthogonal grid used for the calculation,specially near the cylinder wall. This grid was generated by a differential grid-generation procedure developed by Naar & Schoenung[9]. Fig.3(b) shows a comparison between the present prediction,the calculation of Majumdar & Rodi[10]who used an orthogonal polar grid and available experimental results[11] for the pressure distribution around the cylinder at a Reynolds number of 40. Good agreement demonstrates the adequacy of the treatment of the non orthogonal terms and also checks the numerical accuracy of the procedure.

Turbulent flow in an axisymmetric spray-drying chamber. A typical drying chamber geometry,the algebraically generated non-orthogonal grid and the boundary conditions employed are shown in Fig.4(a) wheras Fig.4(b) shows the calculated velocity vectors. The large recirculation zone throughout the chamber due to the sudden expansion near the chamber top and the recirculation created near the bottom due to entry of low-speed fluid at the bottom annulus appear to be physically realistic. The flow pattern near the exit plane,which is perhaps not entirely realistic is found to be sensitive to the kind of exit boundary conditions used. The zero axial gradient condition used in the preliminary calculations presented here are unlikely to do justice to the fairly complex flow behaviour at the exit and experimental data are needed to ascertain better boundary conditions. However, in absence of any experimental data, the realistic flow pattern predicted, in a qualitative sense, confirms the correct treatment of axisymmetric terms and also the correct use of the Wall-function for the non orthogonal control volumes near the wall boundaries.

ACKNOWLEDGEMENTS

The work reported here has been supported by the Deutsche Forschungsgemeinschaft through contract no. Ro 558/3-2. The calculations were carried out in the SIEMENS 7881 of the University of Karlsruhe, Computer Centre.

REFERENCES

[1]. RHIE C.M and CHOW W.L. A numerical study of the turbulent flow past an isolated airfoil with trailing edge separation, AIAA Journal, vol. 21 pp. 1525-1532,1983.

[2]. PERIC M. A finite volume method for the prediction of three dimensional fluid flow in complex ducts,PhD Thesis,Univ. of London,1985.

[3]. VANDOORMAL J.P.,RAITHBY G.D. Enhancements of the SIMPLE method for predicting incompressible fluid flows,Num.Heat Transfer,vol.7,1984.

[4]. DEMIRDZIC I.A. A finite volume method for computation of fluid flow flow in complex geometries,PhD Thesis,Univ. of London,1982.

[5]. LAUNDER B.E.,SPALDING D.B. The numerical computation of turbulent flows, Comp.Meth.Appl. Mech. and Engg.,vol.3,p.269,1974.

[6]. MAJUMDAR S. Development of a finite volume procedure for prediction of fluid flow problems with complex irregular boundaries, Technical Report, SFB 210/T/29, Univ. of Karlsruhe,1986.

[7]. PATANKAR S.V.,SPALDING D.B. A calculation procedure for heat, mass and momentum transfer in three-dimensional parabolic flows,Int.Jour. Heat and Mass Transfer,vol.15,p1787,1972.

[8]. STONE H.L. Iterative solution of implicit approximations of multi-dimensional partial differential equations, SIAM J.Num.Anal. vol.5, p530,1968.

[9]. NAAR M.,SCHOENUNG B. Numerische Gittererzeugung bei Vorgabe von Randwinkeln und Randmaschenweiten,Report No.644, Instt. for Hydromechanics, Univ.of Karlsruhe,1986.

[10]. MAJUMDAR S.,RODI W. Numerical calculation of turbulent flow past circular cylinders, Proceedings 3rd Symp.on Numerical and Physical aspects of Aerodynamic Flows,January,1985,Long Beach,California.

[11]. GROVE A.S., SHAIR F.A., PETERSEN E.E, ACRIVOS A. An experimental investigation of the steady separated flow past a circular cylinder J.Fluid Mechanics,vol.19,1964.

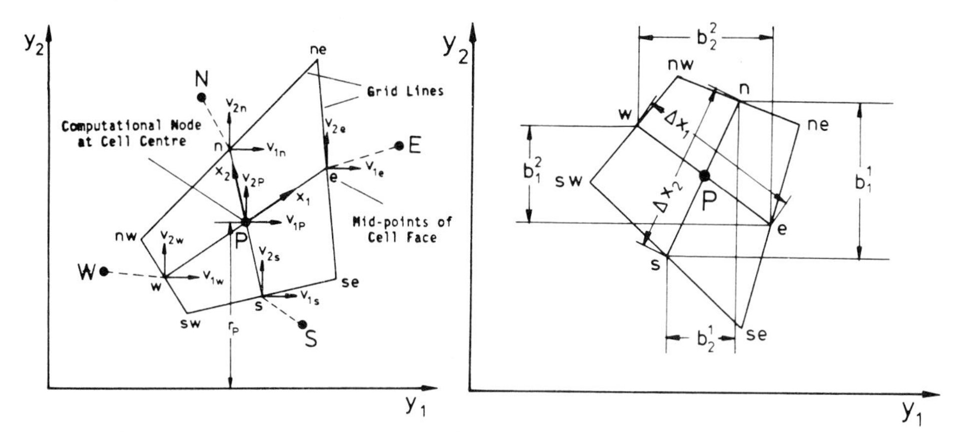

(a) Laying out the control volume (b) Projection lengths for control volume

Fig.1 Grid-related information for non-orthogonal control volumes

Table 1. Coefficients and source terms for the governing equations

General Transport Equation: $\frac{\partial}{\partial x_1}(C_{1\phi}\cdot\phi + D_{1\phi}) + \frac{\partial}{\partial x_2}(C_{2\phi}\cdot\phi + D_{2\phi}) = J\, S_\phi\, r$

ϕ	$C_{1\phi}$	$C_{2\phi}$	$D_{1\phi}$	$D_{2\phi}$	S_ϕ
1	$U_1 r$	$U_2 r$	0	0	0
v_1	$U_1 r$	$U_2 r$	$-\frac{\mu r}{J}\left(B_1^1\frac{\partial v_1}{\partial x_1} + B_2^1\frac{\partial v_1}{\partial x_2} + \beta_1^1\omega_1^1 + \beta_2^1\omega_1^2\right)$	$-\frac{\mu r}{J}\left(B_1^2\frac{\partial v_1}{\partial x_1} + B_2^2\frac{\partial v_1}{\partial x_2} + \beta_1^2\omega_1^1 + \beta_2^2\omega_1^2\right)$	$-\frac{1}{J}\left[\frac{\partial}{\partial x_1}(p\beta_1^1) + \frac{\partial}{\partial x_2}(p\beta_1^2)\right]$
v_2	$U_1 r$	$U_2 r$	$-\frac{\mu r}{J}\left(B_1^1\frac{\partial v_2}{\partial x_1} + B_2^1\frac{\partial v_2}{\partial x_2} + \beta_1^1\omega_2^1 + \beta_2^1\omega_2^2\right)$	$-\frac{\mu r}{J}\left(B_1^2\frac{\partial v_2}{\partial x_1} + B_2^2\frac{\partial v_2}{\partial x_2} + \beta_1^2\omega_2^1 + \beta_2^2\omega_2^2\right)$	Plane: $-\frac{1}{J}\left[\frac{\partial}{\partial x_1}(p\beta_2^1) + \frac{\partial}{\partial x_2}(p\beta_2^2)\right]$ Axisymmetric: $S_\phi = S_\phi - 2\mu\frac{v_2}{r^2}$
k	$U_1 r$	$U_2 r$	$-\frac{\mu r}{J\sigma_k}\left(B_1^1\frac{\partial k}{\partial x_1} + B_2^1\frac{\partial k}{\partial x_2}\right)$	$-\frac{\mu r}{J\sigma_k}\left(B_1^2\frac{\partial k}{\partial x_1} + B_2^2\frac{\partial k}{\partial x_2}\right)$	$G - \rho\varepsilon$
ε	$U_1 r$	$U_2 r$	$-\frac{\mu r}{J\sigma_\varepsilon}\left(B_1^1\frac{\partial \varepsilon}{\partial x_1} + B_2^1\frac{\partial \varepsilon}{\partial x_2}\right)$	$-\frac{\mu r}{J\sigma_\varepsilon}\left(B_1^2\frac{\partial \varepsilon}{\partial x_1} + B_2^2\frac{\partial \varepsilon}{\partial x_2}\right)$	$C_1\frac{\varepsilon}{k}G - C_2\rho\frac{\varepsilon^2}{k}$
ϕ	$U_1 r$	$U_2 r$	$-\frac{\mu r}{J\sigma_\phi}\left(B_1^1\frac{\partial \phi}{\partial x_1} + B_2^1\frac{\partial \phi}{\partial x_2}\right)$	$-\frac{\mu r}{J\sigma_\phi}\left(B_1^2\frac{\partial \phi}{\partial x_1} + B_2^2\frac{\partial \phi}{\partial x_2}\right)$	0

where,

$r = 1$ for plane flows

β_j^i = cofactor of $\frac{\partial y_j}{\partial x_i}$ in J

$$J = \begin{vmatrix} \frac{\partial y_1}{\partial x_1} & \frac{\partial y_2}{\partial x_1} \\ \frac{\partial y_1}{\partial x_2} & \frac{\partial y_2}{\partial x_2} \end{vmatrix}$$

$B_j^i = \beta_1^i\beta_1^j + \beta_2^i\beta_2^j$, $\omega_j^i = \frac{\partial v_i}{\partial x_1}\beta_j^1 + \frac{\partial v_i}{\partial x_2}\beta_j^2$

$U_1 = \rho\cdot(v_1\beta_1^1 + v_2\beta_2^1)$, $U_2 = \rho\cdot(v_1\beta_1^2 + v_2\beta_2^2)$

$\mu = \mu_l + \mu_t$, $\mu_t = \rho\cdot C_\mu\cdot k^2/\varepsilon$

$C_\mu = 0.09$, $C_1 = 1.44$, $C_2 = 1.92$

$\sigma_k = 1.0$, $\sigma_\varepsilon = 1.32$

Plane: $G = \frac{\mu_t}{J^2}\left[2\left(\frac{\partial v_1}{\partial x_1}\beta_1^1 + \frac{\partial v_1}{\partial x_2}\beta_1^2\right)^2 + 2\left(\frac{\partial v_2}{\partial x_1}\beta_2^1 + \frac{\partial v_2}{\partial x_2}\beta_2^2\right)^2 + \left(\frac{\partial v_1}{\partial x_1}\beta_2^1 + \frac{\partial v_1}{\partial x_2}\beta_2^2 + \frac{\partial v_2}{\partial x_1}\beta_1^1 + \frac{\partial v_2}{\partial x_2}\beta_1^2\right)^2\right]$

Axisymmetric: $G = G + 2\mu_t(v_2/r)^2$

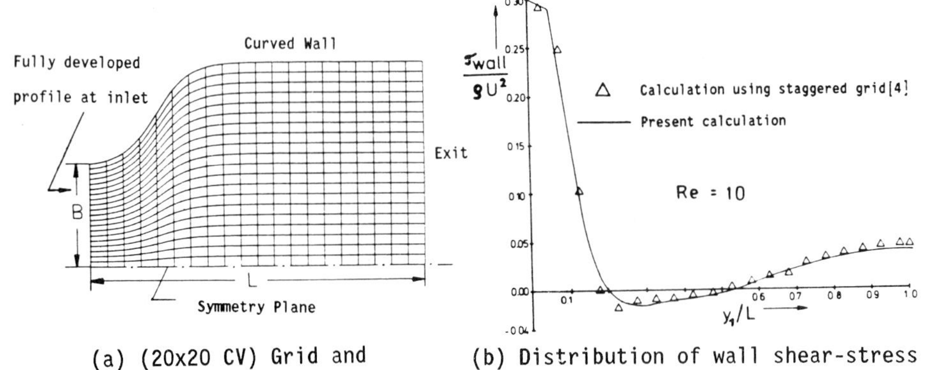

(a) (20x20 CV) Grid and boundary conditions (b) Distribution of wall shear-stress

Fig.2 Laminar flow through a diffuser with curved wall

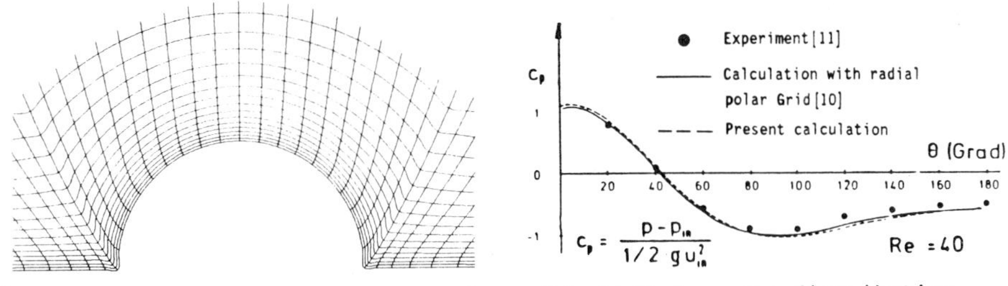

(a)(60x40 CV) Grid near cylinder wall (b) Wall Pressure distribution

Fig.3 Laminar flow around a circular cylinder

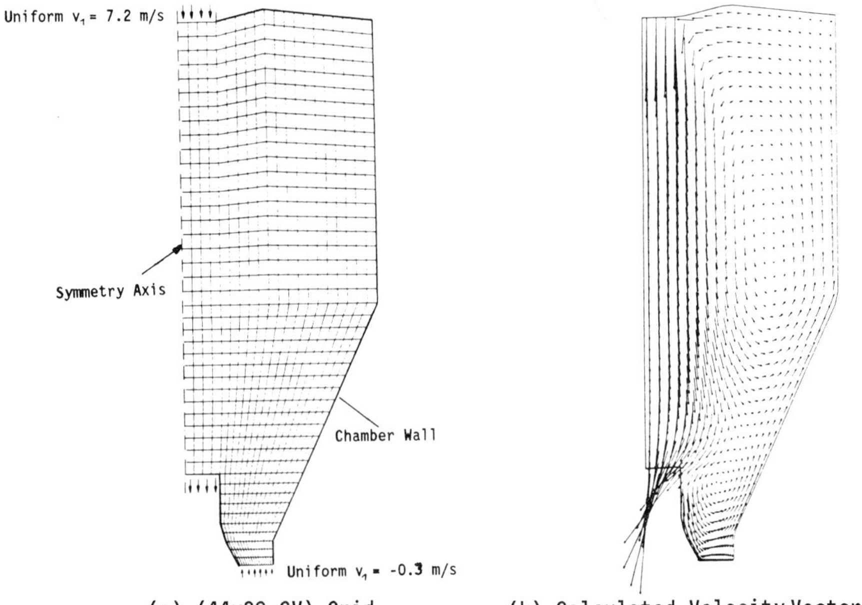

(a) (44x20 CV) Grid (b) Calculated Velocity Vectors

Fig.4 Turbulent flow in an axisymmetric spray-drying chamber

Convergence Acceleration of Finite Element Methods for the Solution of the Euler and Navier Stokes Equations of Compressible Flow

Michel Mallet, Jacques Periaux and Bruno Stoufflet
Avions Marcel Dassault-Breguet Aviation
78 quai Marcel Dassault
92214 Saint Cloud - France

Abstract

Algorithms to solve the Euler and Navier Stokes equations lead to the solution of very large systems of nonlinear equations at each time step. For typical calculations, the cost of this solution is dominant and an increasing research effort is devoted to the development of efficient solvers. The goal of this paper is to describe, test and compare various solvers and to discuss the critical trade-off between the CPU and storage requirements.

1 Introduction

The main points to be addressed are the following:

A transient algorithm is obtained by combining the trapezoidal rule and Newton's method with a linearized Jacobian. At each Newton iteration a linear nonsymmetric system of equations is solved. We apply the recently derived GMRES (Generalized Minimum RESidual) algorithm [1], [2] to this problem. This algorithm has the property of minimizing at each iteration an appropriate norm of the residual over a Krylov space. Various preconditioning of the linear system, including Element by Element factors or block diagonal matrices are proposed.

We introduce a nonlinear extension of GMRES [3], [4] and selected preconditioners. This approach alleviates the storage requirements since the Jacobian matrix is not computed: Gateaux derivatives estimates are used instead.

The above ingredients have been tested in the context of two finite element schemes:

1. a SUPG (Streamline Upwind Petrov Galerkin) method based on an entropy formulation [5],

2. a second order accurate upwind scheme using Osher's approximate Riemann solvers with directional slope limiters [6].

Various numerical experiments including transonic and supersonic Euler and Navier Stokes flow calculations around a NACA0012 airfoil and a BINACA0012 double airfoil and a supersonic Navier Stokes flow calculation over a blunt body show significant improvements in speed and/or storage requirements.

2 Preliminaries

2.1 Approximation methods

The Euler equations in terms of conservation variables U are

$$U_{,t} + F_i^E(U)_{,i} = \mathcal{F} \qquad in\ \Omega \tag{1}$$

where F_i^E is the Euler flux and $\mathcal{F}$ is a source term. The Jacobians of the fluxes are defined by $A_i = F^E_{i,U}$. The Euler equations can be rewritten in terms of entropy variables V (where $V = s_{,U}$):

$$A_0 V_{,t} + \widetilde{A}_i V_{,i} = \mathcal{F} \; . \tag{2}$$

$\widetilde{A}_i$ and A_0 are symmetric matrices and A_0 is positive definite.

The following weighted residual equation defines the Streamline Upwind Petrov-Galerkin (SUPG) finite element method.

$$-\int_\Omega W^h_{,i} \cdot F^E_i(V^h) d\Omega$$

$$+\sum_{e=1}^{n_{el}} \int_{\Omega^e} (\widetilde{\tau}_1 \widetilde{A}_k W^h_{,k} + \widetilde{\tau}_2 \widetilde{A}_{\|k} W^h_{,k}) \cdot (\widetilde{A}_i V^h_{,i} - \mathcal{F} + A_0 V^h_{,t}) d\Omega$$

$$= \int_\Omega W^h \cdot (\mathcal{F} - A_0 V^h_{,t}) d\Omega + \int_\Gamma W^h \cdot (-F^E_i(V^h)) n_i d\Gamma \tag{3}$$

where $\widetilde{W}^h = W^h + (\widetilde{\tau}_1 \widetilde{A}_k W^h_{,k} + \widetilde{\tau}_2 \widetilde{A}_{\|k} W^h_{,k})$ is the perturbed weighting function. See [5] for definition of $\widetilde{\tau}_1$, $\widetilde{\tau}_2$ and $\widetilde{A}_{\|k}$. Γ is the boundary of the domain and n_i its normal vector.

A second-order approximation is defined in the following way. Considering a Finite Element space approximation $\mathcal{V}_\ell$ of continuous functions (currently linear functions by element) and a space of discontinuous functions $\mathcal{W}_\ell$ such that there exists a bijective operator S from $\mathcal{V}_\ell$ onto $\mathcal{W}_\ell$, the formulation can be written :

$$\int_\Omega U^h_{,t} \cdot W^h d\Omega + \int_\Omega F^E_{i,i}(U^h) \cdot S(W^h) d\Omega = \int_\Omega \mathcal{F} \cdot S(W^h) d\Omega \tag{4}$$

where U^h and W^h are in $\mathcal{V}_\ell$. After integration by parts of the second integral, the resulting term is computed with an upwind approximation through a Riemann solver. Boundary conditions are treated in the boundary integral evaluation.

2.2 Semi-discrete equations

Spatial discretization of both formulations leads to the following semi-discrete system of equations:

$$M\dot{X} + N = F \tag{5}$$

where $M = M(X,t)$ is the generalized "mass" matrix, $N = N(X,t)$ is the generalized convection term, $F = F(X,t)$ is the force vector, X is the vector of unknown nodal values and a superposed dot denotes time differentiation. The initial value problem for (5) consists of finding $X = X(t)$ satisfying (5) and the initial condition $X(0) = X_0$ where X_0 is given. The arrays in (5)are assembled from element contributions.

3 Methods with a linearized Jacobian

3.1 Transient algorithm

We consider the following algorithm:

$$R = F_n - N(X_n) \tag{6}$$

$$M^* = M_n(X_n) + \beta \Delta t C_n \tag{7}$$

$$X_{n+1} = X_n + \Delta t M^{*-1} R \,. \tag{8}$$

This algorithm is a member of the family of predictor multi-corrector algorithm which can be derived by combining the trapezoidal rule and Newton's method. (A derivation of the predictor-multicorrector algorithm can be found in [7].) C_n is the linearized jacobian of N at point X_n and Δt the time step. Taking $\beta = 0$ and M^* lumped gives an explicit algorithm (actually a multistep version is used)

Step (8) involves the inversion of a linear system. An iterarive algorithm can be used, for example the GMRES algorithm that we now describe

3.2 Definition of the linear GMRES algorithm

Let us rewrite the system in the following way

$$Ax = b \tag{9}$$

where A is $neq \times neq$, x and b are $neq \times 1$.

Matrix A is nonsymmetric and standard iterative solution methods are not efficient. New algorithms have been proposed recently. We used the GMRES algorithm. It was presented in a number of papers. Implementation issues are discussed in [1]. The work of [3] proves the potential of the GMRES method for industrial applications. A unified formulation for a large class of solvers for nonsymmetric systems can be found in Saad and Schultz [8].

The GMRES algorithm looks for a solution x under the form $x = x_0 + z$ where x_0 is an initial guess and z belongs to the Krylov space $K = < r_0, Ar_0, \ldots, A^{k-1}r_0 >$ $(r_0 = b - Ax)$. The solution x is chosen such that $||b - Ax||$ is minimum.

The algorithm first proceeds to find an orthonormal basis of space K. This is done via Gramm-Schmidt orthonormalization. In this process, a rectangular matrix called the Hessenberg matrix is generated. The following calculations are performed.

$$\hat{v}_1 = r_0, \quad v_1 = \frac{r_0}{||r_0||} \tag{10}$$

for $i = 2$ to k

$$\hat{v}_{i+1} = Av_i - \sum_{j=1}^{i} \beta_{i+1,j} v_j, \qquad \text{where} \quad \beta_{i+1,j} = (Av_i, v_j) \tag{11}$$

$$v_{i+1} = \frac{\hat{v}_{i+1}}{||\hat{v}_{i+1}||} \,. \tag{12}$$

After k steps, the Hessenberg matrix is

$$H_k = \begin{pmatrix} (Av_1, v_1) & (Av_2, v_1) & \ldots & (Av_k, v_1) \\ ||\hat{v}_2|| & (Av_2, v_2) & \ldots & (Av_k, v_2) \\ 0 & ||\hat{v}_3|| & \ddots & \vdots \\ \vdots & \vdots & \ddots & (Av_k, v_k) \\ 0 & 0 & \ldots & ||\hat{v}_{k+1}|| \end{pmatrix} \,. \tag{13}$$

This matrix is rectangular $(k+1 \times k)$. If we denote by $V_k = (v_1, v_2, \ldots, v_k)$, the $neq \times k$ matrix whose columns are the first k basis vectors, the following relation can be checked

$$AV_k = V_{k+1} H_k \ . \tag{14}$$

Having generated the orthonormal basis V_k we proceed to find $x = x_0 + z = x_0 + \sum y_l v_l$ where y_l are reals. Let us define $y = \{y_1, \ldots y_k\}, e_1 = \{1, 0, \ldots, 0\}$ and $\delta = ||r_0||$.

$$\begin{aligned} ||b - Ax|| &= ||b - A(x_0 + \sum_{l=1}^{k} y_l v_l)|| \\ &= ||r_0 - AV_k y|| \end{aligned} \tag{15}$$

using (14)and the notations above

$$\begin{aligned} ||b - Ax|| &= ||V_{k+1}(\delta e_1 - H_k y)|| \\ &= ||\delta e_1 - H_k y|| \text{ (using the orthonormality)} . \end{aligned} \tag{16}$$

The problem of finding y such that

$$||\delta e_1 - H_k y|| = \min_{y_0 \in R^k} ||\delta e_1 - H_k y_0|| \tag{17}$$

is easy due to the small size and "almost" triangular structure of H. (a Q-R algorithm is used)

Remarks:

1 In the Gramm-Schmidt orthonormalization, *if the matrix A is symmetric* it is sufficient to orthogonalize v_{k+1} with respect to v_k and v_{k-1}.

2 The algorithm can be restarted by taking $x_0 = x_p$. Typically, the size of the Krylov space is smaller than 40 even for problems involving 50,000 equations and several restarts are needed.

3.3 Preconditioning

We replace the problem of finding x such that

$$Ax = b \tag{18}$$

by that of finding $\tilde{x}$ such that

$$\tilde{A}\tilde{x} = \tilde{b} \tag{19}$$

where $\tilde{A}$ is better conditionned than A.

Four preconditioning were tried.

1 *diagonal preconditioning*

$\tilde{A} = W^{-1}A$, $\tilde{b} = W^{-1}b$ and $\tilde{x} = x$ where $W = \text{diag}(A)$.

2 *symmetric diagonal preconditioning*

$\tilde{A} = W^{-1/2}AW^{-1/2}$, $\tilde{b} = W^{-1/2}b$ and $\tilde{x} = W^{1/2}x$.

3 *reordered Crout EBE preconditioning*

$\tilde{A} = B^{-1}A$, $\tilde{b} = B^{-1}b$ and $\tilde{x} = x$; with

$$B = W^{1/2}CW^{1/2} \tag{20}$$

where

$$C = \prod_{e=1}^{nel} L(I + \overline{A}_e) \prod_{e=1}^{nel} D(I + \overline{A}_e) \prod_{e=nel}^{1} U(I + \overline{A}_e) \tag{21}$$

$$\overline{A}^e = W^{-1/2}(A^e - D(A^e))W^{-1/2} \tag{22}$$

We use the following notations: A_e is the elementary matrix and $L(M)$, $D(M)$, $U(M)$ are the Crout decomposition factors of matrix M. Heuristically, EBE takes advantage of the fact that the matrices are the result of a finite element assembly procedure. The justification for the EBE preconditioning matrix can be found in [7]. Discussions can be found in [9], [10].

4 *2 pass symmetric Crout EBE preconditioning*

$\tilde{A} = P^{-1}AP^{-T}$, $\tilde{b} = P^{-1}b$ and $\tilde{x} = P^T x$; with

$$P = W^{1/4}CW^{1/4} \tag{23}$$

where

$$C = \prod_{e=1}^{nel} L(I + \frac{1}{2}\overline{A}_e) \prod_{e=1}^{nel} D(I + \frac{1}{2}\overline{A}_e) \prod_{e=nel}^{1} U(I + \frac{1}{2}\overline{A}_e). \tag{24}$$

Remarks

1 With the method based on entropy variables, preconditioning makes GMRES dimensionally consistent: the entries of each matrices have physical dimensions such that matrix-matrix and matrix-vector multiplications result in additions which are physically consistent. In particular, the entries of the Hessemberg matrix are non-dimensional.

2 The SUPG formulation was implemented with a data structure where the element matrices A^e are stored. When an EBE preconditioning was used, L^e, U^e and D^e were also stored.

3 The iterative process stops when $||b - Ax|| \leq \epsilon_l ||b||$.

3.4 Numerical tests

They include both Euler and Navier-Stokes calculations with both attached and detached shocks. The following notations are used: M is the Mach number, α is the angle of attack and Re is the Reynolds number.

The following test problems were chosen with the SUPG algorithm The first test problem is the Euler calculation of the flow around a NACA0012 airfoil at M = .85 and α = 1 degree. The mesh is presented in Figure 1: It is composed of unstructured triangles and includes 800 nodes and 1514 triangles. The solution obtained is presented in Figure 2 (Isomach lines). The second test problem is the Navier-Stokes calculation of the flow around a NACA0012 airfoil at M = .85, α = 0. and Re = 500. The same mesh is used as for the first problem. For both calculations the CFL number was 5 and 150 time steps where enough to converge. The third test problem is the Euler calculation of the flow around a BINACA12 double airfoil at M = .55 and α = 6 degrees. The mesh is presented in Figure 3 : It is composed of unstructured triangles and includes 1728 nodes and 3286 triangles. The solution obtained is presented in Figure 4 (Isomach lines). The fourth test problem is the Navier-Stokes calculation of the flow around a BINACA0012 double airfoil at M = .55, α =0. and Re = 500. The same mesh is used as for the third problem. For both calculations the CFL number was 5 and 200 time steps were required. The fifth test problem is the Navier-Stokes calculation of the flow past a blunt body at M = 3 and Re = 1000. The mesh is presented in Figure 5 : It includes 972 nodes and 1820 triangles. The solution obtained is presented in Figure 6 (Isomach lines). The CFL number was 5 and 300 time steps were required. For all problems, the size of the Krylov space was 14 and ϵ_l was 0.1 .

The ratio of the CPU required by the iterative algorithm and the CPU required by a direct solver are presented in Table 1 below. Three preconditioners are tested.

Table 1 - CPU requirements SUPG

TEST PROBLEM	GAUSS	GMRES+2	GMRES+3	GMRES+4
NACA Euler	1.	.28	.47	.24
NACA N.S.	1.	.28	.47	.27
BINACA Euler	1.	.156	.39	*
BINACA N.S.	1.	.157	.39	*
BLUNT N.S.	1.	.59	1.06	.74

The same data for the storage requirements are presented in Table 2 below.

Table 2 - storage requirements SUPG

	GAUSS	LIN GMRES+1,2	LIN GMRES+3,4
NACA	1.	.50	.83
BINACA	1.	.33	.58
BLUNT BODY	1.	.64	1.11

The second-order upwind method has been applied to the computation of two flow problems around a NACA0012 airfoil : a transonic flow at M = .85 and α=0 degree, a supersonic flow at M = 2. and α = 10 degrees. The mesh presented in Figure 7 is composed of 1360 nodes and 2500 triangles.

The linearized implicit formulation has been tested combined with a Gauss-Seidel relaxation method, the linear GMRES algorithm with and without diagonal preconditioning. For all computations the Krylov space dimension is 6 and ϵ_l is taken to be .1. Because of the presence of stagnation points in the flow, we cannot use a large CFL number at the beginning of the calculation. The CFL number law is given by CFL=minimum(numerous of iteration,50). Comparisons in terms of CPU are presented in Table 3. Efficiency of the GMRES solver is achieved when combined with the preconditioning. The storage requirements are the same for each solver as shown in Table 4. Only the non-zero entries of the matrix are stored.

Table 3 - CPU requirement -upwind

TEST PROBLEM	GAUSS-SEIDEL	GMRES	GMRES+1
NACA M=.85,$\alpha = 0$	1.	1.1	0.8
NACA M=2.,$\alpha = 10$	1.	1.2	0.9

The above methodology has also been used with a modified CFL number law given by CFL=minimum(numerous of iteration, 1./residual). It appears that the linearized implicit algorithm combined with Gauss-Seidel relaxation does not converge while the same algorithm combined with linear GMRES is very efficient. Histories of convergence can be found on Figures 8 and 9 for the same transonic calculation as in Table 3.

Clearly, the linear GMRES algorithm offers a good alternative to direct solve: it alleviates both the storage and CPU requirements; this is true even for meshes with a small number of points and especially so when the mesh is unstructured.

4 The nonlinear GMRES algorithm

4.1 Definition

Implicit algorithms allow the use of large time-steps while the time-step of the explicit algorithm is severaly limited by a stability condition. On the other hand, the block-matrix arising in the implicit integration increases drastically the required memory.The non-linear GMRES algorithm seems to be a good trade-off as a transient algorithm. Let us recall the transient algorithm defined by (6)- (8)and introduce the following functionnal at each time step :

$$G_n(X) = M(X_n)(X - X_n) - \Delta t R(X) . \tag{25}$$

We want to solve the problem :

$$G_n(X) = 0 \tag{26}$$

where X and $G_n(X)$ are $n_{eq} \times 1$.

Let denote by $\overline{G_n}(X; d)$ the Jacobian of G_n at point U in the direction d.

At each iteration p, the non-linear algorithm looks for a solution $X^{(p+1)} = X^{(p)} + Y$ where $X^{(0)}$ is an initial guess and Y belongs to the Krylov space K constructed from successive Jacobian of the function G_n. Y is chosen such that $||G_n(X^{(p+1)})||$ is minimum.

As in the linear case, the algorithm first proceeds to find an orthonormal basis of space K. The following calculations are performed.

For $p \geq 0$,

$$r_1^{(p)} = G_n(X^{(p)}), v_1^{(p)} = \frac{r_1^{(p)}}{||r_1^{(p)}||} . \tag{27}$$

For $i = 1$ to k

$$r_{i+1}^{(p)} = \overline{G_n}(X^{(p)}; v_i^{(p)}) - \sum_{j=1}^{i} b_{i+1,j}^{(p)} v_j^{(p)} \tag{28}$$

$$v_{i+1}^{(p)} = \frac{r_{i+1}^{(p)}}{||r_{i+1}^{(p)}||} \tag{29}$$

where $\overline{G_n}(X^{(p)}; v)$ is evaluated exactly or by the approximate formula

$$\overline{G_n}(X^{(p)}; v) = \frac{G_n(X^{(p)} + \epsilon v) - G_n(X^{(p)})}{\epsilon} \quad \text{with } \epsilon \text{ small} \tag{30}$$

and

$$b_{il}^{(p)} = (\overline{G_n}(X^{(p)}; v_i^{(p)}), v_l^{(p)}) \tag{31}$$

Endly $X^{(p+1)}$ is solution of the minimization problem :

Find $\lambda^{(p)} = \{\lambda_j^{(p)}\}_{j=1}^{k}$ such that $\forall \mu$

$$||G_n(X^{(p)} + \sum_{j=1}^{k} \lambda_j^{(p)} v_j^{(p)})|| \leq ||G_n(X^{(p)} + \sum_{j=1}^{k} \mu_j^{(p)} v_j^{(p)})|| \tag{32}$$

$$X^{(p+1)} = X^{(p)} + \sum_{j=1}^{k} \lambda_j^{(p)} v_j^{(p)} . \tag{33}$$

Remarks

1 We can easily verify that the constructed vector sequence $(v_l^{(p)})_{l=1}^{k}$ is orthogonal.

2 To evaluate $\lambda^{(p)}$, we approximate the minimization problem (32) by a quadratic one replacing the functionnal by

$$||G_n(X^{(p)}) + \sum_{j=1}^{k} \lambda_j^{(p)} \overline{G}_n(X^{(p)}; v_j^{(p)})|| .$$

3 The norm involved in the algorithm must be defined and carefully chosen. Actually, the most simple norm i.e. the l2-norm was taken, which leads to a a-priori non-dimensionally consistent formulation. A suitable preconditioning of the functionnal can overcome this consistency problem.

4.2 Preconditioning

We replace the problem of finding X such that $G_n(X) = 0$ by that of finding $\widetilde{X}$ such that

$$\widetilde{G_n}(\widetilde{X}) = 0 \tag{34}$$

where $\widetilde{G_n}$ is better conditionned than G_n. It seems natural to use preconditioners derived from the linearization of the function G_n at the solution at the previous iteration (X_n) involving the matrix $M^*(X_n)$ described in the previous Section. If $S(X_n)$ denotes a precondioning matrix of $M^*(X_n)$ (for example, one of those described previously) we will take

$$\widetilde{G_n}(\tilde{X}) = S^{-1}(X_n)G_n(X) \tag{35}$$

4.3 Numerical tests

In the context of the SUPG formulation, the nonlinear GMRES algorithm was tested with test problems 1 and 3. The size of the Krylov space was 12, ϵ in (30) was $.510^{-2}$. The CFL number was 1 and 300 time steps were required for convergence. The CPU and memory requirement data are presented in Tables 4 and 5 below using the same format as for Tables 1 and 2 respectively.

Table 4 - CPU requirements SUPG

TEST PROBLEM	GAUSS	NL GMRES
NACA12	1.	1.82
BINACA12	1.	.90

Table 5 - storage requirements SUPG

TEST PROBLEM	GAUSS	NL GMRES
NACA12	1.	.25
BINACA12	1.	.098

We have computed with the second-order upwind method the same test problems as in the previous Section but with various meshes. Table 6 shows a comparison of different solvers :

- The explicit one which is a multistep time integrator

- The non-linear GMRES algorithm with or without diagonal preconditioning. The CFL number used in all calculations is 10. The Krylov space dimension is 4 and 4 iterations are performed at each time step.

- The linearized implicit solver of the previous Section. The CFL law is given by CFL=minimum(numerous of iteration,50).

For both calculations, the preconditioned non-linear GMRES algorithm appears to be more efficient when the number of unknowns increases. Storage requirements (which are linear in the number of mesh nodes) are given in Table 7.

Table 6 - CPU requirements - upwind

NACA0012	EXPL.	NLGMRES	NLGMRES+1	LIN.IMPL.
355nodes M=.85 ,α=0	1.	.80	.60	.20
1360nodes M=.85 ,α=0	1.	.80	.60	.20
2500nodes M=.85 ,α=0	1.	.43	.25	.19
1360nodes M=2. ,α=10	1.	.88	.60	.20
2500nodes M=2. ,α=10	1.	.80	.54	.20

Table 7 - storage requirements - upwind

	EXPL.	NLGMRES	NLGMRES+1	LIN.IMPL.
NACA12(355nodes)	0	.10	.25	1.
NACA12(1360nodes)	0	.10	.25	1.
NACA12(2500nodes)	0	.10	.25	1.

5 Discussion and Conclusion

For implicit matrix solution the linear GMRES algorithm is an efficient solver when combined with an adapted preconditioner. Storage requirements are significantly reduced compared to the direct Gauss solver and CPU efficiency is also increased. In the context of SUPG, the symmetric diagonal preconditioner proved to be very effective. Further research is still needed to find better preconditioners; in particular, the EBE preconditioner in our present version does not seem to be optimal.

On the other hand, the problem of matrix storage is alleviated by the use of the nonlinear GMRES method as a transient algorithm.In terms of efficiency, this method lies between the simple explicit and a linearized implicit one and seems very attractive for 3D problems where the memory required by the matrix storage tends to be prohibitive.

One can advocate a domain decomposition strategy where one would use an explicit method in the near field, a non-linear GMRES method on intermediate regions and an implicit method with a linearized jacobian in the regions where mesh refinement has resulted in small elements for example next to the body or where shocks are present (This implicit method could use either a direct or an iterative solver with various preconditioners).

Acknowledgments

This research was sponsored by DRET under Grant $DRET$86.34.356. The authors would like to thank Q.V. Dinh, B. Mantel and F. Shakib for helpful discussions.

References

[1] Y. Saad and M.H. Schultz, "GMRES: a Generelized Minimal Residual Algorithm for Solving Nonsymmetric Linear Systems", *Research Report YALEU/DCS/RR-254*,(1983).

[2] H.D. Simon, "The Lanszos Algorithm for Solving Symmetric Linear Systems", *Ph.D. Thesis, University of California, Berkeley* (1982).

[3] L.B. Wigton, N.J. Yu and D.P. Young, "GMRES Acceleration of Computational Fluid dynamics Codes", *AIAA 7th Computational Fluid Dynamics Conference, AIAA CP 854*, (1985).

[4] R. Glowinski and J. Periaux, "The GMRES Algorithm for Nonlinear problems", Private Communication, (1987).

[5] M. Mallet, "A Finite Element Method for Computational Fluid Dynamics", *Ph. D. Thesis, Stanford University*, (1985).

[6] B. Stoufflet, J. Periaux, F. Fezoui and A. Dervieux, "Numerical Simulation of 3-D Hypersonic Euler Flows around Space Vehicles using Adapted Finite Elements", AIAA 87-560, (1987).

[7] J.M. Winget, "Element-by-element Solution Procedures for Nonlinear Transient Heat Conduction Analysis", *Ph.D. Thesis, Stanford University*, (1983).

[8] Y. Saad and M.H. Schultz, "Conjugate Gradient-Like Algorithms for Solving Nonsymmetric Linear Systems", *Mathematics of Computation*, Vol 44, (1985), pp 417-424.

[9] T.J.R. Hughes, J. Winget, I. Levit and T.E. Tezduyar, "New Alternating Direction Procedures in Finite Element Analysis based Upon EBE Approximate Factorizations", *Computer Methods for Nonlinear Solids and Structural Mechanics* (eds S.N Atluri and N. Perone) AMD-Vol 45, 75-109, ASME, New-York, (1985).

[10] T.E. Tezduyar and J. Liou, " Element-by-element and Implicit/Explicit Finite Element Formulations in Computational Fluid Dynamics", *Proceedings, 1st International Symposium on Domain Decomposition Methods for Partial Differential Equations*, (1987).

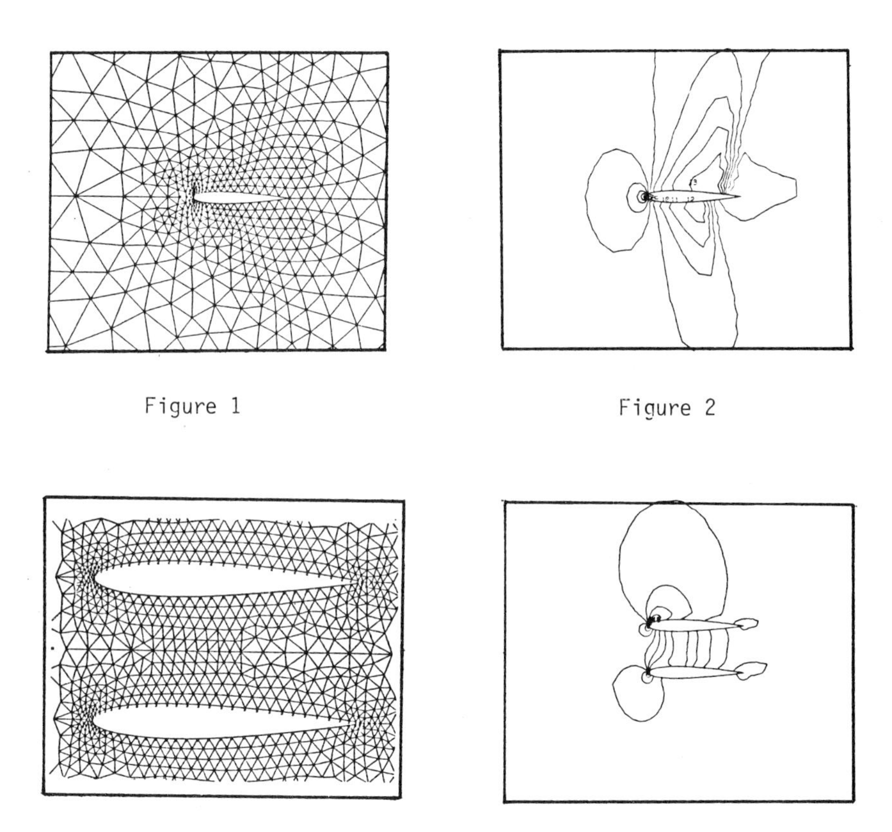

Figure 1

Figure 2

Figure 3

Figure 4

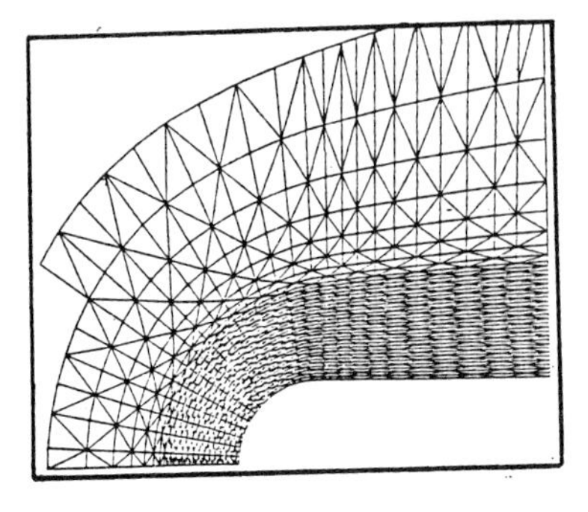

Figure 5

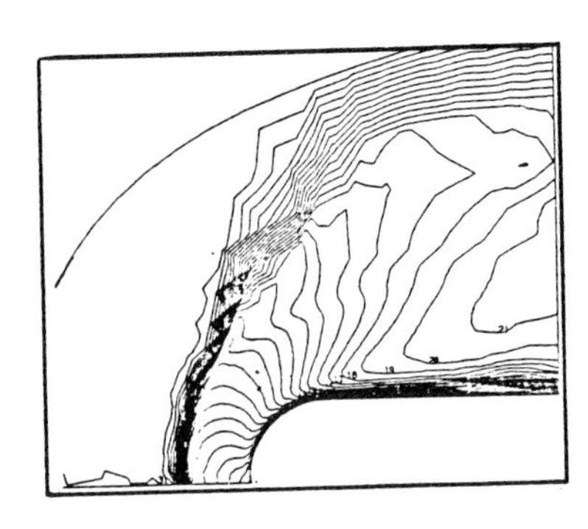

Figure 6

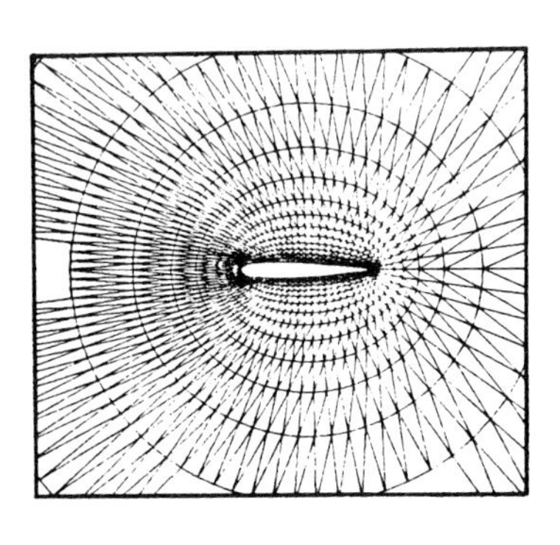

Figure 7

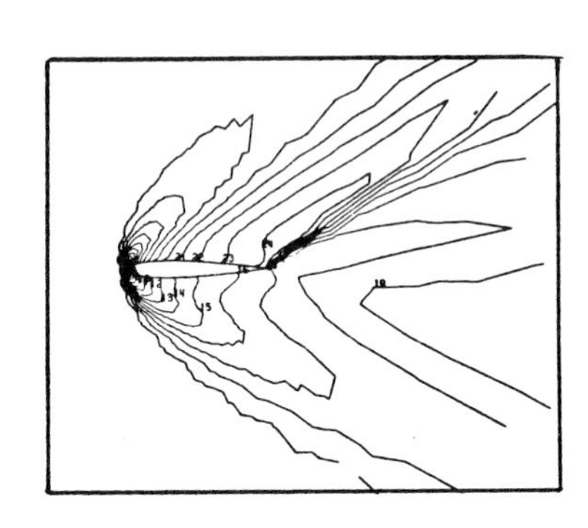

Figure 8

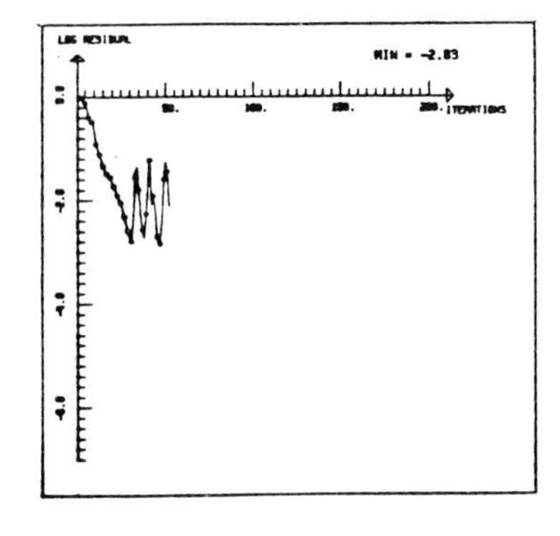

Figure 9

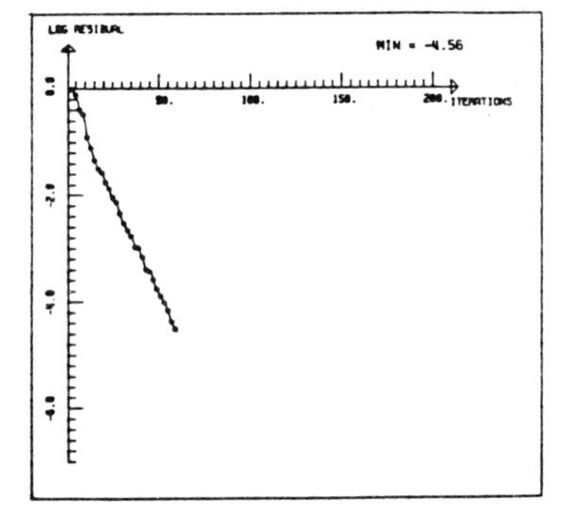

Figure 10

COMPUTATIONS OF UNSTEADY DISSOCIATING NITROGEN FLOWS

Lionel Marraffa
ONERA -BP 72
92322 Chatillon
France

George S. Dulikravich
Pennsylvania State University
State College, PA 16802
U.S.A.

George S. Deiwert
NASA Ames Research
Moffett Field, CA94035
U.S.A.

Abstract

A time-accurate computer program has been developpped to solve flows involving nonequilibrium chemical reactions and vibrational relaxation.The program is based on a multidomain approach and an explicit MacCormack scheme. Results are presented for three simple test cases, involving nitrogen dissociation in a closed box.

Introduction

The growing interest in the field of hypersonic flow computations stimulated the development of a code able to treat nonequilibrium chemically reacting flows. Several approaches [1], [2] are already giving good results for chemical equilibrium, but very few codes that can treat a non equilibrium flow are available [3], and most of them are not time accurate.

A program has been developped to treat unsteady hypersonic flows with non equilibrium chemistry and vibrational nonequilibrium. Even though this program is able to handle any number of chemical species, the results presented here correspond only to the case of dissociating nitrogen. The time accuracy is preserved by the scheme.

Governing equations

the governing equations are of the general form:

$$\frac{\delta \tilde{f}}{\delta t} + \nabla \cdot \approx{F}(\tilde{f}) - \tilde{u}' \cdot \tilde{\nabla} \tilde{f} = \tilde{S}$$

with:

$$\tilde{f} = \begin{bmatrix} X_1 \\ \vdots \\ X_n \\ \rho u_1 \\ \rho u_2 \\ \rho e_t \\ E_{vib_1} \\ \vdots \\ E_{vib_m} \end{bmatrix} \qquad \tilde{S} = \begin{bmatrix} \dot{X}_{1_{react}} \\ \vdots \\ \dot{X}_{n_{react}} \\ 0 \\ 0 \\ \Delta H_{react} \\ \dfrac{E^{(eq)}_{vib_1} - E_{vib_1}}{\tau_{v1}} \\ \dfrac{E^{(eq)}_{vib_m} - E_{vib_m}}{\tau_{vm}} \end{bmatrix}$$

and :

$$
\underset{\approx}{F}(\underset{\sim}{f}) = \begin{bmatrix} \underset{\sim}{u}\, X_1 - \rho\, \mathfrak{D}_1\, \underset{\sim}{\nabla}\, (X_1/\rho) \\ \vdots \\ \underset{\sim}{u}\, X_n - \rho\, \mathfrak{D}_n\, \underset{\sim}{\nabla}\, (X_n/\rho) \\ \rho\, \underset{\sim}{u} \otimes \underset{\sim}{u} + P\, \underset{\approx}{I}_2 - \underset{\approx}{\tau} \\ (\rho\, e_t + P - \underset{\approx}{\tau})\, \underset{\sim}{u} + \underset{\sim}{q} - \sum_{i=1}^{i=n} \rho\, h_i \frac{\rho\, \mathfrak{D}_i}{X_i}\, \underset{\sim}{\nabla}\, (X_i/\rho) \\ E_{vib_1}\, \underset{\sim}{u} \\ \vdots \\ E_{vib_m}\, \underset{\sim}{u} \end{bmatrix}
$$

This system is completed by the thermodynamic relations:

$$
P = \sum_{i=1}^{i=n} X_i\, R\, T \; ; \quad \rho = \sum_{i=1}^{i=n} m_i\, X_i \; .
$$

By associating the diatomic species to the first m indices, and the monoatomic species to indices m+1 to n:

$$
\rho\, e_t = \rho\, u^2/2 + \Big(\sum_{i=1}^{i=m} \frac{5}{2} X_i + \sum_{i=m+1}^{i=n} \frac{3}{2} X_i \Big)\, R\, T + \sum_{i=1}^{i=m} X_i\, E_{vib_i} \; .
$$

The heat of reaction is obtained from the chemical source term:

$$
\Delta H_{react} = \sum_{i=1}^{i=n} \dot{X}_{react_i}\, \Delta H_{f_i} \; .
$$

The details on the computation of thermodynamic and transport properties can be found in reference [4].

As the code uses a multidomain approach, two kinds of boundary conditions have been used for the test case: adiabatic wall and connection between two consecutive domains.

For an adiabatic wall, the velocity $\underset{\sim}{u}$ is imposed equal to the velocity of the wall. The walls are assumed to be impermeable and non ablative. Thus, the boundary conditions can be summarized for a motionless wall as:

$$\underset{\sim}{u} = \underset{\sim}{\emptyset}. \quad \frac{\partial X_i}{\partial \underset{\sim}{n}} = \emptyset. \quad \frac{\partial T}{\partial \underset{\sim}{n}} = \emptyset.$$

For the boundary between two overlapping domains, the values are simply reported from one domain to another.

Chemical source term

Let us assume we have n species, $S_1, \ldots, S_n$ undergoing p reactions:

$$\text{reaction } j: \quad \sum_{i=1}^{i=n} \nu_i^j \, S_i \quad \underset{k'_j}{\overset{k_j}{\rightleftarrows}} \quad \sum_{i=1}^{i=n} \nu'^j_i \, S_i \, .$$

Where the k_j represent the forward rates, and k'_j the backward rates of reaction. These rates are functions of temperature:

$$k'_j(T) = A \, T^B \exp (E/T) .$$

k_j is obtained from k'_j and the constant of equilibrium K_e:

$$k_j(T) = K_e(T) \; k'_j(T)$$

with

$$K_e(T) = (\alpha + \beta \, T + \gamma \, T^2) \exp (-\varepsilon/T) .$$

The rate of production of specie S_i is related to the concentrations X_j of the different species present in the solution by:

$$\frac{dX_i}{dt} = f_i(X_1, \ldots, X_n, T)$$

with:

$$f_i(X_1,\dots,X_n) = \sum_{j=1}^{j=p} (\nu'^j_i - \nu^j_i).(k_i \prod_{l=1}^{l=n} X_l^{\nu^j_l} - k'_i \prod_{l=1}^{l=n} X_l^{\nu'^j_l}).$$

Let us consider a point M in the field at a certain instant t. The chemical composition at M is:

$$X_1(M,t),\ X_2(M,t),\ \dots\ ,\ X_n(M,t)\ .$$

If the point M follows the movements of the fluid, at time t+Δt its new position will be M'.

$$\underset{\sim}{MM'} = \underset{\sim}{u}\ .\ \Delta t\ .$$

The new composition : $X_1(M',t+\Delta t),\ \dots\ ,X_n(M',t+\Delta t)$, is reached

through three different processes: convection, diffusion and

chemical reactions.

This last contribution is modeled by a source term $\Delta t * \dot{X}_{i_{reaction}}$.

This term is the difference of the concentrations X_i of

specie i between time t and time t+Δt, for a reaction taking place at a given temperature T(M,t), with an initial concentration Xi(M,t). Thus, introducing f, define above:

$$\dot{X}_{i_{reaction}} = \int_t^{t+\Delta t} f(X_1(M,\tau),\dots,X_n(M,\tau),T(t))\ d\tau$$

and the initial composition of the mixture is:

$$X_i(M,t) = X_i(M',t) + \underset{\sim}{M'M}\ .\ \underset{\sim}{\nabla}\ X_i(M',t).$$

Application to the case of N_2

For the dissociation of N_2, the chemical reactions are:

$$N_2 + X \rightleftarrows 2N + X$$

where X is a catalyser (N, or N_2). In this case, calling x_1 the concentration of N, and x_2 the concentration of N_2, the following relations apply:

$$\frac{dx_1}{dt} = 2\,(k_1'\,x_1 + k_2'\,x_2)(\,K_e\,x_2 - x_1^{\,2})$$

$$x_1 + 2\,x_2 = x_1^0 + 2\,x_2^0 = C \quad \text{(conservation of atoms N)},$$

In this system, k_1' is the reverse reaction rate for the reaction of dissociation catalysed by N, k_2' for the one catalysed by N_2, and K_e the equilibrium constant. The values of the necessary constants allowing the computation of k_1', k_2' and K_e are taken from reference [5]. This system admits an analytical solution:

$$g(x_1) = g(x_1^0)\ \exp\,(-(t-t_0)(2\,k_1 - k_2))$$

with:

$$g(x_1) = |x_1+a|^{(a-b)(a-d)}\ |x_1+b|^{(b-a)(b-d)}\ |x_1+d|^{(d-a)(d-b)}$$

where:

$$a = \frac{k_1}{2\,k_1 - k_2}\ ;\ b = \frac{K_e/2 + \sqrt{\Delta}}{2}\ ;\ d = \frac{K_e/2 - \sqrt{\Delta}}{2} = -x_1^{\,equilib.}$$

and: $\Delta = K_e^{\,2}/4 + 2\,K_e.C$.

Details of the numerical resolution can be found in reference [4]. This technique can easily be extended to the case of a mixture of dissociating nitrogen and oxygen, whenever the two dissociations are uncoupled.

Results

The code has been applied to three different test cases.In all of them, the geometry is a rectangular box (fig.1), closed, filled with a mixture of molecular nitrogen, N2, and of atomic nitrogen, N.

For the first test case, the box is at a given uniform temperature T and pressure P. The initial concentrations of N2 and N do not correspond to equilibrium composition for the initial pressure and temperature. A very small time step is chosen, adapted to the chemical reactions (10**-18 s). The mixture temperature and pressure increase and in turn the equilibrium composition is modified. An equilibrium is finally reached as shown by the plot of pressure versus time at a point near the walls (fig.2). The recombination of N yield a final pressure and temperature higher than the initial one.

In the second test case, the box is initially partitioned by a membrane. The pressures and temperatures on the two sides of the box are the same and the compositions are different. For this test case, we freeze the chemical reactions, i.e. we impose a species production term always equal to zero. In this situation, the possible physical phenomena are limited to diffusion and convection. These processes are considerably slower than the chemical reactions as the flow is nearly motionless and a much larger time interval is needed to reach the equilibrium. As expected, the concentrations of N2 and N tend to become uniform in the box (fig.3).

The third test case differs from the second case by taking into account the equilibrium chemistry. The diffusion of mass, momentum and energy is a relatively slow process. The convective terms also appear. But the main effect is due to the chemical reactions. The initial conditions correspond to a mixture with a concentration of N over the equilibrium value for the local pressure and temperature. Thus the atoms of N will tend to recombine, as in the first case. On both sides of the membrane, at t=0, we have same pressure and same temperature. But the density and the molar masses are different, and thus correspond to different equilibrium compositions. In turn, the heat released by the chemical reactions on the two sides of the box will be different. Therefore, strong gradients of pressure and temperature appear. An acoustic wave develops and propagates into the box [6], as shown in fig. 4. The frequency of this wave corresponds clearly to the first longitudinal mode of excitation of the box. A strong damping of the wave can also be observed in the fig. 4. This wave is correctly taken into account by the scheme which is time accurate. The last part of the curve corresponds to convection and diffusion processes, tending to produce an uniform temperature and composition. This computation is performed with a quite large time step (2. 10**-7s). One thousand iterations, for 110 points of grid, took 8 minutes on a VAX 8550. Even though this test case corresponds to equilibrium chemistry, the program is computing the source terms as the limit (for a long time step) of nonequilibrium chemistry source terms.

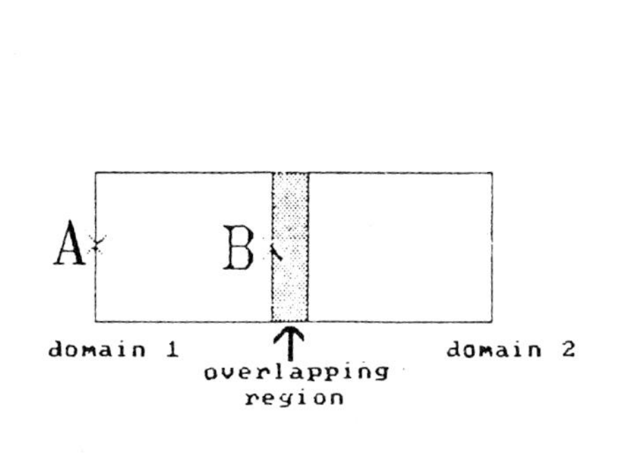

fig. 1: TEST CASE GEOMETRY

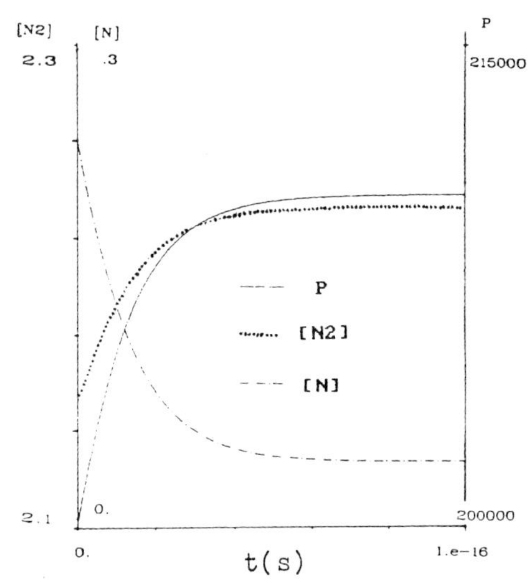

fig. 2: VARIATION OF PRESSURE AND CONCENTRATION OF N VERSUS TIME (REACTING MIXTURE, UNIFORM INITIAL COMPOSITION)

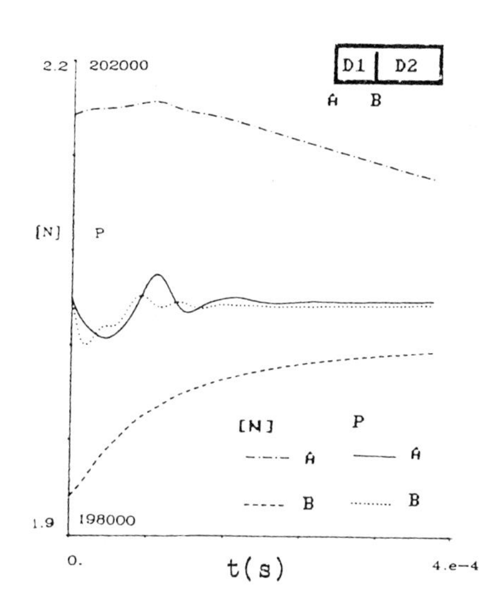

fig. 3: VARIATION OF PRESSURE AND CONCENTRATION OF N VERSUS TIME (FROZEN CHEMISTRY, NONUNIFORM INITIAL COMPOSITION)

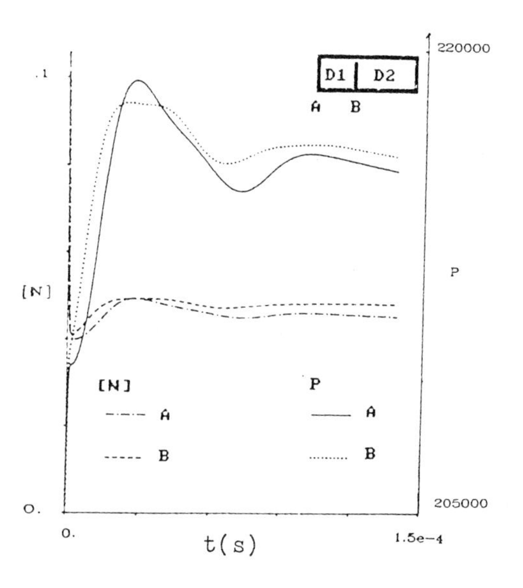

fig. 4: VARIATION OF PRESSURE AND CONCENTRATION OF N VERSUS TIME (REACTING MIXTURE, NONUNIFORM INITIAL COMPOSITION)

Aknowledgement

This research has been supported by NASA Ames Grant NCA 2-120 and by Direction des Recherches, Etudes et Techniques under contract 85.34.1385.

References

[1] Balakrishnan A., Davy W.C. and C.K. Lombard, "Real-Gas Flowfields About Three Dimensional Configurations", J. Spacecraft and Rockets, Vol.22, No1, Jan.-Feb. 1985.

[2]Green, M.J. Davis, W.C. and C.K. Lombard, "CAGI2, a CSCM based Procedure for Flow of an Equilibrium Chemically Reacting Gas", AIAA paper No 85-0927, Williamsburg, Virginia (1985).

[3] Gnoffo, P.A. McCanlers R.S. and H.C. Yee," Enhancements to Program LAURA for Computation of Three-Dimensional Hypersonic Flow" AIAA paper 87-0280, Reno, Nevada (1987).

[4] Marraffa, L. Dulikravich, G.S. and G.S.Deiwert,"Numerical Simulation of Two-Dimensional Viscous Unsteady Dissociating Nitrogen flows", AIAA paper No 87-2549-CP, Monterey, California (1987).

[5] Rakich, J.V., Bailey, H.E. and C. Park, "Computation of Nonequilibrium Three-Dimensional Inviscid Flow over Blunt Nosed Bodies Flying at Supersonic Speeds", AIAA paper 75-835 (1975).

[6] Habiballah, M., Marraffa, L. and Monin, H.," Numerical Simulation of High Frequency Instabilities in a Liquid Propellant Engine Through Combustion-Pressure Coupling", Tenth International Colloquium on Dynamics of Explosions and Reactive Systems, Berkeley, California, aug. 4-9, 1985.

COMPARATIVE STUDY OF HIGH-RESOLUTION SHOCK-CAPTURING SCHEMES FOR A REAL GAS

J.-L. Montagné‡ and H.C. Yee*
NASA Ames Research Center, Moffett Field, CA 94035 USA

and

M. Vinokur**
Sterling Software, Palo Alto, CA 94030 USA

SUMMARY

The recently developed second-order explicit shock-capturing methods of the van Leer, Harten, and Yee types, in conjunction with the generalized flux-vector splittings of Vinokur and Montagné, and a generalized Roe's approximate Riemann solver of Vinokur for a real gas are studied. The comparisons are made on different one-dimensional Riemann problems for equilibrium air with various ranges of Mach numbers, densities, and pressures. The numerical results in the supersonic and low-hypersonic regimes indicate that these approaches produce good shock-capturing capability and that the shock resolution is only slightly affected by the state equation of equilibrium air. The difference in shock resolution between the various methods varies slightly from one Riemann problem to another, but the overall accuracy is very similar. The relative efficiency in terms of operation-count for the different methods is within 30%. The main difference between the methods lies in their versatility in extending to implicit methods with efficient solution procedures, especially for multidimensional steady-state computations.

INTRODUCTION

Several newly developed high-resolution shock-capturing methods [1-5] have been shown to be applicable to many multidimensional fluid dynamics problems for a perfect gas. For problems containing moderate to fairly strong shocks, these methods produce highly accurate solutions near discontinuities [6-10]. These numerical methods belong to the class of total variation diminishing (TVD) schemes.

There exists many ways to achieve higher-order spatial accuracy and at the same time have TVD-type properties. Here, two ways are considered. The first way is due to Harten [3], Roe [4] and Yee [5]; the second way is due to van Leer [1] (it is sometimes referred to as the MUSCL approach). Hereinafter, we will refer to the first way as the non-MUSCL approach. It is emphasized that the basic high-resolution shock-capturing methods for hyperbolic conservation laws are developed for nonlinear scalar hyperbolic conservation laws. There is no identical theory for nonlinear systems or for the multidimensional counterpart. These schemes are formally extended to one- or higher-dimensional nonlinear systems of hyperbolic conservation laws via the so called Riemann solvers and are evaluated by numerical experiments.

There exist three popular ways of extending scalar schemes to nonlinear systems via the

---

‡Ames Associate on leave from ONERA, B.P. 72, 92322 Chatillon Cedex, France.
*Research Scientist, Computational Fluid Dynamics Branch
**Principal Analyst

Riemann-solver approaches: the exact Riemann solvers [11], the approximate Riemann solvers [12-15], and the flux-vector splitting techniques [16-18]. However, these Riemann solvers as originally developed, are only valid for a perfect gas. An approximate Riemann solver [19] and flux-vector splitting approaches [20-22] have recently been generalized to a real gas (see ref. [7] for details). The objective of this paper is to investigate the applicability of second-order explicit shock-capturing methods of the van Leer, Harten, Yee types [1, 3-5] in conjunction with recently developed generalized flux-vector splittings of Vinokur and Montagné [22] and a generalized Roe's approximate Riemann solver of Vinokur [19] for a real gas.

The combination of the three Riemann solvers and of the differencing algorithms considered above yields five different schemes: a symmetric non-MUSCL scheme [5], an upwind non-MUSCL scheme [6,7], and three MUSCL-type schemes, depending on the Riemann solvers. The present study provides a check on the validity of the generalized formulas, since theoretical prediction of their properties appears to be difficult because of the non-analytic form of the state equation. The values of the state equation are obtained using a curve-fit procedure given in reference [23]. Comparisons are made on the accuracy and robustness of the methods. The test cases chosen here are intended to highlight the effect of the high ratios in pressure or density related to shocks, and the effect of departure from perfect gas in the state equation.

DESCRIPTION OF THE NUMERICAL ALGORITHMS

The conservation laws for the one-dimensional Euler equations can be written in the form

$$\frac{\partial U}{\partial t} + \frac{\partial F(U)}{\partial x} = 0, \tag{1}$$

where $U = \left[\, \rho,\ m,\ e \,\right]^T$ and $F = \left[\, \rho u,\ mu + p,\ eu + pu \,\right]^T$. Here ρ is the density, $m = \rho u$ is the momentum per unit volume, p is the pressure, $e = \rho(\epsilon + \frac{1}{2}u^2)$ is the total internal energy per unit volume, and ϵ is the specific internal energy.

Riemann Solvers

The eigenvalues and eigenvectors of the Jacobian matrix $A = \partial F/\partial U$ are in general used in approximate Riemann solvers. Given two states whose difference is ΔU, Roe [12] obtained an average $\overline{A}$ satisfying $\Delta F = \overline{A}\Delta U$ for a perfect gas. The generalization by Vinokur [19] for an arbitrary gas involves the pressure derivatives $\chi = \left(\partial p/\partial \rho\right)_{\tilde{\epsilon}}$ and $\kappa = \left(\partial p/\partial \tilde{\epsilon}\right)_{\rho}$, where $\tilde{\epsilon} = \rho\epsilon$. The relation $c^2 = \chi + \kappa h$ then gives the speed of sound, where $h = \epsilon + p/\rho$. Introducing $H = h + u^2/2$, Vinokur found the same expressions for $\overline{u}$ and $\overline{H}$ as for the perfect gas, and that $\overline{\chi}$ and $\overline{\kappa}$ must satisfy

$$\overline{\chi}\Delta\rho + \overline{\kappa}\Delta\tilde{\epsilon} = \Delta p. \tag{2}$$

Unique values of $\overline{\chi}$ and $\overline{\kappa}$ are obtained by projecting the arithmetic averages of the values for the two states into this relation (see refs. [19] and [7] for the exact formulas).

Flux-vector splitting methods divide the flux F into several parts, each of which has a Jacobian matrix whose eigenvalues are all of one sign. The approach by Steger and Warming [16] made use of the relation $F = AU$, valid for a perfect gas. Van Leer [17] constructed a different splitting, in which the eigenvalues of the split-flux Jacobians are continuous, and one of them vanishes, leading to sharper capture of transonic shocks. Vinokur and Montagné [22] showed that the expressions for both these splittings can be generalized to an arbitrary gas by using the variable $\gamma = \rho c^2/p$, and adding to the split energy flux a term equal to the product of the split mass flux and the quantity $\epsilon - c^2/\left[\gamma(\gamma - 1)\right]$ (see refs. [19] and [7] for the exact formulas).

Numerical Algorithms

Let U_j^n be the numerical solution at $x = j\Delta x$, $t = n\Delta t$, with Δx the spatial mesh size and Δt the time-step. The second-order in time and second-order in space, explicit difference schemes considered here for both the MUSCL and the non-MUSCL approaches can be written in the following form

$$U_j^{n+1} = U_j^n - \lambda(\widetilde{H}_{j+\frac{1}{2}}^n - \widetilde{H}_{j-\frac{1}{2}}^n), \tag{3}$$

where $\lambda = \Delta t/\Delta x$. The vector functions $\widetilde{H}_{j\pm\frac{1}{2}}$ are sometimes referred to as numerical flux functions.

The numerical flux function $\widetilde{H}_{j\pm\frac{1}{2}}$ for a non-MUSCL-type approach for both the upwind and symmetric TVD schemes [6] using an approximate Riemann solver can be expressed as

$$\widetilde{H}_{j+\frac{1}{2}} = \frac{1}{2}\left[F(U_j) + F(U_{j+1}) + R_{j+\frac{1}{2}}\Phi_{j+\frac{1}{2}}\right]. \tag{4}$$

Here $R_{j+\frac{1}{2}}$ is the eigenvector matrix of the Jacobian A evaluated at an average state between U_j and U_{j+1}.

Second-order Symmetric TVD Scheme, non-MUSCL Approach: The elements of the $\Phi_{j+\frac{1}{2}}$ denoted by $(\phi_{j+\frac{1}{2}}^l)^S$ for a general second-order symmetric TVD scheme [5,6] are

$$(\phi_{j+\frac{1}{2}}^l)^S = -\lambda(a_{j+\frac{1}{2}}^l)^2\widehat{Q}_{j+\frac{1}{2}}^l - \psi(a_{j+\frac{1}{2}}^l)\left[\alpha_{j+\frac{1}{2}}^l - \widehat{Q}_{j+\frac{1}{2}}^l\right]. \tag{5.a}$$

The value $a_{j+\frac{1}{2}}^l$ is the lth characteristic speed a^l of the Jacobian A evaluated at an average between U_j and U_{j+1}. The function ψ is

$$\psi(z) = \begin{cases} |z| & |z| \geq \delta_1 \\ (z^2 + \delta_1^2)/2\delta_1 & |z| < \delta_1 \end{cases}. \tag{5.b}$$

Here $\psi(z)$ in equation (5.b) is an entropy correction to $|z|$ where δ_1 is a small positive parameter. Since all of the test problems contain unsteady shocks, δ_1 is set to zero in all of the computations (see ref. [7] for detailed discussion and numerical examples). The limiter function $\widehat{Q}_{j+\frac{1}{2}}^l$ used in the calculations is:

$$\widehat{Q}_{j+\frac{1}{2}}^l = \text{minmod}\left[2\alpha_{j-\frac{1}{2}}^l, 2\alpha_{j+\frac{1}{2}}^l, 2\alpha_{j+\frac{3}{2}}^l, \frac{1}{2}(\alpha_{j-\frac{1}{2}}^l + \alpha_{j+\frac{3}{2}}^l)\right]. \tag{5.c}$$

The minmod function of a list of arguments is equal to the smallest number in absolute value if the list of arguments is of the same sign, or is equal to zero if any arguments are of opposite sign. Here $\alpha_{j+\frac{1}{2}}^l$ are elements of $\alpha_{j+\frac{1}{2}} = R_{j+\frac{1}{2}}^{-1}(U_{j+1} - U_j)$.

Second-Order Upwind TVD Scheme, non-MUSCL Approach: The elements of the $\Phi_{j+\frac{1}{2}}$ denoted by $(\phi_{j+\frac{1}{2}}^l)^U$ for a second-order upwind TVD scheme [7,24], are

$$(\phi_{j+\frac{1}{2}}^l)^U = \sigma(a_{j+\frac{1}{2}}^l)(g_{j+1}^l + g_j^l) - \psi(a_{j+\frac{1}{2}}^l + \gamma_{j+\frac{1}{2}}^l)\alpha_{j+\frac{1}{2}}^l. \tag{6.a}$$

The function $\sigma(z) = \frac{1}{2}[\psi(z) - \lambda z^2]$ and

$$\gamma_{j+\frac{1}{2}}^l = \sigma(a_{j+\frac{1}{2}}^l)\begin{cases} (g_{j+1}^l - g_j^l)/\alpha_{j+\frac{1}{2}}^l & \alpha_{j+\frac{1}{2}}^l \neq 0 \\ 0 & \alpha_{j+\frac{1}{2}}^l = 0 \end{cases}. \tag{6.b}$$

The limiter function g_j^l used in the calculations is

$$g_j^l = \left\{\alpha_{j-\frac{1}{2}}^l\left[\left(\alpha_{j+\frac{1}{2}}^l\right)^2 + \delta_2\right] + \alpha_{j+\frac{1}{2}}^l\left[\left(\alpha_{j-\frac{1}{2}}^l\right)^2 + \delta_2\right]\right\} \Big/ \left[\left(\alpha_{j+\frac{1}{2}}^l\right)^2 + \left(\alpha_{j-\frac{1}{2}}^l\right)^2 + 2\delta_2\right] \quad (6.c)$$

where δ_2 is a small parameter [1]. In all the computations, $\delta_2 = 10^{-7}$ is used.

MUSCL Approach Using an Approximate Riemann Solver: The numerical flux function $\widetilde{H}_{j+\frac{1}{2}}$ for a MUSCL-type approach of an upwind scheme as described in Yee [25] using an approximate Riemann solver can be expressed as

$$\widetilde{H}_{j+\frac{1}{2}} = \frac{1}{2}\left[F(U_{j+\frac{1}{2}}^L) + F(U_{j+\frac{1}{2}}^R) - \widehat{R}_{j+\frac{1}{2}}\widehat{\Phi}_{j+\frac{1}{2}}\right], \quad (7)$$

where the elements of $\widehat{\Phi}_{j+\frac{1}{2}}$ are $\widehat{\phi}_{j+\frac{1}{2}}^l = |\widehat{a}_{j+\frac{1}{2}}^l|\widehat{\alpha}_{j+\frac{1}{2}}^l$ and $\widehat{\alpha}_{j+\frac{1}{2}} = \widehat{R}_{j+\frac{1}{2}}^{-1}(U_{j+\frac{1}{2}}^R - U_{j+\frac{1}{2}}^L)$. Here $\widehat{a}_{j+\frac{1}{2}}^l$ are the eigenvalues and $\widehat{R}_{j+\frac{1}{2}}$ is the eigenvector matrix of A, evaluated using an approximate Riemann solver between the two states $U_{j+\frac{1}{2}}^R$ and $U_{j+\frac{1}{2}}^L$; that is, $\widehat{a}_{j+\frac{1}{2}}^l = a^l(U_{j+\frac{1}{2}}^R, U_{j+\frac{1}{2}}^L)$ and $\widehat{R}_{j+\frac{1}{2}} = R(U_{j+\frac{1}{2}}^R, U_{j+\frac{1}{2}}^L)$.

The variables $U_{j+\frac{1}{2}}^R$ and $U_{j+\frac{1}{2}}^L$ are obtained by interpolating neighboring state variables, using limiters. However, there are options in choosing the types of dependent variables in applying limiters for system cases. For the numerical tests, only one option has been used, imposing the limiters on the set of variables $W = (\rho, u, \rho e)$. Then, if we let P and P^{-1} denote transformation operators such that $U = PW$ and $W = P^{-1}U$, the vectors $U_{j+\frac{1}{2}}^R$ and $U_{j+\frac{1}{2}}^L$ for a second-order in time, second-order in space MUSCL approach can be defined as $U_{j+\frac{1}{2}}^L = P(W_j^{n+\frac{1}{2}} + \frac{1}{2}\widetilde{g}_j)$ and $U_{j+\frac{1}{2}}^R = P(W_{j+1}^{n+\frac{1}{2}} - \frac{1}{2}\widetilde{g}_{j+1})$, with $W_j^{n+\frac{1}{2}} = P^{-1}U_j^{n+\frac{1}{2}}$, where $U_j^{n+\frac{1}{2}} = U_j^n - \frac{\lambda}{2}\left[F\left(P(W_j^n + \frac{1}{2}\widetilde{g}_j)\right) - F\left(P(W_j^n - \frac{1}{2}\widetilde{g}_j)\right)\right]$. Here $\widetilde{g}_j$ is defined as in equation (6.c), except the arguments will be $(W_{j+1}^n - W_j^n)$ and $(W_j^n - W_{j-1}^n)$.

MUSCL Approach Using Flux-Vector Splittings: The numerical flux $\widetilde{H}_{j+\frac{1}{2}}$ for either flux-vector splitting can be expressed as

$$\widetilde{H}_{j+\frac{1}{2}} = F^+(U_{j+\frac{1}{2}}^L) + F^-(U_{j+\frac{1}{2}}^R), \quad (8)$$

where $F^\pm(U_{j+\frac{1}{2}}^{L,R})$ are the split fluxes evaluated using either the generalized Steger-Warming flux-vector splitting or the generalized van Leer flux-vector splitting.

NUMERICAL RESULTS

A detailed description of this study can be found in the expanded version of this paper [26]. Six one-dimensional shock-tube problems were considered. The left- and right-hand-side states of the initial conditions for all six cases are tabulated in table 1. The cases have been ordered in the direction of increasing maximum Mach numbers encountered in the flow.

Table 1. Initial conditions for the test cases.

Case	State	Density, kg/m^3	Pressure, N/m^2	Temp., K	Energy, $(m/sec)^2$	Velocity, m/sec	Mach No.
Case A	Left	0.0660	9.84 10^4	4390	7.22 10^6	0	0.0
	Right	0.0300	1.50 10^4	1378	1.44 10^6	0	0.0
Case B	Left	1.4000	9.88 10^5	2438	2.22 10^6	0	0.0
	Right	0.1400	9.93 10^4	2452	2.24 10^6	0	0.0
Case C	Left	1.2900	1.00 10^5	272	1.95 10^5	0	0.0
	Right	0.0129	1.00 10^4	2627	2.75 10^6	0	0.0
Case D	Left	1.0000	6.50 10^5	2242	2.00 10^6	0	0.0
	Right	0.0100	1.00 10^3	346	2.50 10^5	0	0.0
Case E	Left	0.0100	5.73 10^2	199	1.44 10^5	2200	7.8
	Right	0.1400	2.23 10^4	546	4.00 10^5	0	0.0
Case F	Left	0.0100	5.73 10^2	199	1.44 10^5	4100	14.6
	Right	0.0100	5.73 10^2	199	1.44 10^5	-4000	-14.5

Because of space limitation, only test cases A, D and F, are presented here. Case A is a shock tube problem with moderate density and energy ratios. The resulting shock wave is of moderate strength, but significant real gas effects are found in this temperature range. Case D is a shock tube problem with very large density and energy ratios. It produces a large transonic expansion wave. Case F is a Riemann problem with equal and oppositely directed velocities, and the same initial thermodynamic state on each side. It produces two strong shock waves moving in opposite directions.

The five second-order explicit schemes tested are (a) the symmetric TVD scheme (eqs. 4 and 5), (b) the upwind TVD scheme (eqs. 4 and 6), (c) the upwind TVD scheme (eq. 7), (d) the generalized van Leer flux splitting (eq. 8) and (e) the generalized Steger and Warming flux splitting (eq. 8). Schemes (a) and (b) follow the non-MUSCL approach, while schemes (c)-(e) follow the MUSCL approach. The same approximate Riemann solver is used in the three schemes (a), (b) and (c). The time-step limit is expressed in terms of a CFL number related to the eigenvalues of the numerical fluxes. The CFL number is fixed at 0.9 in cases B and E. In case C, it is fixed at 0.5 for the upwind non-MUSCL and MUSCL scheme and at 0.9 for the symmetric scheme. The actual CFL used for the flux splitting approaches is approximately 80% of the fixed CFL (see ref. [26] for details). The grid is uniform with origin at the initial discontinuity. The number of discretization points is 141 in cases B and C, and only 81 for case E because the expansion fan is replaced by a shock. The time for stopping the computation is chosen for each case in order to use the full computational domain. For a Δx normalized to 0.1 m, these stopping times are $t = 3.5$ $msec$ for case A, $t = 3.2$ $msec$ for case D and $t = 6.0$ $msec$ for case F.

The thermodynamic properties of equilibrium air were obtained from the curve fits of Srinivasan et al. [23]. These curve fits give analytic expressions for $\overline{\gamma} = 1 + p/\rho\epsilon$ in several ranges of density and internal energy. The values of γ, χ, and κ are then calculated from the derivatives of these analytical expressions. The numerical solutions were compared with an "exact solution" computed by solving the Rankine-Hugoniot jump conditions and integrating numerically the characteristic equations in the expansion fan. From the current study, not only the changes in values of γ and $\overline{\gamma}$ but also the differences between them are indications of departure from the perfect-gas case. These differences do not necessarily occur at very high temperatures, but at intermediate temperatures when the vibration is excited and when the dissociation reactions start. Moreover, the resolution of shocks and contact discontinuities do not always behave the same for the different variables. To illustrate the real gas effects, the computed solutions for γ, $\overline{\gamma}$, Mach number, internal energy, velocity and density are shown for completeness. The comparison study can be divided into two aspects, one on the differencing algorithms, and the other on the Riemann solvers.

Comparison of the Differencing Techniques: Parts (a)-(c) of figures 1-3 provide a comparison of the symmetric TVD and the upwind TVD non-MUSCL schemes, and the MUSCL scheme with the same approximate Riemann solver. The three techniques give almost the same results in general and the differences are similar to those found for a perfect gas. The greatest difference occurs in test case C. But this case happens to be already a difficult one when the same initial conditions on density and energy are applied to a perfect gas. The major differences are between schemes (a) and (e), and the three upwind schemes (b)-(d). Although the symmetric scheme (a) behaves similar to scheme (e), and is generally more diffusive than schemes (b)-(d) at the contact discontinuities, the situation is reversed in case E where the main shock is almost stationary and the flow behind it has a very low velocity. Furthermore, scheme (a) yields more stable results in case E (see ref. [26] for details). The influence of the limiters is the same as for a perfect gas as summarized in references [5, 7]. The choice of limiters used corresponds to what we considered as a good compromise between accuracy and stability for the whole set of cases. A comparative study of flux limiters for case B is discussed in reference [7]. The main difference in computational effort lies in the MUSCL and non-MUSCL approaches. The operations count between the non-MUSCL and MUSCL is within 30% for a perfect gas. However, due to extra evaluation in the curve fitting between the left and right states in an equilibrium real gas for the MUSCL formulation, additional computation is required for the MUSCL approach.

Comparison of the Riemann Solvers: Parts (c)-(e) of figures 1-3 compare the three Riemann solvers for the MUSCL scheme. The generalized van Leer splitting yields a sharper capture of the shocks than the generalized Steger and Warming splitting. The results obtained with the approximate Riemann solver are very similar to the ones obtained with the generalized van Leer splitting. Actually, the generalized van Leer splitting seems to be less sensitive to the state equation for the shock resolution while the approximate Riemann solver is more accurate at the contact discontinuities. It is important to note that flux-vector splittings make use of the sound speed only, whereas approximate Riemann solvers of the Roe-type make use of the thermodynamic derivatives χ and κ. These thermodynamic derivatives put more stringent requirements on the curve fit that represents the thermodynamic properties of the gas. In this regard, the curve fits of Srinivasan et al. may be deficient for the approximate Riemann solver as can be seen from the discontinuity of γ and $\tilde{\gamma}$ in the expansion wave of case D. One probably needs more improved curve fits than those of reference [23] before a definite conclusion can be drawn about the accuracy of the different Riemann solvers and schemes.

CONCLUDING REMARKS

In conclusion, for the purpose of calculations in gas dynamics with equilibrium real gases, these numerical tests show that the simple extensions to a real gas for the flux-vector splitting or the approximate Riemann solver presented in this paper are valid. The main effect of using a real-gas equation of state is to exacerbate the problems of the methods for large discontinuities. Test case C is an example of such a situation. Similarly, it seems difficult to give a ranking of the methods. Depending on the case, each one presents some drawbacks or some advantages. The present results also indicate that the state equation does not have a very large effect on the general behavior of these methods for a wide range of flow conditions.

None of the differences observed for the explicit versions seems to be decisive for the one-dimensional tests, but factors such as stability and computational efficiency need further investigation in multidimensional tests. The main differences between the methods lie in their versatility in extending to implicit methods with efficient solution procedures, especially for multidimensional steady-state computations. Preliminary study shows certain advantages of the approximate Riemann solver over the flux-vector splitting approaches.

REFERENCES

[1] B. van Leer, J. Comp. Phys., **32**, 101-136 (1979).

[2] P. Colella and P.R. Woodward, J. Comp. Phys. **54**, 174-201 (1984).

[3] A. Harten, SIAM J. Num. Anal, **21**, 1-23 (1984).

[4] P.L. Roe, Lectures in Applied Mathematics, **22**, 163-194 (Amer. Math. Soc., Providence, R.I., 1985).

[5] H.C. Yee, J. Comput. Phys., **68**, 151-179 (1987); also NASA TM-86775, July 1985.

[6] H.C. Yee, Proc. 10th Int. Conf. on Numerical Methods in Fluid Dynamics, June 1986, Beijing, China; also NASA TM-88325, June 1986.

[7] H.C. Yee, Upwind and Symmetric Shock-Capturing Schemes, NASA TM-89464, May 1987.

[8] Y.J. Moon and H.C. Yee, AIAA Paper 87-0350, Jan. 1987.

[9] W.K. Anderson, J.L. Thomas and B. van Leer, AIAA Paper No. 85-0122, 1985.

[10] M. Borrel and J.-L. Montagné, AIAA Paper No. 85-1497, 1985.

[11] S. K. Godunov, A. V. Zabrodin and G. P. Prokopov, USSR Comp. Math. Phys., **1**, 1187-1219 (1961).

[12] P.L. Roe, J. Comp. Phys., **43**, 357-372 (1981).

[13] L.C. Huang, J. Comp. Phys., **42**, 195-211 (1981).

[14] A. Harten, P. D. Lax and B. van Leer, SIAM Review, **25**, 35-61 (1983).

[15] S. Osher and F. Solomon, Math. Comp., **38**, 339-377 (1981).

[16] J.L. Steger and R.F. Warming, J. Comput. Phys., **40**, 263-293 (1981).

[17] B. van Leer, ICASE Report 82-30; Sept., 1982.

[18] R. H. Sanders and K. H. Prendergast, Astrophysical Journal, **188**, 489-500 (1974).

[19] M. Vinokur, Generalized Roe Averaging for a Real Gas, NASA Contractor Report, in preparation.

[20] J.-L. Montagné , La recherche Aerospatiale, Dec. 1986.

[21] J.-L. Montagné , Rapport de Synthese finale, ONERA, 23/1285 AY, Dec. 1985.

[22] M. Vinokur and J.-L. Montagné, Generalized Flux-Vector Splitting for an Equilibrium Gas, NASA contractor report, in preparation.

[23] S. Srinivasan, J.C. Tannehill, K.J. Weilmunster, ISU-ERI-Ames 86401; ERI Project 1626; CFD15.

[24] H.C. Yee, Computers and Mathematics with Applications, **12A**, 413-432 (1986).

[25] H.C. Yee, On the Implementation of a Class of Upwind Schemes for Systems of Hyperbolic Conservation laws, NASA TM-86839, Sept. 1985.

[26] J.-L. Montagné, H.C. Yee, M. Vinokur, Comparative Study of High-Resolution Shock-Capturing Schemes for a Real Gas, NASA TM-100004, July 1987.

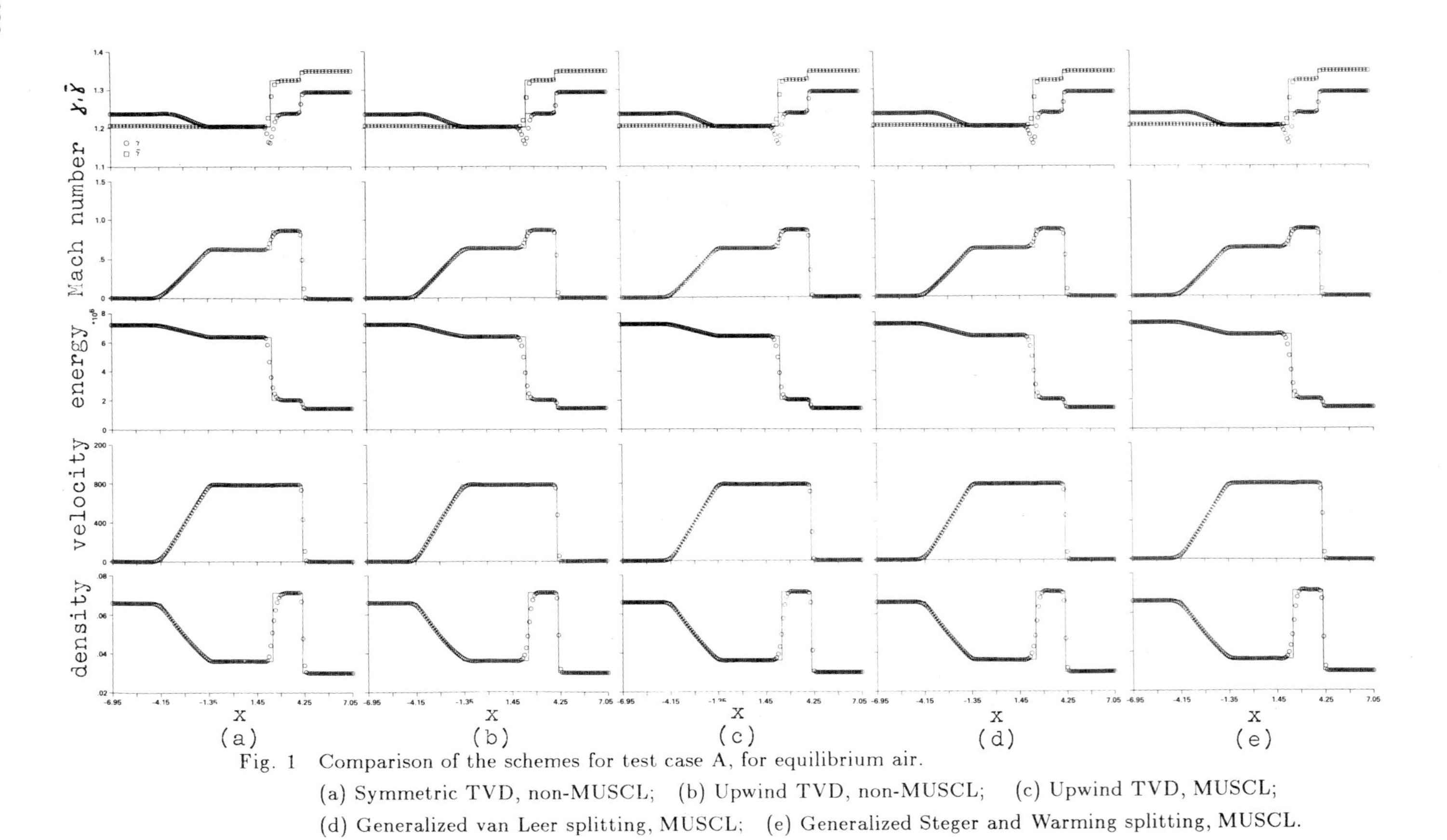

Fig. 1 Comparison of the schemes for test case A, for equilibrium air.

(a) Symmetric TVD, non-MUSCL; (b) Upwind TVD, non-MUSCL; (c) Upwind TVD, MUSCL;
(d) Generalized van Leer splitting, MUSCL; (e) Generalized Steger and Warming splitting, MUSCL.

Note: (a)-(c) use the identical approximate Riemann solver.

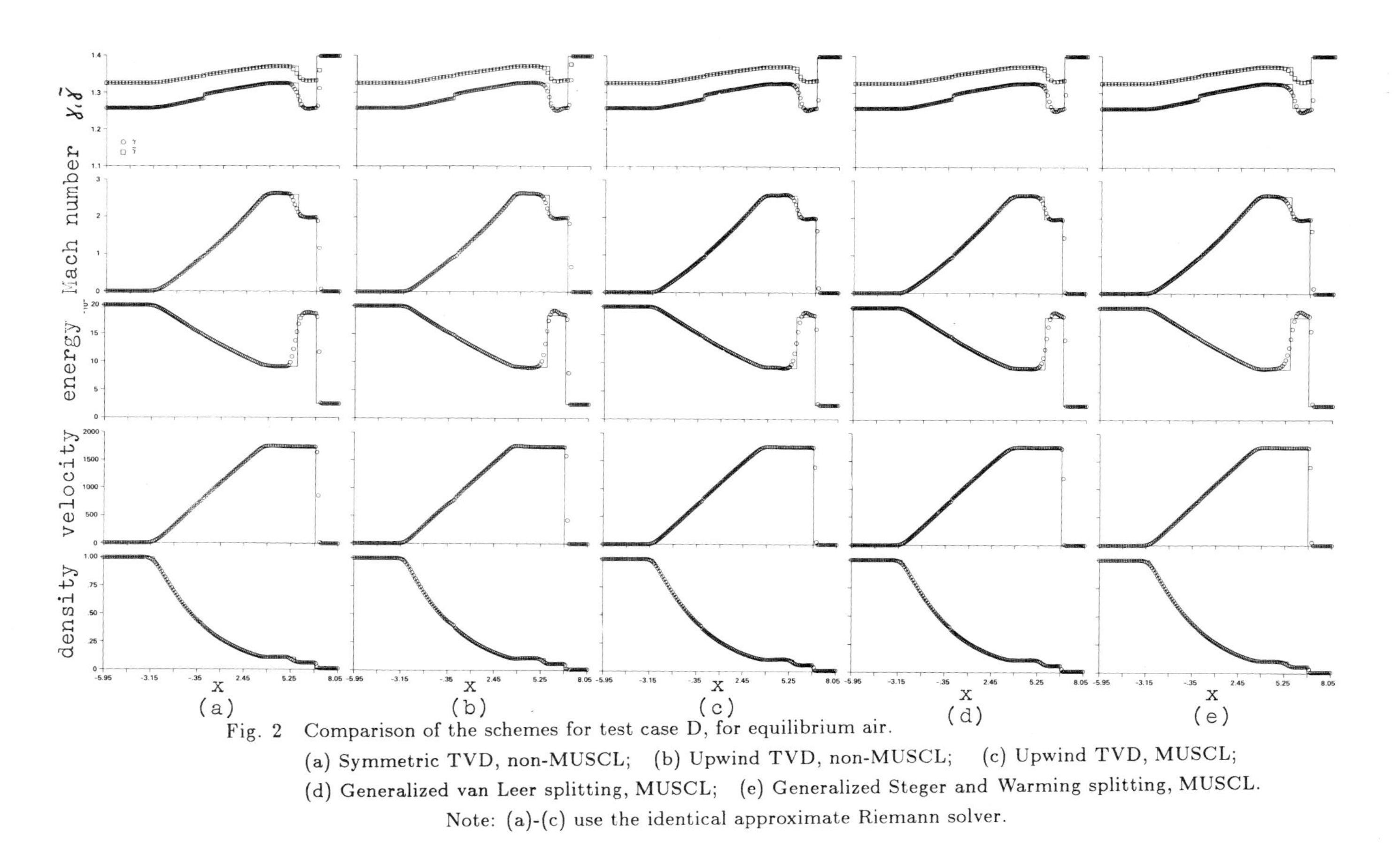

Fig. 2 Comparison of the schemes for test case D, for equilibrium air.
(a) Symmetric TVD, non-MUSCL; (b) Upwind TVD, non-MUSCL; (c) Upwind TVD, MUSCL;
(d) Generalized van Leer splitting, MUSCL; (e) Generalized Steger and Warming splitting, MUSCL.
Note: (a)-(c) use the identical approximate Riemann solver.

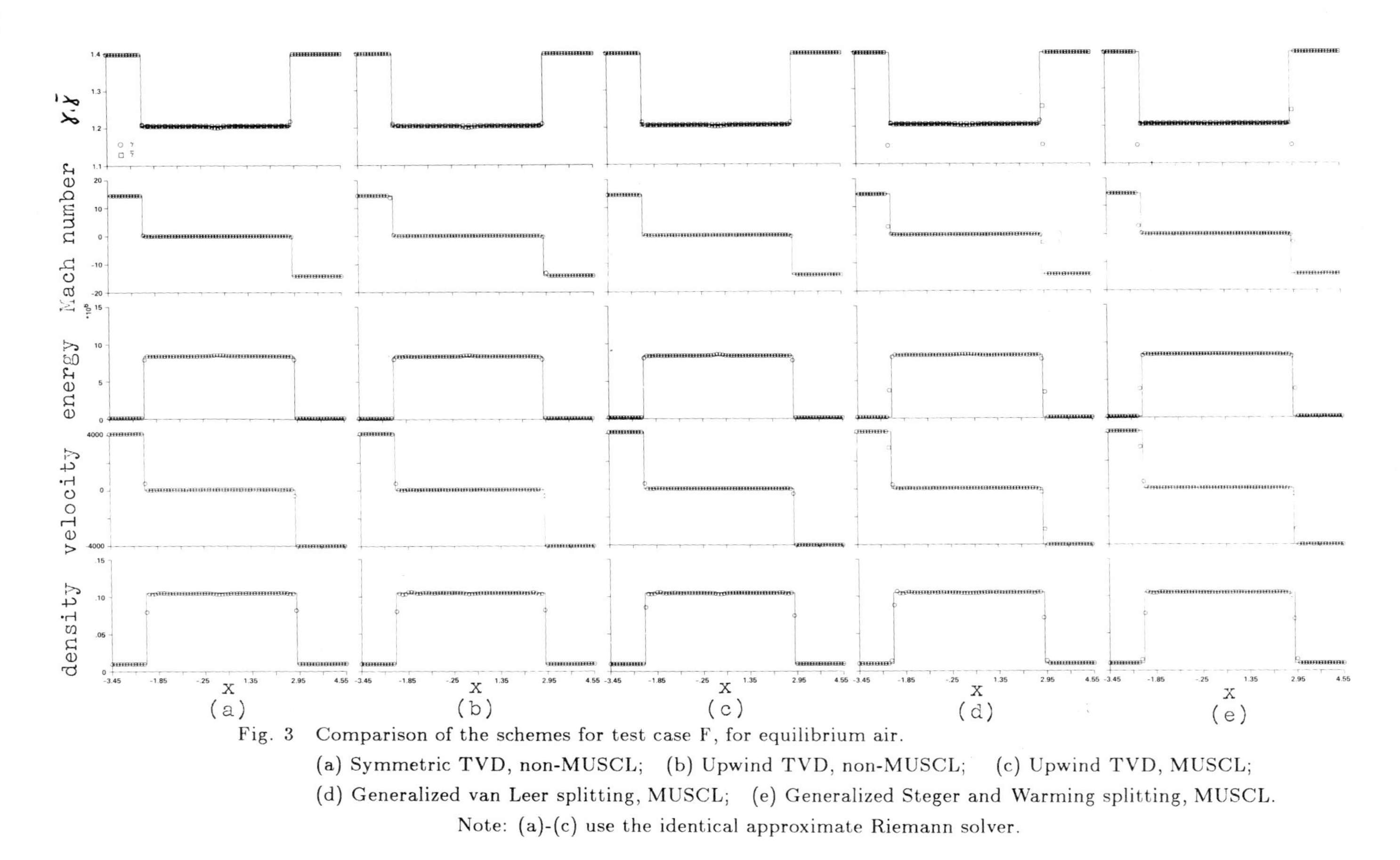

Fig. 3 Comparison of the schemes for test case F, for equilibrium air.
(a) Symmetric TVD, non-MUSCL; (b) Upwind TVD, non-MUSCL; (c) Upwind TVD, MUSCL;
(d) Generalized van Leer splitting, MUSCL; (e) Generalized Steger and Warming splitting, MUSCL.
Note: (a)-(c) use the identical approximate Riemann solver.

CALCULATION OF THE THREE-DIMENSIONAL VISCOUS FLOW PAST ELLIPSOIDS AT INCIDENCE BY ZONAL SOLUTIONS

F. Monnoyer*, K.M. Wanie+, M.A. Schmatz*

SUMMARY

A method is presented for the calculation of three-dimensional viscous flows about general configurations. The zonal solution process is applied, in which the full Navier-Stokes equations are solved only in the regions of strong viscous-inviscid interaction. In the remaining part of the flow, the coupled Euler/boundary layer equations are used. The method is applied to the flow about a 1:6 ellipsoid at incidence.

INTRODUCTION

The importance of the numerical fluid mechanics has been growing significantly in the last few years. In particular, methods for the solution of the complete Navier-Stokes equations have been designed, and the simultaneous development of large vector computers made it possible to compute the viscous flow about realistic configurations. However, systematic application of the Navier-Stokes solvers to the design of new aircrafts is still limited by the extremely high costs in computer time and storage required by these methods.

In practical applications, the flow about three-dimensional configurations is predicted by the use of the boundary layer equations together with the Euler equations, or the potential flow equations when the flow can be assumed irrotational. The influence of viscosity on the inviscid flow solution is then accounted for by a distribution of equivalent sources at the wall. Since the resulting equivalent inviscid flow does not present very steep gradients in the vicinity of the wall, the Euler grid can be coarser than in the presence of viscosity, leading to substantial reduction of the computational effort. On the other hand, the boundary layer equations describe the near-wall viscous flow. They can be solved on very fine meshes, as space-marching procedures are used, which need very small amounts of computer time and storage.

The advantage of the viscous-inviscid coupling procedure on the full Navier-Stokes solution is that it can be applied to configurations of practical interest without leading to unaffordable costs. However, it only applies to regions of weak viscous-inviscid interaction, and the Navier-Stokes equations must be used in regions where strong interaction prevails, such as separated flows or shock-boundary layer interactions.

From these considerations, it is clear that a good compromise can be achieved by combining both procedures in such a way that the Navier-Stokes equations are solved only in the regions where the boundary layer theory does not apply. The so-called "zonal solution" concept was introduced in [1] and successfully applied to two-dimensional flows cases in [2], [3] and [4].

* Messerschmitt-Bölkow-Blohm GmbH, FE122,
Postfach 80 11 60, D-8000 München 80, FRG.
+ Lehrstuhl für Strömungsmechanik, Technische Universität München,
Arcisstr. 21, D-8000 München 2, FRG.

The method has now been extended to three-dimensional flows by Wanie [5]. A computer program has been designed that includes all the parts of the zonal calculation method, as well as the management of the general solution process. This very efficient program is highly vectorizable and easy to handle, and the first results obtained for the subsonic flow about a 1:6 ellipsoid at incidence are presented here.

GENERAL PRINCIPLES

- Governing equations:

The general governing equations are the Reynolds-averaged Navier-Stokes equations for three-dimensional compressible, turbulent flows. The equations are expressed in conservative form, in terms of general curvilinear coordinates [4],[5]. The system is closed with the two-layer algebraic turbulence model of Baldwin and Lomax.

The Euler equations are obtained from the Navier-Stokes equations by simply setting the viscosity coefficients equal to zero.

For the calculation of the equivalent source distribution in the regions of weak interaction, use is made of the second-order boundary layer theory [6]. In the first applications of the zonal solution [2], [3], it turned out that the classical zero- or first-order boundary layer theory was not able to predict with sufficient accuracy the viscous flow in the regions of weak viscous-inviscid interaction. Actually, the assumption of small surface curvature compared to the boundary layer thickness is to restrictive, in particular for calculations at relatively low Reynolds numbers. Furthermore, matching of the boundary layer flow with the external equivalent inviscid flow cannot be achieved satisfactorily when using the classical boundary layer theory.

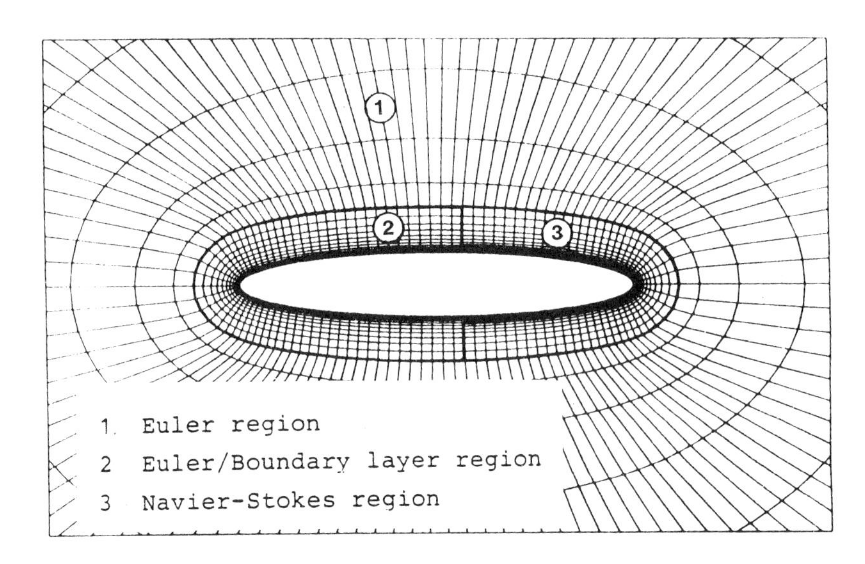

Figure 1 - Part of the grid for the zonal solution on the 1:6 ellipsoid Cut in the plane of symmetry.

The use of second-order boundary layer theory removes these approximations. The governing equations expressed in surface-oriented, locally monoclinic coordinates can be found in [4], [6] and [7]. They differ from the classical equations by the presence of several terms, which all include the surface curvature. In particular, the momentum equation in the wall-normal direction does not reduce anymore to the vanishing normal pressure gradient assumption, as in the first-order theory. Finally, the outer boundary conditions are also influenced by considering second-order effects.

- Close coupling procedure:

Maximal efficiency is obtained in the solution method with the "close coupling" procedure, in which the Euler and Navier-Stokes equations are solved simultaneously in their respective zones of application.

Consider in Figure 1 the mesh used for zonal solutions of the three-dimensional flow past an ellipsoid. The flow field is divided into three regions: the Euler region lies away from the wall where viscosity can be neglected, the weak interaction region where the boundary layer is coupled with the Euler solution is located in the upstream part of the near-wall region, and the strong interaction region fills the downstream part of the body vicinity, where separation takes place for the freestream conditions considered. Note that the second region is a subdomain of the first one, since the Euler solver is applied in both of them. This division of the computational domain introduces an artificial boundary as shown in Figure 2, where boundary conditions have to be provided for both Euler and Navier-Stokes solutions.

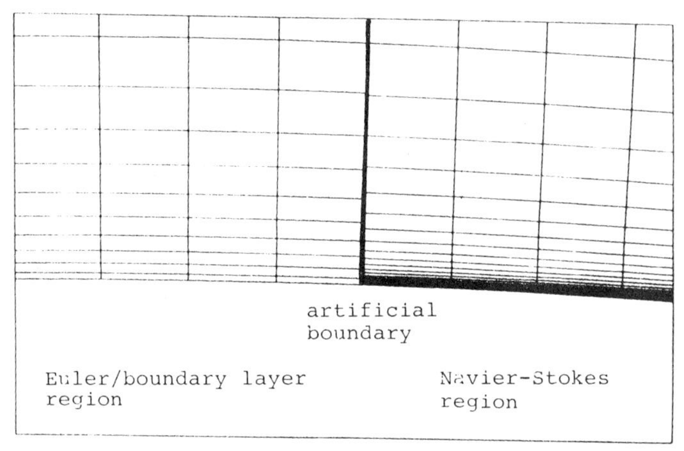

Figure 2 - Detail of the grid in the vicinity of the artificial boundary.

The grid has to satisfy requirements in order to avoid excessive interpolations at the zonal boundaries. The boundary layer equations must be formulated with regard to locally monoclinic surface-oriented coordinates [7], that is, one of the coordinate lines is rectilinear and locally orthogonal to the wall. For this reason, a grid was defined that fits the boundary layer coordinates in the Euler/boundary layer and in the Navier-Stokes regions. Away from the wall, in the pure Euler region, the mesh is defined by an elliptic grid generation algorithm, with the outer part of the surface mesh as boundary condition (Figure 1). The boundary layer equations being solved on the same surface grid as the Euler equations, the equivalent inviscid source distribution can be directly incorporated into the coupling procedure without further interpolation. Concerning the artificial boundary between the Euler/boundary layer and Navier-Stokes regions, the mesh is discontinuous only with respect to the wall-normal direction, as can be seen in Figure 2. Interpolation along this boundary is thus sufficient for determining the zonal boundary conditions.

The close-coupling procedure is started with the solution of the Euler equations in the complete flow field. This calculation is made on the original grid, and boundary values have to be computed at the artificial boundary [4], [5]. It is not necessary to perform the calculation until complete convergence, and about 50 iterations are sufficient for providing a suited initial flow field.

The boundary layer calculation is then activated in the weak interaction region. The necessary information is furnished from the Euler calculation along the corresponding grid lines and the equivalent inviscid source distribution is computed.

The calculation of the viscous flow then proceeds in both weak and strong interaction regions simultaneously. In the Euler/boundary layer region, viscous effects are represented by imposing wall-normal mass fluxes corresponding to the equivalent inviscid sources. In the Navier-Stokes region, the diffusion terms in the Euler/Navier-Stokes solver are activated. At each iteration, new boundary conditions are calculated at the artificial boundary. A new boundary layer calculation needs not to be performed at each iteration, and it is sufficient to keep the inviscid source distribution constant during 10 Euler/Navier-Stokes iterations before recalculating it.

NUMERICAL METHODS

In each region of the computational domain, use is made of a different flow model. For reliability and simplicity reasons, the corresponding numerical methods must be able to interact with each other as far as possible. The same solver is therefore applied to both the Euler and the Navier-Stokes solution, and a new boundary layer code has been developed which is perfectly suited to zonal solutions.

- Euler/Navier-Stokes solution:

An implicit relaxation technique for the unfactored Euler and Navier-Stokes equations is used as main process. The flow solver is based on the NSFLEX code (Navier-Stokes solver using characteristic FLux EXtrapolation), developed by Schmatz [8]. The pseudo-unsteady equations are integrated in time by a Newton method, in which the flux divergence is evaluated with a third-order accurate characteristic flux extrapolation scheme for the inviscid fluxes, with sensors for the detection of shocks, where the accuracy reduces to first order.

A particular feature of the characteristic flux extrapolation scheme is a Godunov-type averaging procedure based on an eigenvalue analysis of the Euler equations, by means of which the inviscid fluxes are evaluated at the cell faces. The viscous fluxes are central-differenced at each cell face.

The discretized equations, linearized in time, are solved with a point Gauss-Seidel relaxation method, in which the use of a red-black strategy allows for high vectorization rates.

- Boundary layer solution:

The boundary layer flow is calculated with the code SOBOL (Second-Order BOundary Layers) of Monnoyer [9]. The parabolic differential equations describing the second-order boundary layer flow are integrated with an implicit finite-difference space-marching method. At each sweep in the downstream direction, the calculation strategy is determined for the calculation in the crossflow direction, such that the best suited difference molecule is selected at each node within a set of five implicit schemes, each of it having typical stability properties. The solution starts with the double zig-zag molecule, so that no boundary conditions have to be prescribed at the lateral boundaries.

In the close coupling procedure, the boundary layer solver is used as a subroutine. Since the boundary layer flow has to be solved in the surface-oriented, locally monoclinic coordinate system, and the Euler/Navier-Stokes program only deals with Cartesian coordinates, a transformation is performed before starting the boundary layer solution. Similarly, the resulting boundary layer flow information is first transformed into Cartesian coordinates before being retrieved to the calling program. These pre- and post-processing procedures being performed within the boundary layer subroutine, both parts of the zonal solution process are thus fully compatible.

- Treatment of the zonal boundaries:

The zonal boundaries are a specific feature of the zonal solution method. They must be handled with extreme care in order to ensure that the information completely and correctly travels from one region to another.

In the weak interaction region, the boundary condition that has to be prescribed to the equivalent inviscid flow in order to take the boundary layer into account reduces to the wall-normal mass flux on the body surface. As the boundary layer is solved on the same surface mesh as the Euler flow, this equivalent source distribution is directly available after the boundary layer has been calculated. The way in which the boundary layer "sees" the inviscid flow is somewhat different. In classical boundary layer theory, the external boundary conditions are simply defined as the inviscid flow at the wall. In second-order theory, however, the effective inviscid flow distribution across the boundary layer must be considered. This insures a smooth blending between the boundary layer profiles and the inviscid flow at the outer edge, as depicted in Figure 3. Furthermore, no adaptation is needed anymore at the Euler/boundary layer boundary, as it is the case when classical boundary layer theory is used, see [3]. It should be noted that taking second-order effects into account also affects the calculation of the equivalent inviscid sources [7].

At the artificial boundary between the Euler/boundary layer and the Navier-Stokes regions, two sets of boundary conditions have to be defined. First, the downstream coupling of the Euler/boundary layer with the Navier-Stokes solution is straightforward when second-order theory is used, since the boundary layer profiles smoothly match the external inviscid flow profiles. The boundary condition for the Navier-Stokes solution is thus realized by simply using the equivalent inviscid flow and the boundary layer profiles to calculate the fluxes at the artificial boundary.

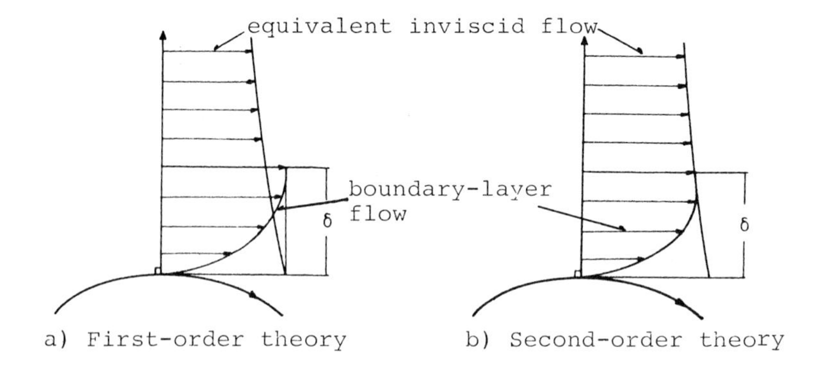

Figure 3 - Matching of the boundary layer with the equivalent inviscid flow.

The upstream coupling of the Navier-Stokes solution with the Euler flow of the weak interaction region needs some more approximation. Equivalent inviscid profiles have to be constructed out of the Navier-Stokes profiles for the Euler fluxes across the artificial boundary. For this purpose, the pressure is assumed equal in both regions along the artificial boundary, and the density is calculated from the assumption of constant entropy in the wall-normal direction. The mass fluxes corresponding to the equivalent inviscid flow at the artificial boundary are linearly extrapolated from those in the neighbouring cells in the inviscid region. It should be noted that the error introduced by this approximation only affects the boundary layer region, that is, a maximum of three to five inviscid cells near the wall.

APPLICATION

As a first three-dimensional application, the zonal solution method is applied to the calculation of the subsonic flow about an ellipsoid of thickness ratio 1:6 at 10° incidence. The freestream Mach number is 0.17, the Reynolds number based on the ellipsoid major axis 7,700,000 , and fixed transition takes place at 10% of the body length. The 1:6 ellipsoid was chosen because it has been extensively studied experimentally, see [10]. In order to facilitate the calculation for this first application, however, transition has been fixed further upstream than in the experiment, where artificial transition takes place at 20% of the body axis, and the sting has not been represented. This should be kept in mind when comparing results with experiment, the exact numerical simulation of the experimentally investigated flow case being planned as the next step in this research program.

A cut of the O-O mesh used for the Euler and the Navier-Stokes solution is shown in Figure 1. It counts 31x54x12 cells in the Euler/boundary layer region, 57x54x15 cells in the pure Euler region, and 26x54x29 cells in the Navier-Stokes region. The embedded boundary layer grid is not drawn in the figure. It has 31x54 surface nodes, fitting the Euler mesh in the corresponding region, and 51 points across the boundary layer thickness at each of them. For safety reasons, the artificial boundary was fixed at 58% of the ellipsoid major axis, although it could be placed further downstream.

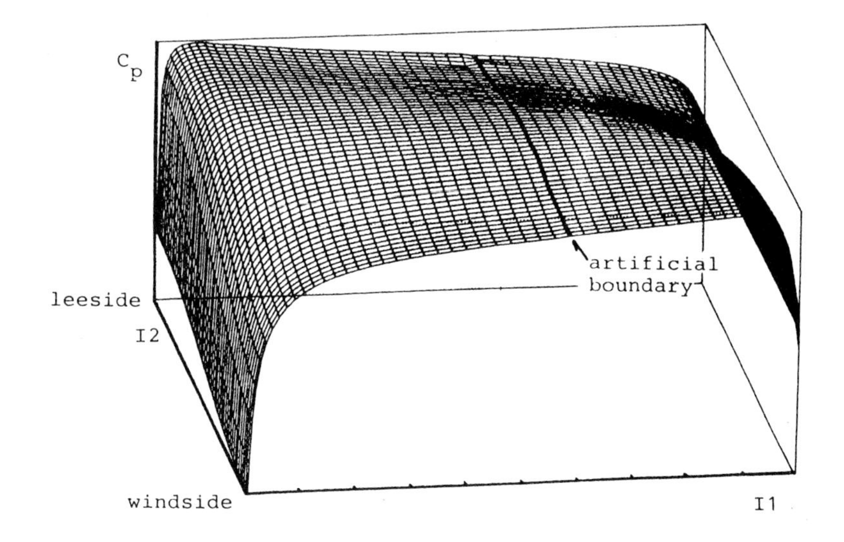

Figure 4 - Surface pressure distribution on the 1:6 ellispoid at 10° incidence.

In Figure 4, the calculated surface pressure distribution is drawn in the computational domain, that is, with respect to the cell indices I1 and I2 in the axial and circumferential directions, respectively. The distribution is perfectly smooth, even at the artificial boundary, which is almost impossible to differentiate from the other grid lines. The calculated and measured

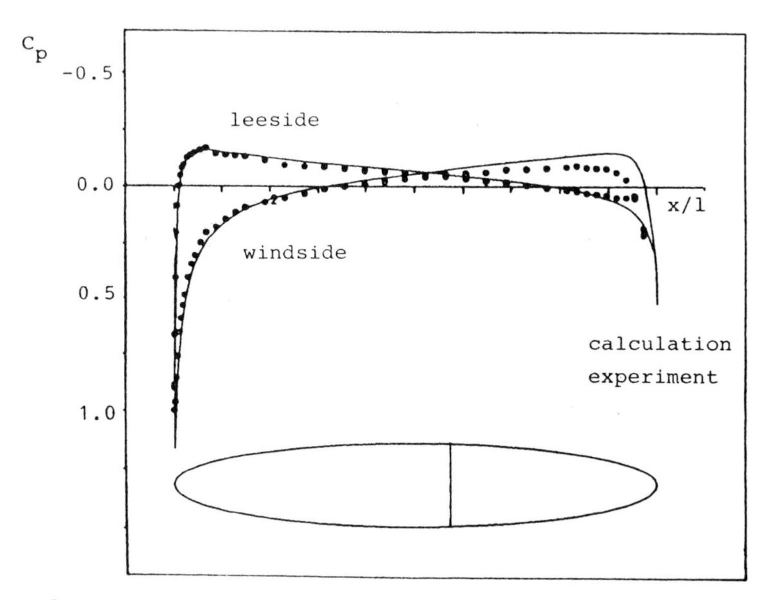

Figure 5 - Comparison of measured and calculated surface pressure in the plane of symmetry.

pressure coefficients in the plane of symmetry are compared in Figure 5. The agreement is extremely good up to about 65% of the ellipsoid major axis, and the discrepancies at the rear part of the body are due to the absence of a sting in the calculation and to the location of transition further upstream than in the experiment.

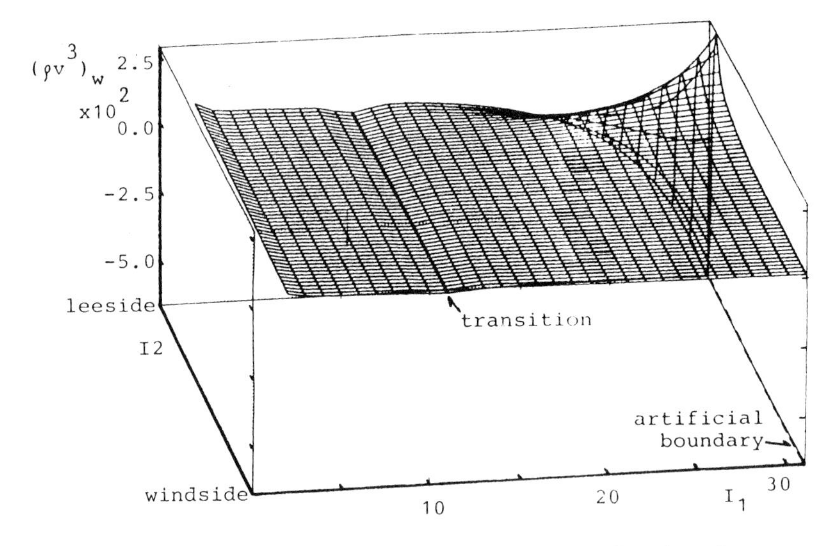

Figure 6 - Equivalent source distribution in the Euler/boundary layer region.

The boundary layer calculation in the weak interaction region is illustrated in Figure 6, which shows the equivalent sources with respect to the mesh indices. Note the local decrease of the equivalent sources as transition to turbulent flow occurs, a phenomenon which is related to the crude modelling of the transitional flow. Strong gradients in the circumferential

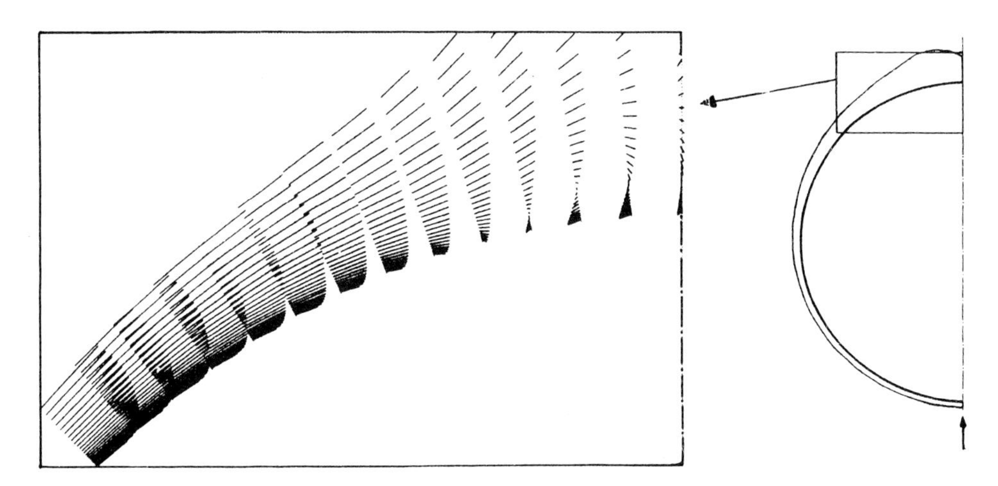

Figure 7 - Boundary layer profiles at the artificial boundary.

direction can also be observed in the rear part of the region; they are due to the strong growing of the boundary layer in the vicinity of the leeside. This situation can be clearly seen in Figure 7, where the boundary layer velocity profiles at the artificial boundary are projected in the plane of separation between the weak and the strong interaction regions.

The limiting streamline patterns are presented in Figure 8. The results of the zonal calculation are compared with the measured data of Meier et al. [10]. The agreement between the calculated and the experimental skin-friction lines is good over most of the body. The different locations of the primary separation lines is thought to be due to the smaller laminar domain in the calculated case than in the experiment; the development of the reverse cross-flow in the laminar region, where the primary separation line originates, is therefore reduced and modifies the behavior of the limiting streamlines further downstream.

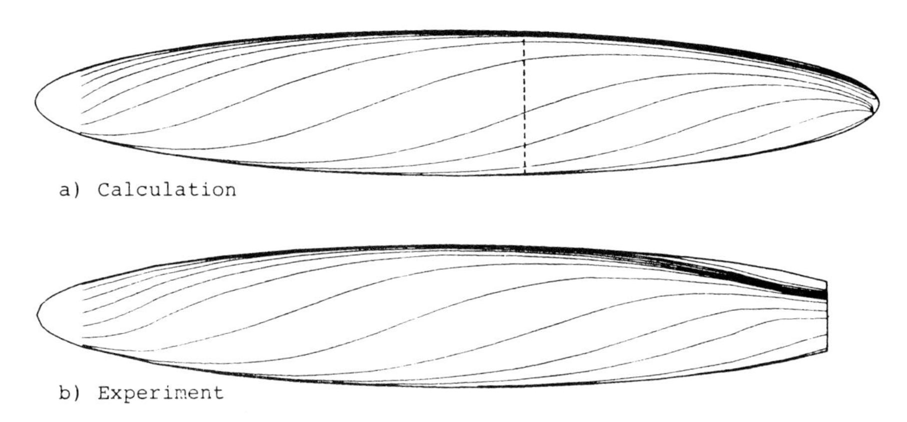

Figure 8 - Limiting streamline patterns.

Finally, the topology of the calculated surface flow on the rear part of the body is presented in Figure 9. Here again, the absence of secondary separation and the presence of a nodal separation point on the windside line of symmetry instead of a combination of a saddle point S′ and a focal point N have to be related to the earlier transition in the calculation than in the experiment [11].

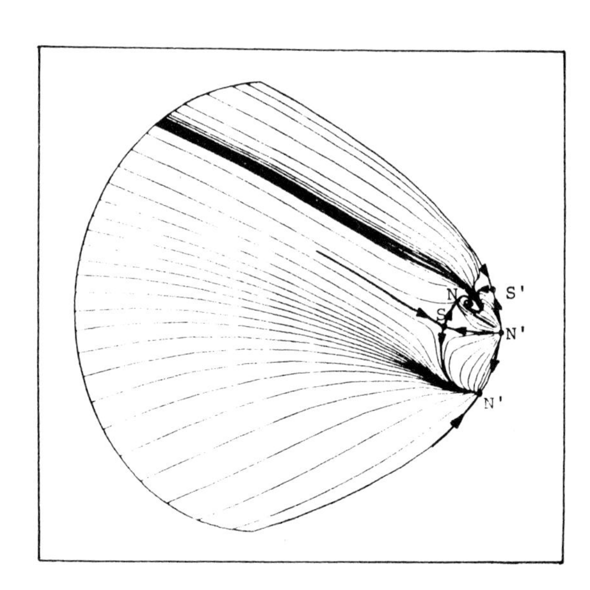

Figure 9 - Topology of the rear separation.

CONCLUSIONS

The zonal solution of the Navier-Stokes equations has been successfully applied to the calculation of three-dimensional viscous flows. The complete solution process is performed by a single, efficient and highly vectorized computer program.

The first application to the 1:6 ellipsoid at 10° incidence demonstrates the feasibility of the method, and further applications are now under progress. The next steps in the present program are the calculation of the ellipsoid with sting and the extension of the method to arbitrary artificial boundaries, in order to optimize the performance of the zonal concept.

This research program is partially supported by the Deutsche Forschungsgemeinschaft (Hi 342/1-5).

REFERENCES

[1] Hirschel E.H., Schmatz M.A.: Zonal solutions for viscous flow problems. E.H. Hirschel (ed.), Finite Approximations in Fl. Mec., DFG-priority Research Program, Results 1983-1985, Notes on Num. Fl. Mec.,Vol.14, Vieweg, Braunschweig-Wiesbaden, 1986, pp.99-112.

[2] Schmatz M.A.: Calculation of strong interaction on airfoils by zonal solutions of the Navier-Stokes equations. In: D. Rues, W. Kordulla (eds.), Proc. of the 6th GAMM-Conference on Num. Methods in Fl. Mec., Notes on Num. Fl. Mec.,Vol.13, Vieweg, Braunschweig-Wiesbaden, 1986, pp.335-343.

[3] Schmatz M.A.: Simulation of Viscous Flows by Zonal Solutions of Euler, Boundary-Layer and Navier-Stokes equations. Z. Flugwiss. Weltraumforsch. 11, 1987, pp.281-290.

[4] Wanie K.M., Schmatz M.A., Monnoyer F.: A close coupling procedure for zonal solutions of the Navier-Stokes, Euler and boundary-layer equations. Z. Flugwiss. Weltraumforsch. 11, 1987, pp.347-359.

[5] Wanie K.M.: Simulation dreidimensionaler viskoser Strömungen durch lokale Lösung der Navier-Stokesschen Gleichungen, Doctoral thesis to be presented at the Technical University München, 1988.

[6] Monnoyer F.: Second-Order Three-Dimensional Boundary Layers. In: D. Rues, W. Kordulla (eds.), Proc. of the 6th GAMM-Conference on Num. Methods in Fl. Mec., Notes on Num. Fl. Mec., Vol.13, Vieweg, Braunschweig-Wiesbaden, 1986, pp.217-278.

[7] Monnoyer F.: The Effect of Surface Curvature on Three-Dimensional, Laminar Boundary Layers. Doctoral Thesis, Université libre de Bruxelles, 1985.

[8] Schmatz M.A.: NSFLEX- A Computer Program for the Solution of the Compressible Navier-Stokes Equations. MBB-LKE122-AERO-MT-778, 1987.

[9] Monnoyer F.: SOBOL, a new Computer Program for the Calculation of Three-Dimensional, Compressible, Second-Order Boundary Layers. Paper in preparation, 1987.

[10] Meier H.U., Kreplin H.-P., Landhäußer A.: Wall pressure measurement on a 1:6 prolate spheroid in the DFVLR 3mx3m Low Speed Wind Tunnel (α=10°, U_∞=55m/s, artificial transition) - Data Report, DFVLR, IB 222-86 A 04, 1986.

[11] Hirschel E.H.: Evaluation of results of boundary-layer calculations with regard to design aerodynamics. In: Computation of three-dimensional boundary layers including separation. AGARD-R-741, 1986. Also: MBB/LKE122/S/PUB/310, 1987.

DETECTION AND FITTING OF TWO-DIMENSIONAL SHOCKS

Gino Moretti
G.M.A.F., Inc.
P.O.Box 184, Freeport, NY 11520, USA
and
Mauro Valorani
University of Rome "La Sapienza", Dpt.of Mech.and Aeronautics
Via Eudossiana 18, Roma, Italy

SUMMARY

Numerical techniques to predict the formation of shocks and to fit them in the framework of a numerical analysis of two-dimensional, inviscid, compressible, unsteady flows are briefly discussed. One application is given, the analysis of a flow expanding into an infinite cavity at constant pressure.

INTRODUCTION

In one-dimensional problems, shocks have been successfully fitted for twenty years, at least [1]. Curiously enough, the publication of [1] occurred three years later than the issue of codes for computing two-dimensional and three-dimensional shocks [2,3]. The reason may be found in the fact that the senior author of this paper and of the above References never foresaw major difficulties in the evaluation of isolated, multidimensional shocks, when treating them as boundaries and, in so doing, using a combination of the Rankine-Hugoniot conditions with compatibility equations along suitable characteristics; on this issue drastically departing from Richtmyer and Morton's pessimism [4]. In addition, whereas the codes described in [2] and [3] were created as working tools for the industry, the code in [1] was issued in an academic environment and in a tutorial spirit. In subsequent years, more studies were made on the fitting of one-dimensional shocks, aiming to an understanding of basic numerical features, extendable to the fitting of complicated two- and three-dimensional shocks [5,6,7,8]. Note also that, in anticipation of forbidding topological difficulties and having experienced them in the building of two-dimensional codes [9], the emphasis was soon shifted from shocks fitted as boundaries to the fitting of shocks, floating within mesh points.

BASIC PROPERTIES OF SHOCKS

Certain properties of one-dimensional shocks should be kept in mind, if we want to create a sound procedure for the fitting of two-dimensional shocks. Using the following notations: u, a, and S for the flow velocity, the speed of sound and the entropy, respectively, λ for $u \pm a$, γ for the ratio of specific heats, δ for $(\gamma-1)/2$, A and B for the low-pressure and high-pressure side of the shock, respectively, W for the shock velocity, M for the relative shock Mach number $[M=(u_A-W)/a_A]$, and

$$\Sigma = [a_B \pm \delta(u_B-u_A)]/a_A \tag{1}$$

(where the + sign is to be used if a decreases with x, and the - sign otherwise), it can be stated that:

a) A necessary condition for the formation of a shock in a mesh interval is a minimum of $d\lambda/dx$,

b) The shock Mach number, M, is a single-valued function of Σ; the functional relation,

$$\Sigma = \{[(\gamma M^2-\delta)(1+\delta M^2)]^{1/2}+\delta(M^2-1)\}/[(1+\delta)M] \qquad (2)$$

allows the Mach number, and consequently all jumps across the shock and the shock velocity itself to be obtained from an inspection of the shock environment, without a need for an a priori knowledge of W.

The numerical exploitation of the first property was foreshadowed in [6], but the property seems not to have been used afterwards. It was recently reproposed [10] under a form which we consider more suitable for our general codes. A further departure from [6] occurs when, after finding condition a) satisfied in an interval, we use property b) as another necessary condition for the appearance of a shock: indeed, a value of Σ slightly larger than 1 implies a value of M slightly larger than 1 and, in this case, we say that the interval contains a shock. Finally, the jump in $\ln a - \delta S$ (which is a measure of the jump in the logarithm of pressure) is evaluated, to avoid defining contact discontinuities as shocks.

SHOCK DETECTION IN TWO-DIMENSIONAL FLOWS

We can extend the above ideas to shocks in two-dimensional flows almost without changes. Briefly, our numerical detection consists of the following steps:

1) All grid lines of one family (the x-lines, say, where x does not necessarily mean a Cartesian coordinate) are scanned for intervals satisfying condition a), with u being the physical velocity component along the x-line. To avoid false starts in accidentally jagged distributions of a and λ, both a_x and λ_{xx} are discretized using a straight distribution of a and λ, obtained as a mean-square approximation over 5 grid nodes. This does not require more statements in the code than a discretization over two points. In Fig. 1, a typical distribution of λ in the proximity of a shock is shown. The vertical scale is immaterial; the figure may shift up and down according to the circumstances (a quasi-steady shock separating a supersonic region on the left from a subsonic region on the right, a shock moving rapidly to the right into a gas at rest, etc.), but the trend is always the same. In the same figure, the mean-square linear approximation to $\Delta\lambda$ and the discretized approximation to $\Delta_2\lambda$, according to our code, are shown. The test on $\Delta_2\lambda$ predicts the appearance of a shock in the same interval where λ has a jump.

2) At all intervals marked as a possible seat for a shock, except when a shock already exists, Σ is computed from (1), again using for u the physical component of the velocity along the x-line. Then a first, linear approximation to $M(\Sigma)$ is used, instead of (2), to decide whether the shock is to be inserted. Two considerations are in order here: (i) on a large range of values of M ($1 \leq M \leq 4$, say), (2) can be well approximated by a straight line, and (ii) the obliquity of the shock with respect to the grid line can be neglected in this preliminary estimate. The jump in $\ln a - \delta S$ is also tested, to discard contact discontinuities.

3) Shocks are then labeled with an index, expressing whether the high-pressure region is to the left or to the right.

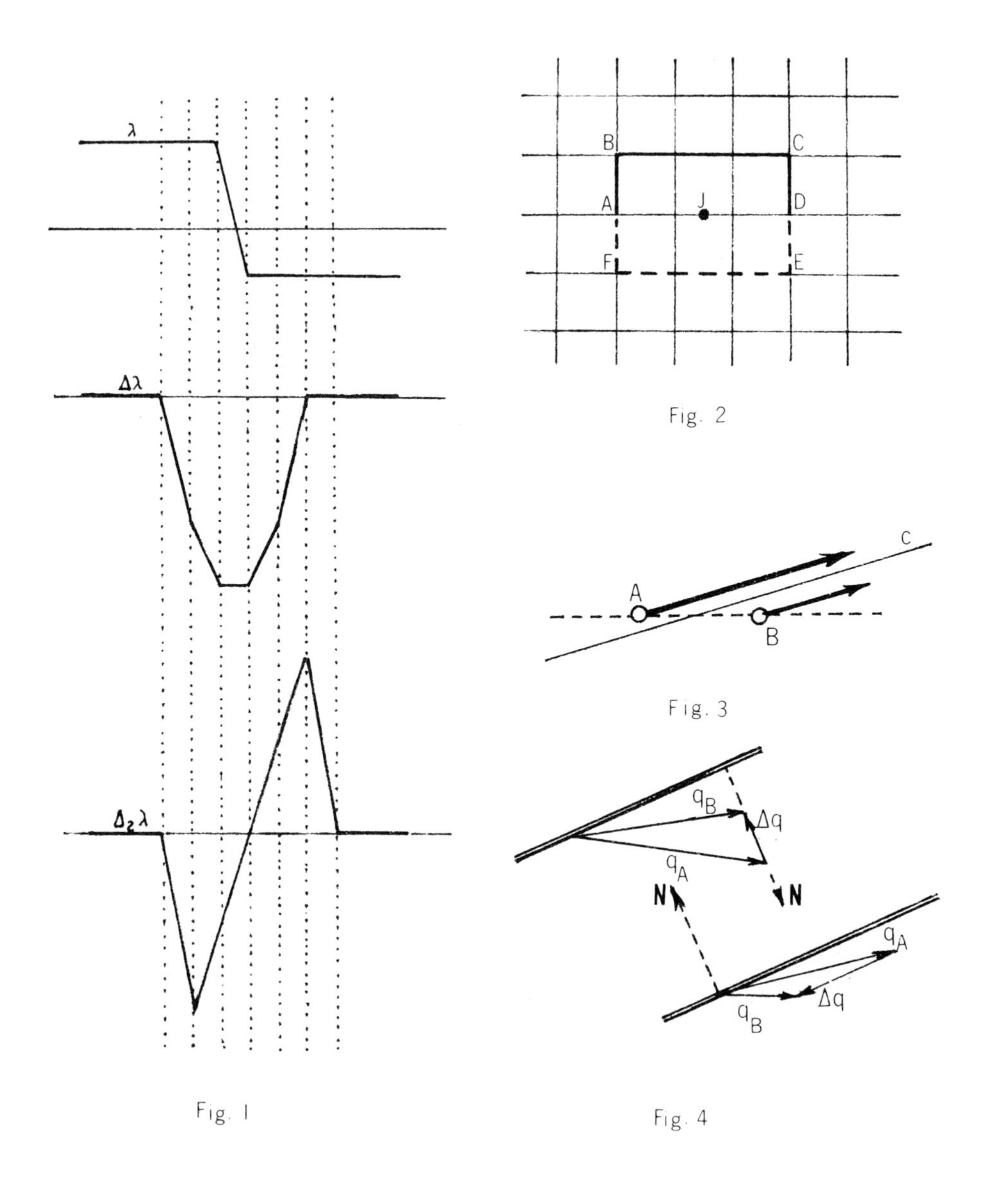

Fig. 1

Fig. 2

Fig. 3

Fig. 4

4) For a shock found at point J (Fig. 2), a search for other shock points located on the intervals AB, BC, and CD ("upper neighbors") and located on the intervals DE, EF, and FA ("lower neighbors") is made. The centers of the upper and lower neighbors are joined by a straight line, m, which is chosen to approximate the slope of the shock at J. The entire step is performed in about 10 statements using some cross-referencing indices to relate shock points to their neighboring grid nodes. If there are no upper neighbors, the slope is defined using point J and its lower neighbors, and vice versa if there are no lower neighbors. If there are no neighbors at all, point J is isolated and it is dropped.

5) The newly formed point, J, is relocated at the intersection of the x-line and line m. In what follows, we will call this shock point an "x-shock".

6) The same procedure is repeated for all grid lines of the other family, using v instead of u and y-derivatives instead of x-derivatives; the search for neighbors is made now on a figure, similar to Fig. 2 but turned 90° counterclockwise. The shock points, thus found, will be called "y-shocks" in what follows.

7) Possible passages of a shock point from one mesh interval to the next are tested; if the test is positive, certain values must be reupdated on one of the nodes bracketing the shock.

8) A final test is taken to avoid treating oblique contact discontinuities as shocks. Consider Fig. 3, where c denotes a contact discontinuity. It may occur that $u_A>u_B$, $a_A<a_B$ in such a way that λ has the same trend as in Fig. 1, and $\Sigma>1$. After a "shock" is detected, and its normal, **N**, has been found, the jump in the velocity vector, **q**, between A and B is evaluated. Ideally, if $\Delta\mathbf{q}$ is parallel to **N** we have a shock; if $\Delta\mathbf{q}$ is normal to **N** we have a contact discontinuity (Fig. 4). In practice, we reach the same conclusions if the angle between $\Delta\mathbf{q}$ and **N** is somewhat less than 20° or greater than 70°, say.

SHOCK COMPUTATION IN TWO-DIMENSIONAL FLOWS

The computation of shocks in two dimensions closely follows the one-dimensional code. Indeed, the only differences between a normal shock and an oblique shock consist of replacing u with the normal velocity component and accounting for the conservation of the tangential velocity component. Once the shock slope, as defined above, is known, obtain M (which is now the <u>normal</u>, relative Mach number:

$$M = [\mathbf{q}_A . \mathbf{N} - W]/a_A \tag{3}$$

where A denotes the low-pressure point and W is the shock velocity) using formulas similar to (2), in which the normal velocity component is used, without having to define it specifically:

$$\Sigma = \{[(\gamma M^2-\delta)(1+\delta M^2)]^{1/2}+\delta(M^2-1)\nu\}/[(1+\delta)M] . \tag{4}$$

The only change with respect to (2) is the presence of the factor, $\nu=|\cos\alpha|$, where α is the angle between **N** and the x-axis or **N** and the y-axis, according to the shock being an x-shock or a y-shock, as defined in the preceding Section. The shock velocity, W, is found from (3) and the displacement of the shock on the x-line or the y-line, respectively, in a time Δt is given by $(W/\nu)\Delta t$. Using the Rankine-Hugoniot equations and the value of M found above, all values on the high-pressure side can be updated.

One important remark is in order. Two-dimensional shocks tend to iron out their wrinkles, not because of viscous effects but because the flow downstream consists of alternating zones, looking like converging and diverging channels [12]. Whether the downstream flow is subsonic or supersonic, the "channel effect" feeds back on the shock, pushing or pulling the shock points to iron out its shape. Such effects are lost in a numerical analysis, if upper and lower neighbors are used to define the slope of the shock, as explained in 4). Devices to account for channel effects have been proposed in [12] and, in the framework of our current

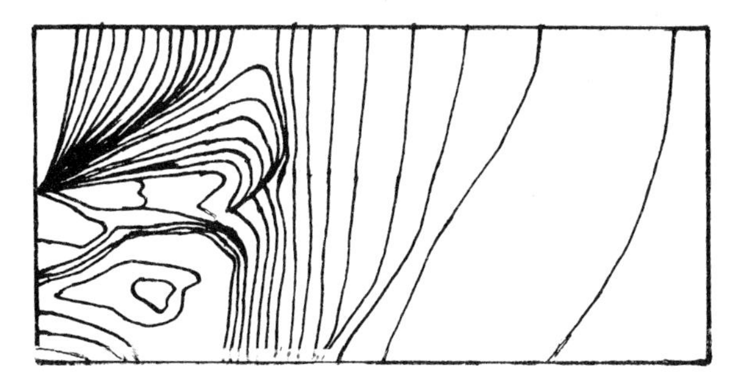

Fig. 5
Constant density lines from an experiment

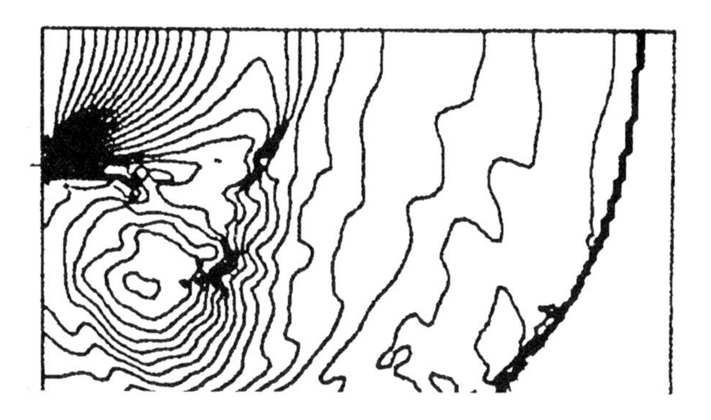

Fig. 6
Constant density lines, as computed

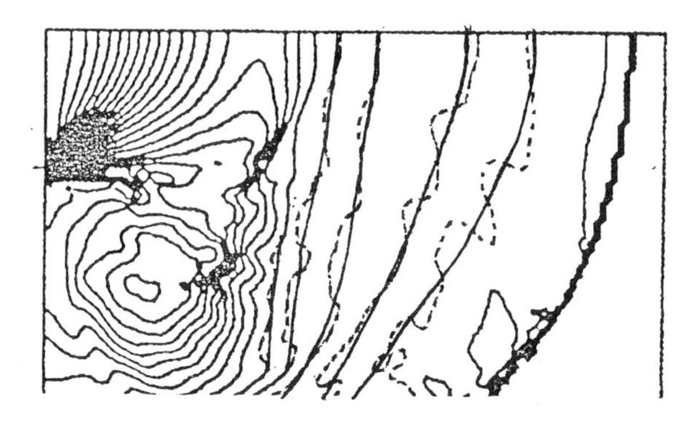

Fig. 7
Computed constant density lines, after rectification

techniques, in [13]. A simplified interpretation of the latter analysis consists of correcting Σ by adding a term:

$$-\kappa\ y''\ q\ \Delta t\ (1+1/M)/\nu \qquad (5)$$

where y'' accounts for the apparent local curvature of the shock, M is the shock Mach number and t is time; κ is a factor, currently taken equal to .005.

AN APPLICATION:
EXPANSION INTO A CAVITY AT CONSTANT PRESSURE

An interesting application of the technique outlined above is the analysis of the flow produced when a normal shock, travelling in a straight duct, reaches the end of the duct and propagates into a gas at rest at constant pressure in an infinite cavity. As the flow expands around the rim of the duct, its pressure falls well below the ambient pressure. Along the wall of the cavity, thus, a sudden recompression must take place. Considering the Mach number of the expanded flow, which is at least 6.819, the recompression may occur only through a normal shock. In an inviscid model, a steady shock would decrease the total pressure to a value, far below the ambient pressure. Therefore, a recirculation initiates along the vertical wall of the cavity, and the flow separates at the rim, forming a plume. As the plume develops in length, a large vortex appears under it, and moves outwards. The vortex should eventually be carried away, leaving a dead water region near the wall and under the plume, at the same pressure as the original pressure in the cavity. The final shape of the plume should be consistent with the pattern computed under the assumption of steady flow, bounded by a constant pressure streamline issuing from the rim of the duct.

Our calculation (on a basic Cartesian square grid with 200 intervals in either direction) may provide a description of the first phase of the expansion or a coarser analysis of the evolution of the flow over a longer period of time, by simply changing the number of intervals across the exit section of the duct. First, we use 21 points across. In 200 computational steps, the perturbed region in the cavity is about the same as an experimental interferogram ([14], as quoted by [15]), the main features of which are sketched in Fig. 5. Density contours, as computed by us, are shown in Fig. 6. The region immediately to the left of the precursor shock is still affected by oscillations, which we intend to correct by a further refinement of the shock detection technique. Nevertheless, if we draw some lines across the wavy constant density lines, we obtain Fig. 7, which compares very well with the experimental results. Using only 7 intervals across the duct, the can proceed farther in time; Fig. 8, 9 and 10 show plots of isomachs and isopycnics and a picture of velocity vectors. The isomach plot reveals the formation of the plume which, in its first portion, is practically stabilized. The slope of the plume edge is the same as the one evaluated by the steady state analysis, mentioned above. The vortex is clearly visible in Fig. 10.

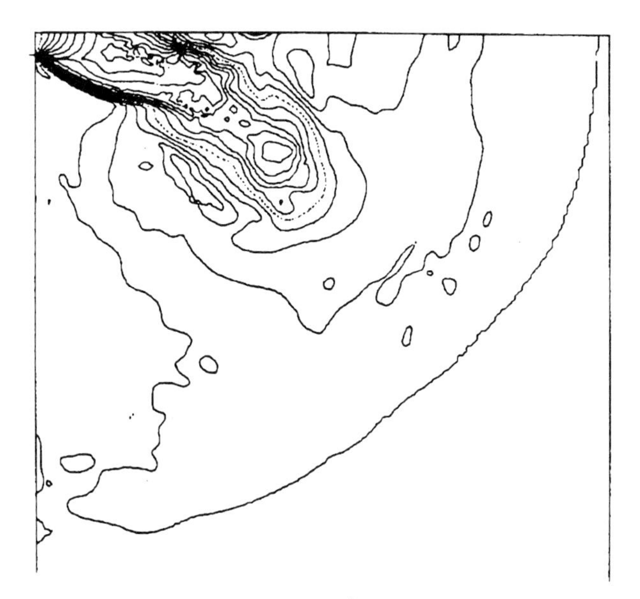

Fig. 8
Constant Mach number lines

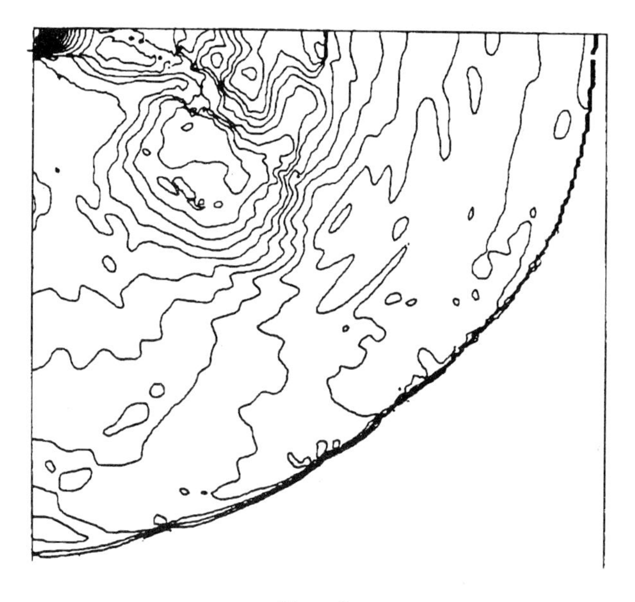

Fig. 9
Constant density lines

REFERENCES

[1]. MORETTI, G., "A critical analysis of numerical techniques: the piston-driven inviscid flow", PIBAL 69-25 (1969).
[2]. MORETTI, G. ABBETT, M., "A time-dependent computational method for blunt body flows", AIAA J., 4 (1966) pp.2136-2141.
[3]. MORETTI, G., BLEICH, G., "Three-dimensional flow around blunt bodies", AIAA J., 5 (1967), pp. 1557-1562.
[4]. RICHTMYER, R.D., MORTON, K.W., "Difference Methods for initial-value problems", Interscience Publ., New York, (1967), p. 378.
[5]. MORETTI, G., "Thoughts and afterthoughts about shock computations", PIBAL Rep. 72-37, (1972).
[6]. SALAS, M.D., "Shock fitting method for complicated two-dimensional supersonic flows", AIAA J. 14 (1976), pp. 583-587.
[7]. MORETTI, G., DI PIANO, M., "An improved λ-scheme for one-dimensional flows", NASA CR 3712 (1983).
[8]. DE NEEF, T., HECHTMAN, C., "Numerical study of the flow due to a cylindrical implosion", Comp. Fluids, 6 (1978), pp. 185-202.
[9]. MARCONI, F., SALAS, M., "Computation of three dimensional flows about aircraft configurations", Comp. Fluids 1 (1973), pp. 185-195.
[10]. DI GIACINTO, M., VALORANI, M., "Shock detection and discontinuity tracking for unsteady flows", to appear in Comp. Fluids (1988).
[11]. MORETTI, G., "A technique for integrating two-dimensional Euler equations", Comp. Fluids, 15 (1987), pp. 59-75.
[12]. MORETTI, G., "Floating shock fitting technique for imbedded shocks in unsteady multidimensional flows", Proc. 1974 Heat Transf. and Fl. Mech. Inst., pp. 184-201.
[13]. MORETTI, G., "Efficient calculation of two-dimensional, unsteady, compressible flows", in Adv. in Comp. Meth. for Part. Diff. Eqs., IMACS, VI: (1987).
[14]. MANDELLA, M., BERSHADER, D., "Quantitative study of the compressible vortices: generation, structure and interaction with airfoils", AIAA Paper 87-0328 (1987).
[15]. YEE, H.C., "Upwind and symmetric shock-capturing schemes", NASA TM 89464 (1987).

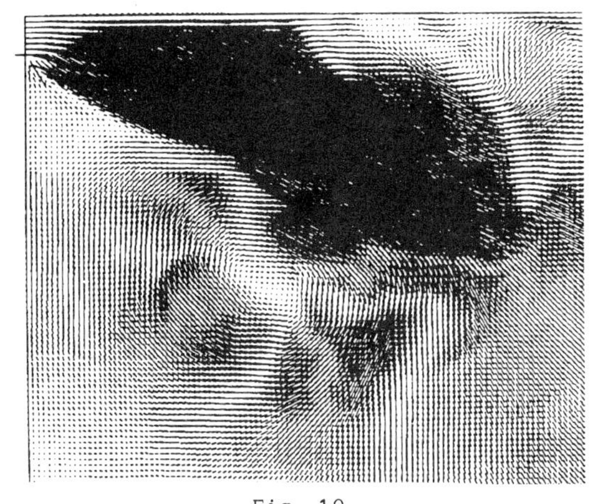

Fig. 10
Velocity vectors

NAVIER-STOKES SOLUTION FOR TRANSONIC FLOW OVER WINGS

Bernhard Müller* and Arthur Rizzi

FFA, The Aeronautical Research Institute of Sweden,
S-161 11 BROMMA, Sweden

SUMMARY

A 3D Navier-Stokes solver has been developed to simulate laminar and turbulent compressible flow over quadrilateral wings. The finite volume technique is employed for spatial discretization with a novel variant for the viscous fluxes. An explicit three-stage Runge-Kutta scheme is used for time integration taking local time steps according to the linear stability condition derived for application to the Navier-Stokes equations. The code is applied to compute laminar primary and secondary separation vortices at transonic speeds over a 65° swept delta wing with round leading edges and cropped tips, and at tached turbulent flow around the ONERA M6 wing. The results are compared with experimental data and Euler solutions.

INTRODUCTION

During the past year we have been developing a method to solve the Navier-Stokes equations for compressible laminar flow over quadrilateral wings. We have applied this method to obtain the solution of transonic flow $M_{\infty}=0.85$, $\alpha=10^{\circ}$, $Re_{\infty}=2.38\times10^{6}$ around a delta wing with a round leading edge [1]. This wing is being studied experimentally, and the results are proposed as a database to verify Euler codes [2]. Ours is the first Navier Stokes solution for this wing, and it is a large-scale computation of over 400.000 mesh points. Here we present new graphical results of that solution which show the presence of a train of vortices between the primary vortex and the leading edge.

We recently developed the code further to treat turbulent flow by the Baldwin-Lomax model. Computed results for the ONERA M6 wing $M_{\infty}=0.84$, $\alpha=3.06^{\circ}$, $Re_{\infty}=11.72\times10^{6}$ are compared with experiment and an Euler solution and serves to verify the code for turbulent flow.

GOVERNING EQUATIONS

Fluid motion is governed by the conservation laws for mass, momentum and energy. The investigated fluid here is a perfect gas obeying Newton's and Fourier's laws. External forces and heat sources are not taken into account. Considering an arbi-

---

*Current address: DFVLR-AVA, SM-TS, Bunsenstr. 10,
D-3400 Göttingen, FRG

trary stationary cell V with the boundary ∂V and the outer normal unit vector $\underline{n}$ in an absolute frame of reference, the Navier-Stokes equations read [1].

$$\int_V \frac{\partial \underline{Q}}{\partial t} dV + \int_{\partial V} \underline{\underline{H}} \cdot \underline{n}\, dA = 0 \,, \tag{1}$$

where

$$\underline{Q} = \begin{pmatrix} \rho \\ \rho \underline{u} \\ e \end{pmatrix} \qquad \underline{\underline{H}} = \begin{pmatrix} \rho \underline{u} \\ \rho \underline{u}\, \underline{u} + p\underline{\underline{I}} - \underline{\underline{\tau}} \\ (e+p)\underline{u} - \underline{\underline{\tau}} \cdot \underline{u} + \underline{q} \end{pmatrix} .$$

$\underline{Q}$ is the vector of the conservative variables, i.e. density, momentum density and total energy density. $\underline{\underline{H}}$ represents the flux tensor, $\underline{\underline{I}}$ the unit tensor. Pressure p and temperature T are related to the conservative variables by the equations of state for perfect gas with the ratio of the constant specific heats γ=1.4. The stress tensor is given by Newton's law:

$$\underline{\underline{\tau}} = \mu[\text{grad } \underline{u} + (\text{grad } \underline{u})^T] + \lambda \text{ div } \underline{u}\, \underline{\underline{I}} \,. \tag{2}$$

Fourier's law states for the heat flux:

$$\underline{q} = -\kappa \text{ grad T}. \tag{3}$$

The viscosity coefficients μ and λ are related by Stokes's hypothesis. The Prandtl number is assumed to be constant, namely Pr=0.72, thus giving a simple relationship between the thermal conductivity coefficient κ and the viscosity coefficient μ. The dependence of μ on the temperature T is obtained from the Sutherland law with the Sutherland constant S/T_∞=0.375.

For turbulent flow, the algebraic turbulence model of Baldwin and Lomax [3] is used to determine the eddy viscosity μ_t. Then, in the Navier-Stokes equations the molecular viscosity coefficient μ is replaced by $\mu + \mu_t$, and in the heat flux terms $\frac{\mu}{Pr}$ is replaced by $\frac{\mu}{Pr} + \frac{\mu_t}{Pr_t}$ with the turbulent Prandtl number Pr_t= 0.9.

NUMERICAL METHOD

Spatial Discretization

The Navier-Stokes equations (1) are discretized in hexahedrons using the finite-volume technique [1]. The surface integral in (1) over the boundary of cell P is approximated by assuming the mean-value of the flux tensor on each side to be equal to the arithmetic average of the flux tensor in the adjacent cells:

$$\int_{\partial V_P} \underline{\underline{H}} \cdot \underline{n} \, dA \cong \sum_{k=1}^{6} \underline{\underline{H}}_{Pk} \cdot \int_{\partial V_{Pk}} \underline{n} \, dA \qquad (4)$$

where

$$\underline{\underline{H}}_{Pk} = 1/2 \, (\underline{\underline{H}}_P + \underline{\underline{H}}_k).$$

∂V_{Pk} denotes the common part of the boundaries of P and its neighbouring cell k.

With the conservative variables given, all terms of the flux tensor are readily available in cell P, except for the gradients of the velocity components and temperature as well as div $\underline{u}$. Following the definition of the conservative variables as cell averages, the gradients in cell P are defined by:

$$\text{grad } \phi_P = \int_{V_P} \text{grad } \phi \, dV / \int_{V_P} dV \qquad (5)$$

where Φ = u, v, w, or T.

Using the gradient theorem, the volume integral in (5) can be expressed by a surface integral, which is approximated similarly to (4):

$$\text{grad } \phi_P = \int_{\partial V_P} \phi \, \underline{n} \, dA / \int_{V_P} dV \cong \sum_{k=1}^{6} \phi_{Pk} \int_{\partial V_{Pk}} \underline{n} \, dA / \int_{V_P} dV \qquad (6)$$

where

$$\phi_{Pk} = 1/2 \, (\phi_P + \phi_k).$$

div $\underline{u}_P$ is evaluated similarly to grad ϕ_P.

Numerical Damping

The spatial discretization constitutes the physical difference operator $\underline{F}_{PH}$ defined by the negative right hand side of (4) divided by the cell volume. These central differences do not damp unphysical oscillations caused by flow discontinuities and waves with short wavelengths. The numerical damping terms $\underline{F}_N(\underline{Q})$, which are therefore added to $\underline{F}_{PH}(\underline{Q})$, comprise nonlinear second-order differences sensed by the discretized second derivative of the pressure, and linear fourth-order differences of the conservative variables [4].

Time Integration

Thus the semi-discrete approximation of the Navier-Stokes equations can be written as:

$$\frac{d\underline{Q}}{dt} = \underline{F}(\underline{Q}) \text{ where } \underline{F} = \underline{F}_{PH} + \underline{F}_N . \qquad (7)$$

Equation (7) represents a large system of first-order ordinary differential equations. It is solved for the steady state by a second-order explicit three-stage Runge-Kutta scheme [4].

Stability

The stability of explicit Runge-Kutta schemes applied to the semi-discretization (7) of the Navier-Stokes equations (1) has been studied for a scalar linear model equation in [1]. If the coefficients of the model equation are obtained from the maximum moduli of the eigenvalues of the coefficient matrix of the linearized Navier-Stokes equations, the von Neumann stability analysis shows that also the mixed derivatives contribute to the time step limitation even on an orthogonal mesh, contrary to the implication in [1]. The refined stability condition for an explicit Runge-Kutta method e.g. the one used here (cf. [4]), applied to the present finite-valume discretization of the Navier-Stokes equations reads:

$$\Delta t \leqslant \min \{ \mathrm{CFL}\ V[|\underline{u}\cdot\underline{S}_I|+|\underline{u}\cdot\underline{S}_J|+|\underline{u}\cdot\underline{S}_K|+c(|\underline{S}_I|+|\underline{S}_J|+|\underline{S}_K|)]^{-1},$$

$$\frac{1}{2}\ |\mathrm{RK}|\, V^2 [\nu(|\underline{S}_I|^2+|\underline{S}_J|^2+|\underline{S}_K|^2)+2\nu(|\underline{S}_I\cdot\underline{S}_J|+|\underline{S}_I\cdot\underline{S}_K|+|\underline{S}_J\cdot\underline{S}_K|)$$

$$+ ((\lambda+\mu)/\rho)(|\underline{S}_I|\,|\underline{S}_J|+|\underline{S}_I|\,|\underline{S}_J|+|\underline{S}_J|\,|\underline{S}_K|)]^{-1}\} \qquad (8)$$

where c is the speed of sound, $\nu = \max\{\mu, \lambda+2\mu, \gamma\mu/Pr\}/\rho$, V the cell volume, $\underline{S}_I$ the surface normal in I-direction, etc.

The stability bounds RK and CFL are chosen such that all complex numbers z with $RK \leqslant Re(z) \leqslant 0$ and $|Im(z)| \leqslant CFL$ lie inside the stability region of the Runge-Kutta method. The factor 1/2 in (8) leaves space on the negative real axis of the stability region to accommodate the numerical damping contribution (cf.[1]).

MESH

A hemisphere with a radius of 3 root chord lengths is chosen as the farfield boundary. Then O-O meshes are generated by the transfinite interpolation method [5]. The fine mesh for the delta wing consists of 129, 49 and 65 grid points in the chordwise I-, near normal J- and spanwise K-directions, respectively, i.e. 410 865 grid points in total. For the M6 wing the size is 97×33×21.

INITIAL AND BOUNDARY DONDITIONS

For the delta wing case $M_\infty=0.85$, $\alpha=10^o$, $Re_{\infty_{CR}}=2.38\cdot10^6$, the calculation is started from freestream on a coarse mesh using a large second-order damping coefficient ($\chi=0.1$), which is subsequently reduced. The converged result is interpolated on the medium mesh and so on to the fine mesh. For the M6 wing the turbulent computation begins with the previously computed laminar solution.

The no-slip condition holds on the wing surface, which is assumed to be adiabatic. The pressure is obtained by neglecting the viscous terms in the wall normal momentum equation:

$$\underline{u}_w = 0; \quad \frac{\partial T}{\partial n}_w = 0 \quad ; \quad \frac{\partial p}{\partial n}_w = 0 \,. \tag{10}$$

The pressure and the stress tensor at the wing interface of the first cell above the wing are approximated by their values in that cell. The boundary conditions at the farfield boundary are based on the theory of characteristics for locally one-dimensional inviscid flow.

RESULTS

Laminar Flow Over Delta Wing

We simulate laminar flow over the cropped delta wing for the conditions $M_\infty=0.85$, $\alpha=10^0$ and $Re_{\infty,CR}=2.38\times10^6$ (based on root chord C_R). The streamlines in Fig. 1a show two distinct vortices which lift before the trailing edge. The inboard one is clearly the primary vortex, but the other is too far outboard to be identified as the secondary vortex. The line towards which the skin friction lines converge (Fig. 1b) indicates the secondary separation at $y/s\approx0.74$. The skin friction lines emanating from points a short distance away from the tip and trailing edge turn upstream and apparently terminate in a nodal point of separation on the secondary separation line.

The pressure minimum (Fig. 2a) is just inboard of the secondary separation line, around which large gradients of the modulus of vorticity (Fig. 2b) can be seen where the boundary layer lifts off the surface. Pressure and vorticity are nearly conical up to $x/C_R\sim0.55$.

The comparison of pressure coefficients (Fig. 3) at the stations $x/C_R=0.3$, 0.6 and 0.8 with experimental data verifies the realismof the main features simulated here, and points out the contrast with the Euler solution. The position of the primary vortex core and the pressure level under it are predicted in good agreement with the experiment, whereas the pressure minimum in the Euler solution lies too close to the leading edge and its suction is much higher. Two distinct minima occur in the viscous simulation which is typical of laminar vortex flow. The second minimum usually is not seen in turbulent flow, and this is consistent with the experimental results here, which are transitional.

Figure 4 presents contour curves of constant static C_p, total pressure coefficient $C_{p_t}=1-p_t/p_{t_\infty}$ and vorticity modulus drawn in the true plane $x/c_R=0.80$. Secondary separation occurs under the primary vortex where the contour lines of C_{p_t} and vorticity are most dense. Further outboard both sets of contours show another uplifting from the wing surface and could be called a tertiary separation. And even a fourth island of vorticity appears very near the tip. Velocity vectors bring out the sense of the flow direction. Figure 5 displays them drawn at equispaced intervals (not grid points) in a true plane normal to the leading edge at x/c_R=0.70. Here (Fig. 5a) we see clearly the primary separation, the vortex core, and the primary attachment point. The secondary and tertiary vortices are seen only crudely because of the coarse spacing of the vectors. With a tighter spacing over a region close to the upper surface, Fig. 5b brings out the velocity profile over the whole upper surface of the wing section and now clearly indicates the secondary and tertiary vortices. In the region just surrounding the tip of the wing the velocity vectors in Fig. 5c, even reveal a fourth vortex which might be called a "roller" vortex because it is sandwiched between the shed shear layer and the wing surface.

Turbulent Flow Over M6 Wing

In the second application, turbulent flow $M_\infty=0.84$, $\alpha=3.06^o$ and $Re_\infty=1.72\times10^6$ (based on mean aerodynamic chord) over the ONERA M6 wing is considered. A steady solution based on root chord is reached but we found that convergence is slower for the turbulent calculation than for the laminar one.

The lift and drag coefficients are predicted to C_L=0.2728 and C_D=0.0157 by the present Navier-Stokes solution, whereas the Euler computation [4] yields C_L=0.286 and C_D=0.0116.

The skin friction lines show attached flow (Fig. 6). They are turned in the spanwise direction, especially near the trailing. The flow around the tip at ~45% local chord is depicted by Fig. 7 showing the velocity vectors in the cells I=24 and I=73 taken in the cell centers and seen in the negative χ-direction.

The isobars on the upper surface of the wing (Fig. 8) reflect an expansion and subsequent pressure increase near the leading edge as well as a compression region on the wing. However, a lambda shock as in the Euler result [4], is not resolved by the present solution.

The surface pressure distribution at three chordwise sections is compared with experimental data and the Euler solution of [4] obtained with a 97×21×21 0-0 mesh (Fig. 11). On the lower side, the agreement is rather good although at the section y/S=0.95 (half span denoted by S) near the tip, the surface pressure between leading edge and midchord predicted by the present approach is higher than the Euler result and the measurements. On the upper surface of the wing, the suction peaks are not reached by the present solution and the compression regions are smeared out in particular near the tip, whereas the Euler solution is closer to the experimental data. Finer grids are needed for the Navier-Stokes simulation [6]. In addition, the present upper surface pressure is predicted too low between midchord and trailing edge at y/S=0.95. Although at each of the three sections the pressure predicted here drops at the trailing edge, the present solution there is closer to the measurements than the Euler result.

CONCLUSIONS

A new Navier-Stokes analysis code for laminar and turbulent compressible flow over quadrilateral wings has been developed. The finite-volume technique is employed with a larger difference molecule than the conventional compact differencing of the viscous fluxes. The method has been applied to simulate transonic flow over the cropped delta and the ONERA M6 wings using O-O meshes. For the delta wing, the computation with a $129 \times 49 \times 65$ O-O mesh shows that the location of and pressure level under the primary vortex are predicted well compared with experimental data for $M_\infty = 0.85$, $\alpha = 10^\circ$ and $Re_{\infty, c_R} = 2.38 \times 10^6$. However, the pressure level between the primary and the secondary vortices is overpredicted. For the M6 wing the computed surface pressure distribution is in satisfactory agreement with experimental and Euler results for the supercritical flow case $M_\infty = =0.84$, $\alpha = 3.06^\circ$ and $Re_\infty = 11.72 \times 10^6$ using a $97 \times 33 \times 21$ grid. The resolution of the suction peaks and compression regions on the upper surface, however, requires a finer mesh.

ACKNOWLEDGEMENTS

The extension of the Euler solver to a Navier-Stokes code for quadrilateral wings was supported by the Swedish Board for Technical Development, STU. The computing time on the CYBER 205 was provided by Control Data Corporation, CDC.

REFERENCES

[1] Müller, B., and Rizzi, A.: "Navier-Stokes Computation of Transonic Vortices Over a Round Leading Edge Delta Wing", AIAA Paper No. 87-1227, 1987.

[2] Elsenaar, A., and Eriksson, G. (eds), Proc. Intl. Vortex Flow Study, FFA TN, Stockholm, July 1987.

[3] Baldwin, B.S. and Lomax, H.: "Thin Layer Approximation and Algebraic Model for Separated Turbulent Flows", AIAA Paper 78-257, Jan. 1978.

[4] Rizzi, A., and Eriksson, L.-E.: "Computation of Flow around Wings Based on the Euler Equations". J. Fluid Mech., Vol. 148, 1984, pp. 45-71.

[5] Eriksson, L.-E.: "Generation of Boundary Conforming Grids Around Wing-Body Configurations Using Transfinite Interpolation". AIAA J., Vol. 20, No. 10, Oct. 1982, pp. 1313-1320.

[6] Vatsa, V.N.: "Accurate Numerical Solutions for Transonic Viscous Flow over Finite Wings". J.Aircraft, Vol.24, No.6, June 1987, pp. 377-385.

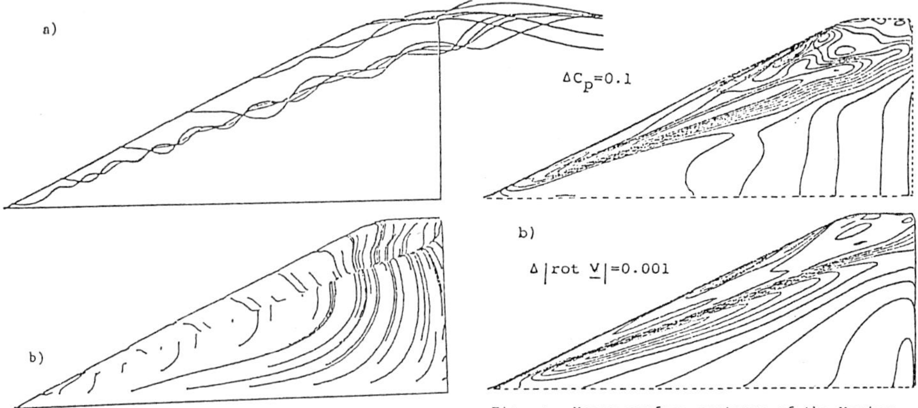

Fig. 1 a) streamlines, and b) skin friction

Fig. 2 Upper surface contours of the Navier-Stokes solution. a) C_p b) vorticity

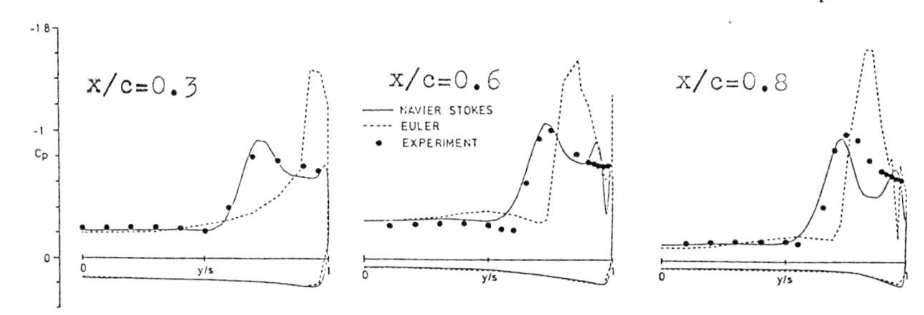

Fig. 3 Comparison of measured and computed surface C_p distributions Euler and Navier Stokes solutions

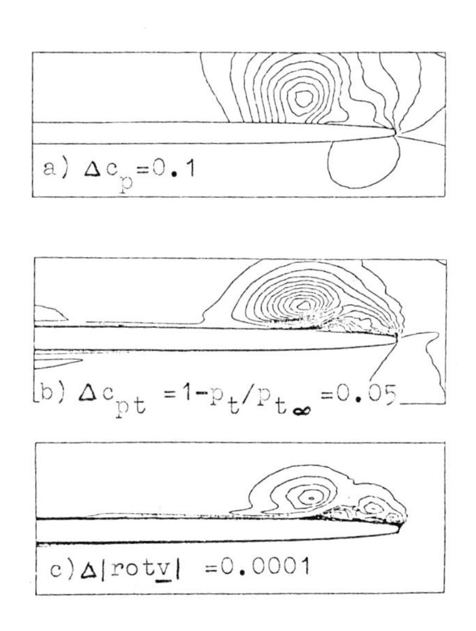

Fig. 4 Contour lines in the plane x/c_R=0.8 Navier Stokes solution

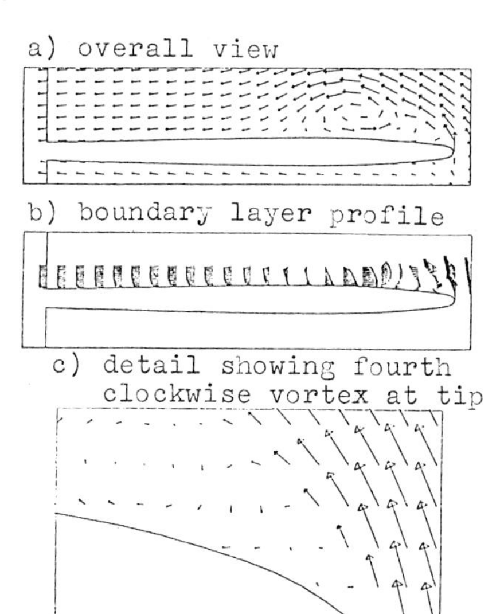

Fig. 5

Velocity vectors drawn in the palne normal to the leading edge at x/c_R=0.7 Navier Stokes solution

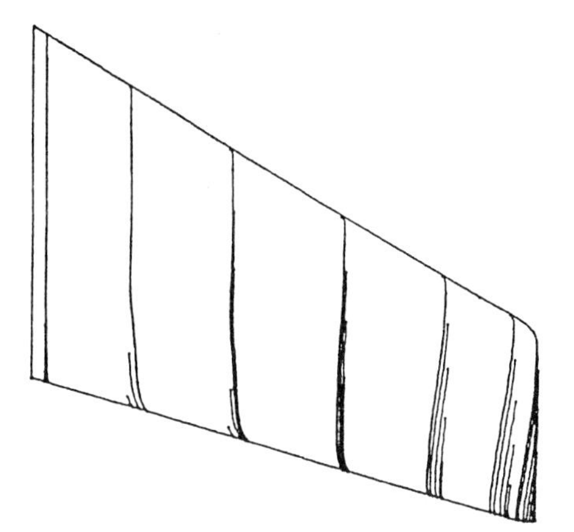

Fig 6 Skin friction lines over M6 wing.

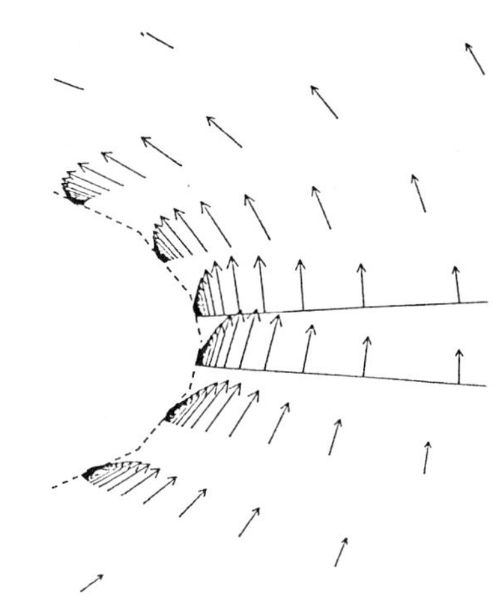

Fig 7 Attached flow around tip edge.

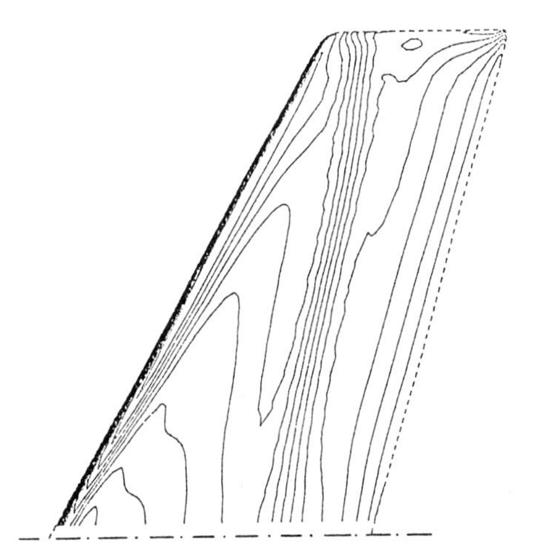

Fig 8 Isobars on wing upper surface

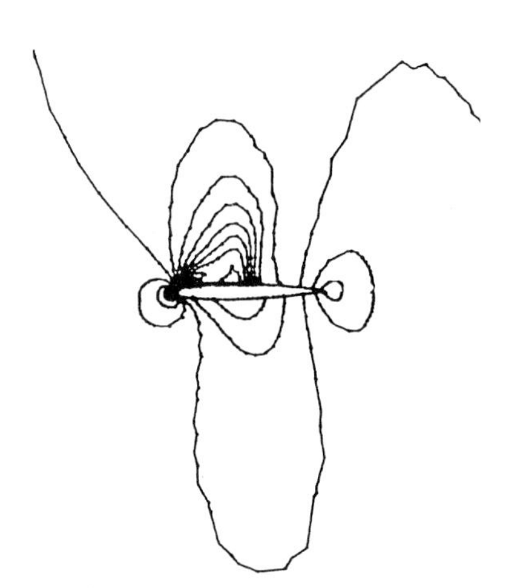

Isobars around wing at midspan

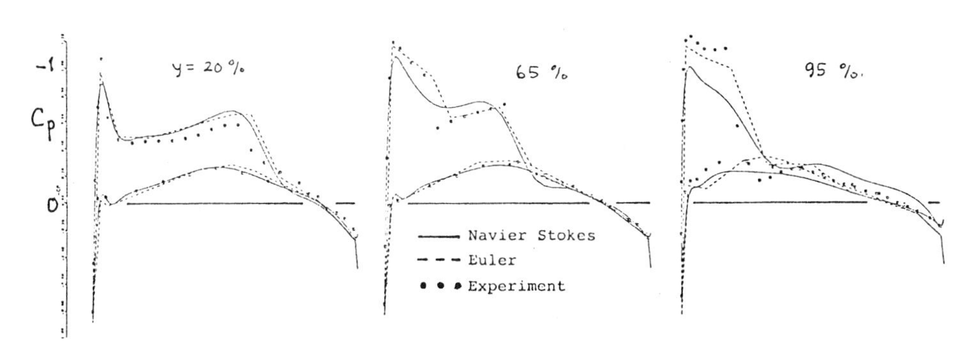

Fig 9 Comparison of the surface C_p distribution in chordwise sections

NUMERICAL SIMULATIONS OF FLOW FIELD AROUND THREE-DIMENSIONAL COMPLEX CONFIGURATIONS

Satoru OGAWA+, Tomiko ISHIGURO+ and Yoko TAKAKURA++

+ Computational Sciences Division, National Aerospace Laboratory, Japan
++ Scientific Systems Department, Fujitsu Limited, Japan

SUMMARY

Numerical simulations of flow around three-dimensional complex configurations are performed by solving both the Euler equations and the Navier-Stokes equations. Recent TVD schemes are used for two problems; (1) Transonic and supersonic flow around shuttle orbiter, (2) Hypersonic flow in scramjet inlet. It is shown the results of numerical flow simulation using the present methods is reasonable.

INTRODUCTION

Recently, several shock-capturing methods [1-3] have been poposed to solve the hyperbolic conservation laws. In [4], we have developed the three-dimensional TVD codes and solved the typical transonic flows around ONERA-M6 wing. From the numerical experiment, it was known that slight modifications are needed in TVD schemes to obtain clear solutions in three-dimensional general coordinate systems, and that our modified schemes have been proved to produce excellent solutions with very few numerical oscillations when applied to the three dimensional Euler equations. The present computations are mainly carried out to estimate the capability of the TVD codes to the applications of complex flow problems.

Two problems are numerically solved in this paper. The one is the flows around a shuttle orbiter, and the other is the flow in scramjet inlet. Regarding the first case in which the grid generation is the most difficult and important problem, Rizk et al.[5,6] have already solved the supersonic flow around NASA shuttle orbiter dividing the flow fields into multiple regions, however, the configuration used by them lacks the rear part. In [7], we have proposed a modification of hyperbolic grid generation procedure and have successfully generated a numerical grid around the whole configuration of shuttle orbiter. Using the grid, we solve several flows around a shuttle orbiter and examine the capability of TVD codes. The computations of the second case are mainly performed to examine the capability of the codes for hypersonic flow condition. The hypersonic internal flow in scramjet inlets [8-10] are solved at various freestream Mach numbers.

The Harten-Yee type TVD scheme [2,4] and the Chakravarthy-Osher type TVD scheme [3,4] are used in the computations of the Euler equations and the Navier-Stokes equations. In the Navier-Stokes equations, the Baldwin-

Lomax algebraic turbulence model [11] is used in the first case, and the subgrid scale model [12] is used in the second case.

GOVERNING EQUATIONS AND DIFFERENTIAL SCHEMES

The three-dimensional Euler equations and Navier-Stokes equations in conservation law forms in general curvilinear coordinates are written as

$$\partial_\tau \tilde{Q} + \partial_\alpha \tilde{F}^\alpha = \partial_\alpha \tilde{\sigma}^\alpha,$$

where σ^α denotes the viscous stress tensor and in the Euler equations the r.h.s. equals to zero.

In applying the TVD scheme to the system equations, the flux of convective term F^α must be diagonalized, and the quantities used in diagonalization are fully described in [13]; for example, so called the Jacobian matrices $A^\alpha = \partial F^\alpha / \partial Q$, and the similarity transformation for diagonalization $A^\alpha = T^\alpha{}_\xi \Lambda^\alpha{}_\xi T^\alpha{}_\xi{}^{-1}$, $\Lambda^\alpha{}_\xi = \mathrm{diag}(a^\alpha)$, where a^α is the eigenvalue of matrix A^α.

The implicit approximate factorization algorithm [13] is mainly used to solve the finite-difference forms of the above equations. The TVD numerical flux is used for the convective terms, and the central differencing is used to approximate the viscous term. The numerical procedures employed such as the treatments of the boundary are similar to [14]. The local time stepping $\Delta t = \Delta t_{ref}/(1+J^{1/3})$ is used in the computation of Euler equations, and the local time stepping CFL~5.0 is also used for the Navier-Stokes equations. We briefly descrive the numerical fluxes of TVD schemes in the followings:

Harten-Yee TVD scheme

The numerical flux of the Harten-Yee scheme is given as follows;

$$F_\kappa = (1/2)[F_i + F_{i+1} + (T\Phi)_\kappa],$$

where the elements of the Φ denoted by ϕ^m are given by

$$\phi^m{}_\kappa = (1/J_\kappa)[(1/2)\psi(a^m{}_\kappa)(g^m{}_i + g^m{}_{i+1}) - \psi(a^m{}_\kappa + \gamma^m{}_\kappa)\alpha^m{}_\kappa]$$

$$g^m{}_i = \mathrm{minmod}[\alpha^m{}_\kappa, \alpha^m{}_{\kappa-1}],$$

$$\alpha_\kappa = T^{-1}{}_\kappa \Delta_\kappa Q$$

where the subscript κ denotes i+1/2. Note that $(1/J_\kappa)$ is outside of minmod function in our expression, which is differ from the original form of Harten-Yee (for detail see[4]).

Chakravarthy-Osher(van Leer) TVD scheme

The numerical flux of the Chakravarthy-Osher(van Leer) TVD scheme is given by

$$F_\kappa = h_\kappa-(1-\phi)/4\{d\bar{F}^-_{\kappa+1}\}-(1+\phi)/4\{d\bar{\bar{F}}^-_\kappa\}+(1+\phi)/4\{d\bar{F}^+_\kappa\}+(1-\phi)/4\{d\bar{\bar{F}}^+_{\kappa-1}\},$$

where h_κ is the first-order accurate flux;

$$h_\kappa = (1/2)(F_{i+1} + F_i) - (1/2)(dF^+_\kappa - dF^-_\kappa),$$

and $d\bar{F}^\pm$ and $d\bar{\bar{F}}^\pm$ are flux limited values of $dF^\pm$ defined by

$$d\bar{F}^-_{\kappa+1} = (1/J)_\kappa T_\kappa \mathrm{minmod}[\Lambda^-_\kappa T^{-1}_\kappa \Delta_{\kappa+1}Q,\ \beta\Lambda^-_\kappa T^{-1}_\kappa \Delta_\kappa Q],$$

$$d\bar{\bar{F}}^-_\kappa = (1/J)_\kappa T_\kappa \mathrm{minmod}[\Lambda^-_\kappa T^{-1}_\kappa \Delta_\kappa Q,\ \beta\Lambda^-_\kappa T^{-1}_\kappa \Delta_{\kappa+1} Q],$$

$$d\bar{F}^+_\kappa = (1/J)_\kappa T_\kappa \mathrm{minmod}[\Lambda^+_\kappa T^{-1}_\kappa \Delta_\kappa Q,\ \beta\Lambda^+_\kappa T^{-1}_\kappa \Delta_{\kappa-1} Q],$$

$$d\bar{\bar{F}}^+_{\kappa-1} = (1/J)_\kappa T_\kappa \mathrm{minmod}[\Lambda^+_\kappa T^{-1}_\kappa \Delta_{\kappa-1} Q,\ \beta\Lambda^+_\kappa T^{-1}_\kappa \Delta_\kappa Q],$$

where β is the compression parameter. Note that $\phi=1/3$ corresponds to the third-order accurate flux.

NUMERICAL EXAMPLES

Transonic and supersonic flow around a shuttle orbiter

The numerical grids are generated for almost complete configuration of NASA shuttle orbiter by using the hyperbolic grid generation scheme [15,16]. The data of shuttle are taken from [5,17]. For convex simple body, it is well known that the hyperbolic grid generation procedure is efficient, and that the application of it to the complex configurations is difficult as the coordinate lines easily overlap in concave region. To increase the applicability of the scheme we propose two strategies; (1) determine the cell volume of coordinates by using the second fundamental tensor [18], (2) relax the orthogonality condition of coordinate lines by considering one-dimensional elastic model (see [7] for detail). Figure 1(a) shows the grid points distributed on the surface of shuttle orbiter, and Fig. 1(b) and Fig 1(c) show the grid on symmetric plane and the various ξ=constant grid surface projected into y-z plane, respectively. Correspondence between the physical space and computational space is shown in Fig.2. Computations are performed using the Harten-Yee type TVD scheme under various freestream Mach number (0.1~4.0). As the examples of numerical flow simulation, we show a few results of both the Euler equations and the Navier-Stokes equations. Figure 3 shows the isobaric contours obtained by solving the Euler equations under the freestream Mach number $M_\infty=0.6$ and the angle of attack $\alpha=0°$, where the flow behind of the body is also computed by imbedding another grid. Figure 4 is the iso-Mach contours for $M_\infty=3.0$ and $\alpha=0°$. For about 4×10^5 grid points, the computational time for Euler equations using FACOM VP-400 is within an hour. The numerical results of Navier-Stokes equations under $M_\infty=1.4$ are shown in Fig. 5 ($\alpha=0°$) and Fig. 6 ($\alpha=25°$). Figure 5(a) shows the iso-Mach contours on the symmetrical plane, and Fig. 5(b) is the oil-flow pattern on the surface of orbiter. The separation occurs near the root of the vertical tail, and the velocity vectors in that region is shown in Fig. 5(c). Figures 6(a)-(c) show the oil-flow pattern, the total pressure contours at various vertical planes and the isobaric contours, respectively, at $\alpha=25°$. Computations are performed for the wide range of angle of attack (0~35°), and it is known the various separations occur everywhere of the surface. Though the experimental data of shuttle orbiter are few, the Cp distribution at $\alpha=0°$ shows a good agreement with the experimental data [5].

Hypersonic flow in scramjet inlet

The TVD schemes have high resolution capability of shock wave, and we examine the applicability of the TVD codes by solving the hyperbolic flow in scramjet inlets developed in NASA Langley [8-10]. The chemical reactions as the combustion and dissociation are not considered yet. Both the Euler equations and the Navier-Stokes equations are solved for two models, the one is single-strut scramjet inlet, and the other is three-struts inlet. To make the grid generation easy, the rectangular shapes without sweep are used. The computations of Euler equations are easily carried out without any trouble by the Chakravarthy-Osher scheme. Figure 7 shows the isothermal contours for single-strut inlet model in a plane located at mid-inlet height obtained by solving Euler equations. The isothermal contours and the isobaric contours for three-struts model are shown in Figs. 9(a) and 9(b), respectively. From the figures, it is known the resolution of shock wave is marvelously clear. However, in the computations of Navier-Stokes equations, the compression parameter in TVD scheme must be decreased as the freestream Mach number increases. For example, if the maximum value of compression parameter for $M_\infty=8.0$ case is used, the solution never converges in our experiences. Further, the local time stepping to obtain the steady solution is bad in viscous cases. In our computation of Navier-Stokes equations, the first computations are performed by using implicit local time stepping technique for several hundred time steps, and then the next computations are continued by simple Euler explicit method with CFL~1.0. In this computations, full viscous terms are considered and the computation using implicit scheme takes more time to obtain convergent solutions. Figure 8 shows the isobaric contours obtained by solving the Navier-Stokes equations, where the Harten-Yee type TVD schemes is used and 1.2 million points are distributed in the half space of internal region of inlet. The computational time in this case is about 10 hours using FACOM VP-400. We have computed the same problem using the Chakravarthy-Osher scheme, however, the solutions did not converge in spite of many efforts. The SGS eddy viscosity [12] was used in the computations, however, the turbulence model in hypersonic flow is ambiguous at present.

CONCLUSIONS

The invisid and viscous flow simulations around three-dimensional complex configuration have been carried out by solving the Euler equations and the Navier-Stokes equations. Recent TVD schemes have been successfully used for this study.

Two flow problems have been solved; the one is the transonic and supersonic flow around the shuttle orbiter and the other is the hypersonic flow in scramjet inlets. It becomes clear that the reasonable solutions are obtained for wide range of freestream Mach number using the TVD codes.

ACKNOWLEDGEMENT

The authers would like to express their sincere gratitude to Mr. Hajime Miyoshi, Director of Computational Sciences Division of NAL, for his advice and encouragements. The authors would also like to thank Mr. Yasuhiro Wada at NAL for his help of preparing this paper.

REFERENCES

[1] van Leer, B., "Towards the ultimate conservative difference scheme, V", J. Comp. Phys., Vol. 32, 101(1979).

[2] Yee, H.C., Warming, R.f. and Harten, A., "Implicit total variation diminishing (TVD) schemes for steady-state calculations", J. Comp. Phys., Vol.57, 327(1985).

[3] Chakravarthy, S.R. and Osher, S., "A new class of high accuracy TVD schemes for hyperbolic conservation laws", AIAA Paper 85-0363, 1985.

[4] Takakura, Y., Ishiguro, T. and Ogawa, S., "On the recent difference schemes for the three-dimensional Euler equations", AIAA Paper 87-1151-CP, 1987.

[5] Rizk, Y.M., and Ben-shmuel, J.L., "Computation of the viscous flow around the shuttle orbiter at low supersonic speeds", AIAA Paper 85-0168, 1985.

[6] Rizk, Y.M., Steger, J.L. and Chaussee, D.S., "Use of a hyperbolic grid generation scheme in simulating supersonic viscous flow about three-dimensional winged configurations", Proceedings ISCFD-Tokyo, in: Oshima, K. (ed), (ISAS, 1986), pp. 392-403.

[7] Ogawa, S., Ishiguro, T. and Takakura, Y., "Hyperbolic grid generation scheme for simulating flow about three-dimensional complex configuration" Proceedings ISCFD-Sydney, in: Fletcher, C. (ed), 1987.

[8] Drummond, J.P. and Weidner, E.H., "Numerical study of a scramjet engine flowfield", AIAA J. Vol.20, No.9, 1182(1982).

[9] Trexler, C.A. and Souders, S.W., "Design and performance at local Mach number of 6 of an inlet for an integrated scramjet concept", NASA TN D-7944, 1975.

[10] Kumar, A., "Numerical simulation of flow through scramjet inlets using three-dimensional Navier-Stokes code", AIAA Paper 85-1664,

1985.
[11] Baldwin, B.S. and Lomax, H., "Thin layer approximation and algebraic model for separated turbulent flows", AIAA Paper 78-257, 1978.
[12] Deardoff, J.W., "A numerical study of three-dimensional turbulent channel flow at large Reynolds number", J. Fluid Mech., Vol.41, 453(1970).
[13] Pulliam,T.H. & Chaussee,D.S., "A Diagonal Form of an Implicit Approximate Factorization Algorithm, " J.Comp.Phys. Vol.39, 347 (1981).
[14] Pulliam,T.H. & Steger,J.L., "Recent Improvements in Efficiency, Accuracy, and Convergence for Implicit Approximate Factorization Algorithms," AIAA Paper 85-0360, 1985.
[15] Steger, J.L. and Chaussee, D.S., "Generation of body fitted coordinates using hyperbolic partial differential equations", SIAM J.Sci.Stat .Comput. 1, 431(1980).
[16] Steger, J.L. and Sorenson, R.L., "Use of hyperbolic partial differential equations to generate body fitted coordinates", in: Smith, R.E., (ed), Numerical Generation Techniques (NASA-CP, 1980) pp. 463-479.
[17] "Jane's all the world's aircraft 1985-1986", (Jane's Pub. Co., Ltd, 1986).
[18] Eisenhart, L.P., "Riemannian geometry", (Princeton Univ. Press, 1950).

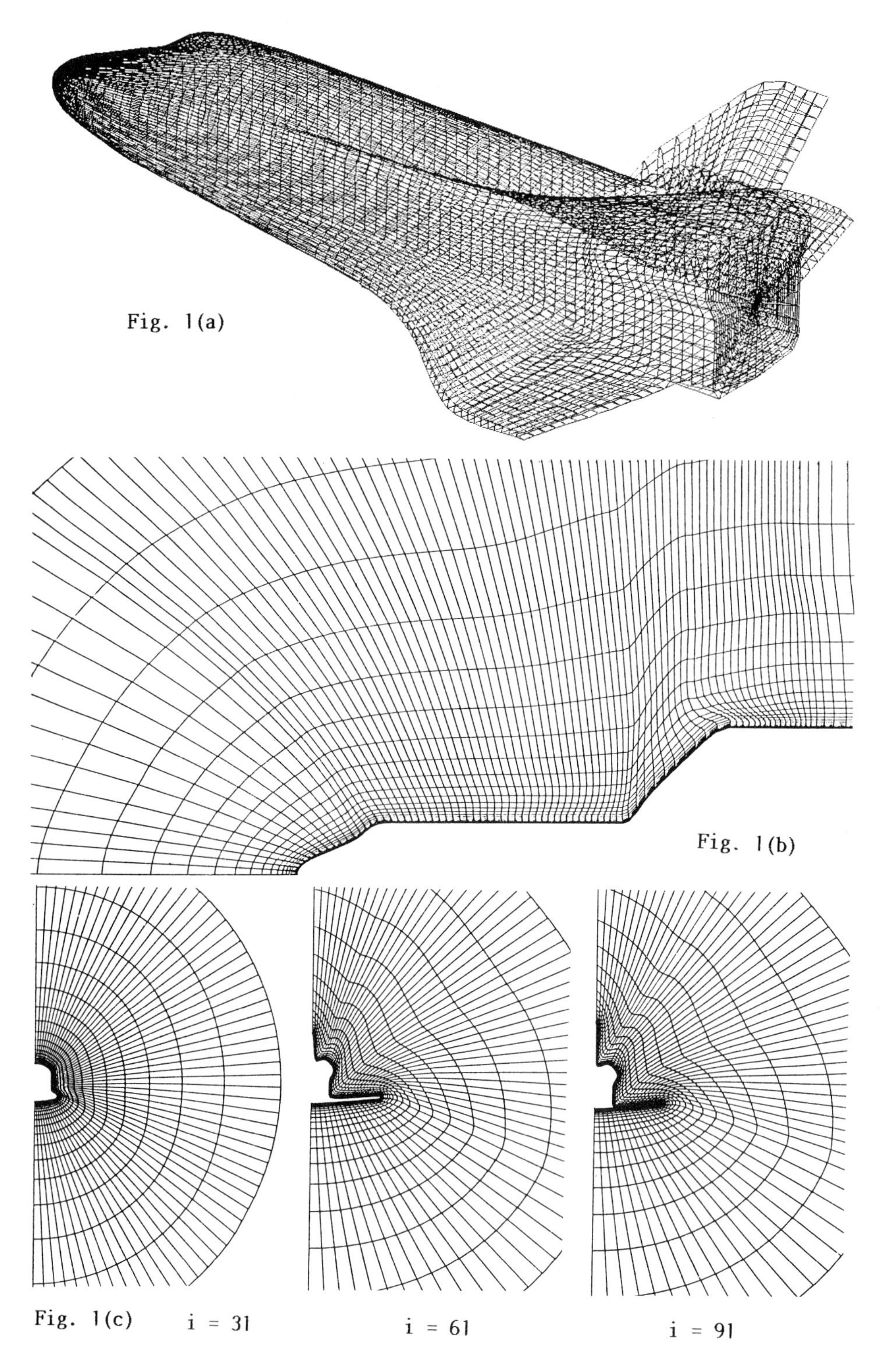
Fig. 1(a)
Fig. 1(b)
Fig. 1(c)
i = 31
i = 61
i = 91

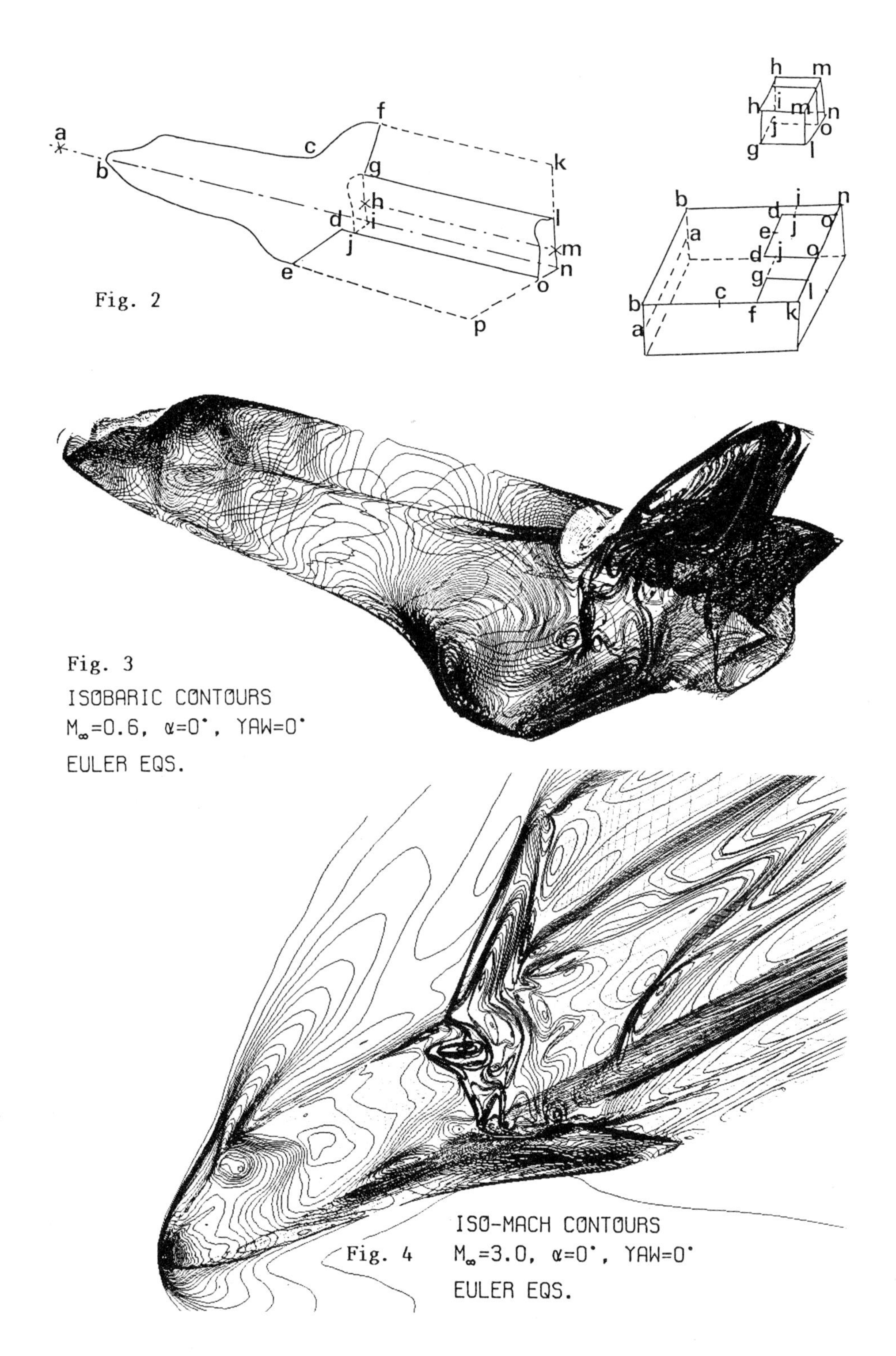

Fig. 2

Fig. 3
ISOBARIC CONTOURS
M_∞=0.6, α=0°, YAW=0°
EULER EQS.

ISO-MACH CONTOURS
Fig. 4 M_∞=3.0, α=0°, YAW=0°
EULER EQS.

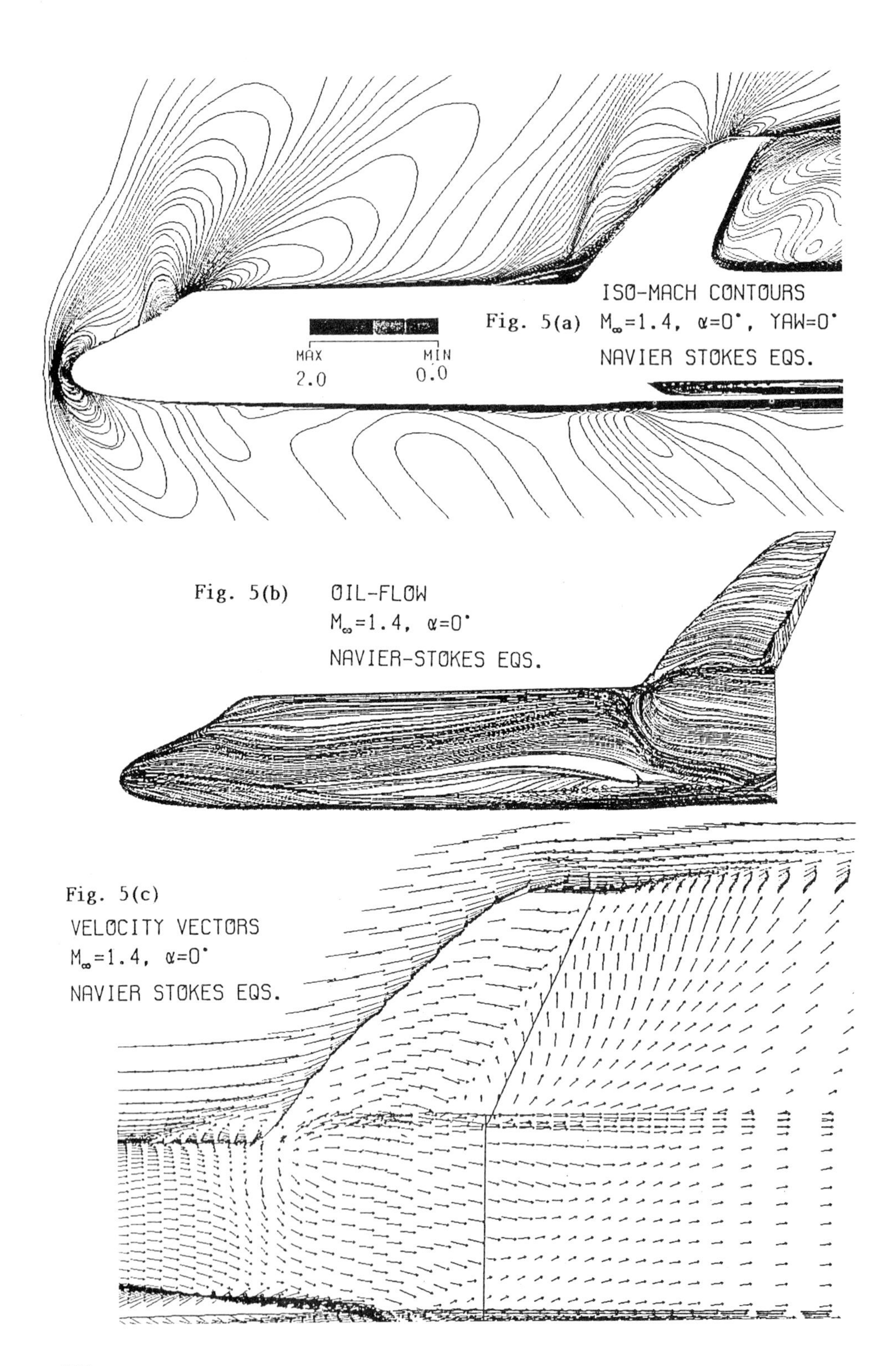
ISO-MACH CONTOURS
Fig. 5(a) M_∞=1.4, α=0°, YAW=0°
NAVIER STOKES EQS.
MAX
2.0
MIN
0.0
Fig. 5(b) OIL-FLOW
M_∞=1.4, α=0°
NAVIER-STOKES EQS.
Fig. 5(c)
VELOCITY VECTORS
M_∞=1.4, α=0°
NAVIER STOKES EQS.

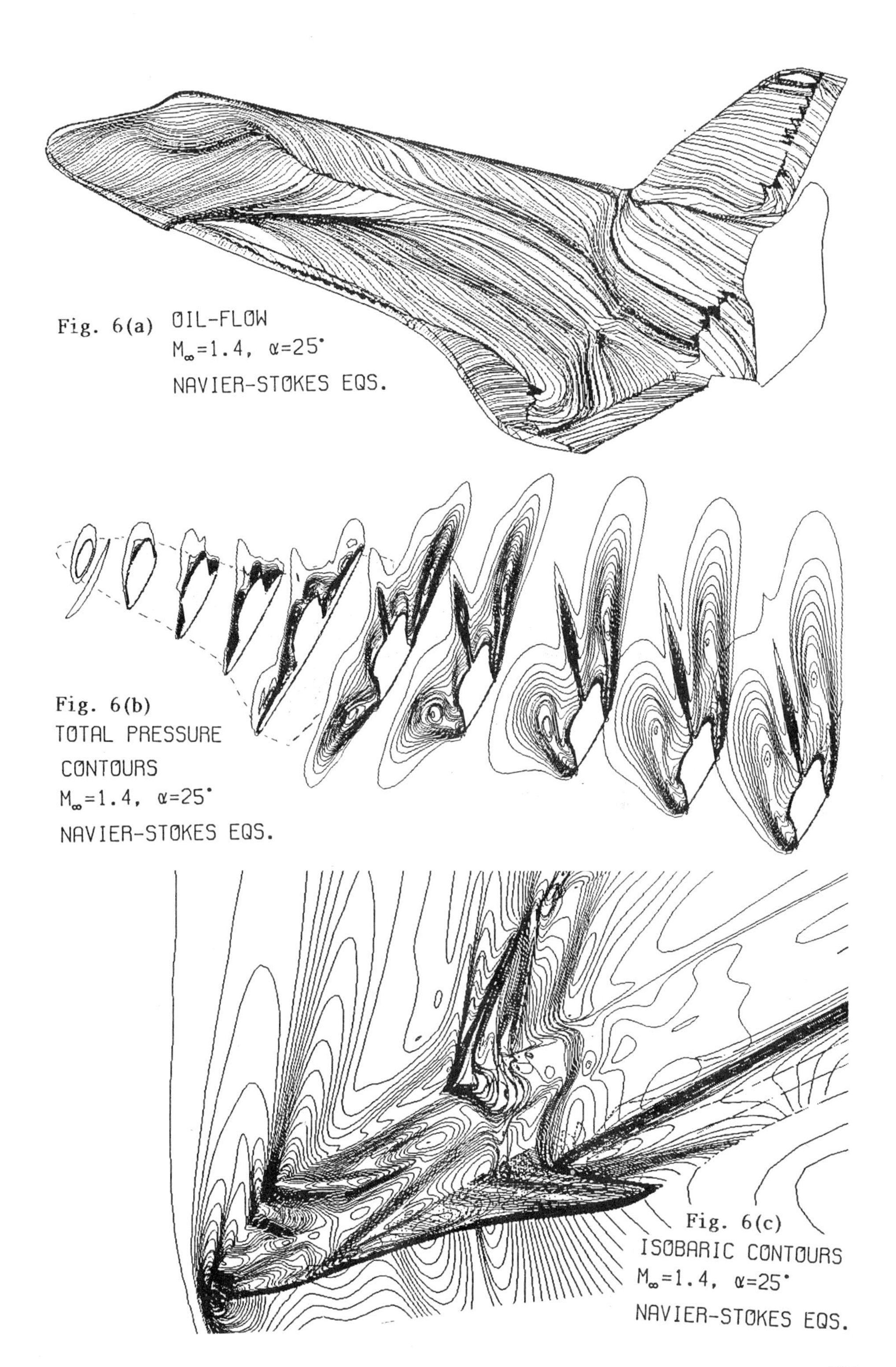

Fig. 6(a) OIL-FLOW
$M_\infty=1.4$, $\alpha=25^\circ$
NAVIER-STOKES EQS.

Fig. 6(b)
TOTAL PRESSURE CONTOURS
$M_\infty=1.4$, $\alpha=25^\circ$
NAVIER-STOKES EQS.

Fig. 6(c)
ISOBARIC CONTOURS
$M_\infty=1.4$, $\alpha=25^\circ$
NAVIER-STOKES EQS.

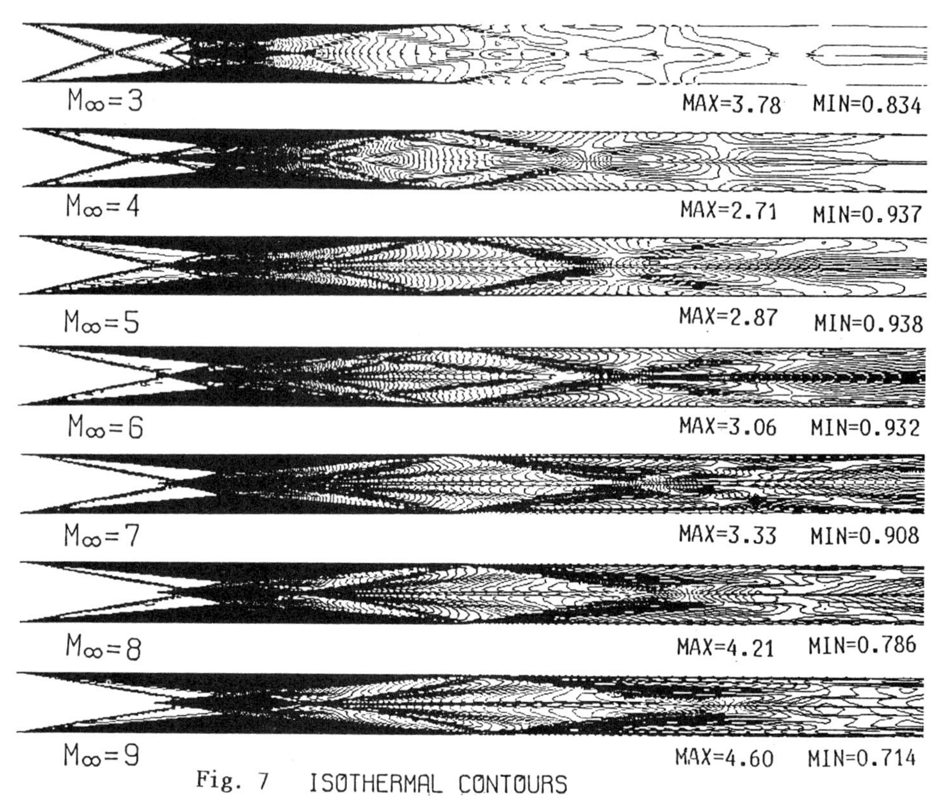

Fig. 7 ISOTHERMAL CONTOURS

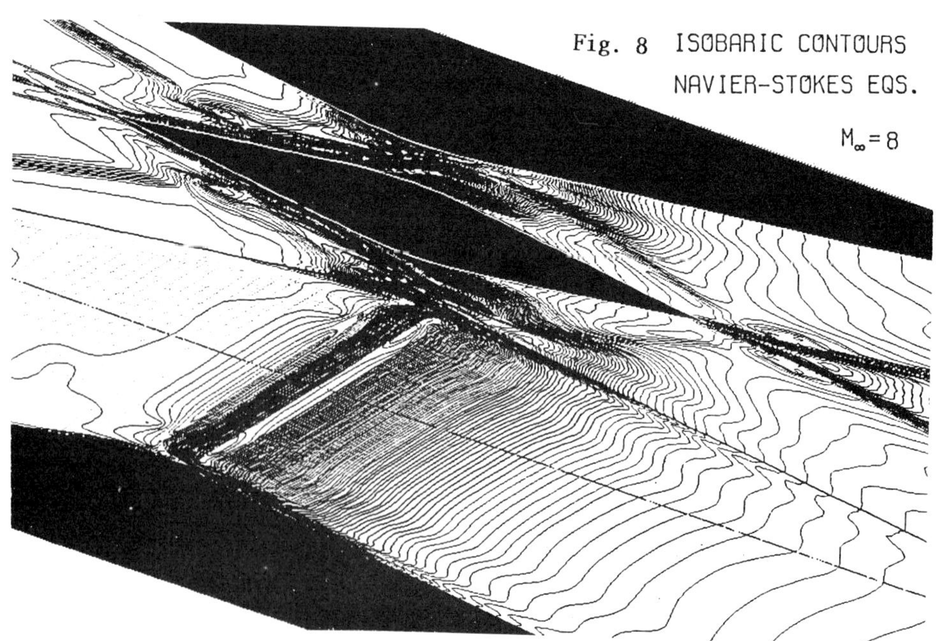

Fig. 8 ISOBARIC CONTOURS
NAVIER-STOKES EQS.

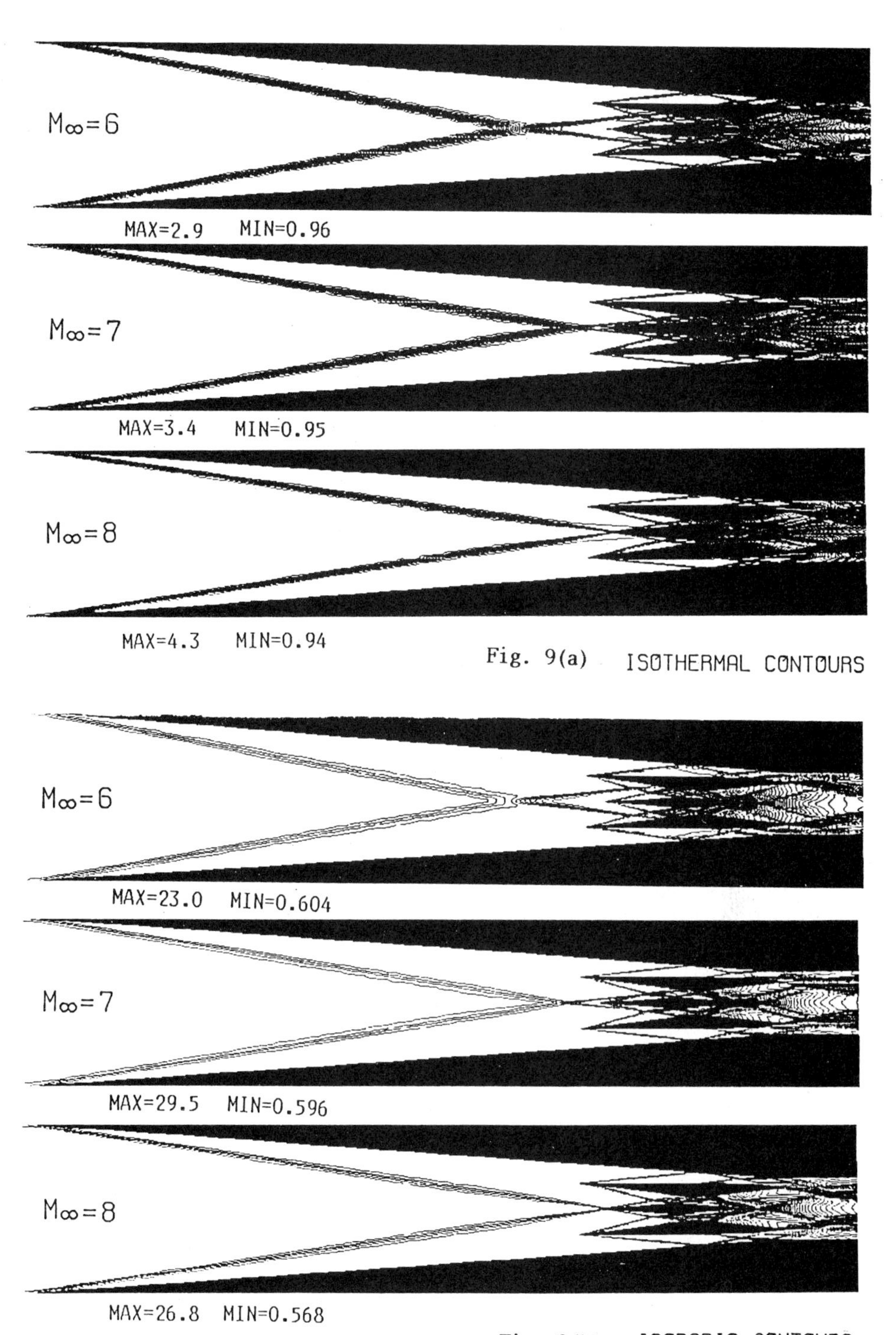

Fig. 9(a) ISOTHERMAL CONTOURS

Fig. 9(b) ISOBARIC CONTOURS

INVISCID HYPERSONIC BLUNT BODY FLOWS WITH FINITE-RATE CHEMICAL KINETICS

M. Onofri - D. Lentini

Dipartimento di Meccanica e Aeronautica
Universita' degli Studi di Roma "La Sapienza"
Via Eudossiana 18, I-00184 Rome, Italy

ABSTRACT

Aim of this study is to show that an accelerating technique can successfully be used to solve the reacting flows about space vehicles during the re-entry phase, downstream of a curved bow shock, where a large flow region about the body is in conditions of chemical nonequilibrium, in both subsonic and supersonic regime. In particular, a fast numerical solver for 2-D hypersonic blunt body flows undergoing finite-rate chemical kinetics is proposed. The solution method is based on the non-conservative Lambda formulation and the bow shock is treated by a "shock fitting" technique.

NOMENCLATURE

a_f	frozen sound speed
c_{pf}	specific heat at constant pressure and frozen composition
c_{vf}	specific heat at constant volume and frozen composition
e_i	internal energy of species i
$\vec{q}$	velocity
R	$\bar{R}/\bar{W}$
$\bar{R}$	universal gas constant
s	entropy
y	distance from the axis
Y_i	mass fraction of species i
$\bar{W}$	average molecular weight
γ_f	specific heat ratio of the mixture at frozen composition
ρ	density
θ	temperature
δ_f	$(\gamma_f - 1)/2$
μ_i	chemical potential of species i

INTRODUCTION

In order to obtain accurate predictions of the flow about re-entry space vehicles a numerical solution of a nonequilibrium chemically reacting inviscid flow, operating from subsonic to supersonic conditions, is needed.

The early studies on this argument date back to the '60s. At that time many efforts were devoted to carry out solutions of multidimensional inert blunt body flows [1] and to circunvent the numerical difficulties connected with the solution of nonequilibrium reacting flows.

As regards the latter aspect, the studies were mainly devoted to cope with the stiffness of the species continuity equations. In particular, 1-D flow solutions were carried out [2-5], in which the different techniques proposed had, as common basis, the adoption of a local linearization of the species production terms as a function of the mass fractions.

Further studies developed later have introduced this result in more general methodologies for multidimensional flows, based on a separate numerical integration of the fluid dynamic equations and of the species continuity equations. In particular, the fluid equations are integrated by the current schemes commonly adopted to compute inert flows, while the species equations are integrated by different implicit algorithms, based on the early studies.

A significant representation of these studies is given by [6], in which Rakich, Bailey and Park use the method of characteristics for the fluid equations, therefore limiting the computing field to the supersonic region only, while the species equations are integrated by an implicit algorithm.

Another significant approach is the time-split explicit method [7,8] in which the fluid and species equations are integrated separately by explicit techniques, using different time step according to the stability criteria of each equation set. Even if treating stiff equations implicitly seems to be the most effective and commonly practiced way, these results are widely used as a basis for setting relaxation strategies in more recent studies.

Moving from this stage, studies in the last two year have been performed [9-11] in order to get:

a) higher accuracy from the gasdynamic point of view;
b) easy approach to describe the flow from subsonic to supersonic conditions, with no limitation on the geometry;
c) significant saving of computing time, that is heavy when solving the species continuity equations with a large number of species and reactions.

In the same spirit this study is conceived. Two main goals are pursued:

a) To extend to blunt body flows in chemical nonequilibrium conditions a very accurate gasdynamic formulation, as the non conservative Lambda scheme. This method is, in fact, based on an upwind discretization, consistent with the physical dominion of dependence of each unknown, that results in a high accuracy in the description of wave propagation phenomena. Moreover, in the spirit of the non conservative form, a "shock fitting" technique can easily be adopted, that is more effective than "shock capturing" methods in treating blunt body flows.

b) To introduce a numerical fast solver, enable to reduce the over all computing time. This technique is an extension of the Euler fast solver proposed for 1-D flows by Moretti in [12] and further developed by Moretti and Onofri in [13] for blunt body flows and by Onofri and Lentini in [14] for internal flows. The occurrence of finite-rate chemical reactions is accounted for by extending the scheme already developed in [15,16] for nozzle flows. The technique is implicit, second order accurate. It reaches a convergence with a residual of 1.E-10 in a limited amount of steps, thus showing particularly useful in problems where each iteration is highly time consuming. Moreover, the formulation is also efficient from a computational point of view, because the integration is simply reduced to the inversion of bidiagonal matrixes.

MATHEMATICAL MODEL

Equations of motion

Dimensionless quantities are obtained by assuming the following reference values for length, pressure, temperature and velocity:

$$\tilde{x} = x_b \qquad \tilde{p} = p_0 \qquad \tilde{\theta} = \theta_0 \qquad \tilde{q} = \sqrt{\bar{R}\,\theta_0}$$

where x_b is the body lenght, p_0 and θ_0 define the reference state, and $\bar{R}$ is the universal gas constant; further, the kilomole is assumed as molar unit.

The governing equations are obtained extending the Lambda formulation for inert flows, described in detail in [14], to nonequilibrium chemically reacting flows, as also reported in [15,16]. In particular, the gas is assumed to be a mixture of N components, each both thermally and calorically perfect, in nonequilibrium conditions. Thus, the equations of motion for an adiabatic inviscid non heat-conducting flow, are:

continuity: $$\rho_t + \nabla.(\rho \vec{q}) = 0 \tag{1a}$$

momentum: $$\vec{q}_t + \vec{q}.\nabla\vec{q} + \frac{1}{\rho}\nabla p = 0 \tag{1b}$$

energy: $$s_t + \vec{q}.\nabla s = \varepsilon \tag{1c}$$

species: $$Y_{it} + \vec{q}.\nabla Y_i = \tau_i \quad (i=1,2,...,N) \tag{1d}$$

where $\varepsilon = -\frac{1}{\theta}\sum_{i=1}^{N} \mu_i \tau_i$ and τ_i is the species source function per unit mixture mass of species i; it accounts for the contributions of all reactions, which consist of both the forward and the backward steps, with reaction rates related by the equilibrium constants [17,18].

In the spirit of the Lambda formulation [19,20], the equations of motion are recast in the form of compatibility equations, written in terms of the variables $\vec{q}$, a_f, s, Y_i $(i=1,2,...,N)$, along characteristic lines. Whereas eqs. (1c) and (1d) are already in the form of compatibility equations, in order to attain a similar form for the other equations the procedure outlined in [19, 20] for non-reacting flows is here extended to nonequilibrium chemically reacting flows. Defining a unit vector $\vec{w}$ the generic compatibility equation is:

$$b_t + \vec{w}.\vec{q}_t + \vec{q}.\nabla b + \vec{w}.(\vec{q}.\nabla\vec{q}) + a_f\nabla.\vec{q} + a_f\vec{w}.\nabla b - \theta\vec{w}.\nabla s + \sum_{i=1}^{N}\pi_i\vec{w}.\nabla Y_i = \beta \quad (2)$$

with:

$$b = \frac{a_f}{\delta_f}$$

$$\beta = a_f \sum_{i=1}^{N} \left[\frac{1}{\gamma_f - 1} \frac{\partial \ln(\gamma_f R/\delta_f^2)}{\partial Y_i} - \frac{e_i}{R\theta}\right] \tau_i \quad (3)$$

$$\pi_i = \theta \left[s_i - c_{pf} \frac{\partial \ln\ (\gamma_f R/\delta_f^2)}{\partial Y_i}\right] \qquad (i=1,2,...,N)$$

where e_i and s_i are the internal energy and entropy of species i.

Following an idea of Butler [21], the wave propagation phenomena can be simulated as propagations along only four orthogonal bicharacteristic directions. To this end a suitable choice of vector w must be done. The set of orthogonal vectors $\vec{n}$, $\vec{\tau}$, $\vec{k}$, with $\vec{n}$ and $\vec{\tau}$ lying in the x-y plane and $\vec{n}$ forming an angle α_0 with the x-axis, are defined at any node of the computational grid. From eq.2, the following set results, with j = 1,2,3,4:

$$R_{jt} + \vec{\Lambda}_j.\nabla R_j - \theta\vec{w}_j.\nabla s + \sum_{i=1}^{N}\pi_i\vec{w}_j.\nabla Y_i + a_f\ \vec{w}'_j.\nabla(\vec{w}'_j.\vec{q}) =$$
$$= \alpha'_0\vec{w}'_j.\vec{q} - \alpha''_0 - \delta_{ax}\ a_f\ \frac{V}{y} + \beta \quad (4)$$

where the four bicharacteristic directions and the associated Riemann variables can be defined as

$$\vec{\Lambda}_1 = \vec{q}+a_f\vec{n} \qquad R_1 = b+\vec{n}.\vec{q} \qquad \vec{w}_1 = \vec{n} \qquad \vec{w}'_1 = \vec{\tau}$$

$$\vec{\Lambda}_2 = \vec{q}-a_f\vec{n} \qquad R_2 = b-\vec{n}.\vec{q} \qquad \vec{w}_2 = -\vec{n} \qquad \vec{w}'_2 = -\vec{\tau}$$

$$\vec{\Lambda}_3 = \vec{q}+a_f\vec{\tau} \qquad R_3 = b+\vec{\tau}.\vec{q} \qquad \vec{w}_3 = \vec{\tau} \qquad \vec{w}'_3 = -\vec{n}$$

$$\vec{\Lambda}_4 = \vec{q}-a_f\vec{\tau} \qquad R_4 = b-\vec{\tau}.\vec{q} \qquad \vec{w}_4 = -\vec{\tau} \qquad \vec{w}'_4 = \vec{n}$$

$$\alpha'_0 = \vec{q}.\nabla\alpha_0 \qquad \alpha''_0 = a_f\vec{k}\times\vec{q}.\nabla\alpha_0\ .$$

Computational grid

A region limited by the shock and the body has been considered, mapped by a non orthogonal grid. A correspondence between this physical region and an auxiliar computational plane (X,Y) mapped with an orthogonal grid has been obtained by a coordinate transformation.

In particular, by means of a conformal mapping, with metric denoted by G, a ξ-η orthogonal grid is firstly generated. In the present case as a simple shape like a cylinder is considered, a polar frame is used. Then, a non-orthogonal computing grid (fig. 1) is defined by the transformations

$$w = \xi \qquad\qquad z = \frac{\eta - \eta_b}{\eta_s - \eta_b}$$

where η_s and η_b are the η values at the shock and the body, respectively. If α_0 is chosen as $\alpha_0 = \tan^{1}(z\ \eta_{s_\xi})$ and C, S, T denote the cosine, sine and tangent of α_0; then

$$z_\xi = -Tz_\eta \qquad\qquad z_\eta = \frac{1}{\left(\eta_s(\xi) - \eta_b(\xi)\right)} .$$

Let U and V be the cartesian components of $\vec{q}$, and u and v those in the $\vec{n}$ and $\vec{T}$ directions, it results

$$u = CU + SV \qquad\qquad v = -SU + CV .$$

Since for any quantity ψ is $\nabla\psi = \vec{i}\,(\psi_\xi + z_\xi\psi_z) + \vec{j}\ z_\eta\psi_z$ by substituting in eqs. (4), (1c), (1d), and expanding the dot-products, the set of equations of motion takes on the form

$$R_{1t} + G\lambda_1 R_{1w} + GV_0 R_{1z} - GC\theta s_w + GC\sum_{i=1}^{N}\pi_i Y_{iw} = \alpha_1 + \beta_1 \tag{5a}$$

$$R_{2t} + G\lambda_2 R_{2w} + GV_0 R_{2z} + GC\theta s_w - GC\sum_{i=1}^{N}\pi_i Y_{iw} = \alpha_2 + \beta_1 \tag{5b}$$

$$R_{3t} + GU_3 R_{3w} + G\lambda_3 R_{3z} + GS\theta s_w - G\frac{z_y}{C}\theta s_z - GS\sum_{i=1}^{N}\pi_i Y_{iw} + G\frac{z_y}{C}\sum_{i=1}^{N}\pi_i Y_{iz} = \alpha_3 + \beta_2 \tag{5c}$$

$$R_{4t} + GU_4 R_{4w} + G\lambda_4 R_{4z} - GS\theta s_w + G\frac{z_y}{C}\theta s_z + GS\sum_{i=1}^{N}\pi_i Y_{iw} - G\frac{z_y}{C}\sum_{i=1}^{N}\pi_i Y_{iz} = \alpha_4 + \beta_2 \tag{5d}$$

$$s_t + GUs_w + GV_0 s_z = \epsilon \tag{5e}$$

$$Y_{it} + GUY_{iw} + GV_0 Y_{iz} = \tau_i \qquad (i=1,2,..N) \tag{5f}$$

where

$$\lambda_1 = U + Ca_f \qquad\qquad \alpha_1 = \alpha_0' v - \alpha_0'' - \delta_{ax}\ a_f\ \frac{V}{y}$$

$$\lambda_2 = U - Ca_f \qquad\qquad \alpha_2 = -\alpha_0' v - \alpha_0'' - \delta_{ax}\ a_f\ \frac{V}{y}$$

$$\lambda_3 = \frac{z_y}{C}(v + a_f) \qquad\qquad \alpha_3 = -\alpha_0' u - \alpha_0'' - \delta_{ax}\ a_f\ \frac{V}{y}$$

$$\lambda_4 = \frac{z_y}{C}(v - a)_f \qquad\qquad \alpha_4 = \alpha_0' u - \alpha_0'' - \delta_{ax}\ a_f\ \frac{V}{y}$$

$$V_0 = \frac{z_y}{C}v \qquad\qquad U_3 = U - Sa_f \qquad\qquad U_4 = U + Sa_f$$

$$\beta_1 = \beta + GSa_f v_w - G\frac{z_y}{C}a_f v_z \qquad\qquad \beta_2 = \beta - GCa_f u_w .$$

NUMERICAL INTEGRATION

The above equations are solved on a region limited by the shock and the body. The shock is fitted and the fluid dynamic quantities behind the shock are determined by considering frozen flow conditions across it. As regards the other boundary conditions, they are imposed just assuming v=0 at the wall and simmetry conditions along the symmetry line.

Eqs. (5e,5f) describe the propagation along $\vec{q}$ of the quantities s and Y_i (i=1,2,..,N), respectively. Once these equations have been solved, the derivatives of s and τ_i appearing in (5a-5d) can be evaluated and regarded as known term. Thus each equation (5a-5d) can be regarded as describing the propagation of a scalar term R_i along a bicharacteristic direction $\vec{\Lambda}_i$ (i=1,2,3,4), which is independent of the propagation of the other three R_j carried by the other $\vec{\Lambda}_j$ (j≠i).

As a consequence, when solutions of steady flows are considered, the integration of the above equation set can be obtained as the result of an iterative procedure in which each equation is solved over the whole field independently of the others. In particular, R_1 and R_2 proceed essentially along the $\vec{n}$ direction, whereas R_3 and R_4 proceed essentially along the radial direction. Therefore, R_1 is integrated sweeping in the positive ξ-direction, according with the positive sign of λ_1, R_2 is integrated in the same direction but in the negative or positive way according to the sign of λ_2, R_3 and R_4 are integrated in the radial direction, in the positive way the former and in the negative one the latter, consistently with the sign of λ_3 and λ_4.

Thus the integration of the equations can be performed by treating a two-dimensional problem as two alternate quasi one-dimensional problems and the numerical scheme can be patterned on the very efficient one-dimensional formulation of the fast solver technique [12].

The integration procedure at each step of the iterative process can therefore be summarized as follows:

a) Integration of species continuity equations.
At each computational step, the species continuity equations, are integrated in steady form in order to accelerate the convergence by an implicit algorithm. The integration is performed by an implicit algorithm, sweeping along η=const. lines, in the direction of increasing ξ assumed to be the main direction of propagation. In order to attain second order accuracy, the transport quantities and the source term τ_i must be averaged in the direction of incoming velocity components. η- and ξ-derivatives are approximated by upstream differences, as U and V are positive. The source term τ_i is averaged in the ξ-direction. However, in order to overcome the numerical stiffness inherent in these equations, $\tau_{i\,m,n}$ is expressed by means of a Taylor expansion with respect to all Y_j (j=1,2,...N)

$$\tau_{i\,m,n} \cong \tau_i\left[a_{f\,m,n}, s_{m,n}, Y_{j\,m-1,n} \quad (j=1,2,...N)\right] + \sum_{j=1}^{N} \frac{\partial \tau_i}{\partial Y_j} (Y_{j\,m,n} - y_{j\,m-1,n})$$

where derivatives are computed for the same values of the state variables appearing as arguments in τ_i. By using this expression a linear system of N equations in the un knowns $y_{jm,n}$ $(j=1,2,...N)$ results; their solution has then to be carried out simultaneously. Therefore, the set of eqs. (5f) can be integrated apart from the remaining ones, even though the single species continuity equations cannot be integrated separately.

b) Integration of entropy equation.
The entropy equation is integrated in steady form by an implicit algorithm. The source term ϵ is computed by considering the values of the state variables at the end of the previous computational step.

c) Integration of the equations in the Riemann variables.
At this stage, the Riemann eqs. (1-4) can be solved by an implicit scheme, which computes the specific R_i at each grid point as a function of the updated values of s and Y_i and of the values of R_i at the contiguous grid point just computed. Thus, the integration algorithm results very simple, based on the simple solution of bidiagonal matrixes.

All the derivatives are approximated by upwind differences, and the transport terms are averaged in the same way to attain second order accuracy.

COMPUTATIONAL ASPECTS

Some preliminary results are reported in the following figures, showing the behaviour of the above referred method when implemented on different test cases.

The results reported in Figs. 2,3 are referred to calculations performed for a cylinder in a supersonic inert flow with an infinity Mach number equal 4. A computational grid with 13x25 points is used. Fig. 2 shows the convergence hystory of the solution. A convergence with a residual of 10E-7 is reached after 180 steps. Nevertheless, as referred in [14], for more pratical purposes the steady state can be considered reached after 50 steps. Fig. 3 shows the iso-Mach lines computed. The standoff distance is in good agreement with the values reported in the literature.

So far, the technique has been implemented on multidimensional non-equilibrium flows only on nozzles with a H/O reacting mixture. The results are reported in [15,16] and show that the decay of the residual versus the number of iterations reaches 10E-9 in less than 150 steps.

As regards applications of the technique to a blunt body flow, aim of the tests performed in this study is to analyze the numerical problems connected with the coupling of the two separate solutions of fluid and species equations. As a matter of fact, even if an implicit scheme is used to solve the species equations, numerical instabilities can still occur as result of the coupling of this solution with the solution of the fluid equations.

A test has been considered in which the path of a generic particle about the blunt body is simulated by a quasi-1D flow in a duct. The configuration includes a shock in a constant area duct followed by a convergent divergent shape. As in a blunt body region the particles cross the shock and then flow through a nonequilibrium region, passing from subsonic to supersonic conditions.

A first numerical difficulty is connected with the initialization. As shown in Fig. 4, the trend of the temperature initial values produces at the first step a behaviour of the specie mass fraction very far from the aspected. Instability can originate from that. To circunvent this problem a relaxation factor has been introduced in the integration of species equations to allow a gradual changing of the species mass fraction from the frozen initial conditions.

Since this numerical behaviour originates from the strong dependence of the compatibility equations from the values of the mass fractions, the same trend occurs as flows with high total temperature and significant dissociation phenomena are considered. In these cases, an effective solution is the adoption of the "successive specie solution", according with the species equations are solved only after a certain ammount of steps of the "fluid equations" solution (20 steps have been found the optimum delaying). In fact, the flow gradients can be sufficiently smoothed in this way, before a new variation of the fluid mixture composition occurs.

REFERENCES

[1]. Moretti G., Abbett M., "A Time Dependent Computational Method for Blunt-Body Flows". AIAA J., Vol. 4, 12, 1966.

[2]. Moretti G., "A New Technique for the Numerical Analysis of Nonequilibrium Flows". AIAA J., Vol. 3, Feb. 1965.

[3]. Treanor C.E., "A Method for the Numerical Integration of Coupled First-Order Differential Equations with Greatly Different Time Constants". Math. Comp., Vol. 20, Jan. 1966.

[4]. Lomax H., Bailey H.E., "A Critical Analysis of Various Numerical Integration Methods for Computing the Flow of a Gas in Chemical Nonequilibrium". NASA TN D-4109, 1967.

[5]. Bailey H.E., "Numerical Integration of the Equations Governing the One-Dimensional Flow of a Chemically Reacting Gas". The Physics of Fluids, Vol. 12, 11, November 1969.

[6]. Rakich J.V., Bailey H.E., Park C., "Computation of Non equilibrium, Supersonic Three-Dimensional Inviscid Flow over Blunt-Nosed Bodies", AIAA J., Vol. 21, June 1983.

[7]. Li, C.P., "Time-Dependent Solutions of Nonequilibrium Airflow past a Blunt Body". J. Spacecraft, Vol. 9, no. 8, Aug. 1972, pp. 571-572.

[8]. Rizzi A.W., Bailey H.E., "Reacting Nonequilibrium Flow around the Space Shuttle Using a Time-Split Method", Aerodynamic Analyses Requiring advanced Computers, NASA SP-347, March 1975, pp. 1327-1376.

[9]. Yee H.C., Shinn J.L., "Semi-Implicit and Fully Implicit Shock-Capturing Methods for Hyperbolic Conservation Laws with Stiff Source Terms". NASA TM 89415, Dec. 1986.

[10]. Li C.P., "Implicit Methods for Computing Chemically Reacting Flows". 10th ICNMFD, Beijing (China). Also NASA TM 58274, Sept. 1986.

[11]. Gnoffo P.A., McCandless R.S., "Three-Dimensional AOTV Flow fields in Chemical Nonequilibrium". AIAA Paper 86-230, Jan. 1986.

[12]. Moretti G., "Fast Euler Solver for Steady, One-Dimensional Flows". Computers and Fluids, 13, 1985.

[13]. Moretti G., Onofri M., "A Fast Euler Solver for Blunt Body Flows". Poly M/AE Report 86/13, Polytechnic University, New York, 1986.

[14]. Onofri M., Lentini D., "Fast Numerical Technique for Transonic Flows". Symp. on Physical Aspects of Numerical Gas Dynamics, Polytechnic University, Farmingdale, N.Y., August 1987.

[15]. Lentini D., Onofri M., "A Fast Numerical Solver for Non-Equilibrium Chemically Reacting Flows in Two-Dimensional Nozzles", SIAM Conf. on Numerical Combustion, S. Francisco, CA, March 1987.

[16]. Lentini D., Onofri M., "Fast Numerical Technique for Nozzle Flows with Finite Rate Chemical Kinetics". ISCFD, Sidney (Australia), Aug. 1987.

[17]. Vincenti W.G., Kruger C.H., "Introduction to Physical Gas Dynamics". Wiley, New York 1965.

[18]. Williams F.A., "Combustion Theory". Benjamin/Cummings, Menlo Park 1985.

[19]. Moretti G., Zannetti L., "A New Improved Computational Technique for Two-Dimensional Unsteady Compressible Flows". AIAA J., Vol. 22, Jun. 1984.

[20]. Moretti G., "A Technique for Integrating Two-Dimensional Euler Equations". Computers and Fluids, Vol. 15, 1, 1987.

[21]. Butler D.S., "The Numerical Solution of Hyperbolic Systems of Partial Differential Equations in Three Independent Variables", Proc. Royal Soc., Series A, pp.232-252, 1960.

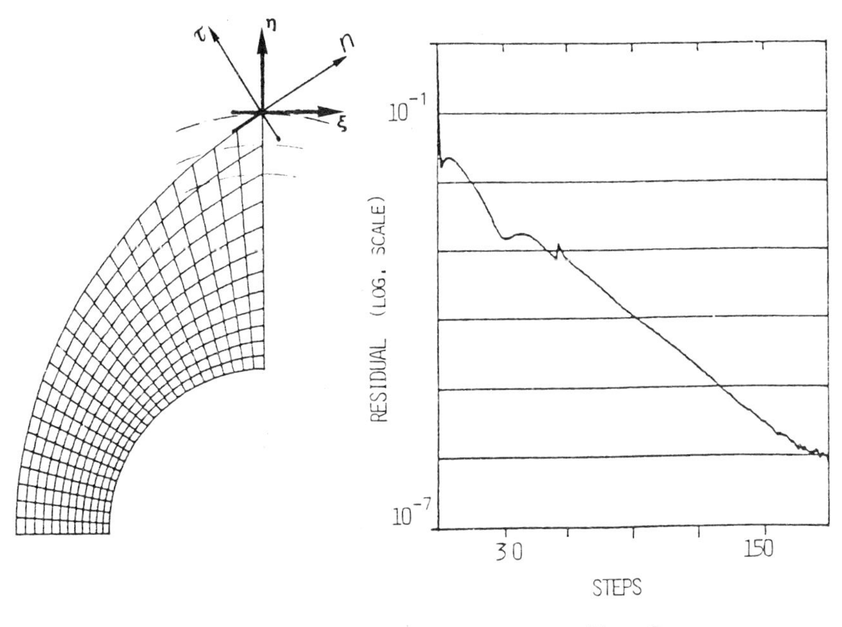

FIG. 1

FIG. 2

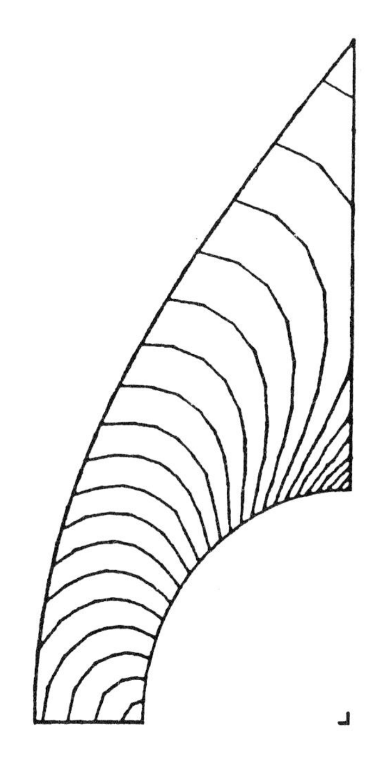

FIG. 3

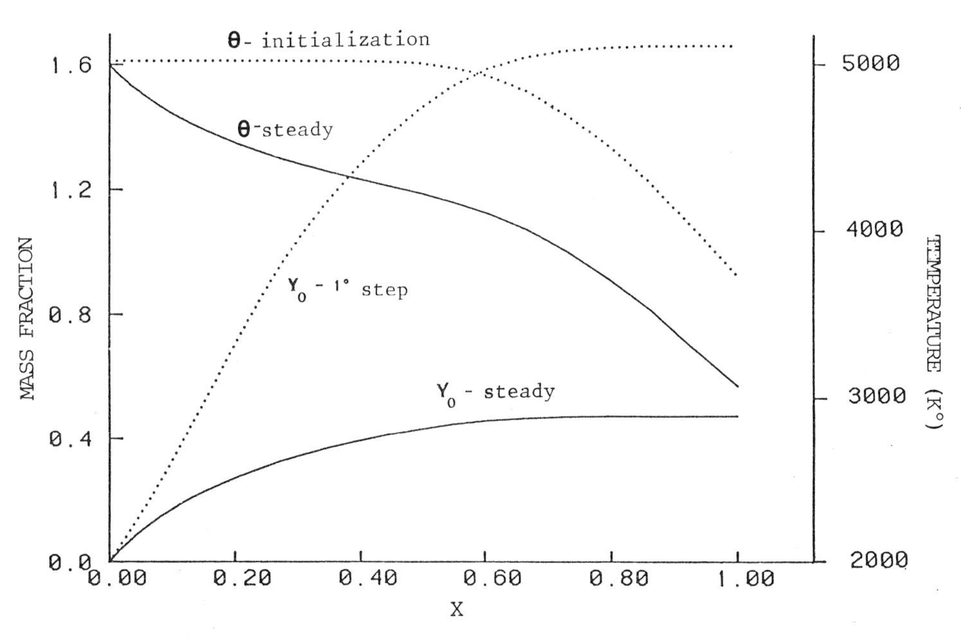

FIG. 4 $Mach_{\infty}$=10.10 - H=60 Km.

FINITE ELEMENT SOLUTION OF THE EULER EQUATIONS IN TWO AND THREE DIMENSIONS

J. Peiro, L. Formaggia, J. Peraire and K. Morgan

Institute for Numerical Methods in Engineering
University of Wales SWANSEA SA2 8PP

ABSTRACT

We present a finite element procedure for solving the equations of compressible flow over bodies of arbitrary geometry. The numerical solution algorithm employed is an explicit two-step version of a second order Taylor-Galerkin scheme. The discretization of the computational domain into unstructured meshes of triangles in 2D and tetrahedra in 3D is performed by an automatic mesh generator. In our approach, the mesh generator is coupled to the finite element solver to produce an adaptive remeshing procedure.

1. INTRODUCTION

Finite element methods for the solution of the equations of compressible flow implemented on unstructured grids are, in general, computationally more expensive than traditional methods using structured meshes when solving problems with the same number of unknowns. However, this overhead is largely compensated in simulations of two- and three-dimensional industrial flows by the use of unstructured grids, together with the utilization of mesh generators and adaptive techniques, which can provide representations of geometry and flow features of an accuracy that is not easily attainable with traditional approaches.

In this paper we present an adaptive remeshing procedure that combines directional adaptivity with mesh generation to produce a powerful tool for the simulation of complicated flows over general geometries.

The performance of the adaptive procedure in 2D is shown by solving a shock interaction problem over a cylindrical leading edge in a Mach 8 flow. The preliminary results in 3D show the mesh generated for a wing tip and the computation of a Mach 2 flow past an engine intake.

2. SOLUTION OF THE EULER EQUATIONS

The unsteady Euler equations of compressible flow are written in the conservation form as

$$\frac{\partial U}{\partial t} + \frac{\partial F^j}{\partial x_j} = 0 \,. \tag{1}$$

Our numerical algorithm for the solution of this equation is a two-step version of the explicit second order Taylor - Galerkin scheme introduced by Donea [1]. The elimination of oscillations near discontinuities is accomplished by the introduction of a novel finite element based artificial viscosity.

2.1. Two-step Taylor-Galerkin scheme.

The Taylor-Galerkin scheme employed for the solution of the equation (1) involves a two-step explicit evaluation of the advective terms. In the first step, given a piecewise linear distribution of the solution vector U^n at time t^n we can obtain a piecewise linear discontinuous representation of $U^{n+1/2}$ by writing

$$U^{n+1/2} = U^n + \delta U. \tag{2}$$

δU is computed by means of the weighted residual statement

$$\int_\Omega \delta U \, P_e \, d\Omega = - \int_\Omega \frac{\Delta t^e}{2} \left. \frac{\partial F^j}{\partial x_j} \right|^n P_e \, d\Omega \tag{3}$$

where P^e is the constant shape function associated to element e, Ω represents the computational domain and Δt^e is an element local timestep.

In the second step, the solution is updated according to

$$U^{n+1} = U^n + \Delta U \tag{4}$$

The Galerkin weighted residual statement for calculating ΔU is

$$\int_\Omega \Delta U \, N_i \, d\Omega = -\Delta t_i \left(\int_\lambda n_j \, F^j |^{n+1/2} N_i \, d\lambda - \int_\Omega F^j |^{n+1/2} \frac{\partial N_i}{\partial x_j} d\Omega \right). \tag{5}$$

Here, N_i is the linear shape function associated with node i, λ denotes the boundary of Ω and n_j are the components of the unit outward normal to λ. Δt_i is a local timestep for node i computed as the minimum of the timesteps for the elements surrounding i.

Boundary conditions are directly applied to the boundary integrals appearing in equation (5) by performing a one-dimensional linearised characteristic analysis in the direction normal to the boundary.

2.2. Artificial viscosity.

Before advancing to the next step, some artificial viscosity needs to be applied to stabilize the solution. In the present method this is accomplished by smoothing the values at time t^{n+1} according to

$$U_s^{n+1} = U^{n+1} + \Delta t \, D(U^{n+1}) \tag{6}$$

D being the diffusion term. A Laplace type artificial viscosity employing a diffusion coefficient depending on pressure derivatives [2,3] is used.

The expensive computation of the second derivatives is avoided by using the fact that in one dimension

$$\left(M_L^{-1} \sum_e (M^e - M_L^e) \, U^e \right)_i = \left(\frac{d}{dx} \left(h^2 \frac{dU}{dx}\right) \right)_i \tag{7}$$

where e stands for element and i for node, M represents the consistent mass matrix and M_L the diagonal lumped mass matrix. The straightforward extension of the expression on the left of equation (7) to more dimensions and the inclusion of a pressure switch provides the final form of the numerical smoothing term

$$D = M_L^{-1} \sum_e \frac{C \, S^e}{\Delta t_e} \left(M^e - M_L^e \right) U^e \tag{8}$$

where C is a user defined parameter. The element pressure switch S_e is the mean of the element nodal switches S_i given by

$$S_i = \frac{\{ \sum_e (M^e - M_L^e) P^e \}_i}{| \sum_e (M^e - M_L^e) P^e |_i}. \tag{9}$$

In this formula P denotes the vector of nodal pressures and |.| indicates absolute value. The values of S_i always lie between 0 and 1, giving a value close to 1 in the proximity of discontinuities regardless of their relative strength.

3. MESH GENERATION

One of the main difficulties encountered when using structured mesh generators is the representation of geometries and flow features in multiple connected domains with closely coupled boundaries.

In order to overcome this flaw and fully exploit the advantages of unstructured grids, our mesh generator introduces the possibility of generating elements (triangles or tetrahedra) of variable size which can be stretched along certain directions. This approach allows us to produce meshes that can easily adapt to complex geometries and features of the flow.

A complete description of the 2D version of the mesh generator can be found elsewhere [4]. Therefore, in what follows, our exposition will be mainly centred upon the extension of the 2D concepts and procedures to the 3D case.

3.1. Mesh parameters.

Directionality in an element of the mesh is taken into account by introducing the concepts of spacing and stretching. The spacing δ in the direction α is defined as the maximum distance between projections, on a straight line parallel to α, of the nodes of the element and the stretching s will be the positive ratio between spacings in two perpendicular directions.

The mesh parameters needed to define an element are: the value δ of nodal spacing and two stretchings s_1 and s_2 associated to two orthogonal directions α_1 and α_2 (figure 1).

3.2. Mesh generator strategy.

The computational domain is represented by a geometrical model in which we can construct a hierarchical structure, with different levels of geometrical complexity, of one-, two- and three-dimensional components. In our convention, a component is a closed region defined in a support geometry: a curve, a surface or a volume. The boundary of this region is formed by a set of components of the next inferior level of geometrical complexity. The representation of the support geometry for curves and surfaces is accomplished by fitting, imposing curvature continuity, bicubic segments and patches, respectively, through a set of data coordinates. Each of these 1D, 2D and 3D components is, independently, discretized into straight, triangular and tetrahedral elements whose nodal points lie in the support geometry of the component.

In the discretization process, elements and points are generated simultaneously according to the grid characteristics specified by a background grid [4] of linear triangles in 2D and tetrahedra in 3D in which the values of the mesh parameters are specified at the nodes.

For 1D components, the position of the generated nodes is calculated by integrating the distribution of spacings along the length of the curve. This distribution is obtained by interpolating, from the background grid, the spacing at sampling points in the support geometry.

The methodology for the discretization of 2D and 3D components is based upon the advancing front concept. The initial front is the set of orientated items (segments or triangles) that form the discretized boundary of the component to be meshed and which will be used as bases for the construction of new elements. Once a base for the element has been chosen, the value of the mesh parameters δ, s_i and α_i are obtained by interpolation from the background grid. Then, the coordinates of the points relevant for the generation of an element are transformed, using a linear mapping $T(\delta, s_i, \alpha_i)$, into an unstretched space in which the element will look 'equilateral' with a typical dimension of 1. After that, a regular element is generated in the new space and its coordinates transformed back to the original space. Finally, the validity of the element is checked by verifying that none of its sides intersects the front and that its area/volume is positive. Every time a new element is created, the front is updated by deleting the boundary faces or sides of the element that belong to the front and including the newly created ones. The end of the discretization is reached when the number of items in the front is zero.

In the case of 2D components, the definition of the support geometry by bicubic patches provides a one-to-one mapping $R=R(x_1, x_2)$ between a 2D space x_1-x_2 and the 3D space. In our approach, the surface mesh is obtained as the image by $R(x_1, x_2)$ of a triangular mesh in the x_1-x_2 space that is generated using the 2D procedure [4]. The values of the 2D mesh parameters that ensure compatibility of the resulting discretization with the 3D mesh parameters are obtained from the mapping Q between the unstretched space and the x_1-x_2 plane given by

$$Q(x_1, x_2) = T(\delta, s_i, \alpha_i)\ R(x_1, x_2). \tag{10}$$

The spacing δ to be used in any direction defined by the unit vector α in the plane x_1-x_2 is found as

$$\left(\frac{1}{\delta}\right)^2 = \frac{\partial Q}{\partial x_i} \frac{\partial Q}{\partial x_j} \alpha_i \alpha_j. \tag{11}$$

From this expression, the calculation of the 2D parameters reduces to an eigenvalue problem.

The construction of the mesh representing the computational domain is performed in a bottom-up manner. The first step is the discretization of all 1D components. Then, for each 2D component, we build the initial front as an assembly of the discretized 1D components forming its boundary and a triangulation is generated afterwards in the way previously described. Finally, the 3D mesh of tetrahedra is generated from an initial front obtained by assembling discretized 2D components.

3.3. Adaptive remeshing.

For the solution of steady state problems, an adaptive remeshing procedure has been proposed [4] as a more economic alternative to standard mesh enrichment and mesh movement methods.

In this approach the mesh is completely regenerated at each mesh adaptation stage. The information about the distribution of mesh parameters is obtained from the computed solution in the current mesh by an error estimation process based upon interpolation theory. In this process, an extension to several dimensions of the one-dimensional criterion of equidistribution of the root mean square value of the local error [4] suggests that the spacing δ in the direction of the unit vector α should be computed according to

$$\delta^2 \quad \alpha_i \, \alpha_j \quad \frac{\partial^2 \xi}{\partial x_i \, \partial x_j} = \text{constant} \tag{12}$$

where ξ is a key variable chosen as representative of the features of the flow. Then, the mesh parameters are obtained from the solution of an eigenvalue problem for the matrix of second derivatives in (12). The new grid is finally constructed, following the procedure described above, with the current mesh acting as background grid.

4. EXAMPLES

As an illustration of the performance of the adaptive remeshing method we show the 2D computation performed for the shock wave interaction problem on a cylindrical leading edge. The free stream Mach number is 8 and the angle of the shock generator is 12.5 degrees. The schlieren photograph of the experimentally observed interference pattern is displayed in figure 2d. Figures 2a, 2b and 2c show the sequence of meshes used in the calculations and the Mach number contours for their respective computed solutions. The wall and outflow values of the Mach number and the residual curve for the third mesh are displayed in figures 2e and 2f.

Three-dimensional examples of mesh generation and flow solution are shown in figures 3 and 4. Figure 3 shows the initial surface discretization of a winglet with the planes defining the boundary of the computational domain. Figure 4 shows the 3D computation of a supersonic inviscid flow through an engine intake at zero angle of attack. The flow conditions correspond to a free stream Mach number of 2 and a prescribed Mach number at the engine entrance of 0.4. Due to the symmetry of the problem only a quarter of the engine has been computed. The surface triangulation of the computational domain used is shown in figure 3a. Figures 3b and 3d show the pressure and Mach number solutions obtained.

The flow computations have been carried out on a CRAY XMP/48 computer using a fully vectorized code. Typical CPU times in milliseconds required per iteration and point are 0.067 in 3D and 0.015 in the 2D case.

5. CONCLUSIONS

We have presented a method for the generation of unstructured meshes of triangles and tetrahedra. The mesh generator allows the generation of elements which can exhibit large variations in size and be stretched along certain prescribed directions. Therefore, it is specially suited to be combined with a finite element solver to produce an adaptive remeshing method. This adaptive approach has shown to be very effective in the solution of 2D problems and dramatic savings compared with traditional methods are expected in the solution of 3D problem.

6. ACKNOWLEDGMENTS

This work has been partially supported by the Aerothermal Loads Branch of the NASA Langley Research Center under NASA Research Grant NAGW-478. The first author also acknowledges the support by the 'Generalitat de Catalunya' in the form of a CIRIT research studentship.

7. REFERENCES

[1] J. Donea, 'A Taylor-Galerkin method for convective transport problems', Int. J. Num. Meth. Engng., 20, pp. 101-119, (1984).

[2] R.W. MacCormack and B.S. Baldwin, 'A numerical method for solving the Navier-Stokes equations with applications to shock-boundary layer interactions', AIAA paper 75-1, (1975).

[3] A. Jameson, 'Multigrid algorithms for compressible flow computations', MAE Report 1743, (1985).

[4] J. Peraire, M. Vahdati, K. Morgan and O.C. Zienkiewicz, 'Adaptive remeshing for compressible flow computations', accepted for publication in J. Comp. Phys., (1987).

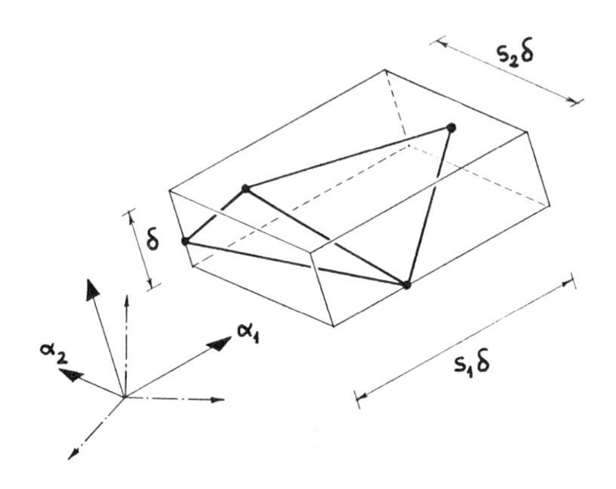

FIGURE 1. Mesh parameters.

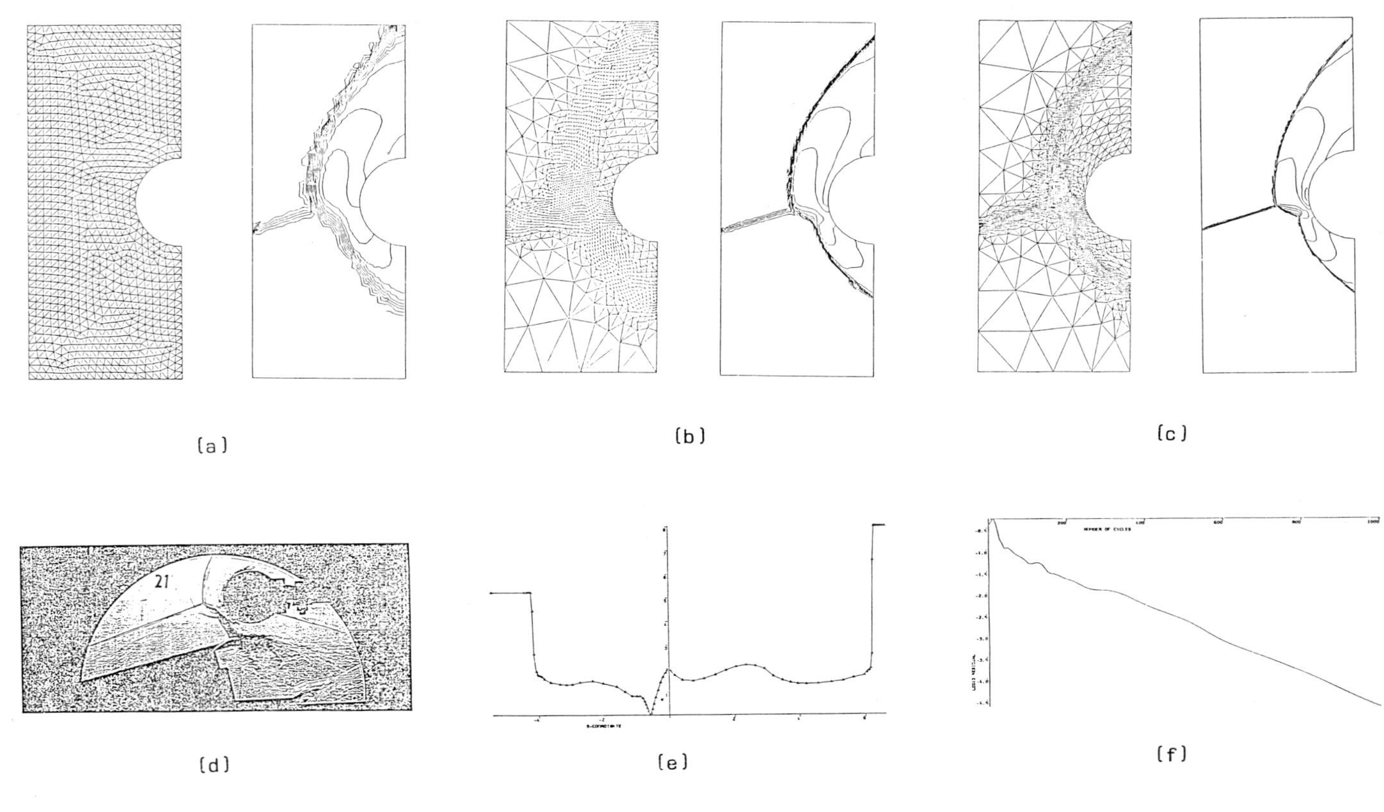

FIGURE 2. Mach 8 shock interaction problem in a cylindrical leading edge.
(a,b,c) Mesh and Mach number contours for 1st, 2nd and 3rd meshes,
(d) Schlieren photograph, (e) Mach number at cylinder and outflow,
(f) Residual curve.

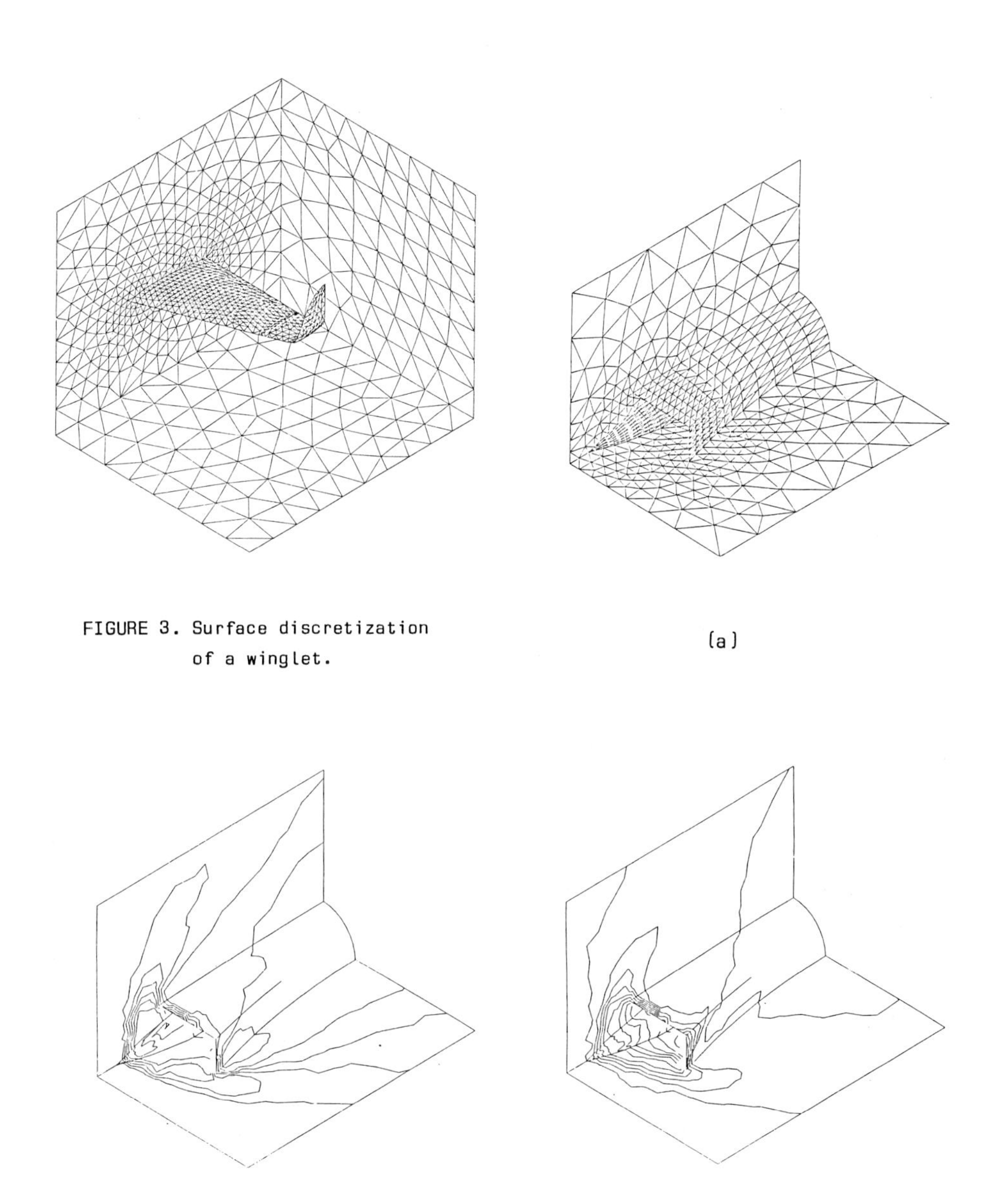

FIGURE 3. Surface discretization of a winglet.

(a)

(b)

(c)

FIGURE 4. Three-dimensional engine air intake.
(a) Surface triangulation of the computational domain.
(b,c) Pressure and Mach number contours.

HIGH ORDER SCHEME AND MULTIDOMAIN TECHNIQUE FOR 2.D. NAVIER-STOKES EQUATIONS : SHOCK - TEMPERATURE SPOT INTERACTION

by Bernadette PERNAUD-THOMAS (*)

ABSTRACT

For compressible viscous flow, the shock-turbulence interaction is characterized by interferences between different modes : entropy, vorticity and sound. As a first approach, the shock-temperature spot interaction is studied numerically in two dimensions. The flow is considered periodic in the direction parallel to the shock line. The complete Navier-Stokes equations are discretized by means of a high-order method (fourth order finite differences) using two domains of computation which overlap over one mesh along the shock line. Production of vorticity is evidenced as a result of the interaction between pressure and temperature gradients. The calculation has been done in two cases : a normal shock and an oblique shock.

1. Introduction

Some fundamental mechanisms of the interaction of a shock wave with an inhomogeneous medium have been studied by solving the complete 2-D. unsteady Navier-Stokes equations for a compressible fluid. This inhomogeneity can come from density or velocity variations. These studies are aimed to shed some light on turbulent processes involved in shock boundary layer [1, 4] or shock flame interactions [6, 7].

The Chu - Kovasnay the ry [2] on turbulent compressible flows described the principal modes to be taken into account to understand interaction mechanisms. For weak enough pertubations, the linear theory shows that the three basic modes (entropy, vorticity, acoustic) are governed by three independent differential equations. But, for larger pertubations, second order interactions are no longer negligible and can generate other new pertubations.

The model case of shock-temperature spot interaction considered here gives an exemple of interaction between two of the three basic modes : the entropy and acoustic modes, involved by larger values of their space gradients.

Existing results about shock-temperature spot interaction concern only inviscid flow (Euler equations solved by Shock-Fitting technique [6] or a Flux Corrected Transport scheme [11]). Here, the viscous case is considered and the complete Navier-Stokes equations are solved in two dimensions by a high order scheme and a multidomain technique.

Correct treatment of the inhomogeneous medium (temperature spots, vortices, fine grained turbulence, ...) requires a highly accurate numerical method (spectral or high-order finite differences). Because of boundary conditions, centered fourth-order finite differences have been preferred to spectral schemes.

Inviscid calculation techniques of shock-capturing [8] or shock-fitting [3] must be adapted to viscous flows to handle the high space gradients associated with shocks. Subdomains overlapping over one mesh are defined to fit high gradient zones.

Section 2 describes the continous problem and Section 3 presents the numerical method. Section 4 studies the physical problem and Section 5 gives results for the interaction of a temperature spot with a normal or oblique shock.

2. Continuous problem

Let

- ρ : volumic mass
- $\vec{V}$: velocity (components u, v)
- p : pressure
- e : total energy per unit volum
- ϵ : internal energy
- T : temperature

(*) Doctoral Student at O.N.E.R.A.

Ω : calculation domain.

A viscous compressible flow without external forces is governed by the Navier-Stokes equations :

. mass conservation $$\frac{\partial \rho}{\partial t}+\vec{\nabla}.(\rho \vec{V})=0$$

. momentum conservation $$\frac{\partial}{\partial t}(\rho \vec{V})+\vec{\nabla}.(\rho \vec{V}\otimes \vec{V}+p\bar{\bar{I}})=\vec{\nabla}.\bar{\bar{\tau}}$$

where $\bar{\bar{\tau}}$ is the viscous stress defined as follows : $\bar{\bar{\tau}}=\lambda(\vec{\nabla}.\vec{V})\bar{\bar{I}}+\mu\ \{\vec{\nabla}\vec{V}+(\vec{\nabla}\vec{V})^t\}$

and p is given by the state equation : $p=(\gamma-1)(e-\rho\frac{V^2}{2})$ where $V=||\vec{V}||$

. energy conservation $$\frac{\partial e}{\partial t}+\vec{\nabla}.\ \{(p+e)\vec{V}\}=\vec{\nabla}.\ \{\vec{V}.\bar{\bar{\tau}}-\vec{q}\}$$

where $\vec{q}$ is the heat flux $\vec{q}=-\kappa\vec{\nabla}T$.
To complete those equations, we need the initial condition on Ω and the boundary conditions on $\partial\Omega$.
The normalized equations using the Reynolds number Re and the Prandtl number Pr defined from reference values U_0, ρ_0, L_0, μ_0 are :

$$\begin{cases} \dfrac{\partial \vec{W}}{\partial t}+\dfrac{\partial}{\partial x}\{\vec{F_1}(\vec{W})\}+\dfrac{\partial}{\partial y}\{\vec{F_2}(\vec{W})\}=\vec{0} \\ \text{with } \vec{W}(X,0)=\vec{W_0}(X) \text{ where } \vec{W}=(\rho,\rho u,\rho v,e) \end{cases}$$

and the fluxes $\vec{F_1}$ and $\vec{F_2}$ are :

$$\vec{F_1}(\vec{W})=\begin{pmatrix} \rho u \\ \rho u^2+p-\tau_{xx}/\mathrm{Re} \\ \rho uv-\tau_{yx}/\mathrm{Re} \\ u(p+e)-(u\tau_{xx}+v\tau_{yx}+\dfrac{\gamma\mu}{Pr}\dfrac{\partial\epsilon}{\partial x}) \end{pmatrix}$$

$$\vec{F_2}(\vec{W})=\begin{pmatrix} \rho v \\ \rho uv-\tau_{xy}/\mathrm{Re} \\ \rho v^2+p-\tau_{yy}/\mathrm{Re} \\ v(p+e)-(u\tau_{xy}+v\tau_{yy}+\dfrac{\gamma\mu}{Pr}\dfrac{\partial\epsilon}{\partial y}) \end{pmatrix}$$

The problem consists in the determination of $\vec{W}(X,t)$ for $X\epsilon\Omega$ and $t\epsilon[t_0, t_1]$.

3. Numerical method

Such a computation requires a highly accurate numerical method. The high order finite difference method used is described in 3.1. Shock-capturing or shock-fitting techniques are necessary to compute high gradient zones and to avoid divergence of the high order space discretization for reasons analogous to the Gibbs phenomenon.
Inviscid calculation techniques of shock-capturing [8] and shock-fitting [3] have been adapted

to viscous flow. The shock-fitting technique described in 3.3. gives better results for a much lower cost.

3.1. *Space scheme.*
The choice of a 4th order finite difference scheme has been guided by two considerations :
- mesh can be adapted to the problem : clustered in shock zones and stretched near boundaries.
- treatment of boundary conditions stays local with a finite difference method : with a global method, as a spectral method, errors at the boundaries will be spread to the whole domain.

The classical centered 4th order finite difference scheme is recalled here : For an internal point x_j, the x derivative of w at point x_j is :

$$\frac{\partial w}{\partial x}(x_j)=\frac{1}{12\delta x}(-w_{j+2}+8w_{j+1}-8w_{j-1}+w_{j-2}) \qquad (j=2,...,N-2).$$

For inhomogeneous boundary, non centered formulae are used ($j=0$, 1 or $j=N-1$, N) and for homogeneous boundary, a periodic formula is used.
A second order derivative is calculated by using twice the above formula.

3.2. *Time scheme.*
To simplify the implementation, an explicit time scheme is used. Stability studies [9] have conducted to the following choice : for the convective flux, a second order Adams-Bashforth scheme, for the diffusive flux, a first order Euler explicit scheme.
At mesh points $\{x_j\}_{j=0,...,N}$, and time $t_{n+1}=(n+1)\delta t$, $\vec{W}_j^{n+1}=\vec{W}(t_{n+1},x_j)$ is calculated by

$$\left\{\begin{array}{l}\vec{W}_j^{n+1}=\vec{W}_j^n-\delta t\frac{\partial}{\partial x}\{\vec{F}_1(\vec{W}_j^{*})\}-\delta t\frac{\partial}{\partial x}\{\vec{F}_2(\vec{W}_j)\}\\ \text{with}\\ \vec{W}_j^{*}=1.5\,\vec{W}_j^n-0.5\,\vec{W}_j^{n-1}\end{array}\right.$$

where $\vec{W}_j^{n-1}$ and $\vec{W}_j^n$ are the values of $\vec{W}$ at times $t_{n-1}=(n-1)\delta t$ and $t_n=n\delta t$ respectively.

3.3. *Two- domain technique.*
Because of viscosity, the "shock" is in fact a high space gradient zone. Calculation is done with two subdomains overlapping on one mesh corresponding to the high gradient zone. The computation is done on each subdomain separately and only the internal values are taken into account (figure 1).
Such a two-domain method is in fact a finite difference one-domain method with non centered formula in the high gradient zone. The two domain treatment is easy to implement and is amenable to parallelisation.

3.4. *Initial and boundary conditions.*
Figure 2 displays the initial and boundary conditions in two dimensions. At initial time, the flow field is separated in two domains by a discontinuity parallel to the mesh lines. The lower pressure side is pertubated by a circular quasi Gaussian temperature spot :

$$\frac{\Delta T}{T_0}=\frac{(\rho^2-\alpha^2)^2}{\alpha^4}\exp(-\frac{\rho^2}{\sigma^2}) \text{ where } \rho=((x-x_0)^2+(y-y_0)^2)^{\frac{1}{2}}.$$

In all cases, periodic boundary conditions are imposed in spanwise direction. The computation is performed in a frame evolving at the shock line mean velocity. So, the flow is supersonic upstream of the shock and subsonic downstream. Consequently, values of all quantities are imposed at the upstream boundary (1). Downstream (2), different tests have been done and finally, classical Euler conditions are used. With an adapted mesh, which allows transfer of boundary conditions to infinity, the influence of the boundary conditions is negligible for calculation.

4. Results

Comparisons with the Chu-Kovasnay theory [2] for compressible turbulent flows will be made such a theory that vortices appear in case of interaction between an entropy pertubation mode (from the temperature spot) and a pressure pertubation mode (from the shock). Those pertubations are not negligible compared to the mean value of $||\vec{W}||$ and second order terms must be taken into account. The second order term in power serie development of the vorticity satisfies the following equation :

$$\frac{\partial \overrightarrow{\omega^{(2)}}}{\partial t} - \nu \Delta \overrightarrow{\omega^{(2)}} = \vec{\nabla} \times \overrightarrow{f^{(2)}} + \text{viscous terms}$$

where $\vec{\nabla} \times \overrightarrow{f^{(2)}}$ is a source term of order $\vec{\nabla} \times (S_s \frac{\partial \vec{u}_p}{\partial t}) = - a_0^2 \vec{\nabla} S_s \times \vec{\nabla} P_p$.

(S_s is an entropy pertubation, P_p a pressure pertubation and $\vec{u}_p$ the convection velocity of the pressure pertubation).

4.1. *Normal shock.*

To obtain interactions between a real " Navier-Stokes shock ", we calculate first a 1-D.-field without temperature spot. The initial discontinuity is regularised by a fourth degree polynomial (thus, the shock is at least stretched on 5 points) and computation is made till steady solution is obtained, i. e. :

$$|| \frac{\rho^n - \rho^{n-1}}{\rho^1 - \rho^0} ||_{L^2} \leq 10^{-3} \text{ (figure 3).}$$

The " shock " velocity is nearly one-dimensional and can be compute only in the x direction. The shock is weak ($\frac{\Delta P}{P_0} = 0.4$) and the two-dimensional effects are due to the circular temperature spot. The spot intensity is given by $\frac{\Delta T}{T_0} = 1$. The Reynolds number, based on the domain length (10 times spot dimension) is Re=2000 .

4.1.1. *Results for regular mesh.*

The calculation domain is divided in two subdomains using 2x(50x33) discretization points. Time step is $2.\ 10^{-5}$. Computing cost is $0.7\ 10^{-5}$ s/point/time step on CRAY 1S.

Figure 4 displays the time evolution of the spot which is convected at the fluid.

Figure 5 displays vorticity time evolution. Figures 6 and 7 are given for comparison with one domain calculation using 151x51 points. The symmetry of the calculation leads to vorticity symmetry and in agreement with the Chu - Kovasnay theory [2], counter rotating vortices appear.

To evaluate the interaction source term, figure 8 gives the evolution in time of $||\nabla S \times \nabla P||_{L^2}$. The peak of this function corresponds to the meeting of the spot with the high pressure gradient zone and its width to the time during which the spot crosses the shock zone.

Much care must be taken not to induce spurious vorticity production due to an insufficient number of mesh points. Accuracy tests are shown in figures 9 and 10 with 2x(33x33) and 2x(50x33) mesh points respectively. In the first case (figure 8) due to numerical oscillations, vortices are produced after the interaction has ended. In the second case (figure 9) the growth of the vorticity intensity $||\vec{\omega}||_{L^2}$ takes place during interaction. As expected, the vorticity then decreases slowly due to viscosity. such a result has been verified with more refined meshes.

4.1.2. *Results for an adapted refined mesh.*

Results obtained with the regular mesh show a large oscillation on the isovorticity lines which is not physical. These results can be improved with a refined mesh adapted to the high gradient zone. Figure 11 displays part of such a refined mesh using 2x(60x33) points. The technique of refined mesh uses a geometric progression in x with a reason of 1.04.

The refinement is chosen so as to allow at least five points in the high gradient zone.

An evaluation of the thickness of the steady shock can be obtained as a function of the viscosity in looking at a model viscous steady equation.

$$\begin{cases} u\dfrac{\partial u}{\partial x} - \nu\dfrac{\partial^2 u}{\partial x^2} = 0 \\ u(-\infty) = u_0 \\ u(+\infty) = -u_0 \end{cases}$$

where u_0 is a real positive number.
For the Navier-Stokes equations, the viscosity is obtained as a function of the global Reynolds number. This refinement allows local grid Reynolds number close to 1.
Temperature spot is then superimposed to the field. A comparison of isovorticity lines obtained with or without refinement is given on figure 12 : the non physical oscillations of the vorticity have disappeared.
On figure 13, the adapted refined mesh is shown after passage of the spot through the shock.

4.2. *Oblique shock.*
A supersonic case is studied with Mach number 2.9 and incidence angle of 29^0 . Upstream pressure is $p_0=0.714$ and high pressure is $p_1=1.528$. Boundary conditions are :
. Dirichlet conditions upstream
. Euler conditions downstream
The mesh is oblique and the y - axis is parallel to shock-line. It allows oblique periodic boundary conditions at top and bottom of the domain. Figure 14 displays a refined adapted mesh using 2x(60x47) points with a geometric progression coefficient of 1.035.
As in [3], it can be shown that " shock " velocity which is approximated by

$$\frac{1}{\delta t}\left(\max\left|\frac{\partial \rho^{n+1}}{\partial x}\right| - \max\left|\frac{\partial \rho^{n}}{\partial x}\right|\right) \vec{x}$$

in the one-dimensional case.
It is possible to calculate a new mesh at each time step, but it is necessary to interpolate the different functions. This method would be very expensive. Instead, equations are modified by introducing a convection term corresponding to shock velocity [3] and calculation is then done on a fixed mesh.
Same initialisation technique as in 4.1. is used. Reynolds number based on the spot size is equal to 1200. During passage of the spot through the high gradient zone, non symmetric counter rotating vortices appear. They are convected by the downstream velocity flow. Figures 15 and 16 display time evolution of density and vorticity.

5. Conclusion

A two-domain method has been developed for 2.D. viscous compressible flow. This method is a generalization of the shock-fitting technique to the viscous case. It has been applied to shock/temperature spot interactions, for normal and oblique shocks. Main advantages of this method are :
- implementation of refined mesh without interpolation at each time step.
- treatment high gradient zones with a method of order greater than 2.

Due to such advantages, the method is more accurate and less expensive than the one-domain technique for which smaller time and space steps are necessary.
Future work will consist in extending this technique to 3.D flow and computing interaction between turbulence spot and shock.

Acknowledgements

The author wishes to thank K. Dang, P. Loisel, Y. Maday and Y. Morchoisne for their contributions.

BIBLIOGRAPHY

[1] Anyiwo J. C. and Bushnell D. M. : " Turbulence amplification in shock wave boundary layer interactions. " AIAA J. Vol. 20 (1982), pp. 893-899.
[2] Chu B. T. and Kovasznay L. S. G. : " Non linear interactions in a viscous heat-conducting compressible gas. " J. Fluid Mec. 3, 494-514. (1957)

[3] Dang K., Maday Y., Pernaud-Thomas B., Vandeven H. : " Shock-fitting techniques for solving hyperbolic problems with spectral methods." La Recherche Aérospatiale. 1986-5 .French and English Editions.

[4] Delery J. M. : " Experimental investigation of turbulence properties in transonic shock/boundary layer interactions. " AIAA J. Vol. 21 (1983), pp. 180-185.

[5] Delorme P., Loisel P., Morchoisne Y., Pernaud-Thomas B. : " Résolution par méthodes spectrales des équations de Navier-Stokes pour un fluide compressible. " CEA-INRIA-EDF Course on spectral methods. 1985.

[6] Hussaini M. Y., Kopriva D. A., Salas M. D., Zang T. A. : " Spectral Methods for Euler equations Chebyshev Methods and shock-fitting technique. " NASA Contractor Report N^0 172295 . ICASE. Jan. 1984.

[7] Hussaini M. Y., Collier Fa,. Bushnell D. M., : " Turbulence alteration due to shock motion " . IUTAM Symposium Palaiseau/France 1985 (turbulent Shear-Layer/Shock waves interactions).

[8] Loisel P. : " A hybrid finite difference-spectral method for computing compressible fluid flows ". La Recherche Aérospatiale. 1984-6. French and English translation.

[9] Pernaud-Thomas B. " Time stability of a few schemes using high-order spatial discretization in the case of a convection equation ". La Recherche Aerospatiale. 1987-3. French and English translation.

[10] Pernaud-Thomas B. " Calcul d'écoulements compressibles par des méthodes d'ordre élevé ". Paris VI University Thesis (to appear).

[11] Picone J. M., Oran E. S., Boris J. P. and Young Jr T. R. : " Theory of vorticity generation by shock wave and flame interactions in Dynamics of Shock waves ", Explosions and detonations, J. R. Bowen, N. Manson, A. K. Oppenheim, and R. I. Soloukhin (eds.), Vol. 94, Progress in Astronautics and Aeronautics. (1983)

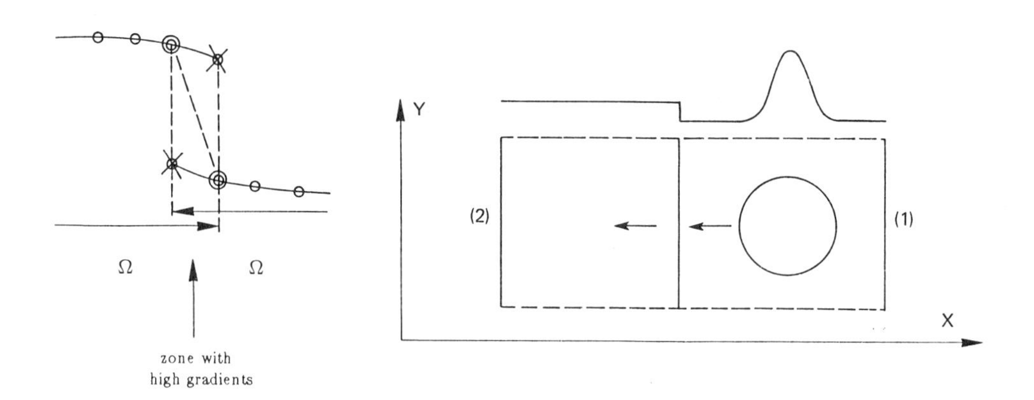

figure 1.
Shock – fitting technique.

figure 2. *Initial condition.*

figure 3.
Two – domain technique.
Steady solution without temperature spot.
Isodensity lines.
2x(60x33) *points.*

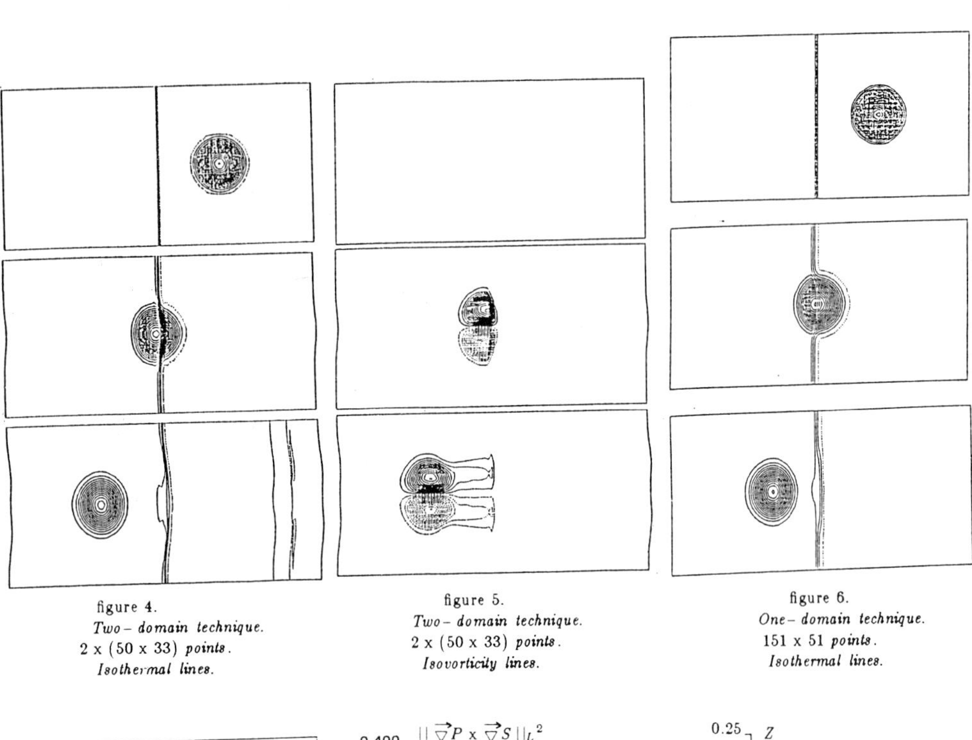

figure 4.
Two – domain technique.
2 x (50 x 33) *points*.
Isothermal lines.

figure 5.
Two – domain technique.
2 x (50 x 33) *points*.
Isovorticity lines.

figure 6.
One – domain technique.
151 x 51 *points*.
Isothermal lines.

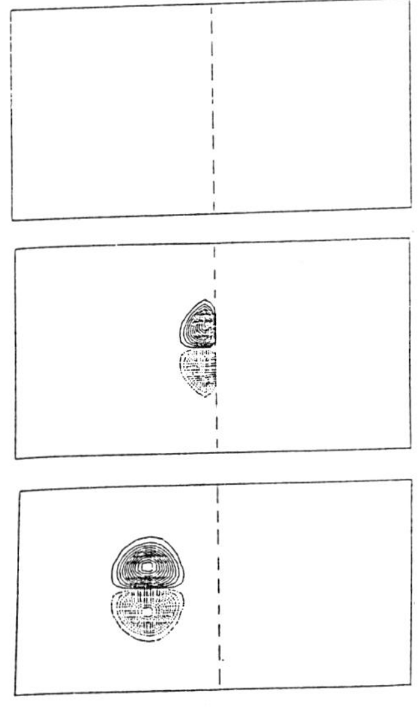

figure 7.
One – domain technique.
151 x 51 *points*.
Isovorticity lines.

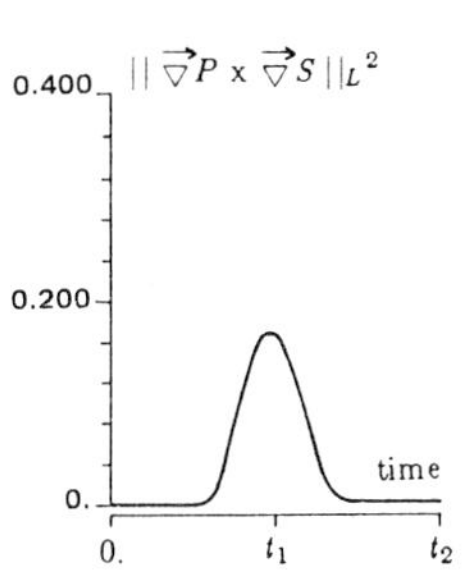

figure 8.
Source term $|| \vec{\nabla} P \times \vec{\nabla} S ||_{L^2}$.

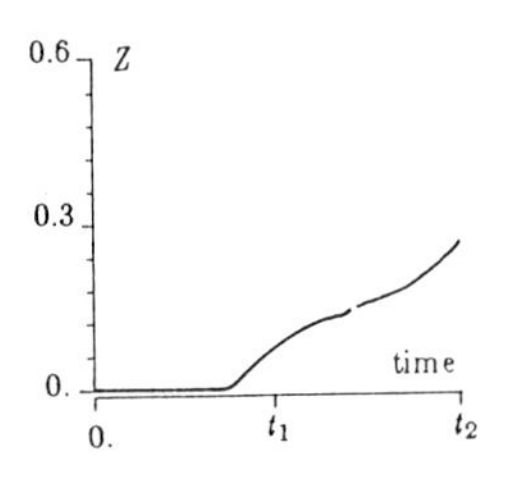

figure 9. $Z = || \vec{\omega} ||_{L^2}$.
2x(33x33) *points*.

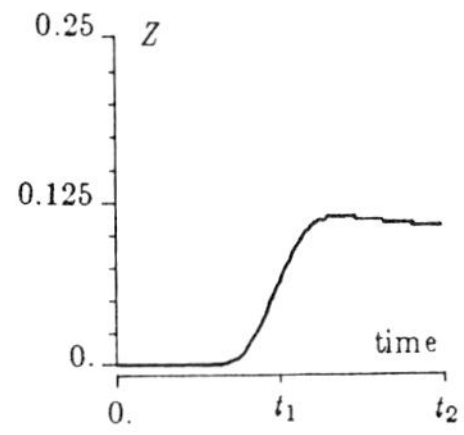

figure 10. $Z = || \vec{\omega} ||_{L^2}$.
2x(50x33) *points*.

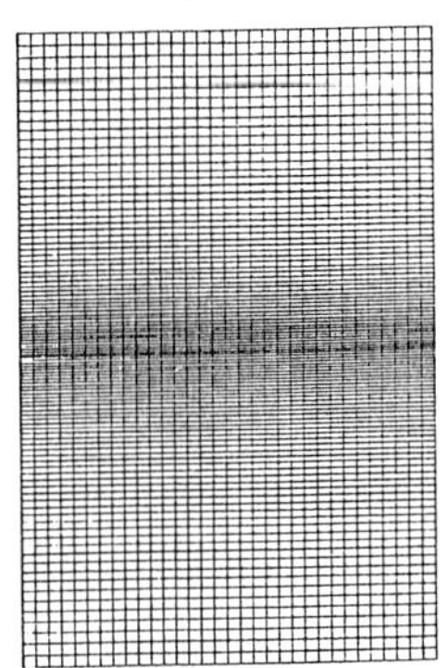

figure 11.
Refined mesh adapted to high gradient zone.
2x(60x33) *points*.

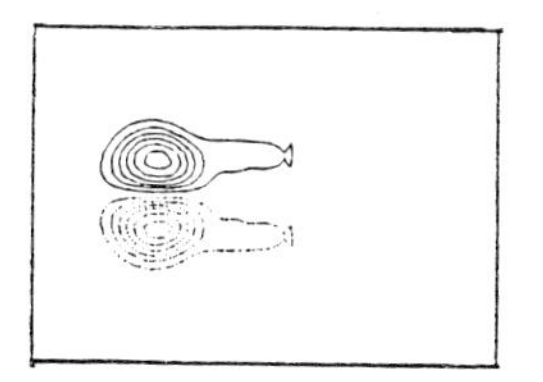

figure 12-a.
Isovorticity lines with regular mesh.

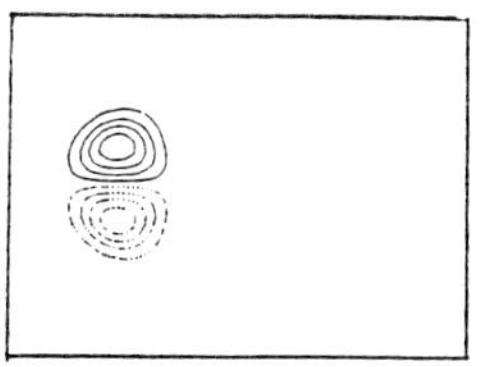

figure 12-b
Isovorticity lines with refined mesh.

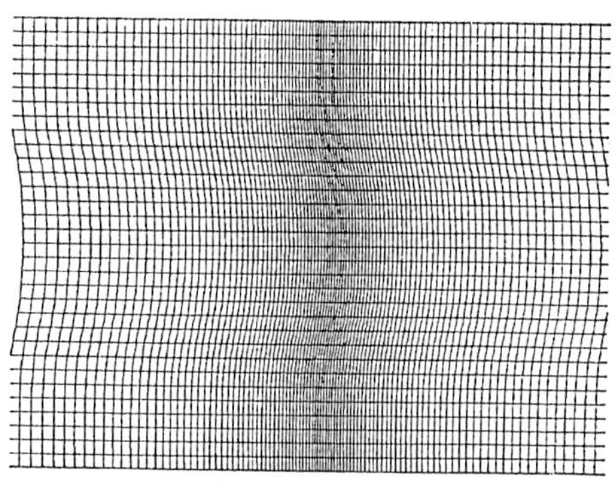

figure 13.
Refined mesh adapted to high gradient zone.
After passage through the shock.
2x(60x33) *points.*

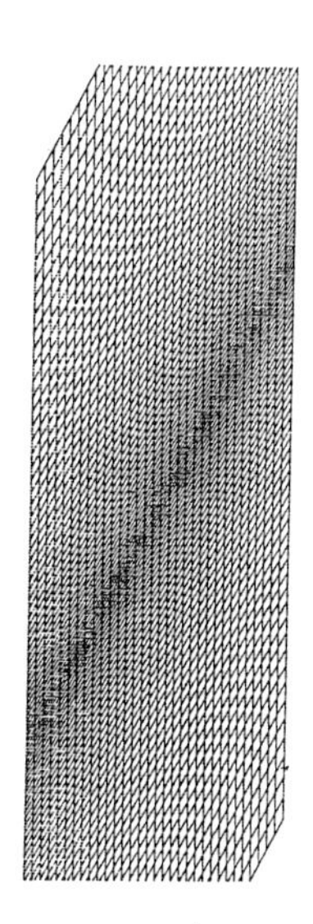

figure 14.
Refined mesh adapted to high gradient.
Oblique shock case.
2x(60x47) *points.*

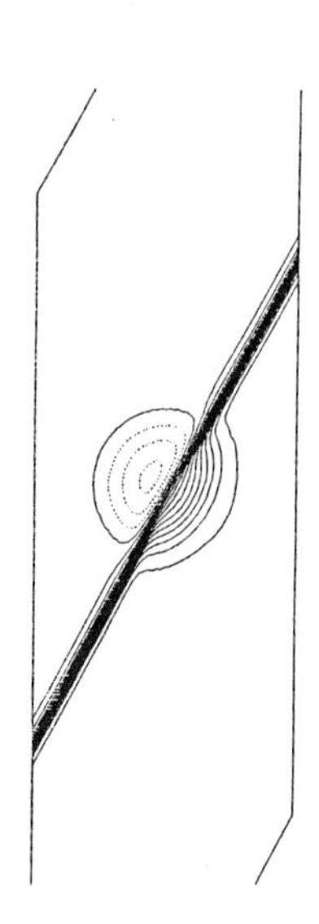

figure 15.
Two – domain technique.
Oblique shock.
2x(60x47) *points.*
Isodensity lines.

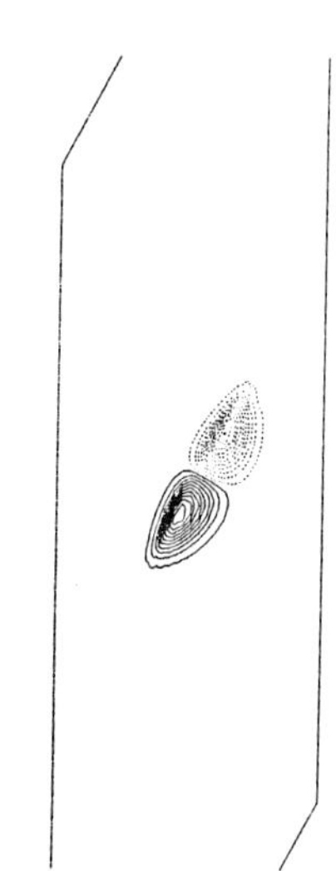

figure 16.
Two – domain technique.
Oblique shock.
2x(60x47) *points.*
Isovorticity lines.

STEADY THREEDIMENSIONAL VISCOUS FLOW PAST A SHIPLIKE HULL

J. PIQUET & M. VISONNEAU
CFD Group, ENSM (UA1217 CNRS)
1 Rue de la Noe, 44072 Nantes, France

SUMMARY

The computation of the threedimensional viscous flow on the stern part and in the wake of a ship hull is investigated. A marching iterative technique is applied to the Reynolds-averaged Navier-Stokes equations (RANSE) written in a nonorthogonal curvilinear body-fitted coordinate system. The potentiality of the method is demonstrated on the SSPA 720 ship liner hull.

INTRODUCTION

Computations as well as experiments indicate that the thin boundary layer assumptions are no more valid in the stern region of a shiplike hull because of the dramatic thickening of the viscous zone across which important variations of pressure occur. Other significant features of this problem include strong viscous-inviscid interaction, significant hull surface curvature effects, abnormally low levels of turbulence in the greatest (outer) part of the viscous layer and occurrence of a large longitudinal vorticity component which occasionally leads to a free vortex separation without noticeable flow reversal in the direction of the ship motion.

Because of the limited success of thick boundary layer theories, the RANSE in their pressure-velocity formulation are considered for the solution of this problem. The turbulent problem is closed by means of a newtonian viscosity assumption - here of $K-\epsilon$ type -. The equations are written in a body-fitted curvilinear coordinate system, the grid being classically generated using the solution of three Poisson equations relating the numerical body-fitted coordinates to the physical cartesian coordinates, rather than stacking numerical conformal transformations of each plane section. A fully implicit technique is used where both the advection and the diffusion are treated implicitly. The pressure-velocity coupling is satisfied iteratively by a successive segregated updating of each variable. In order to build-in the compatibility with the multistructured boundary-layer theories, a partially-parabolic mode of iteration is retained in which the ellipticity of the momentum equations in the longitudinal direction is treated as a source term. The solution can therefore be marched in the downstream direction from some initial plane, provided, in some cases, there is no recirculation or reverse flow in the upstream direction. The pressure is regarded as an unknown which provides the mechanism of "upstream influence" allowing the propagation of information upwards (pressure ellipticity). The viscous-inviscid interaction then results from the multiple sweep approach.

The developped procedures which rest on this approach can be classified into two categories : those in which the continuity equation is the dominant ingredient of the solution procedure in that one attempts to satisfy it exactly at each sweep (as in boundary-layer computations) [6][7][8][19][24] and those in which the continuity constraint is satisfied only indirectly via a pressure-correction procedure [1][4][5][12][15][16][18][23].

Once the curvilinear coordinate system is selected, it then remains to transform the governing equations of motion in the chosen system. Two approaches are possible: the first is to transform both the independent and dependent variables (as for boundary layer equations) using either the contravariant components of the velocity [6][7][23] or the covariant components. Such an approach has never been used in its full generality since it involves a vast amount of geometric coefficients -the Christoffel symbols- and therefore leads to an increased computer storage if the coefficients are computed once for all or to an increased computer time if they are computed whenever necessary. For this reason, only orthogonal coordinates have been used in each plane cross section. Apart from the resulting limitations, such an approach may produce divergences in the keel region and in the wake because the strong curvature of coordinate lines and the large gradients of geometric quantities lead to severe variations of contravariant velocity components although cartesian components do not vary so much. The alternative which is used here is to transform only the independent variables leaving the dependent variables in the original physical coordinates, as in [1][4][5][16]. The use of Christoffel symbols is then avoided but there is a price to pay which will be discussed further.

Mean momentum equations and turbulent transport equations are then solved using standard numerical schemes for advection-diffusion type equations. Here, the finite analytic method [3] is retained, following [4][5]. A nine point stencil results in each plane conjuguate to the marching direction and the resulting stiffness matrix is usually solved using an extrapolated Gauss-Seidel technique resting on the standard tridiagonal matrix algorithm. Because the method is found to lead to bad convergence results in the wake, ad-hoc strongly implicit procedures have been proposed [6][19] while, here, a ad-hoc LU decomposition is available (direct solver). The pressure-correction approach "SIMPLER" [17] is used to enforce the continuity constraint, but it has to be modified to account for the curvilinear and non orthogonal character of the coordinate system. In all these aspects, the method to be presented here follows the work described in [4][5], but it departs from it in several significant aspects which will be detailed further.

The paper is outlined as follows. The equations to be considered are written down in the curvilinear coordinate system and the specific aspects of the method used for the grid generation around the shiplike hull are detailed in §1. §2 deals with the numerical aspects of the discretization, with the pressure correction procedure and focusses on the technical details associated to its implementation. Endly, preliminary results concerning the SSPA 720 shipliner hull are presented in §3 together with remaining points which deserve further attention.

1. THE EQUATIONS

1.1. The equations

Using the partial transformation of only the independent variables ($x^1 \equiv x$, $x^2 \equiv y$, $x^3 \equiv z$) to the curvilinear coordinates ($\lambda^1 \equiv \lambda$, $\lambda^2 \equiv \eta$, $\lambda^3 \equiv \zeta$), the RANSE can be conveniently written under the following generic form [4]

$$\tilde{\nabla}^2 \Phi = \sum_{j=1}^{3} A_\Phi^j \frac{\partial \Phi}{\partial \lambda^j} + E_\Phi \frac{\partial \Phi}{\partial t} + s_\Phi \tag{1}$$

where Φ represents any of the advected quantities U, V, W, K or ϵ and subscripts denote λ, η, ζ and t-derivatives.

$$\tilde{\nabla}^2 \Phi \equiv g^{11}\Phi_{\lambda\lambda}+g^{22}\Phi_{\eta\eta}+g^{33}\Phi_{\zeta\zeta}+2g^{12}\Phi_{\lambda\eta}+2g^{13}\Phi_{\lambda\zeta}+2g^{23}\Phi_{\eta\zeta}$$

$$A_\Phi^j = -f^j + J^{-1} Reff \left\{ b_1^j [a_\Phi U - b_\Phi \nu_{Tx}] + b_2^j [a_\Phi V - c_\Phi \nu_{Ty}] + b_3^j [a_\Phi W - d_\Phi \nu_{Tz}] \right\}. \tag{2}$$

$E_\Phi = a_\Phi Reff$; coefficients a_Φ, b_Φ, c_Φ, d_Φ are equal to 1 except $b_U = c_V = d_W = 2$, $a_K = \sigma_K$; $a_\epsilon = \sigma_\epsilon$. s_Φ coefficients are given by (3)

$$\begin{aligned} s_U &= Reff\ [p_x + 2K_x/3 - \nu_{T,y} V_x - \nu_{T,z} W_x] \\ s_V &= Reff\ [p_y + 2K_y/3 - \nu_{T,x} U_r - \nu_{T,z} W_y] \end{aligned} \tag{3}$$

$$s_W = \text{Reff}\ [p_z + 2K_z/3 - \nu_{T,x}U_y - \nu_{T,y}V_x] \tag{3}$$

$$s_K = -\sigma_K \text{Reff}\ (\ G-\epsilon\ )\ ;\ s_\epsilon = -\sigma_\epsilon \text{Reff}\ (\ C_{\epsilon 1}\frac{\epsilon}{K}G - C_{\epsilon 2}\frac{\epsilon^2}{K}\ ).$$

where x, y, z derivatives are expressed with respect to λ, η, ζ-derivatives using the chain-rule relation $\partial\zeta^i/\partial x^\ell = b_i^\ell/J$. The effective Reynolds number Reff has been defined by $\text{Reff}^{-1} = \text{Re}^{-1} + \nu_T$; $\nu_T = C_\mu K^2/\epsilon$. G is the turbulent production term and the constants of the turbulent model are fixed to their standard values. Other quantities which remain to be defined are computed once the curvilinear body-fitted coordinate system is obtained : g^{ij} are the contravariant components of the metric tensor and the area measures b_i^j are defined from the covariant base vectors $\mathbf{a}_i = \partial\mathbf{r}/\partial\lambda^i$ by $b_\ell^i = (\mathbf{a}_j x \mathbf{a}_k)_\ell$ where i, j, k are in cyclic order. J is the jacobian of the transformation $\lambda^i \to x^i$:
$J = D(x,y,z)/D(\lambda,\eta,\zeta)$.

The continuity equation is written $\text{div}\mathbf{V} = J^{-1}\partial(JV^i)/\partial\lambda^i$ which gives

$$(b_1{}^1U+b_2{}^1Vb_3{}^1W)_\lambda + (b_1{}^2U+b_2{}^2V+b_3{}^2W)_\eta + (b_1{}^3U+b_2{}^3V+b_3{}^3W)_\zeta = 0. \tag{4}$$

1.2. The grid generation procedure

The coordinates of the physical domain are generated numerically from a solution of three Poisson equations relating the numerical coördinates (λ,η,ζ) to the physical orthogonal coordinates (x, y, z) through

$$\nabla^2 x^i \equiv \tilde{\nabla}^2 x^i + f^1 x^i{}_\lambda + f^2 x^i{}_\eta + f^3 x^i{}_\zeta = 0\ ;\ i=1,2,3. \tag{5}$$

These equations involve the specification of the grid control functions f^i. For ship stern flow computations, the comparison with experiments is made possible choosing $x^1 \equiv x \equiv x(\lambda)$ so that constant λ sections are the transverse sections [4]. This implies $2a = -f^1/g^{11} = x_{\lambda\lambda}/x_\lambda$ while (5) is solved only for i=2,3. Dirichlet boundary conditions are used with (5) and girthwise boundary points in each crosssection are defined from a curvilinear abcissa equirepartition. Grid control functions f^2 and f^3 differ from those introduced in [5] in that they both depend on η and ζ. For the fore part of the ship, plane transverse cross sections cannot be used. Instead of Neumann boundary conditions, a surface mesh generation procedure similar to that used in [25] is used to specify grid points on the ship, once its surface has been parametrically defined. In the wake, the ship lines corresponding to λ=const. and $\eta=0$ become vertical segments of the plane of symmetry and define the so-called fictive rudder with its singularity located at $\eta=\zeta=0$. Typical grids are shown in figs. 1 and 2 while fig. 3 details the mesh in transverse crosssections close to the stern.

2. THE NUMERICS

2.1. The momentum equations

The partially parabolic mode consists in solving eqn (6) in place of (1)

$$g^{22}\Phi_{\eta\eta} + g^{33}\Phi_{\zeta\zeta} = 2A_\Phi\Phi_\zeta + 2B_\Phi\Phi_\eta + D_\Phi\Phi_\lambda + E_\Phi\Phi_t + s_\Phi \tag{6}$$

where $2A_\Phi \equiv A_\Phi{}^1$; $2B_\Phi \equiv A_\Phi{}^2$; $D_\Phi \equiv A_\Phi{}^3$ and $S_\Phi = s_\Phi - g^{11}\Phi_{\lambda\lambda} - 2g^{12}\Phi_{\lambda\eta} - 2g^{13}\Phi_{\lambda\zeta} - 2g^{23}\Phi_{\eta\zeta}$.

Marching in increasing λ is possible for $D_\Phi \geq 0$. The next step is to discretize (6) with respect to t and λ and to linearize the resulting equation in each rectangular finite volume of the computational plane. Then

$$g_P^{22}\ \Phi_{\eta\eta} + g_P^{33}\ \Phi_{\zeta\zeta} = 2A_{\Phi P}\Phi_\zeta + 2B_{\Phi P}\Phi_\eta + g \quad \text{where} \quad g = D_{\Phi P}(\Phi_P - \Phi_U) + \tau^{-1}E_{\Phi P}(\Phi_P - \Phi_P{}^{n-1}) + S_{\Phi P}. \tag{7}$$

The subscript P denotes the center of the element, $\tau = t^n - t^{n-1}$ is the time increment and the superscript t^n corresponding to the "unknown time" is omitted

; thus t and λ derivatives are lagged and U denotes the upstream value (fig. 4). The finite analytic method [3] is endly applied to a normalized form of (7). The result is a set of simultaneous algebraic equations involving a nine point molecule in the crosssection. The corresponding linear system is solved by a accelerated line Gauss-Seidel technique for which a condition of diagonal dominance is $D_{\phi P} + a_{\phi} Reff \geq 0$.

2.2. The iterative method for the pressure

The "simpler" method of [17] is retained. In order to avoid the classical chequerboard uncoupling problem, a MAC staggered mesh is used. This choice, while optimal when cartesian and contravariant coordinates coincide, reintroduce the possibility of decoupling: the cartesian velocity components (U, V, W) are neither orthogonal to the control surfaces, nor parallel to the direction of the coordinate lines. In order to discuss the consequences of this choice, it is necessary to write down the projection phase of the procedure:

$$U_d = U^*_d - d_d P_{xd} \ ; \ V_n = V^*_n - d_n P_{yn} \ ; \ W_e = W^*_e - d_e P_{ze} \text{ where}$$

$$P_x = J^{-1}[b_1^1 P_\lambda + b_1^2 P_\eta + b_1^3 P_\zeta] \ ; \ P_y = J^{-1}[b_2^1 P_\lambda + b_2^2 P_\eta + b_2^3 P_\zeta] \ ; \ P_z = J^{-1}[b_3^1 P_\lambda + b_3^2 P_\eta + b_3^3 P_\zeta].$$

A first possibility is to drop the underlined contributions in order to retain the structure of the five-point approximation [4][5]. Because of the non coincidence of contravariant and cartesian velocity components, it has been found that some of the dropped derivatives could have more influence than those undropped. The structure of the five-point approximation is retained if these terms are treated as source terms but this technique is prone to divergence [20]. What is done here consists in using $U_d = U^*_d - d_d{}^u(P_D - P_P)$ while $V_n = V^*_n - (d_\eta{}^v P_\eta + d_\zeta{}^v P_\zeta)$; $W_e = W^*_e - (d_\zeta{}^w P_\zeta + d_\eta{}^w P_\eta)$ so that the pseudo velocities V^* and W^* do not contain pressure contributions. Substituting these expressions in the continuity equation and performing necessary consistent interpolations [2], a nine point pressure molecule results in the unknown crosssection. Here, an extrapolated line Gauss-Seidel technique is used to start the resolution of the resulting system for the pressure and for the pressure correction which is achieved by means of an ad-hoc direct (block tridiagonal) solver.

3. NUMERICAL RESULTS

3.1. Technical difficulties

Fig. 5 specifies the grid layout in a cross plane. Mesh nodes are pressure nodes. At $\eta=0$ (j=1) which gives the wall for U and V and at j=3/2 for V, the pressure has to be computed, U and V are fixed by the no slip condition and $V_{3/2}$ is unknown. Therefore the pressure scheme takes a particular form here. Moreover, while the wall pressure is never needed with the five-point molecule in the crosssection, it becomes necessary with the nine-point molecule and it is evaluated here from $\partial P/\partial N = 0$ treated as implicitly as possible. The pressure scheme must also be written in the far field (where W_n need not be corrected).

The last important difficulty is connected with the fictive rudder and with the geometric singularity S(1,2) at its edge where j=0 (fig. 6). The topology is such that points (1,1) and (1,3) coincide in the physical space while the symmetry condition imposes that (2,1) and (2,3) correspond each other. The pressure node (1,2) and the V node (3/2,2) have therefore to be treated specifically.

3.2. Results

The SSPA 720 has been used extensively as a test case for boundary layer calculation methods, all of which break down some distance ahead of the stern. Meanflow data are found in [9] and corresponding Reynolds stress data are reported in [11]. For this hull, the domain solution extends from x/L = 0.5 (midship station) to x/L = 2. so that the fictive rudder starts at x/L=.95 and the hull disappears at x/L=1. In the cross sectional planes, the domain extends from the hull surface and wake centreline to a cylindrical boundary 2 ship lengths away from the ship axis where the application of uniform flow conditions U=1 ; P=0 is valid.

The pressure distribution along the waterline and the keel line are given in fig. 7, using the experimental values corrected for blockage [9][5]. The projections of the calculated vectors in the transverse sections shown in fig. 8 provide a picture of the crossflow and normal velocities. While the crossflow is directed from the keel to the waterline at upstream stations, a crossflow reversal appears very early (x/L=.35) and it is responsible for the local thickening of the boundary layer, firstly close to the so-called C potential line and further downstream, in a more important part of the hull. Axial velocity profiles along the keel and below the waterline are also available. From these figures, it is evident that the first numerical point in the thick boundary layer along the waterline lies at a reasonable distance from the wall. This is a major difference with respect to computations presented in [5] where the near-wall problem is left unresolved. Here the mesh is far more clustered close to the wall and the wall function approach has been abandonned in favor of a zonal mixing length model which is systematically used in an inner region and which is patched with the $K-\epsilon$ model in the log region. The details of the technique and its need were clearly demonstrated in [18] for axisymmetric flows.

4. CONCLUSION

A numerical method for the calculation of complex threedimensional turbulent flows over the stern and in the wake of arbitrary ship forms has been described. Some more work is needed both to improve the numerical method, and to assess in more details the results by more systematic comparisons with experiments. The former aspect is made necessary as it is considered that the convergence of the solution is low, especially in the wake. A serious reduction in the needed CPU (actually of the order of 15' Cray2 for 44x30x13 points along x, η, ζ) is possible through the reduction of the number of sweeps using accelerating techniques, as well as through the reduction of the computational effort, improving the technique of resolution in each crosssection (for the abovementionned results, one mesh point needs .6μs on Cray2 per iteration -three times more than for the axisymmetric case-). The later aspect needs some more systematic comparisons with turbulent measurements [11] as well as the investigation of other ship hulls where a clearly identifiable longitudinal vortex motion imbedded in the stern flow is present.

Acknowledgments: Authors gratefully acknowledge Pr. Patel who provided them with an earlier source listing of his partially parabolic code with wall functions (polar case afterbody 3:1). Computations have been performed on the VP200 (CIRCE) and with a dotation of 8 hours on the Cray 2 (CCVR) by the scientific Committee of CCVR. Financial support of DRET through contract 86.34.104 is also gratefully acknowledged.

REFERENCES

[1] ABDEL MEGUID, A.M. ; MARKATOS, N.C. ; SPALDING, D.B. & MURAOKA, K. [1978] Proc 1st. Int. Symp. Ship Viscous Resistance ; SSPA Goteborg, Sweden.

[2] BRAATEN, M. & SHYY, W. [1986] Num. Heat Transfer, Vol.9 ; pp. 559-574.

[3] CHEN, J. & CHEN H.C., [1984] J. Comp. Phys., Vol.53 ; N°2, pp.209-226

[4] CHEN, H.C. & PATEL, V.C. [1985a] IIHR Rept N°285 (Iowa City, Ia)

[5] CHEN, H.C. & PATEL, V.C. [1985b] Proc. 4th. Num. Conf. on Numerical Ship Hydrodynamics (Washington D.C.) pp.492-511.

[6] HOEKSTRA, M. & RAVEN, H.C. [1985] Proc. Osaka Int. Colloq. on Ships Viscous Flow, pp.125-142.
[7] HOEKSTRA, M. & RAVEN, M.C. [1985] Proc. 4th Int. Conf. on Numerical Ship Hydrodynamics (Washington, D.C.) pp.470-491.
[8] ISRAELI, M. & LIN, A. [1985] Comput. & Fluids, Vol.13, N°4, pp. 397-410.
[9] LARSSON, L. [1975] "Boundary Layers on Ships" ; PHD Thesis Chalmers Univ. Goteborg.
[10] LARSSON, L [1980] Proc. SSPA-ITTC Workshop on Ship Boundary Layers, SSPA Rept N°90 (1981).
[11] LOFDAHL, L. & LARSSON, L. [1984] J. Ship Research Vol.28, N°3.
[12] MALIN, M.R. ; ROSTEN, H.I. ; SPALDING, D.B. & TATCHELL, D.G. [1985] Proc. 2nd Int. Symp. Ship Viscous Resistance (SSPA Goteborg), paper 16.
[13] MALISKA, C.R. & RAITHBY, G.D. [1984] Int. J. Num. Methods in Fluids, Vol.4, pp. 519-537.
[14] MARX, Y. ; PIQUET, J. & VISONNEAU, M. [1986] in Num. Grid Generation in Comp. Fluid Dynamics (Hauser,J. & Taylor, C. eds.) Pineridge Press ; pp.773-784.
[15] MORI, K. & ITO, N. [1985] Proc. 4th. Int. Conf. on Numerical Ship Hydrodynamics (Washington D.C.) pp.512-528.
[16] MURAOKA, K. [1982] Proc. 13th. ONR. Symp. Naval Hydrodynamics (Tokyo) Ed. Shipbuilding Res. Ass. of Japan ; pp.601-616.
[17] PATANKAR, S.V. [1980] "Numerical Heat Transfer and Fluid Flow" ; Hemisphere Publishing Comp. New-York.
[18] PIQUET, J. ; QUEUTEY, P. & VISONNEAU, M. [1987] Proc 5th. Int Symp. Num. Meth. in Laminar & Turbulent Flows (Montreal) (Ed. Taylor, C.) Pineridge Press ; pp. 644-655
[19] RAVEN, H.C. & HOEKSTRA, M. [1985] Proc. 2nd Int. Symp. Ship Viscous Resistance, SSPA (Goteborg) Paper 14.
[20] SHYY, N. ; TONG, S.S. & CORREA, S.M. [1985] Num. Heat Transfer, Vol.8 ; pp. 99-113.
[21] THOMPSON, J.F. ; THAMES, F.C. & MASTIN, C.N. [1974] J. Comp. Phys. Vol.15 ; p. 2-99.
[22] THOMPSON, J.F. [1982] ed. Numerical Grid Generation, North Holland Publishers, New-York.
[22] TZABIRAS, G.D. & LOUKAKIS, T.A. [1983] Int. Shipb. Progress, N°345.
[23] VANKA, S.P. & LEAF, G.K. [1984] AIAA Paper 84-1244 (AIAA/SAE/ASME 20th Joint Propulsion Conf.).
[24] WARSI, Z.U.A. [1982] in Numerical Grid Generation, Ed. Thompson, J.F. (North Holland) pp. 41-77.

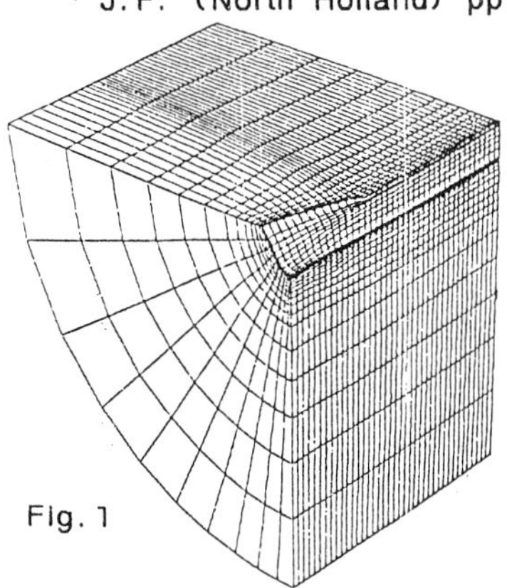
Fig. 1

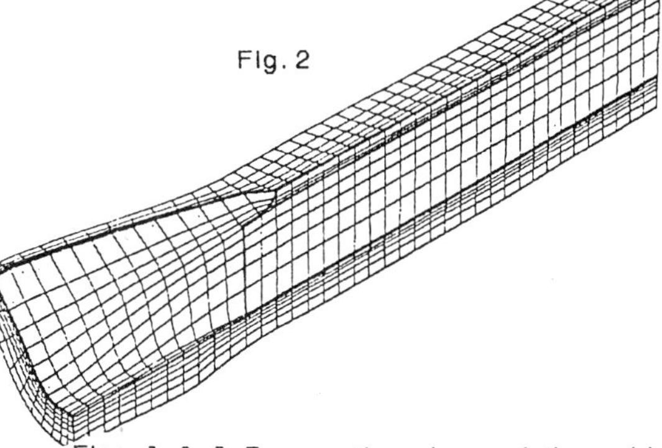
Fig. 2

Figs. 1 & 2 Perspective views of the grid for the SSPA 720 ship liner.

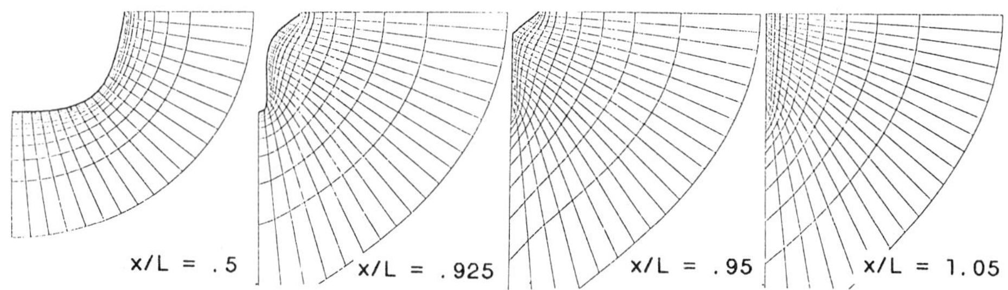

Fig. 3 Grid for the SSPA 720 ; transverse crosssections

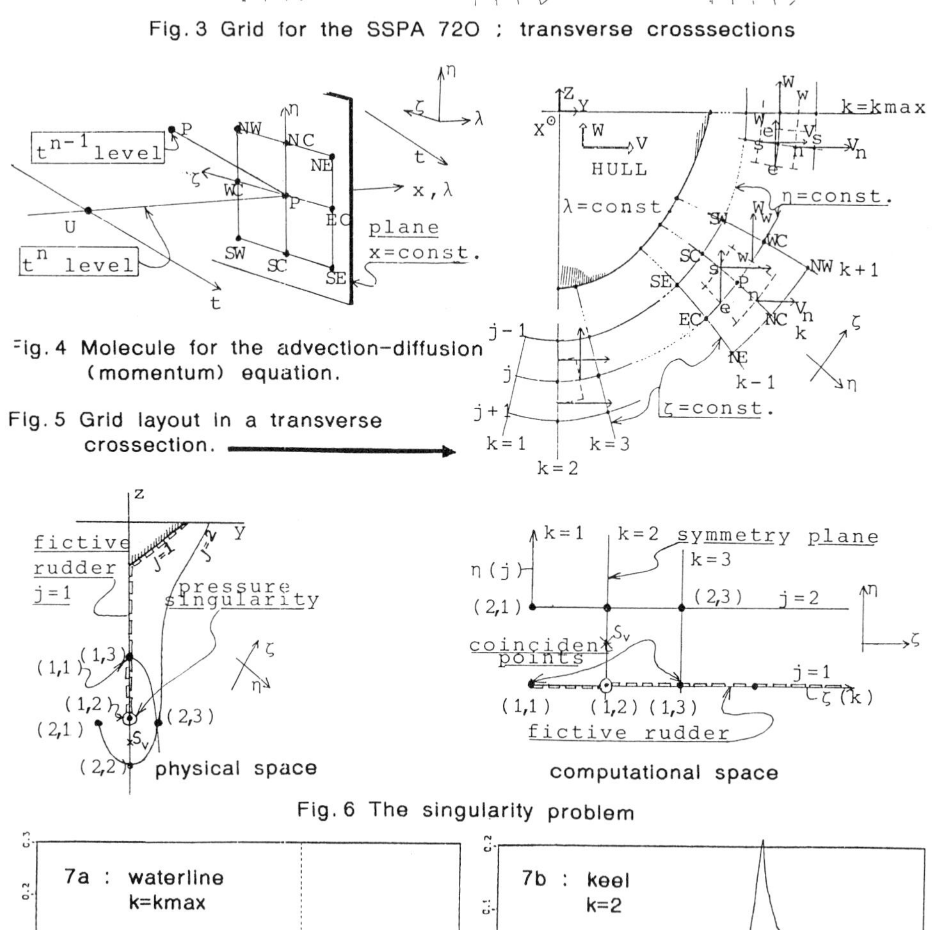

Fig. 4 Molecule for the advection-diffusion (momentum) equation.

Fig. 5 Grid layout in a transverse crossection.

Fig. 6 The singularity problem

Fig. 7 SSPA 720 : Pressure distribution along the waterline and along the keel
$\bar{x} = (x+L)/2L$ (2L = ship length along the waterline)

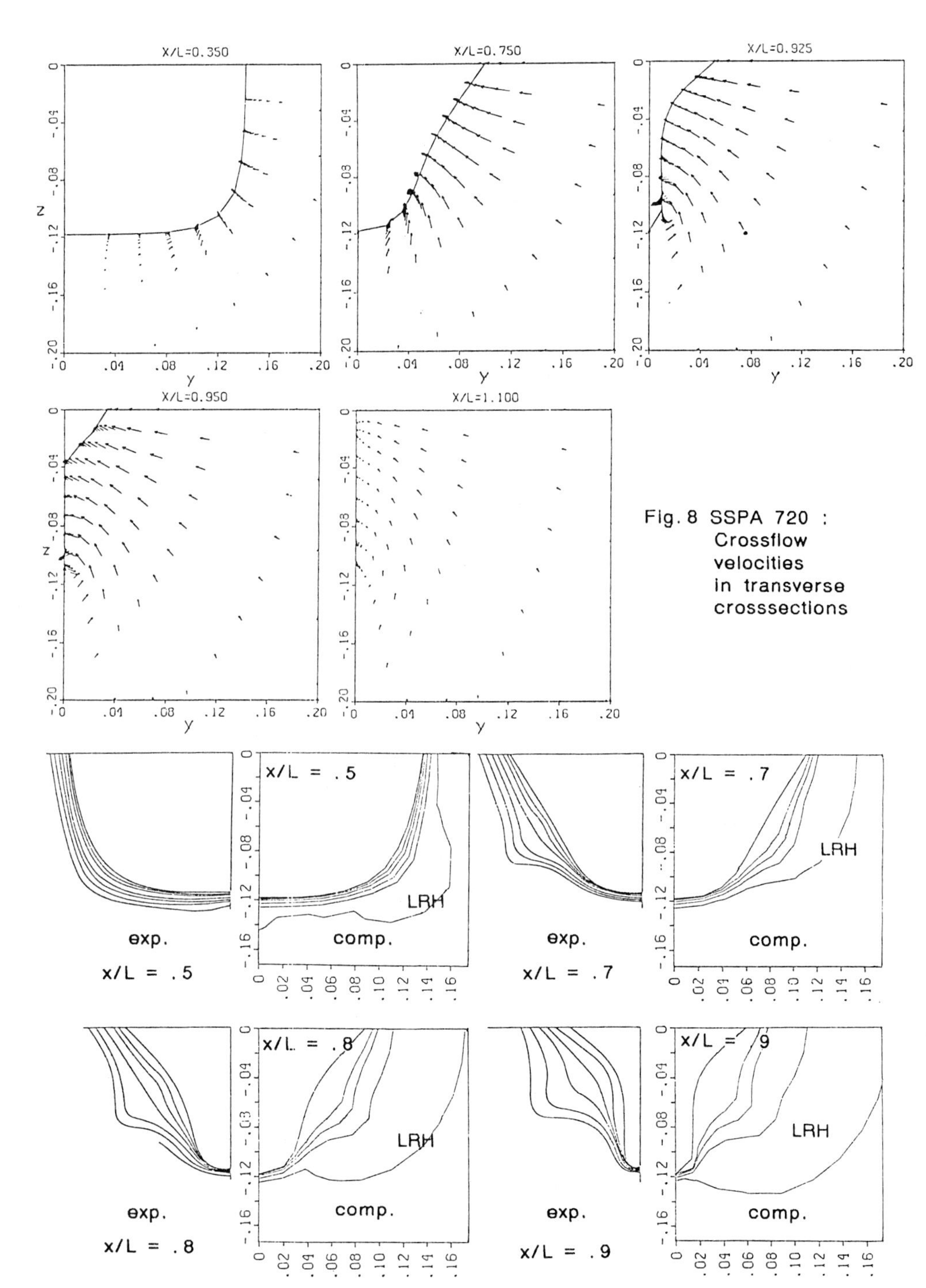

Fig. 8 SSPA 720 : Crossflow velocities in transverse crosssections

Fig. 9 SSPA 720 ; Isowakes ; experiments U = .7, .8, .9, .95, 1., 1.05
computations : U = .7, .8, .9, 1.
LRH *means : low resolution & need to refine the mesh here*

A STUDY OF THE VORTEX SHEET METHOD FOR SOLVING THE PRANDTL BOUNDARY LAYER EQUATIONS

Elbridge Gerry Puckett[†]
Lawrence Livermore National Laboratory
Livermore, California 94550

SUMMARY

The vortex sheet method for solving the Prandtl boundary layer equations is the most widely used means of satisfying the no-slip boundary condition in a random vortex method solution of the Navier-Stokes equations. It is a fractional step method in which random walks are used to solve the diffusion equation and particles, called 'sheets', are created at the boundary in order to satisfy the no-slip boundary condition. A review of those theoretical results currently known is presented including the consistency of the random walk/sheet creation process. The convergence of the method to Blasius flow is then demonstrated numerically and from the data presented rates of convergence in terms of the computational parameters are conjectured.

§1 INTRODUCTION

One of the fundamental outstanding problems in computational fluid mechanics today is that of computing fluid flows at very large Reynolds numbers. Many methods (such as finite difference methods) which do well at small Reynolds number suffer from numerical diffusion as the Reynolds number is increased. Thus, in order to maintain a given degree of accuracy they require an increasingly expensive refinement of the computation, eventually reaching a point that the calculation cannot be performed even on today's fastest computers. For this reason Chorin [8] developed the random vortex method as a way to solve the Navier-Stokes equations at arbitrarily high Reynolds number. This grid free particle method has the property that the work required to approximate a given flow at a given accuracy is independent of the Reynolds number. After subsequent work Chorin concluded that the random vortex method performed well for free space boundary conditions but that his original technique of creating vorticies on the boundary lead to certain difficulties such as the creation of too many vorticies at each time step. Therefore, in [9], he presented a method of coupling a random vortex solution of the Navier-Stokes equations away from boundaries with a vortex sheet solution of the Prandtl equations next to boundaries. The goal of this work is to study the vortex sheet method in some detail.

In vorticity formulation the Prandtl equations are

$$\omega_t + u\omega_x + v\omega_y = \nu\omega_{yy} \ , \tag{1.1a}$$

$$\omega = -u_y \ , \tag{1.1b}$$

$$u_x + v_y = 0 \ . \tag{1.1c}$$

Here (x,y) are the coordinates which are parallel and perpendicular to the boundary respectively, (u,v) are the corresponding velocity components, ω is the vorticity, and ν is the viscosity. The solution of these equations is evolved subject to the following constraints:

$$u(x,0,t) = 0 \ , \tag{1.2a}$$

$$v(x,0,t) = 0 \ , \tag{1.2b}$$

[†] Part of this work was performed while the author was a GSRA at Lawrence Berkeley Laboratory under contract DE-AC03-76SF00098 and part was performed while he was employed at Lawrence Livermore Laboratory under contract W-7405-Eng-48.

$$\lim_{y\to\infty} u(x,y,t) = U_\infty(x,t), \tag{1.2c}$$

where it is assumed the boundary is at $y = 0$ and U_∞ is some prescribed 'velocity at infinity' in the direction parallel to the boundary imposed on the flow from outside the boundary layer. See [14] for a review of the mathematical aspects of these equations.

The vortex sheet method is a fractional step method. One step is the solution of the advection equation

$$\omega_t + u\,\omega_x + v\,\omega_y = 0 \tag{1.3}$$

subject to boundary conditions (1.2b,c) and the other is the solution of the diffusion equation

$$\omega_t = \nu\omega_{yy} \tag{1.4}$$

subject to the no-slip boundary condition (1.2a). It is also a particle method in which the particles, called 'vortex sheets', are line segments parallel to the boundary and carry concentrations of vorticity. The tangential velocity component is recovered from the vorticity with the aid of (1.1b) and (1.2c),

$$u(x,y,t) = U_\infty(x,t) + \int_y^\infty \omega(x,s,t)\,ds\,. \tag{1.5a}$$

The normal velocity component is determined from (1.1c) and (1.2b),

$$v(x,y,t) = -\int_0^y u_x(x,s,t)\,ds\,. \tag{1.5b}$$

Note that the velocity field given by (1.5a,b) automatically satisfies (1.2b,c). The numerical solution of (1.3) is found by advecting the sheets according to this velocity field and the numerical solution of (1.4) is obtained by allowing the sheets to undergo a random walk in the y direction. During this random walk new sheets are created at the boundary in order to satisfy the no-slip boundary condition (1.2a). One of the attractions of the vortex sheet method is that this sheet creation process mimics the creation of vorticity at the boundary and hence it provides a convenient mechanism for introducing vorticity into the flow.

The coupling to the random vortex method solution is accomplished in the following manner. The computational domain is divided into two regions: an interior region in which the random vortex method is used to solve the Navier-Stokes equations and a sheet layer next to the boundary in which the vortex sheet method is used to solve the Prandtl equations. These two solutions are matched by letting U_∞ be the velocity tangent to the boundary due to the random vortex method and by allowing sheets to become vortices when they exit the sheet layer. See [10] and [21] for further details. The natural manner in which sheets become vorticies, thereby introducing vorticity into the interior of the flow from the boundary, is the key aspect of the sheet method. Such hybrid random vortex/vortex sheet methods have been successfully used to model such problems as flow past a circular cylinder (Cheer [5,6], Tiemroth [25]), driven cavity flow (Choi, Humphrey, and Sherman [7]), instability of the two dimensional boundary layer (Chorin [10]), flow past a backwards-facing step (Sethian and Ghoniem [20]), turbulent combustion (Ghoniem, Chorin and Oppenheim [11], Sethian [21,22]), and wind flow over a building (Summers, Hanson, and Wilson [24]).

§2 THE METHOD

In the vortex sheet method the vorticity at the kth time step is represented as the sum of linear concentrations of vorticity,

$$\tilde{\omega}^k(x,y) = \sum_{j=1}^{N_k} \omega_j\, b_h(x - x_j^k)\, \delta(y - y_j^k), \tag{2.1}$$

where (x_j^k, y_j^k) is the center of the jth sheet at the kth time step, ω_j is the 'weight' or 'strength' of the jth sheet, $b_h(x)$ is the 'smoothing' or 'cutoff' function, $\delta(y)$ is the Dirac delta function, and N_k is the number of sheets in the flow at the kth time step. It is usually most efficient to choose all sheet strengths to have the same magnitude, $|\omega_j| = \omega_{max}$ where ω_{max} is some small, positive parameter. Most workers use the piecewise linear smoothing originally proposed by Chorin,

$$b_h(x) = \begin{cases} 1 - |x/h| & |x| \le h, \\ 0 & otherwise. \end{cases} \quad \textbf{(2.2)}$$

Typically the support of b_h is of length nh for some small integer n.

From (1.5a) and (2.1) we see that the computed tangential velocity $\tilde{u}^k$ is given by

$$\tilde{u}^k(x,y) = U_\infty(x,k\Delta t) + \sum_{j=1}^{N_k} \omega_j \, b_h(x - x_j^k) \, \mathrm{H}(y_j^k - y) \quad (2.3a)$$

where $\mathrm{H}(y)$ is the Heaviside function,

$$\mathrm{H}(y) = \begin{cases} 1 & y \ge 0, \\ 0 & otherwise. \end{cases}$$

By approximating $\tilde{u}_x$ with a centered divided difference and using (1.5b) we now obtain the computed perpendicular velocity component,

$$\tilde{v}^k(x,y) = -\partial_x U_\infty(x,t)\, y - \frac{1}{h} \sum_{j=1}^{N_k} \omega_j \, [b_h(x + \frac{h}{2} - x_j^k) - b_h(x - \frac{h}{2} - x_j^k)] \min(y, y_j^k) \, . \quad (2.3b)$$

Given the positions (x_j^k, y_j^k) and strengths ω_j of the sheets at some time $t = k\Delta t$ the flow field at the next time step is determined as follows. First the sheets are advected by evaluating (2.3a,b) at the center of each sheet and moving the sheet one time step of length Δt in this direction,

$$(x_j^{k+1/2}, y_j^{k+1/2}) = (x_j^k, y_j^k) + \Delta t(\tilde{u}^k(x_j^k, y_j^k), \tilde{v}^k(x_j^k, y_j^k)) \, .$$

The tangential velocity $\tilde{u}^{k+1/2}(x,0)$ induced on the boundary by these new sheet positions is then evaluated at r equally spaced points on the boundary with grid spacing h. Denote the ith gridpoint on the boundary by a_i and the tangential velocity at this point by $\tilde{u}_i^{k+1/2}$. For $i = 1, \ldots, r$ let $m_i = [|\tilde{u}_i^{k+1/2}|/\omega_{\max}]$ where $[x]$ is the greatest integer less than or equal to x. One creates m_i new sheets, each of strength $sign(\tilde{u}_i^{k+1/2}) \cdot \omega_{\max}$, at the ith gridpoint. The new flow field obtained by this sheet creation process is identical to $(\tilde{u}^{k+1/2}, \tilde{v}^{k+1/2})$ except that (1.2a) is now satisfied up to $O(\omega_{\max})$ at each of the r gridpoints on the boundary. Finally every sheet (including those just created) is given a random displacement in the y direction, reflecting those sheets which go below the boundary. Thus,

$$(x_j^{k+1}, y_j^{k+1}) = (x_j^{k+1/2}, |y_j^{k+1/2} + \eta_j|)$$

where the η_j are independent and chosen from a Gaussian distribution with mean 0 and variance $2\nu\Delta t$.

Note that there are three computational parameters in the sheet method: the time step Δt, the sheet length/grid spacing h, and the maximum sheet strength $\omega_{\max}$. (Actually, for the cutoff (2.2), sheets have length $2h$.) The principal goal of this work is to establish guidelines for choosing these parameters effectively. The only widely agreed upon criterion for governing parameter selection is the so called CFL condition, $\Delta t \cdot \max U_\infty \le h$. It is usually argued that this condition is necessary in order to ensure that sheets created at a given gridpoint influence each subsequent grid point downstream before passing on. (Also see [15].)

Our approach has been to establish estimates for the random walk/sheet creation process separately from the advection step with the view point that the most efficient choice of parameters will be achieved by balancing the error from the random walk step with that from the advection step. In what follows we will only present theoretical results for the random walk sheet creation process. Recall that the error in solving the heat equation is $O(\sqrt{N^{-1}})$ where N is the number of particles (e.g. see [16], §7) and note that that for a fixed a_i on the boundary the number of sheets above a_i will be $O(\omega_{\max}^{-1})$. Consequently, it is plausible that the error in our numerical solution of (1.4) with boundary condition (1.2a) is $O(\sqrt{\omega_{\max}})$. Evidence will be presented below to substantiate this conjecture.

§3 A SUMMARY OF THEORETICAL RESULTS

To date there is no proof that vortex sheet method converges to solutions of the Prandtl equations. Recent related work includes Benfatto and Pulvirenti [4], Goodman [12], and Hald [13]. In [4] however, the authors do not attempt to establish the rate of convergence in terms of the computational parameters while the convergence proof in [12] holds only for free space boundary conditions thus avoiding the difficulty inherent in proving the convergence of a method which creates particles at the boundary. Hald [13] has addressed this issue directly by proving the convergence of a one dimensional particle method with creation of vorticity. The convergence of both the vortex sheet method and the random vortex method with boundaries present and/or in three dimensions remain open and difficult problems. In this section we present several results about the random walk/sheet creation algorithm including the consistency of the random walk/sheet creation solution of (1.4) with boundary condition (1.2a). In the following section we demonstrate numerically the convergence of the method to Blasius flow and obtain (experimental) estimates of the convergence rate.

We review now those theoretical results which are known about the vortex sheet method. (For more detail see [15].) We begin with the sheet creation process. Assume the boundary is the line segment ($x \in [a,b]$,0). Let b_h be given by (2.2), let $t = (k+1)\Delta t$, and let

$$g(x) \equiv \tilde{u}^{k+1/2}(x,0) = U_\infty(x,t) + \sum_{j=1}^{N_k} \omega_j \, b_h(x - x_j^{k+1/2})$$

be the tangential velocity on the wall after the $(k+1)$st advection step. We wish to create sheets with centers at the gridpoints $(a_i,0)$ and strengths ω_{il} so that

$$g(x)+\sum_{i=1}^{r} \sum_{l=1}^{q_i} \omega_{il} \, b_h(x - a_i) \approx 0 \,. \tag{3.1}$$

Note that the functions $b_h(x-a_i)$, $i = 1, \ldots, r$ form a basis for the space of piecewise linear functions on $[a,b]$ ([19]). Thus, the no-slip boundary condition (1.2a) may be approximately satisfied by choosing the function on the right hand side of the plus sign in (3.1) to be the piecewise linear interpolant of $g(x)$. This is easily accomplished if one lets $\sum_{l=1}^{q_i} \omega_{il} = -g(a_i)$. However, since the amount of work required to compute the velocity from (2.3a) and (2.3b) increases with the number of sheets in the flow we have found that it is more effective to choose $\omega_{il} = -sign(g(a_i))\, \omega_{\max}$ and set $q_i = [|g(a_i)|/\omega_{\max}]$. For this choice of q_i and ω_{il} one can show that

$$\| g + \sum_{i=1}^{r} \sum_{l=1}^{q_i} \omega_{il} \, b_h(\cdot - a_i) \|_{L^1_{[a,b]}} \leq const. \, (\| \partial_x g \|_{L^\infty} h + \omega_{\max}) \tag{3.2}$$

where the constant depends on the length of the interval $[a,b]$ and where the sup norm of the first derivative of g is suitably defined for piecewise linear functions. It is natural to inquire what happens if one chooses the smoothing function in (2.2) so that the functions $b_h(x - a_i)$ form a basis for piecewise C^2 functions on $[a,b]$, say cubic splines. In this case it is possible to show that the error in (3.2) is $O(h^2 + \omega_{\max})$ but numerical experiments have not revealed an improvement over the piecewise linear smoothing defined by (2.2) ([15], §6).

Now consider the random walk and sheet creation algorithm together. Given

$$\tilde{u}^{k+1/2}(x,y) = U_\infty(x,t) + \sum_{j=1}^{N_k} \omega_j \, b_h(x - x_j^{k+1/2}) \, \mathrm{H}(y_j^{k+1/2} - y)$$

where $t = (k+1)\Delta t$ let $D_{\Delta t} \, \tilde{u}^{k+1/2}$ denote the exact solution at time Δt to the heat equation

$$u_t = \nu u_{yy}$$

with initial data $\tilde{u}^{k+1/2}$ and boundary condition (1.2a). Similarly, let $\tilde{D}_{\Delta t} \, \tilde{u}^{k+1/2}$ represent the computed solution to this problem obtained by creating sheets and random walking all sheets with reflection as described above,

$$\tilde{D}_{\Delta t} \, \tilde{u}^{k+1/2}(x,y) \equiv U_\infty(x,t) + \sum_{j=1}^{N_{k+1}} \omega_j \, b_h(x - x_j^{k+1}) \, \mathrm{H}(y_j^{k+1} - y) \,.$$

It is possible to find an exact expression for $D_{\Delta t} \, \tilde{u}^{k+1/2}$ from which it follows that

$$D_{\Delta t} \, \tilde{u}^{k+1/2}(x,y) = U(x,t) + \mathrm{E}[\sum_{j=1}^{N_k} \omega_j \, b_h(x - x_j^{k+1/2}) \, \mathrm{H}(|y_j^{k+1/2}+\zeta_j|-y) - \tilde{u}^{k+1/2}(x,0) \, \mathrm{H}(|\zeta| - y) \,] \tag{3.3}$$

where E denotes expected value taken over the independent, Gaussian distributed random variables ζ, $\zeta_1, \ldots, \zeta_{N_k}$. Combining (3.2) and (3.3) one can then prove the consistency of the random walk/sheet creation algorithm in the L^1 norm,

$$\|E\tilde{D}_{\Delta t}\, u^0 - D_{\Delta t}\, u^0\|_1 \le const.\ (\ \|\partial_x u^0(\cdot,0)\|_{L^\infty}\, h + \omega_{max})\ \sqrt{4\nu\Delta t}\ . \tag{3.4}$$

It is possible to prove other, similar results for one time step of the random walk/sheet creations process. In particular, it can be shown that

$$P(\ \|\tilde{D}_{\Delta t}\, \tilde{u}^{k+1/2} - D_{\Delta t}\, \tilde{u}^{k+1/2}\|_{L^2} \ge \gamma\sqrt{\omega_{max}}) \le const.\ \sqrt{2\nu\Delta t}\ \gamma^{-2} \tag{3.5}$$

where γ is an arbitrary positive constant and P denotes probability. In other words, the error in the L^2 norm is $O(\sqrt{\omega_{max}})$ lending credence to our earlier conjecture that the error due to the random walk is $O(\sqrt{\omega_{max}})$. Also note the favorable dependence of the estimates in (3.4) and (3.5) on the viscosity ν. The random vortex method has also been observed to have such a favorable dependence on ν ([3, 17]).

§4 A NUMERICAL DEMONSTRATION OF CONVERGENCE TO BLASIUS FLOW

In order to demonstrate the validity of the sheet method and to estimate its rate of convergence in terms of the computational parameters we used it to model Blasius flow over a semi infinite flat plate ([18, 26]). In this case

$$U_\infty(x,t) \equiv 1$$

and there exists a well known (stationary) similarity solution

$$u(x,y) = f'(\eta) \tag{4.1}$$

where

$$\eta = y \,/\, \sqrt{\nu x} \tag{4.2}$$

and f satisfies the ordinary differential equation (ODE)

$$ff'' + 2f''' = 0$$

$$f(0) = 0, \quad f'(0) = 0, \quad \text{and } f'(\infty) = 1\ .$$

The function f may be determined to arbitrarily high accuracy using any standard numerical ODE solver (e.g. [26], p. 262.)

In our test we computed over the portion of the plate which extends from $3h$ to $1 + 3h$ and identified the endpoints of this interval with appropriately rescaled y coordinates. Thus, whenever the center of a sheet left the interval $[3h, 1 + 3h]$ its y coordinate was rescaled according to (4.2) and it was then placed at the opposite end of the computational domain. Initially there are no sheets so that

$$u(x,y,0) = \begin{cases} 1 & y > 0, \\ 0 & y = 0. \end{cases}$$

We computed until time $t = 2$ and then examined the error. Note that the following results are obtained at one instant in time rather than being the errors averaged over many time steps. All errors were measured in the similarity coordinates $(x,\eta) = (x, y/\sqrt{\nu x})$. This has the effect of making the error independent of the viscosity (see [15], p. 39). We examined the errors in the discrete L^1, L^2 and L^∞ norms. The errors in the L^1 norm were normalized by dividing by

$$\|1 - u\|_{L^1([3h,1+3h]\times[0,\infty))} \approx 1.7208$$

where u is the exact solution given by (4.1). All of the following experiments were made with Chorin's smoothing function (2.2) and a value of $\nu = 10^{-4}$.

The numbers in Table 1 represent an estimate of the expected value of the error obtained by averaging the error over 25 trials. Table 2 contains an estimate of the standard deviation of this error based on our sample of 25 runs. The dagger (†) indicates that the sample size was 5 instead of 25 and the double dagger (‡) indicates a sample size of 1.

It is apparent from Table 1 that the error decreases as ω_{max}, h and $\Delta t \to 0$. Note that for fixed h and decreasing ω_{max} the expected value of the error eventually reaches a plateau but that the standard deviation continues to diminish. This indicates that the statistical errors depend primarily (if not exclusively) on ω_{max} while the overall error also depends on Δt and h. In the sequence of runs shown here the time step Δt was set equal to h. To test the dependence of these results on Δt we made identical runs but with $\Delta t = h/2$ and $\Delta t = h/4$. We observed no significant change in the error. We conjecture that, at least for this simple test case, when $\Delta t \leq h$ the errors due to time discretization are dominated by those due to the parameter h.

An examination of the diagonal that begins at $\omega_{max} = 0.05$ and $h = 0.2$ reveals that the error decreases like $O(\sqrt{\omega_{max}})$. Along diagonals above this one the rate is slower. Furthermore, for decreasing h with fixed $\omega_{max} \geq h$ the error actually increases. We conjecture that the error depends in some manner on the ratio ω_{max}/h, perhaps because

$$\frac{d}{dx}\omega_j\, b_h(x - x_j) \;=\; O(\omega_{max}/h)$$

and errors in interpolating piecewise C^1 functions with piecewise linear functions depend on the first derivative of the function being interpolated (e.g. (3.2)). Along diagonals such that $\omega_{max} \leq h/4$ however, the error decreases at the expected rate of $\sqrt{\omega_{max}}$.

Table 1 *Convergence Rate in the Discrete L^1 Norm*

	h ($\Delta t = h$)				
ω_{max}	0.2	0.1	0.05	0.025	0.0125
0.2	0.4002	0.4538	0.5029	0.6755	-
0.1	0.2989	0.2983	0.3290	0.4060	0.6606
0.05	0.2580	0.2239	0.2230	0.2891	0.4091
0.025	0.2663	0.1773	0.1657	0.1903	0.2766
0.0125	0.2483	0.1636	0.1267	0.1346	0.1864
0.00625	0.2529	0.1594	0.1088	0.0990	0.1159†
0.003125	0.2473	0.1528	0.1007	0.0778†	-
0.0015625	0.2511	0.1534	0.0857†	0.0669‡	-

Table 2 *Standard Deviation of the Errors in Table 1*

	h ($\Delta t = h$)				
ω_{max}	0.2	0.1	0.05	0.025	0.0125
0.2	0.0734	0.0649	0.0519	0.0511	-
0.1	0.0638	0.0521	0.0350	0.0373	0.0274
0.05	0.0475	0.0399	0.0279	0.0175	0.0180
0.025	0.0473	0.0239	0.0112	0.0165	0.0102
0.0125	0.0243	0.0202	0.0126	0.0101	0.0069
0.00625	0.0252	0.0104	0.0103	0.0075	0.0060†
0.003125	0.0146	0.0093	0.0069	0.0077†	-
0.0015625	0.0088	0.0068	0.0078†	-	-

Now consider decreasing h with $\omega_{max} \ll h$ fixed. For $\omega_{max} \leq h/8$ the error decreases like $O(h^{2/3})$. We conjecture that for $\Delta t \leq h$ and $\omega_{max} \leq h/8$ the L^1 error in computing Blasius flow is $O(\sqrt{\omega_{max}} + h^{2/3})$. Table 3 shows the results of a sequence of computer runs designed to test this conjecture. Note that here the number reported is the error after one trial rather than an estimate of the expected value of this error obtained by averaging the error over several trials.

The column labeled 'sheets' indicates the number of sheets in the computation at time $t = 2$. The column labeled 'time' contains the time, in minutes, it took for 1 trial on a CRAY X-MP. The column labeled $\delta_1 av$ is the average error in the displacement thickness defined by

$$\delta_1 av = \frac{1}{r}\sum_{i=1}^{r}|\delta_1(a_i) - \tilde{\delta}_1(a_i)|\,.$$

Here $\delta_1(a_i)$ is the displacement thickness ([18], p. 140) above the ith gridpoint on the wall and $\tilde{\delta}_1(a_i)$ is a trapezoid rule approximation to $\delta_1(a_i)$. Similarly, $\delta_2 av$ is the average error in the momentum thickness ([18], p. 141). We have not scaled out the effect of ν in our calculation of $\tilde{\delta}_1$ and $\tilde{\delta}_2$ and hence $\delta_1 av$ and $\delta_2 av$ are $O(\sqrt{\nu}) = O(10^{-2})$.

Table 3 (1 trial per row) *A Convergence Study with* $h = O(\omega_{max}^{3/4})$ *and* $\Delta t = h$

	($\Delta t = h$)						
ω_{max}	L^1 norm	L^2 norm	L^∞ norm	sheets	time	$\delta_1$av	$\delta_2$av
0.025	0.2817	0.2489	0.2708	226	0.002	0.0044	0.00161
0.0125	0.1617	0.1756	0.2677	939	0.022	0.0024	0.00070
0.00625	0.1044	0.1070	0.1536	2954	0.218	0.0013	0.00037
0.003125	0.0823	0.0915	0.2719	13755	3.638	0.0009	0.00029
0.0015625	0.0592	0.0631	0.1141	37206	31.131	0.0007	0.00019

It is clear that for the integrated quantities (L^1, L^2, $\delta_1 av$ and $\delta_2 av$) the error decreases at the anticipated rate of $1/\sqrt{2}$. The error in the L^∞ norm is not so well behaved. This is to be expected however since it is likely that with a random walk method the error may be large on small sets. (ex. [13], §5.)

§5 CONCLUSIONS

The foregoing results demonstrate that the vortex sheet method converges to Blasius flow. This establishes the fundamental validity of the algorithm and makes it plausible that the vortex sheet method will converge in the case of more general solutions to the Prandtl equations. On the basis of the data above we have conjectured that for $\omega_{max} \le h/8$ and $\Delta t \le h$ the error in the discrete L^1 norm is $O(\sqrt{\omega_{max}} + h^{2/3})$. It has not been possible to determine the dependence of the method on the time step Δt. This may be because Blasius flow is stationary or because Δt and h must satisfy the CFL condition $\Delta t \le h$ making it difficult to isolate the time discretization errors. We note that many of the conclusions drawn here are similar to

Two unusual aspects of the vortex sheet algorithm are the combination of a deterministic step and a random walk step into a two part fractional step method and the creation of particles (sheets) at boundaries in order to satisfy the no-slip boundary condition. We have shown how the sheet creation process and random walk together constitute the numerical solution of the diffusion equation (1.4) subject to the no-slip boundary condition (1.2a) and that this numerical solution is is consistent in the L^1 norm,

$$\| E\tilde{D}_{\Delta t}\, \tilde{u}^{k+1/2} - D_{\Delta t}\, \tilde{u}^{k+1/2} \|_{L^1} = O((h + \omega_{max})\sqrt{\nu \Delta t}) .$$

We have also attempted to demonstrate that the most efficient choice of computational parameters is obtained when the error due to the random walk (roughly $O(\sqrt{\omega_{max}})$) is balanced with that due to the advection step (roughly $O(h^{2/3})$ assuming $\omega_{max} \le h/8$ and $\Delta t \le h$). This balancing of errors should work as a rule of thumb for most fractional step methods in which one step is random and the other is deterministic (e.g. [1, 23]). Similar conclusions have been drawn by Ghoniem and Sethian in [20].

Currently the sheet method is the only viable way to satisfy the no-slip boundary condition in a random vortex method solution of the Navier-Stokes equations. Such hybrid methods have the advantage that the error is independent of the viscosity enabling one to compute at arbitrarily large Reynolds numbers. In fact, the statistical error decreases with decreasing ν. On the other hand, for fixed ν, the sheet method converges slowly, constrained by the $O(\sqrt{\omega_{max}})$ rate at which the random walk solution of the diffusion equation converges. This is particularly evident when one notes from Table 3 that it takes roughly 10 times as long to reduce the error by a factor of $\sqrt{2}$. Thus, it is unlikely to be the most efficient way to solve the Prandtl equations alone. The sheet method would be improved by the replacement of the random walk with a random walk type algorithm with higher order convergence. Alternatively, one can attempt to use a finite difference method to solve the flow near the boundary but then the problem remains of how to couple such a scheme to the vortex method. (See Anderson [2] for related work.) However, in spite of its slow rate of convergence the vortex sheet method remains widely used because of the very natural manner in which vorticity is created at the boundary and subsequently released into the flow.

References

1. B. J. Alder and D. M. Ceperley, "Quantum Monte Carlo," *Science*, vol. 231, pp. 555-560, 1986.
2. C. R. Anderson, "Vorticity Boundary Conditions and Vorticity Boundary Generation for Two Dimensional Viscous Incompressible Flows," *submitted to JCP* , UCLA Math Dept, 1987.
3. T. J. Beale and A. Majda, "Rates of Convergence for Viscous Splitting of the Navier-Stokes Equations," *Math Comp*, vol. 37, pp. 243-259, 1981.
4. G. Benfatto and M. Pulvirenti, "A Diffusion Process Associated to the Prandtl Equations," *J. Funct. Anal.*, vol. 52, pp. 330-343, 1983.
5. A. Y. Cheer, "A Study of Incompressible 2-D Vortex Flow Past a Circular Cylinder," *SIAM J. Sci. Stat. Comput.*, vol. 4, pp. 685-705, 1983.
6. A. Y. Cheer, "Unsteady Separated Wake Behind an Impulsively Started Cylinder in Slightly Viscous Fluid," *manuscript*, U. C. Davis, 1986.
7. Y. Choi, J. A. C. Humphrey, and F. S. Sherman, "Random Vortex Simulation of Transient Wall-Driven Flow in a Rectangular Enclosure," *submitted to JCP*, UC Berkeley ME Dept, 1986.
8. A. J. Chorin, "Numerical Study of Slightly Viscous Flow," *J. Fluid Mech.*, vol. 57, pp. 785-796, 1973.
9. A. J. Chorin, "Vortex Sheet Approximation of Boundary Layers," *J. Comp. Phys.*, vol. 27, pp. 428-442, 1978.
10. A. J. Chorin, "Vortex Models and Boundary Layer Instability," *SIAM J. Sci. Stat. Comput.*, vol. 1, pp. 1-21, 1980.
11. A. F. Ghoniem, A. J. Chorin, and A. K. Oppenheim, "Numerical Modeling of Turbulent Flow in a Combustion Tunnel," *Philos. Trans. Roy. Soc. London*, vol. A304, pp. 303-325, 1982.
12. J. Goodman, "Convergence of the Random Vortex Method in Two Dimensions," *Comm Pure and Applied Math*, to appear.
13. O. H. Hald, "Convergence of a Random Method With Creation of Vorticity," *Siam J. Sci. Stat. Comput.*, vol. 7 , pp. 1373-1386, 1986.
14. K. Nickel, "Prandtl's Boundary-Layer Theory from the Viewpoint of a Mathematician," *Annual Review of Fluid Mechanics*, 1973.
15. E. G. Puckett, "A Study of the Vortex Sheet Method and Its Rate of Convergence," *Center for Pure and Applied Mathematics preprint* PAM-369 , U. C. Berkeley, 1987.
16. E. G. Puckett, "Convergence of a Random Particle Method to Solutions of the Kolmogorov Equation," *Lawrence Berkeley Laboratory Preprint* 22420 , 1987.
17. S. G. Roberts, "Accuracy of the Random Vortex Method for a Problem with Non-Smooth Initial Conditions," *J. Comp. Phys.*, vol. 58, pp. 29-43, 1985.
18. H. Schlichting, *Boundary-Layer Theory*, McGraw-Hill, New York, 1968.
19. M. H. Schultz, *Spline Analysis*, Prentice-Hall, Inc., Englewood Cliffs, N.J., 1973.
20. J. A. Sethian and A. F. Ghoniem, "Validation Study of Vortex Methods," *J. Comp. Phys. (to appear)*.
21. J. A. Sethian, "Turbulent Combustion in Open and Closed Vessels," *J. Comp. Phys.*, vol. 55, pp. 425-456, 1984.
22. J. A. Sethian, "Vortex Methods and Turbulent Combustion," in *Lectures in Applied Mathematics*, vol. 22, Springer Verlag, New York, 1985.
23. A. S. Sherman and C. S. Peskin, "A Monte Carlo Method for Scalar Reaction-Diffusion Equations," *SIAM J. Sci Stat Comp*, 1986.
24. D. M. Summers, T. Hanson, and C. B. Wilson, "A Random Vortex Simulation of Wind-Flow Over a Building," *Int. J. for Num. Meth. in Fluids*, vol. 5, pp. 849-871, 1985.
25. E. Tiemroth, "The Simulation of the Viscous Flow Around a Cylinder by the Random Vortex Method," *Ph. D. Thesis*, U. C. Berkeley Naval Arch. Dept, 1986.
26. F. M. White, *Viscous Fluid Flow*, McGraw-Hill, New York, 1974.

SPARSE QUASI–NEWTON METHOD FOR HIGH RESOLUTION SCHEMES

Ning QIN and Bryan E. RICHARDS

Department of Aeronautics and Fluid Mechanics
The University of Glasgow, Glasgow G12 8QQ, Scotland, U.K.

1. INTRODUCTION

To speed up the convergence to steady state solution in CFD, there is a tendency to use implicit time marching schemes. Generally, as compared with explicit time marching schemes, the penalty that implicit methods take more computing time per time step and more coding work is amply offset by the large improvement in convergence to the stationary solution.

Linearization is generally used in an implicit scheme to avoid the difficult task of solving the nonlinear system at each time step. This gives the name *non–iterative implicit method*. If the linearization is *exact*, the implicit scheme will, as $\Delta t \to \infty$, reduce to a Newton iteration method for the steady state equations. Using the exact linearization, Mulder and van Leer[1] found a quadratic convergence to the steady state in the solution for a nozzle problem. This fast convergence can only be obtained if an applicable procedure exists for the exact linearization (i.e. the evaluation of the Jacobian).

Unfortunately this is seldom the case in practice. The differential equations can be discretized in space by various methods. Some sophisticated schemes have been developed for high resolution of shock waves. Furthermore more complicated mechanisms such as viscous effects and turbulence may involve more complicated schemes. In all these cases, the evaluation of the Jacobian is often too difficult to apply so that different simplification in constructing the implicit operator are made for a specific discretization scheme or for a specific problem. This process brings about an irretrievable loss of information that exhibits itself in the degradation of the convergence rate.

The motivation of the present research stems from the desire to develop an iterative scheme to greatly increase the efficiency for a steady state solution. For this purpose, we present a new strategy which updates the Jacobian by a sparse quasi– Newton method for the solution of nonlinear system. To form a successful nonlinear algorithm, we combine this fast locally convergent method with some time marching approach, which results in a globally convergent procedure. The analysis of numerical results of a nozzle problem with van Leer's flux splitting[2,3] and Harten's TVD[4] scheme reveals that the present procedure outperforms the implicit schemes by a huge margin for steady state solution.

2. GENERAL PROCEDURE OF THE SPARSE QUASI– NEWTON METHOD

We consider the nonlinear system,

$$R(U) = 0, \tag{1}$$

which may arise from the finite difference discretization of the steady Euler or Navier– Stokes equations.

This nonlinear system is usually large while the Jacobian $J = \partial R/\partial U$ generally has a sparse structure. Specifically if a FD discretization scheme is used, the resulting nonlinear system often has a Jacobian with a regular banded sparsity pattern, e.g. block tridiagonal, block pentadiagonal, and so on. Each single equation can be very complicated and, therefore, the analytic expression of the Jacobian is generally unavailable and the function evaluation is usually very expensive.

2.1. *Sparse quasi–Newton method*

Since the Jacobian is unavailable, the Newton method cannot be applied. However making use of the sparsity, we can devise a sparse finite difference Newton method and the usual n additional function evaluations can be reduced significantly. Although the reduction is remarkable, many times of the function evaluation per iteration may still be too time consuming. To avoid these extra function evaluation, we follow a quasi– Newton strategy.

The basic idea of the quasi– Newton method is to approximate the Jacobian of the nonlinear system using only function values that we have already calculated. To make a full use of sparsity in the quasi– Newton updating of the Jacobian, we introduce here the sparse quasi– Newton method, which was first proposed by Schubert[5] and Broyden[6]. Also see Dennis and Schnabel[7] for details.

We define the matrix projection operator $P_J : R^{n\times n} \rightarrow R^{n\times n}$ by

$$(P_J(M)) = \begin{cases} 0, & \text{if } J(U)_{i,j} = 0 \text{ for all } U \in R^n \\ M_{i,j}, & \text{otherwise.} \end{cases} \tag{2}$$

Similarly we define $S_i \in R^n$ by

$$(S_i)_j = \begin{cases} 0, & \text{if } J(U)_{i,j} = 0 \text{ for all } U \in R^n \\ S_j, & \text{otherwise.} \end{cases} \tag{3}$$

The procedure of the sparse quasi– Newton method may be written as following:

$$\begin{array}{l} \text{Given } R: R^n \rightarrow R^n,\ U^0 \in R^n,\ A^0 \in R^{n\times n} \\ \quad \text{DO for } k = 0,1,2,\ldots \\ \qquad \text{Solve} \quad A^k S^k = -R(U^k) \text{ for } S^k \\ \qquad\qquad\quad U^{k+1} = U^k + S^k \\ \qquad\qquad\quad Y^k = R(U^{k+1}) - R(U^k) \\ \qquad\qquad\quad A^{k+1} = A^k + P_J[D^+(Y^k - A^k S^k)(S^k)^T]. \end{array} \tag{4}$$

Here $D^+ \in R^{n\times n}$ is a diagonal matrix with

$$(D^+)_{ii} = \begin{cases} 1/(S_i)^T(S_i), & \text{if } (S_i)^T(S_i) \neq 0 \\ 0, & \text{otherwise.} \end{cases} \tag{5}$$

Under standard assumptions, this procedure has been proved to be locally q– superlinearly convergent[8]. Although constructing the above sequence imposes no analytical assumption on R and no requirement on U^0 and A^0, we have to exercise care to enable a successful application. Since R in question is generally complicated, we have little analytical information about it. Thus it is usually impossible to check all the analytical properties of R before we use the method so that numerical experiments are needed. However it is important to note that the fast convergence is a local property. The basic idea in forming a successful nonlinear algorithm is to combine a fast local convergence strategy with a global convergence strategy in a way that derives benefit from both.

2.2. *Initialization and global convergence*

1) Initial value U^0 – Time dependent approaches

For problems in CFD, a natural and robust way to initialize the procedure is the time marching approach, although as an iterative method for steady state solutions it may be extremely slow. The time marching is switched to the sparse quasi– Newton iteration as soon as the solution goes into the convergent region. Because the theory on the convergent region for the sparse quasi– Newton method is absent, this switching point has to be determined by experimentation.

2) Initial approximation A^0 to the Jacobian – Sparse finite difference Newton method
If in the above initialization procedure, an explicit time differencing is used, the initial approximation A^0 to the Jacobian $J(U^0)$ is evaluated by a sparse finite difference(FD) Newton method[9] to start the sparse quasi–Newton procedure. While if an implicit time differencing is used in the initialization, we have another way for the initial A^0 by using the implicit operator. The latter approach is simpler, while, if the implicit operator is far away from $J(U^0)$, the convergence may be greatly degraded. Since implicit methods are used in the following calculations for initialization, both approaces are tested for comparison.

3. FORMULATION FOR THREE–POINT AND FIVE–POINT SCHEMES

In the last section, we have discussed the general formulation and application procedure. Here we formulate it for three–point and five–point schemes not only for its practical importance but also to provide a clarified presentation of the application of the sparse quasi–Newton method and the sparse FD Newton method.

Suppose Eq.(1) results from discretization of a steady state problem by a three–point or five–point finite difference scheme. We introduce the notation

$$U = \begin{bmatrix} u_1 \\ u_2 \\ \cdot \\ \cdot \\ \cdot \\ u_I \end{bmatrix}, \quad R = \begin{bmatrix} r_1 \\ r_2 \\ \cdot \\ \cdot \\ \cdot \\ r_I \end{bmatrix}, \quad S = \begin{bmatrix} s_1 \\ s_2 \\ \cdot \\ \cdot \\ \cdot \\ s_I \end{bmatrix}, \tag{6}$$

where $u_i, r_i, s_i \in \mathbf{R}^m$ and

$$r_i = r_i(u_{i-1}, u_i, u_{i+1}) \text{ or } r_i = r_i(u_{i-2}, u_{i-1}, u_i, u_{i+1}, u_{i+2}),$$

Therefore the Jacobian has a block tridiagonal or pentadiagonal structure

$$J(U) = \frac{\partial R}{\partial U} = \begin{bmatrix} D_1 & C_1 & & & \\ B_2 & D_2 & C_2 & & \\ & \cdot & \cdot & \cdot & \\ & & \cdot & \cdot & \cdot \\ & & & B_I & D_I \end{bmatrix} \text{ or } \begin{bmatrix} D_1 & C_1 & C1_1 & & & \\ B_2 & D_2 & C_2 & C1_2 & & \\ B1_3 & B_3 & D_3 & C_3 & C1_3 & \\ & \cdot & \vdots & \vdots & \vdots & \vdots & \cdot \\ & & & B1_I & B_I & D_I \end{bmatrix} \tag{7}$$

where D_i, B_i, C_i, $B1_i$, $C1_i \in \mathbf{R}^{m \times m}$. We suppose that the boundary conditions at $i = 1$ and $i = I$ can be embedded into the above structures.

3.1. *Block tridiagonal and pentadiagonal quasi–Newton updating*

Now we describe the sparse quasi–Newton update for the tridiagonal or pentadiagonal Jacobian. The approximation A^k to $J(U^k)$ will have the same sparse structure as J so we only need to update the approximation D_i^k, B_i^k and C_i^k to $D_i(U^k)$, $B_i(U^k)$ and $C_i(U^k)$. According to the procedure (4), the formula may then be written as:

Solve
$$A^k S^k = -R(U^k) \text{ for } S^k$$
$$U^{k+1} = U^k + S^k$$
Update A^k to A^{k+1} by
$$\begin{aligned} D^{k+1}{}_i &= D^k{}_i - d_{ii} r^{k+1}{}_i (s^k{}_i)^T && i=1,\dots,I \\ B^{k+1}{}_i &= B^k{}_i - d_{ii} r^{k+1}{}_i (s^k{}_{i-1})^T && i=2,\dots,I \\ C^{k+1}{}_i &= C^k{}_i - d_{ii} r^{k+1}{}_i (s^k{}_{i+1})^T && i=1,\dots,I-1 \end{aligned} \tag{8}$$
and for a five-point scheme add
$$B1^{k+1}{}_i = B1^k{}_i - d_{ii} r^{k+1}{}_i (s^k{}_{i-2})^T \qquad i=3,\dots,I$$

$$Cl^{k+1}{}_i = Cl^k{}_i - d_{ii} r^{k+1}{}_i (s^k{}_{i+2})^T \qquad i=1,\dots,I-2$$

where for a three-point scheme

$$d_{ii} = [\ (s_{i-1})^T(s_{i-1})+(s_i)^T(s_i)+(s_{i+1})^T(s_{i+1})\ ]$$

and for a five-point scheme

$$d_{ii} = [(s_{i-2})^T(s_{i-2})+(s_{i-1})^T(s_{i-1})+(s_i)^T(s_i)+(s_{i+1})^T(s_{i+1})+(s_{i+2})^T(s_{i+2})]$$

where s_{-1}, s_0, s_{I+1} and s_{I+2} are equal to zero.

3.2. *Block tridiagonal and pentadiagonal FD Newton method*

In the sparse FD Newton method, we only need 3×m or 5×m additional evaluations of R(U) for a finite difference approximation of J(U) for a block tridiagonal or pentadiagonal sparsity pattern respectively.

For the block tridiagonal case, we can evaluate this approximation by

$$\begin{aligned}(D_i)\cdot_n &= [r_i(u_{i-1},u_i+h^n{}_i e^n{}_i,u_{i+1})-r_i(u_{i-1},u_i,u_{i+1})]/h^n{}_i \\ (B_i)\cdot_n &= [r_i(u_{i-1}+h^n{}_{i-1}e^n{}_{i-1},u_i,u_{i+1})-r_i(u_{i-1},u_i,u_{i+1})]/h^n{}_{i-1} \\ (C_i)\cdot_n &= [r_i(u_{i-1},u_i,u_{i+1}+h^n{}_{i+1}e^n{}_{i+1})-r_i(u_{i-1},u_i,u_{i+1})]/h^n{}_{i+1}.\end{aligned} \tag{9}$$

Similarly we can derive formulas for the block pentadiagonal structure.

For real calculations, boundary conditions should be embedded into the above formulation according to R(U).

4. APPLICATION TO FLUX–SPLITTING AND TVD SCHEMES FOR THE NOZZLE PROBLEM

4.1. *The test problem*

The governing equation for the nozzle problem can be written as

$$\frac{\partial F(u)}{\partial x} + H(u) = 0 \tag{10}$$

with

$$u = \begin{bmatrix} \rho\kappa \\ m\kappa \\ e\kappa \end{bmatrix}, \qquad F = \begin{bmatrix} m\kappa \\ (m^2/\rho+p)\kappa \\ (e+p)m\kappa/p \end{bmatrix}, \qquad H = \begin{bmatrix} 0 \\ -p(\partial\kappa/\partial x) \\ 0 \end{bmatrix}, \tag{11}$$

where $m=\rho v$, κ is the area of the nozzle, a function of x

$$\kappa(x) = 1.398 + 0.374 \tanh(0.8x - 4), \tag{12}$$

and

$$p = (\gamma-1)(e-m^2/2\rho). \tag{13}$$

In the calculations to be presented the computational domain is $0 \leqslant x \leqslant 10$. We use the spacing $\Delta x = 0.5$. The initial and boundary conditions are treated as in Yee et al[10].

4.2. *Time dependent approach – backward Euler implicit operator*

We use an implicit formulation as proposed by Mulder and van Leer[1],

$$[\ (I/\Delta t) - M^k\ ]\ (U^{k+1}-U^k) = R(U^k), \tag{14}$$

which can reduce to the Newton method as $\Delta t \to 0$ if the linearization is exact, i.e. if $M = J = \partial R/\partial U$. Therefore it provides a possibility of quadratic convergence.

The convergence is monitored by the maximum residuals

$$RES = \max(|\ r^1{}_i\ |/|\ u^1{}_i\ |) \tag{15}$$

and the time step with the implicit operator is

$$\Delta t^k = \epsilon/RES^k. \tag{16}$$

The implicit operators used below are of this evolution/relaxation type with $\Delta t \to \infty$ as RES $\to 0$

4.3. *van Leer's flux vector splitting*

1) First order accuracy
A flux vector splitting method is used to Eq(10) for the upwind differencing in space[2]. The resulting nonlinear system can be written as

$$r_i = r_i(u_{i-1}, u_i, u_{i+1}) = -(F_i^+ - F_{i-1}^+ + F_{i+1}^- - F_i^-)/\Delta x - H_i = 0. \tag{17}$$

This flux–splitting scheme has the advantage that the splitted fluxes F^+, F^- are continuously differentiable and the analytic expression of the Jacobian for this problem can easily be derived.

Implicit operator. After the exact linearization of the implicit backward Euler time differencing the implicit operator can be written as

$$\begin{aligned} D_i &= I/\Delta t + (E^+{}_i - E^-{}_i)/\Delta x - G_i \\ B_i &= -E^+{}_{i-1}/\Delta x \\ C_i &= -E^-{}_{i+1}/\Delta x. \end{aligned} \tag{18}$$

Corresponding to the explicit boundary conditions, the implicit boundary treatment is

$$\begin{aligned} \bar{D}_I &= D_I + (2/\Delta x)E^-{}_{I+1} I_1 \\ \bar{B}_I &= B_I - (1/\Delta x)E^-{}_{I+1} I_1 \end{aligned} \tag{19}$$

where

$$I_1 = \begin{bmatrix} 0 & & \\ & 1 & \\ & & 1 \end{bmatrix}.$$

This correspondence on the boundary is important for the implicit operator. Otherwise the implicit operator will not reduce to the Jacobian as $\Delta t \to \infty$.

Sparse quasi–Newton method. In the sparse quasi–Newton method we use formula (8) to update approximations D^k, B^k, C^k to D^{k+1}, B^{k+1}, C^{k+1}. In contrast to the implicit operator no special treatment on the boundary is needed.

Fig. 1 shows the convergence history against the iteration number (convergence) and the work unit (efficiency) respectively. We define one work unit as the CPU time per iteration needed by the corresponding implicit method. As shown in c) the sparse quasi–Newton approach has almost the same efficiency as a Newton approach. The FD Newton method is approximately identical to a Newton method if the increments are properly chosen according to the machine zero and R(U). The switching point is clearly shown in the figures (b) and (c) at iteration point five.

2) Second order accuracy – MUSCL approach
The MUSCL approach[1,3] is used for a second order flux vector splitting scheme

$$\begin{aligned} r_i &= r_i(u_{i-2}, u_{i-1}, u_i, u_{i+1}, u_{i+2}) \\ &= -[F^+(u^-{}_{i+1/2}) - F^+(u^-{}_{i-1/2}) + F^-(u^+{}_{i+1/2}) - F^-(u^+{}_{i-1/2})]/\Delta x - H(u_i). \end{aligned} \tag{20}$$

Implicit operator. The implicit operator is the same as the first order case, which does not affect the accuracy of the right hand side but prevent the method from becoming a

Newton– type as $\Delta t \to \infty$.

Sparse quasi–Newton method. Since the scheme is a five– point scheme, the block pentadiagonal quasi– Newton update presented in Section 3 is used.

Fig.2 Shows the result for the second order case. The convergence by the implicit operator is heavily degraded as compared to the first order case. While sparse quasi Newton approach still exhibits high convergence rate as compared to the FD Newton method. Althouth the sparse FD Newton method has a slightly higher convergent rate (shown in (b)), the sparse quasi– Newton with sparse FD Newton method evaluating the initial Jacobian is the most efficient (shown in (c)).

4.4. *Harten's TVD scheme*

If first or second order accurate (in space) TVD scheme[4,10] is used in the discretization, the resulting non– linear system may be written as

$$r_i = -(1/\Delta x)(\tilde{F}_{i+1/2} - \tilde{F}_{i-1/2}) - H_i = 0. \tag{21}$$

Implicit operator. To accelerate convergence to steady state, Harten[4] extended his explicit TVD scheme to an implicit method by an so– called TVD linearization. The resulting implicit operator can be written as

$$\begin{aligned} D_i &= (1/\Delta t)I + (1/\Delta x)(\tilde{E}^-_{i+1/2} + \tilde{E}^+_{i-1/2}) - G_i \\ B_i &= (1/\Delta x)\,\tilde{E}^+_{i-1/2} \\ C_i &= -(1/\Delta x)\,\tilde{E}^-_{i-1/2}. \end{aligned} \tag{22}$$

At the downstream boundary B_I and D_I are modified as in (19).

This "TVD linearization" is clearly not an exact linearization. Hence the implicit operator will not reduce to a Newton iteration method as $\Delta t \to \infty$.

Sparse quasi–Newton approach. The above scheme (21) is a five– point second order accurate (in space) scheme. Therefore the block pentadiagonal sparse quasi– Newton method formulated in Section 3 is used.

To get the three– point first order accurate TVD scheme, we simply set $g=\gamma=0$ in the above second order form. The corresponding sparse quasi– Newton method is block tridiagonal updating.

Fig. 3 and 4 show the results for first and second order TVD scheme respectively. The convergent rate is improved greatly and the CPU time is reduced marginally by the sparse quasi– Newton approach. Again the most efficient is the sparse quasi– Newton scheme with sparse FD Newton Jacobian initialization.

5. CONCLUDING REMARKS

A sparse quasi– Newton iterative method is introduced for solving steady state problems in fluid dynamics. Compared with implicit time marching approaches, this approach exhibits a much faster convergence to steady state. The robustness of the time marching approach is made use of to initialize the sparse quasi– Newton procedure. The algorithm treats the system as a black box and seeks derivative information from the solution process. Therefore, in addition to its fast convergence, it is general, simple and not sensitive in implementation to different space differencing schemes or different physical problems.

Extension to multidimensional problems is underway, where the major task is to combine an efficient iterative linear system solver for sparse linear system $A^k s^k = -R(U^k)$.

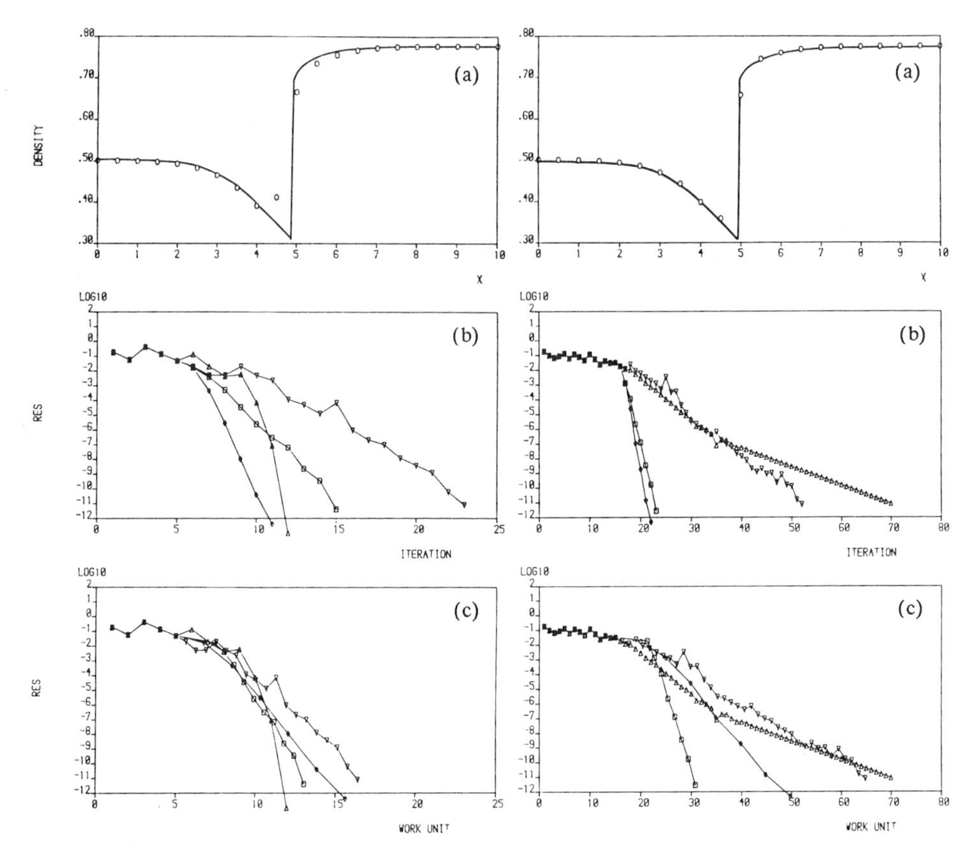

FIG.1. FLUX SPLITTING– FIRST ORDER FIG.2. FLUX SPLITTING– SECOND ORDER

(a) density distribution; —— exact; ◦ numerical
(b) convergence against iteration number
(c) convergence against work unit

△ implicit
□ sparse quasi– Newton with sparse FD Newton Jacobian initialization
▽ sparse quasi– Newton with imlicit operator as initial Jacobian
◇ sparse FD Newton

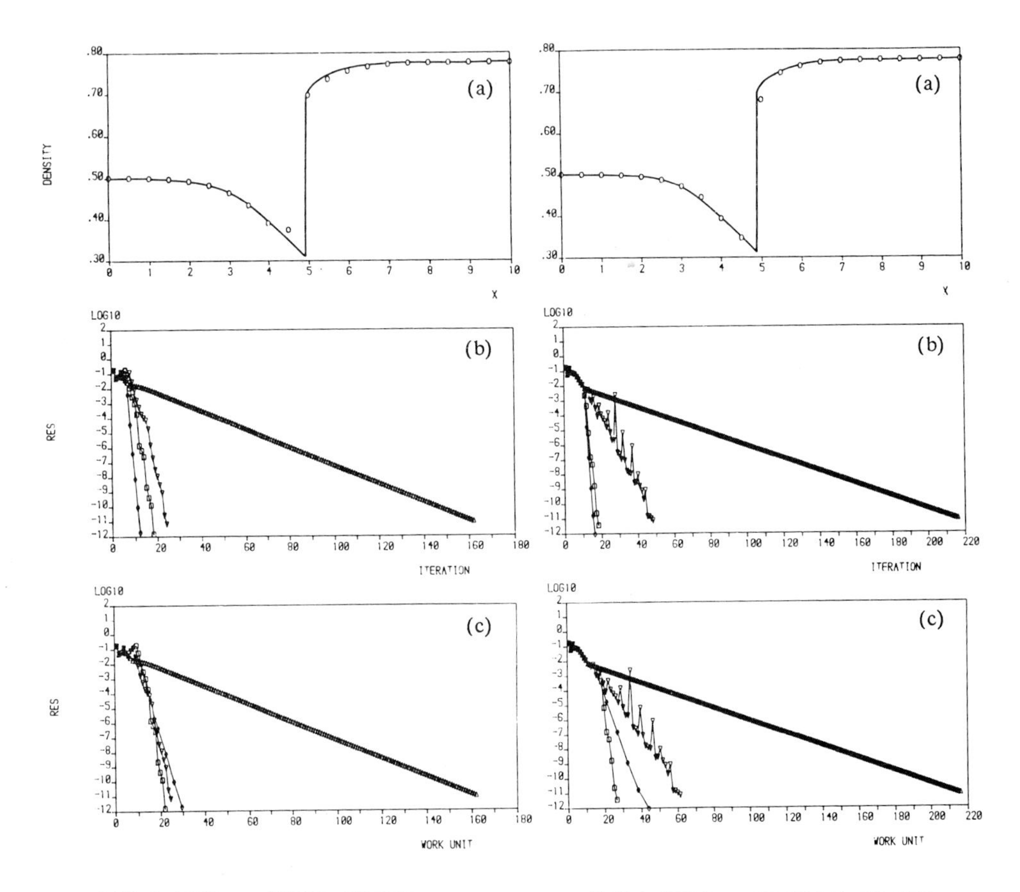

FIG.3. TVD – FIRST ORDER

FIG.4. TVD – SECOND ORDER

REFERENCES

1. W.A. MULDER AND B. VAN LEER, *J. Comput. Phys.* **59**(1985), 232.
2. B. VAN LEER, *Lecture Notes in Phys.* **170**(1982), 507.
3. B. VAN LEER, *J. Comput. Phys.* **32**(1979), 101.
4. A. HARTEN, *SIAM J. Numer. Anal.* **21**(1984), 1.
5. L.K. SCHUBERT, *Math. Comp.* **24**(1970), 27.
6. C.G. BROYDEN, *Math. Comp.* **25**(1971), 285.
7. J.E. DENNIS, JR. AND R.B. SCHNABEL, Numerical Methods for Unconstrained Optimization and Nonlinear Equations (Prentice Hall, Englewood Cliffs, N.Y., 1983).
8. E.S. MARWIL, *SIAM J. Numer. Anal.* **16**(1979), 588.
9. A. CURTIS, M.J.D. POWELL AND J.K. REID, *J.I.M.A.* **13**(1974), 117.
10. H.C. YEE, R.F. WARMING AND A. HARTEN, *J. Comput. Phys.* **57**(1985), 327.

Acknowledgement
The first author was supported by the Research Scholarship of Glasgow University and an ORS award during this research.

A LEGENDRE SPECTRAL ELEMENT METHOD FOR THE INCOMPRESSIBLE NAVIER-STOKES EQUATIONS

Einar M. Rønquist and Anthony T. Patera
Department of Mechanical Engineering
Massachusetts Institute of Technology
Cambridge, Massachusetts 02139
U.S.A.

ABSTRACT

Spectral element methods are high-order weighted residual techniques for partial differential equations that combine the geometric flexibility of finite element methods with the rapid convergence rate of spectral techniques. In this paper we describe Legendre spectral element methods for the simulation of incompressible fluid flow, with special emphasis on optimal Stokes discretizations, iterative Stokes solvers, and incomplete iteration.

STEADY STOKES

Discretizations. We consider here the two-dimensional steady Stokes problem: Find a velocity $\mathbf{u}$ and a pressure p in a rectangularly decomposable domain Ω such that

$$-\nu\Delta\mathbf{u} + \nabla p = \mathbf{f} \quad \text{in } \Omega, \tag{1a}$$

$$-\mathrm{div}\,\mathbf{u} = 0 \quad \text{in } \Omega, \tag{1b}$$

subject to homogeneous velocity boundary conditions on the domain boundary $\partial\Omega$. Here $\mathbf{f}$ is the prescribed force, ν is the kinematic viscosity, and the density of the fluid is taken to be unity. The spectral element discretization for solving (1) is based on the equivalent variational formulation: Find $(\mathbf{u},p)$ in $X\times M$ such that

$$\nu(\nabla\mathbf{u},\nabla\mathbf{w})-(p,\mathrm{div}\,\mathbf{w}) = (\mathbf{f},\mathbf{w}) \;, \quad \forall\mathbf{w}\in X \;, \tag{2a}$$

$$-(q,\mathrm{div}\,\mathbf{u}) = 0 \;, \quad \forall q\in M \;, \tag{2b}$$

where $(\phi,\psi) = \int_\Omega \phi\psi d\Omega$. The velocity space is given by $X = [H_0^1]^2$ where H_0^1 is the space of all functions which satisfy the homogeneous boundary conditions, which are square integrable, and whose first derivatives are also square integrable. The space for the pressure is given by $M=L_0^2(\Omega)$, the space of functions which are square integrable and have zero average over Ω.

The spectral element discretization proceeds by breaking up the domain into K quadrilateral elements $\Omega_1,\ldots,\Omega_K$, and searching for a solution in polynomial subspaces of X and M. However, choosing the polynomial degree N to be the same for the velocity and the pressure leads to spurious modes in the pressure. It can be shown that an optimal strategy [1-3] is to use a staggered mesh for the pressure for which the associated polynomial degree is N-2. The discrete formulation corresponding to (2) can then be stated as: Find $\mathbf{u}_h$ in $X_h=[H_0^1(\Omega)\cap\mathbf{P}_{N,K}(\Omega)]^2$ and p_h in $M_h=\mathbf{P}_{N-2,K}(\Omega)$ such that

$$\nu(\nabla\mathbf{u}_h,\nabla\mathbf{w}_h)_{h,GL}-(p_h,\mathrm{div}\mathbf{w}_h)_{h,G} = (\mathbf{f},\mathbf{w}_h)_{h,GL}, \quad \forall\mathbf{w}_h\in X_h, \tag{3a}$$

$$-(q_h,\mathrm{div}\mathbf{u}_h)_{h,G} = 0, \quad \forall q_h\in M_h. \tag{3b}$$

Here $\mathbf{P}_{N,K}(\Omega)=\cup_k\mathbf{P}_N(\Omega_k)$, and $\mathbf{P}_N(\Omega_k)$ is the space of all polynomials over Ω_k of degree $\leq N$ with respect to each spatial variable x_1 and x_2. The subscripts h,G and h,GL in (3) refer to numerical quadrature based on the Gauss and the Gauss-Lobatto points, respectively. For an error analysis of the discretization (3) we refer to [2,4], where it is shown that spectral convergence obtains as $N\Rightarrow\infty$ for fixed K.

We now define the bases for the spaces X_h and M_h. In each element Ω_k ($(x_1,x_2)\in\Omega_k\Rightarrow(r_1,r_2)\in\Lambda\times\Lambda$, $\Lambda=]-1,1[$), the velocity $\mathbf{u}_h$ is expanded in tensor-product form

$$\mathbf{u}_h^k(r_1,r_2) = \sum_{i=0}^{N}\sum_{j=0}^{N} \mathbf{u}_{ij}^k h_i(r_1)h_j(r_2) , \tag{4}$$

where the $h_i(r)$ are the one-dimensional N^{th} order Lagrangian interpolants through the Gauss-Lobatto points ξ_i, and $\mathbf{u}_{ij}^k$ is the velocity at the local point $(\xi_i,\xi_j)\in\Lambda\times\Lambda$ within the mapped element Ω_k. The pressure p_h is represented at the Gauss points,

$$p_h^k(r_1,r_2) = \sum_{i=1}^{N-1}\sum_{j=1}^{N-1} p_{ij}^k \tilde{h}_i(r_1)\tilde{h}_j(r_2) , \tag{5}$$

where $\tilde{h}_i(r)$ are the one-dimensional $(N-2)^{th}$ order Lagrangian interpolants through the Gauss points ς_i, and p_{ij}^k is the pressure at the local point $(\varsigma_i,\varsigma_j)\in\Lambda\times\Lambda$ within the mapped element Ω_k.

The velocity $\mathbf{u}_h$, the test functions $\mathbf{w}_h$, and the data f are all expressed in terms of the nodal basis (4), while the pressure p_h and the test functions q_h are expressed in terms of the nodal basis (5). Choosing appropriate "delta" testfunctions $\mathbf{w}_h$ and q_h we arrive at the discrete saddle problem,

$$\underline{A}\,\underline{u}_i - \underline{D}_i^T\underline{p} = \underline{B}\,\underline{f}_i , \quad i=1,..,d, \tag{6a}$$

$$-\underline{D}_i\underline{u}_i = \underline{0} . \tag{6b}$$

Here d refers to the space dimension, the matrix $\underline{A}$ is the discrete Laplacian, $\underline{B}$ is the diagonal mass matrix, $\underline{D}=(\underline{D}_1,..,\underline{D}_d)$ is the discrete gradient operator, and underscore refers to nodal values. In (6) we assume that the homogeneous boundary conditions are imposed by eliminating appropriate rows and columns.

Using the idea of tensor-product expansions, the spectral element discretization for the two-dimensional steady Stokes problem can be readily extended to solve three-dimensional problems in rectilinear geometry. We have further extended the method to allow for treatment of general deformed geometries by using subparametric mappings. The geometry is expanded in a space $G_h = [\mathbf{P}_{N-2,K} \cap H^1(\Omega)]^d$ which insures the preservation of the (physical) hydrostatic mode, that is $(p_h = \text{Const}, \text{div}\mathbf{w}_h)_{h,G} = 0 \ \forall \mathbf{w}_h \in X_h$ in *all* geometries.

Solvers. In this section we consider solution of the algebraic system of equations (6) resulting from the spectral element discretization of the Stokes problem (1). It should be noted that the algorithms presented here are equally appropriate for other types of variational discretizations, and are in fact extensions of the classical Uzawa algorithm used in finite element analysis [5,6].

Our approach [7] to solving (6) is a global iterative procedure in which the original coupled saddle problem is decoupled into two positive-definite symmetric forms,

$$\underline{A}\,\underline{u}_i - \underline{D}_i^T \underline{p} = \underline{B}\,\underline{f}_i \quad , \ i=1,..,d, \tag{7a}$$

$$\underline{S}\,\underline{p} = -\,\underline{D}_i \underline{A}^{-1} \underline{B}\,\underline{f}_i \quad , \tag{7b}$$

where

$$\underline{S} = \underline{D}_i \underline{A}^{-1} \underline{D}_i^T \quad . \tag{8}$$

In the solution process (7b) is first solved for $\underline{p}$, and then (7a) is solved for $\underline{u}=(\underline{u}_1,..,\underline{u}_d)$ with $\underline{p}$ known.

The equations for the velocities $\underline{u}_i$ correspond to standard Laplacian solves, and the inversion of the matrix $\underline{A}$ is done by conjugate gradient iteration using tensor-product sum factorization [4]. Although the matrix $\underline{S}$ in the equation for the pressure is full due to the presence of $\underline{A}^{-1}$, the matrix is extremely well-conditioned. In particular, it can be shown [7] that the matrix $\underline{S}$ is spectrally close to the variational equivalent of the identity matrix $\underline{I}$, namely the mass matrix $\tilde{\underline{B}}$ defined on the pressure mesh. This suggests an inner/outer conjugate gradient iteration procedure in which the outer iteration corresponds to inverting the matrix $\underline{S}$ preconditioned with the diagonal mass matrix $\tilde{\underline{B}}$, and the inner iteration corresponds to inverting the discrete Laplacian, $\underline{A}$. As the condition number for the matrix $\tilde{\underline{B}}^{-1}\underline{S}$ is of order unity, the above algorithm requires only order unity elliptic solves, and is therefore an ideal decoupling of the Stokes problem.

UNSTEADY STOKES

Discretizations. The discretization and solution of the steady Stokes equations (1) can be readily extended to the unsteady Stokes equations [4],

$$\frac{\partial \mathbf{u}}{\partial t} - \nu \Delta \mathbf{u} + \nabla p = \mathbf{f} \quad \text{in } \Omega, \tag{9a}$$

$$-\text{div}\mathbf{u} = 0 \quad \text{in } \Omega, \tag{9b}$$

with appropriate boundary conditions and initial conditions.

Applying temporal discretization by the implicit Euler backward method, and spatial discretization by the spectral element spatial operators developed previously, we arrive at the following set of discrete equations:

$$\frac{1}{\Delta t}\,\underline{B}(\underline{u}_i^{n+1} - \underline{u}_i^{n}) + \underline{A}\,\underline{u}_i^{n+1} - \underline{D}_i^T \underline{p}^{n+1} = \underline{B}\,\underline{f}_i^{n+1} \quad , \quad i=1,..,d, \qquad (10a)$$

$$-\underline{D}_i \underline{u}_i^{n+1} = \underline{0} \quad , \qquad (10b)$$

to be solved for the velocity $\underline{u}_i^{n+1}$ and the pressure $\underline{p}^{n+1}$ at time $t=(n+1)\Delta t$, where Δt is the time step. The scheme (10) can easily be shown to be unconditionally stable for all time steps Δt.

Solvers. In a similar fashion to the steady Stokes problem, the original set of discrete equations (10) can be decoupled into two positive-definite symmetric forms:

$$\sigma \underline{B}\,\underline{u}_i + \underline{A}\,\underline{u}_i - \underline{D}_i^T \underline{p} = \underline{g}_i \quad , \qquad (11a)$$

$$\underline{S}_t \underline{p} = -\,\underline{D}_i (\underline{A} + \sigma \underline{B})^{-1} \underline{g}_i \quad , \qquad (11b)$$

where

$$\underline{S}_t = \underline{D}_i (\underline{A} + \sigma \underline{B})^{-1} \underline{D}_i^T \qquad (12)$$

is the unsteady Stokes pressure operator analogous to the steady operator $\underline{S}$ defined in (8). Here $\sigma=1/\Delta t$, $\underline{u}_i=\underline{u}_i^{n+1}$, $\underline{p}=\underline{p}^{n+1}$, and $\underline{g}_i$ represents all inhomogeneities.

Although the matrix $\underline{S}_t$ is not well-conditioned for small Δt, $\underline{S}_t$ can nevertheless be efficiently inverted by the following Richardson iteration,

$$\underline{E}\,\tilde{\underline{p}}^m = \underline{E}\,\underline{p}^m - \alpha\sigma\,\underline{S}_t \underline{p}^m - \alpha\sigma\,\underline{D}_i (\underline{A} + \sigma \underline{B})^{-1} \underline{g}_i \quad , \qquad (13a)$$

$$\underline{S}\,\underline{p}^{m+1} = \underline{S}\,\tilde{\underline{p}}^m - \alpha\,\underline{S}_t \tilde{\underline{p}}^m - \alpha\,\underline{D}_i (\underline{A} + \sigma \underline{B})^{-1} \underline{g}_i \quad , \qquad (13b)$$

where α is a relaxation parameter, and $\underline{E}$ is the pseudo-Laplacian $\underline{E} = \underline{D}_i \underline{B}^{-1} \underline{D}_i^T$. The scheme can formally be written in the form

$$\underline{p}^{m+1} = \underline{Q}\,\underline{p}^m + \underline{q}^m \quad , \qquad (14)$$

where

$$\underline{Q} = (\underline{I} - \alpha \underline{S}^{-1} \underline{S}_t)(\underline{I} - \alpha\sigma\,\underline{E}^{-1} \underline{S}_t) \quad , \qquad (15)$$

$\underline{I}$ is the identity matrix, and $\underline{q}^m$ is the inhomogeneity vector. The Q-iteration (13-15) can be viewed as the product of the unsteady (explicit) and steady Stokes operators, thus providing uniformly rapid convergence for all Δt. Using Fourier analysis it can be shown [4,7] that an optimal relaxation parameter $\alpha=1.1716$ gives an upper bound for the spectral radius of Q equal to $\rho^Q=0.1716$, *independent of the time step Δt and the spatial resolution.*

Incomplete Iteration. Although the two-level Q-iteration (13) provides rapid convergence of the pressure, the iterative procedure involves nested iteration in which systems in $\underline{E}$, $\underline{S}$, and $\underline{A}$ need to be solved. In order for any nested iterative solver to be efficient, it is crucial that the correct criteria be used to stop the various iterations. If all the iterations are continued until complete convergence, the scheme will become very inefficient; on the other hand, stopping at arbitrarily chosen residuals is not appropriate in high accuracy spectral calculations.

One way to set appropriate tolerances is to estimate the error in the velocity and the pressure due to incomplete iteration. This can be done by considering the solution of the perturbed momentum and continuity equations,

$$\sigma\underline{B}(\delta\underline{u}_i^{n+1}-\delta\underline{u}_i^n) + \underline{A}\delta\underline{u}_i^{n+1} - \underline{D}_i\delta\underline{p}^{n+1} = \underline{B}\delta\underline{f}_i^{n+1} , \quad (16a)$$

$$-\underline{D}_i\delta\underline{u}_i^{n+1} = \widetilde{\underline{B}}\delta\underline{g}^{n+1} , \quad (16b)$$

where $\delta\underline{u}$ and $\delta\underline{p}$ are the perturbed velocity and pressure due to incomplete iteration. We assume that for all times $t=n\Delta t$ the residuals $\delta\underline{f}$ and $\delta\underline{g}$ satisfy the tolerances

$$||\delta\underline{f}||_0 = \left(\sum_{j=1}^{d} \delta\underline{f}_j^T\underline{B}\ \delta\underline{f}_j\right)^{\frac{1}{2}} \leq \epsilon , \quad (17a)$$

$$||\delta\underline{g}||_0 = \left(\delta\underline{g}^T\widetilde{\underline{B}}\ \delta\underline{g}\right)^{\frac{1}{2}} \leq \mu , \quad (17b)$$

where $\underline{B}$ and $\widetilde{\underline{B}}$ are the diagonal mass matrices defined on the velocity mesh and pressure mesh, respectively. It can then be shown [7,8] that the discrete L^2 in time of H^1 in space velocity error due to incomplete iteration is bounded in terms of ϵ and μ as follows:

$$\overline{||\delta\underline{u}||}_1 = \left(\frac{1}{n\Delta t}\sum_{m=0}^{n}\Delta t\sum_{j=1}^{d}(\delta\underline{u}_j^m)^T(\underline{A}+\underline{B})\delta\underline{u}_j^m\right)^{\frac{1}{2}}$$

$$< \gamma^2\epsilon + \left[\frac{\gamma^3(1+\frac{2}{\Delta t})+\gamma}{\alpha_S^{1/2}}\right]\mu , \quad (18)$$

where $\gamma=(1+1/\sqrt{\alpha_A})$, and α_A and α_S are the minimum eigenvalues of the discrete Laplacian $\underline{A}$ and the steady pressure matrix $\underline{S}$, respectively. We emphasize that $\delta\underline{u}$ is the difference between two discrete solutions.

The effect of incomplete iteration on the convergence rate ρ^Q of the Q-iteration can also be estimated, and an optimal iteration strategy for (13) involving dynamic setting of the tolerances can be established [7,8]. It should be noted that the error estimates following from the incomplete iteration analysis [7,8] involve the minimum and maximum eigenvalues of the matrices $\underline{A}$, $\underline{E}$, $\underline{S}$ and $\underline{S}_t$; these eigenvalues must be computed or estimated before the time-stepping begins.

The implicit Stokes algorithm can be further extended to solve the unsteady Navier-Stokes equations by explicit treatment of the nonlinear term using a third-order Adams-Bashforth scheme [4]. The resulting Navier-Stokes algorithm can be viewed as an implicit Stokes scheme with an "augmented" force which includes the explicit convective contributions.

NUMERICAL RESULTS

To demonstrate the accuracy of the spectral element method we consider the following exact two-dimensional solution to the time-dependent, incompressible Navier-Stokes equations [9]:

$$\mathbf{u} = (u_1, u_2) = (-\cos x_1 \cdot \sin x_2 \cdot e^{-2t},\ \sin x_1 \cdot \cos x_2 \cdot e^{-2t}) \qquad (19a)$$

$$p = -\frac{1}{4} \cdot (\cos 2x_1 + \cos 2x_2) \cdot e^{-4t} \quad . \qquad (19b)$$

The problem is solved on a domain $\Omega =]-1,1[^2$, imposing time-dependent velocity boundary conditions according to (19a), and with initial conditions corresponding to the exact velocity (19a) at time t=0. The domain is broken up into K=4 similar spectral elements, each element being of order N. First, we simulate until a fixed time t=1.0 using different time steps Δt, while keeping N large enough such that the spatial errors are negligible compared to the temporal errors. As can be seen from Fig. 1, the scheme is clearly first-order in time. Next, we simulate until a fixed time t=.5 using different spatial resolutions (N=3,4 and 5), but now keeping the time step small enough in order to insure that temporal errors are negligible. The results shown in Fig. 2 indicate that the spectral element method gives exponential convergence as the order of the elements, N, is increased.

To show the effect of incomplete iteration we plot in Fig. 3 the discretization error $||\mathbf{u}-\mathbf{u}_h||_1$ obtained at (effectively) complete convergence ($\epsilon=2\cdot 10^{-6}$, $\mu=4\cdot 10^{-9}$), together with the incomplete iteration error $||\delta\underline{u}||_1(t)$ obtained for $\epsilon=2\cdot 10^{-1}$, $\mu=4\cdot 10^{-4}$ (the predicted upper bound of $||\delta\underline{u}||_1$ from (18) for these parameters is 1.0). The calculations are carried out to a final time t=1.0, using a time step Δt=0.01 and a spatial discretization K=4, N=6. From Fig. 3 and other numerical experiments we note that the error bound (18) is generally too conservative (this is particularly true for (19) due to the very smooth solution), however it does give the correct dependence when varying the spatial and temporal discretization parameters. Fig. 3 demonstrates how careful incomplete iteration can result in virtually no deterioration in the solution accuracy, yet yield significant increases in computational efficiency (e.g., factors of ten).

As an example of a three-dimensional deformed geometry solution we consider buoyancy-induced steady Stokes flow in the geometry shown in Fig. 4. A heat source is placed on one side of the domain that then drives the flow through Boussinesq coupling. The velocity vectors are shown in Fig. 5, in which the expected circulation pattern is clearly seen.

ACKNOWLEDGEMENT

The authors would like to thank Yvon Maday of Université Pierre et Marie Curie for his many contributions to this work. This work was supported by ONR and DARPA under Contract N00014-85-K-0208, and by ONR under Contract N00014-87-K-0439.

REFERENCES

1. Bernardi, C. and Maday, Y., "A collocation method over staggered grids for the Stokes problem", submitted to Int. J. Num. Meth. in Fluids.
2. Maday, Y., Patera, A.T. and Rønquist, E.M.,"Optimal Legendre spectral element methods for the Stokes semi-periodic problem", submitted to SIAM J. Numer. Anal.
3. Maday, Y., Patera, A.T. and Rønquist, E.M.,"Optimal Legendre spectral element methods for the multi-dimensional Stokes problem", in preparation.
4. Maday, Y. and Patera, A.T., "Spectral element methods for the incompressible Navier-Stokes equations", in State-of-the-Art Surveys in Computational Mechanics (ed: A.K. Noor), ASME, 1987.
5. Girault, V. and Raviart, P.A., Finite element approximation of the Navier-Stokes equations, Springer-Verlag, 1986.
6. Bristeau, M.O., Glowinski, R. and Periaux, J., "Numerical methods for the Navier-Stokes equations. Applications to the simulation of compressible and incompressible viscous flows", Computer Physics Report, to appear, 1987.
7. Maday, Y., Meiron, D.I., Patera, A.T. and Rønquist, E.M., "Saddle-decomposition methods for spectral discretization of the steady and unsteady Stokes problem", submitted to J. of Comp. Phys.
8. Rønquist, E.M., "Legendre spectral element methods for the incompressible Navier-Stokes equations", Ph.D. Thesis, M.I.T., in progress.
9. Kim, J. and Moin, P., "Application of a fractional-step method to incompressible Navier-Stokes equations", J. Comp. Phys., **59**, p. 308, 1985.

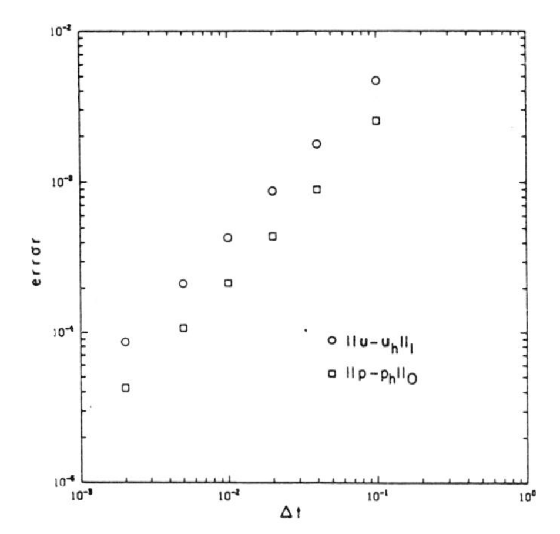

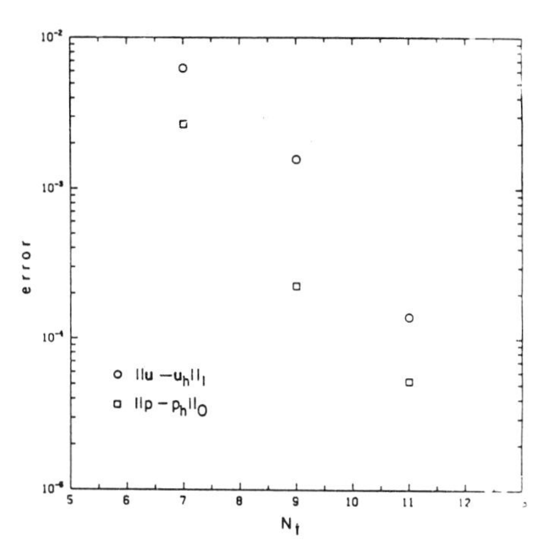

Figure 1. A plot of the discretization errors $\| u-u_h \|_1$ (○) and $\| p-p_h \|_0$ (□) as a function of the time step Δt when solving the two-dimensional, unsteady, incompressible Navier-Stokes equations with exact solution given by (19). Here the spatial errors are negligible compared to the temporal errors. The results indicate that the scheme is first-order accurate in time.

Figure 2. A plot of the discretization errors $\| u-u_h \|_1$ (○) and $\| p-p_h \|_0$ (□) as a function of the total number of degrees-of-freedom, N_t, in one spatial direction when solving the two-dimensional, unsteady, incompressible Navier-Stokes equations with exact solution given by (19). Here the temporal errors are negligible compared to the spatial errors. The results indicate that the scheme gives exponential convergence as the order of the elements, N, is increased.

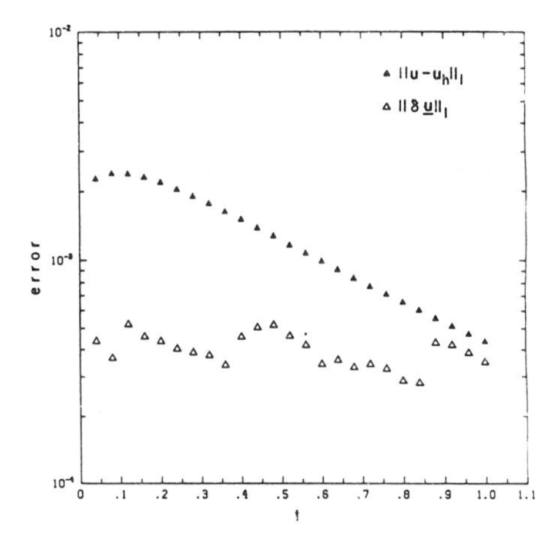

Figure 3. A plot of the discretization error $\| u-u_h \|_1$ (▲) obtained by iterating until complete convergence, together with the H^1 velocity error $\| \delta \underline{u} \|_1 (t)$ due to incomplete iteration (△), as a function of time t.

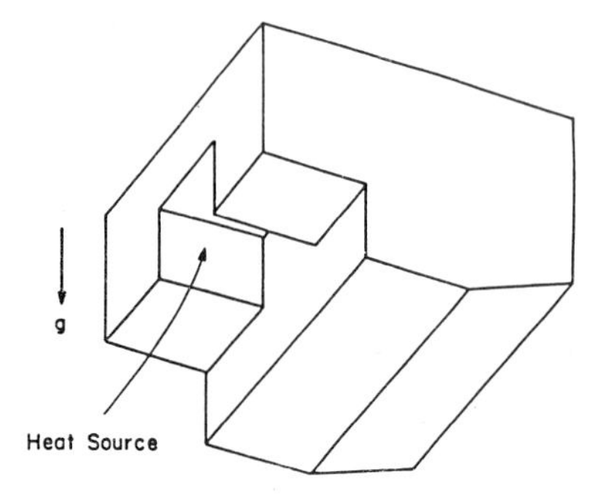

Figure 4. A plot of the three-dimensional non-rectilinear furnace geometry. The steady Stokes problem is forced by the Boussinesq term through the temperature field arising from a localized heat source.

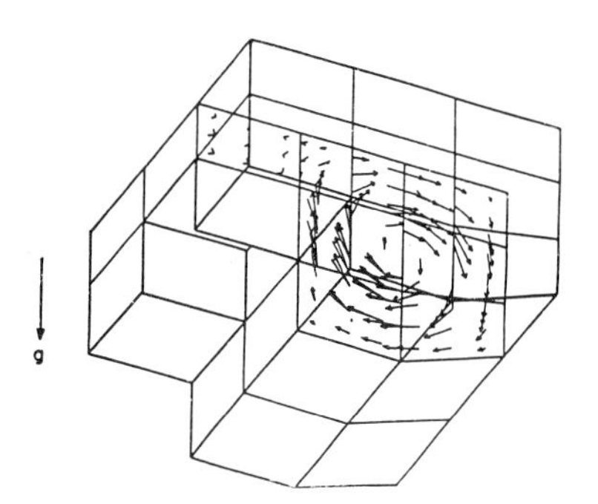

Figure 5. Velocity vectors for the steady Stokes problem shown in Figure 4. The spectral element mesh is indicated by the lighter lines on the domain surface.

Comparison of Cell Centered and Cell Vertex Finite Volume Schemes

C. Rossow

DFVLR, Institute of Design Aerodynamics
Braunschweig, F.R. Germany

Summary

Three finite volume discretization methods for the unsteady Euler equations are investigated. One method is a cell centered scheme while the other two are cell vertex schemes. An error analysis using Taylor series expansion is applied to estimate the theoretical order of accuracy of these schemes. Numerical results on different meshes confirm the theoretical statement that only the cell vertex schemes perform accurately on grids with slope-discontinuities. On smooth meshes the cell centered scheme and one of the cell vertex schemes deliver nearly identical results and exhibit an accuracy of almost second order.

Introduction

In the numerical analysis of fluid flow the finite volume technique for discretizing the governing equations is well known. There exists a variety of schemes implementing this technique. For the spatial discretization, however, generally two basic strategies are applied: a cell centered scheme in which the flow variables are associated with the center of a cell in the computational mesh, and a cell vertex scheme where the flow quantities are ascribed to the corners of a cell. In the present study we limit our investigations to methods which discretize the unsteady Euler equations in a cartesian coordinate frame. In order to establish a fair comparison for all schemes under investigation, the modeling of artificial dissipation, the time stepping scheme, and the acceleration techniques are identical and calculations are performed on the same meshes. The present study is split into two major parts. In the first part the different discretization methods are introduced and their accuracy is explored by using a Taylor series expansion. The second part of the study contains a brief description of the numerical method used, and a discussion of the results gained from the application of the different discretization techniques.

Governing equations

The two-dimensional Euler equations for unsteady compressible inviscid flows may be written in integral form using a cartesian coordinate system as:

$$\iint_V \frac{\partial}{\partial t}\vec{W}\, dV = -\int_{\partial V} \vec{F}^1 \cdot n_x dS - \int_{\partial V} \vec{F}^2 \cdot n_y dS \qquad (1)$$

where

$$\vec{W} = \begin{bmatrix} \rho \\ \rho u \\ \rho v \\ \rho E \end{bmatrix}, \quad \vec{F}^1 = \begin{bmatrix} \rho u \\ \rho u^2 + p \\ \rho uv \\ \rho Hu \end{bmatrix}, \quad \vec{F}^2 = \begin{bmatrix} \rho v \\ \rho uv \\ \rho v^2 + p \\ \rho Hv \end{bmatrix}, \quad \vec{n} = \begin{bmatrix} n_x \\ n_y \end{bmatrix}.$$

In equation (1) p, ρ, u, v, E and H are the pressure, density, cartesian velocity components, total energy and total enthalpy. V denotes an arbitrary control volume fixed in time and space and ∂V is the closed boundary of the volume. $\vec{F}^1$ and $\vec{F}^2$ are the vectors of flux density and $\vec{n}$ is the outward normal along ∂V. Applying the integral mean value theorem equation (1) can be converted to

$$\left(\frac{\partial}{\partial t}\vec{W}\right) = - \frac{\int_{\partial V} \vec{F}^1 \cdot n_x dS}{\iint_V dV} - \frac{\int_{\partial V} \vec{F}^2 \cdot n_y dS}{\iint_V dV} . \qquad (2)$$

The term on the left hand side of equation (2) represents the integral mean value of the rate of change of $\vec{W}$ in the control volume V, and the right hand side is the flux per volume of mass, momentum and energy through the surface of V. Together with the equation of state

$$p = \rho R T \qquad (3)$$

which relates the pressure to the components of $\vec{W}$, equation (2) forms a system of four partial differential equations of first order for the four unknows ρ, u, v and E.

Finite Volume discretization

In the following three methods for the spatial discretization will be investigated:

- a cell centered scheme according to Jameson [1]
- a cell vertex scheme according to Ni [2]
- a cell vertex scheme according to Hall [3].

These schemes will be denoted schemes A, B and C, respectively. In order to apply the discretization the physical domain is divided into quadrilateral cells by generating a body fitted grid. The integral mean value of the rate of change of $\vec{W}$ in each grid cell is estimated by the approximation of equation (2). As this equation is valid for any arbitrary control volume, it holds also locally for each cell $V_{i,j}$ in the mesh. Hence

$$\left(\frac{\partial}{\partial t}\vec{W}\right)_{i,j} = - \frac{1}{V_{i,j}} \cdot \left[\int_{\partial V_{i,j}} \vec{F}^1 \cdot n_x dS + \int_{\partial V_{i,j}} \vec{F}^2 \cdot n_y dS \right] = \vec{Q}_{i,j} . \qquad (4)$$

To evaluate the flux $Q_{i,j}$ numerically, the surface integrals are split into a sum of integrals over the four faces of a cell and these integrals are approximated by the respective finite volume technique.

First the cell centered scheme is considered. The flow-quantities are located at the cell centers (Fig. 1). The mean values of the flux vectors on a face are assumed to be functions of the flow quantities at the midpoints of that faces. The flow quantities there are approximated by the average of the quantities of those cells having the considered face in common. On this way equation (2) can be approximated for each cell. The further approximation that the integral mean value of the change is similar to the change of the flow variables located at the cell centers

$$\frac{\partial}{\partial t}\vec{W}\bigg|_{i,j} \approx \vec{Q}_{i,j} \qquad (5)$$

leads to a system of ordinary differential equations for $\vec{W}$ in all cells of the mesh.

Next the cell vertex schemes will be outlined. In Fig. 2 four control volumes are sketched, which have point A as a common vertex. The flow quantities are located at the vertices of the control volumes. The flow quantities at the midpoints of the faces are obtained by linear interpolation of the quantities at the endpoints, and the rate of change in each control volume can be approximated. In order to relate these cell averaged quantities to the rate of change at a vertex a further approximation is made by the use of a distribution formula. The application of this formula assumes the change at a vertex to be a function of the changes in all cells having the particular vertex in common. The special formulation of the distribution formula causes the difference between schemes B and C:

Scheme B uses an arithmetic average of the changes in the neighboring cells. For point A in Fig. 2 for example the distribution formula reads:

$$\frac{\partial}{\partial t}\vec{W}\Big|_A \approx \frac{1}{4}\left\{ \vec{Q}_{i,j} + \vec{Q}_{i-1,j} + \vec{Q}_{i,j-1} + \vec{Q}_{i-1,j-1} \right\}. \tag{6}$$

In Scheme C the changes in the cells surrounding a vertex are weighted by the volumes of the corresponding cells, and this average is taken to be the change at the common vertex:

$$\frac{\partial}{\partial t}\vec{W}\Big|_A \approx \frac{\vec{Q}_{i,j}V_{i,j} + \vec{Q}_{i-1,j}V_{i-1,j} + \vec{Q}_{i,j-1}V_{i,j-1} + \vec{Q}_{i-1,j-1}V_{i-1,j-1}}{V_{i,j} + V_{i-1,j} + V_{i,j-1} + V_{i-1,j-1}}. \tag{7}$$

The application of the distribution formula leads to a system of ordinary differential equations for the flow quantities at all nodal points of the grid.

Error analysis

The integral form of the equations as given in (2) requires no assumption about the differentiability of the flow quantities. In the following error analysis the differential form of the equations will be considered:

$$\frac{\partial \vec{W}}{\partial t} = -\frac{\partial \vec{F}^1}{\partial x} - \frac{\partial \vec{F}^2}{\partial y}. \tag{8}$$

The integral form (2) can be converted to (8) by the use of Gauss' theorem, provided the derivatives of $\vec{F}^1$ and $\vec{F}^2$ exist. Note that the rate of change is no longer an integral mean value but the rate of change of $\vec{W}$ at a certain point. In order to simplify the error analysis the numerical approximation of the first derivative of an arbitrary function ϕ will be investigated. If ϕ is differentiable the first derivative of ϕ in the x-direction is given by Gauss' theorem as:

$$\frac{\partial \Phi}{\partial x} = \Phi_x = \lim_{V\to 0} \frac{\int_{\partial V} \Phi n_x dS}{\iint_V dxdy}. \tag{9}$$

This derivative can be approximated for a finite volume:

$$\Phi_x \approx \frac{\int_{\partial V} \Phi n_x dS}{\iint_V dxdy}. \tag{10}$$

In the present analysis the finite volume schemes outlined before will be used to approximate the surface integrals. This leads to finite difference formulae for ϕ_x at a particular point in the considered mesh. Provided the derivatives of the function ϕ and the function itself are known at the chosen point, the expansion point, the values at neighboring points can be calculated by a Taylor series expansion. Using now the finite volume schemes to approximate ϕ_x in the expansion point allows to compare the expressions from the approximation with the exact value of ϕ_x at this point, and the accuracy of the approximation can be analysed. This analysis will be performed on two types of grids. The first grid is stretched in x-direction whereas the spacing in the y-direction is constant. The second grid exhibits a slope-discontinuity given by an angle β. The spacing in x- and y-direction is constant and all volumes have the same size in this grid.

Stretched grid. First the cell centered scheme is considered. ϕ_x has to be approximated at the center of the cell $V_{i,j}$ (Fig. 3). Scheme A provides for the numerical approximation of ϕ_x at the center of $V_{i,j}$:

$$\begin{aligned} \Phi_x\big|_{num} &= \Phi_x\big|_{i,j} + \Phi_x\big|_{i,j}\frac{\Delta x_+ + \Delta x_- - 2\Delta x_o}{2\Delta x_o} \\ &+ \frac{1}{2}\Phi_{xx}\big|_{i,j}\frac{\Delta x_+^2 - \Delta x_-^2}{2\Delta x_o} + O(\Delta x^2)\,. \end{aligned} \tag{11}$$

The subscripts i,j denote the exact values of the derivatives at the cell center. Equation (11) shows the deviation of the approximation from the exact value of ϕ_x by the appearance of certain error terms. The first error term is independent of the size of Δx and therefore of zeroth order, it will not drop by a mesh refinement which keeps the geometrical relations between Δx_o, Δx_- and Δx_+ constant. The following error terms depend on Δx or higher orders of Δx and will therefore be reduced by reducing the size of Δx. Note that the numerators of the error terms given in (11) are differences of Δx. On a cartesian grid the differences in the numerators are zero causing the zeroth and first order errors to vanish and the scheme will be second order accurate. If the differences in the numerators are not zero but drop when the mesh is refined, the scheme can behave as being second order accurate. This is usually the case on smooth grids, as already shown by Hoffmann [4].

Considering the cell vertex schemes the first derivative will be approximated at the point A shown in Fig. 4. Application of scheme B leads to

$$\begin{aligned} \Phi_x\big|_{num} &= \Phi_x\big|_A \\ &+ \frac{1}{4}\Phi_{xx}\big|_A(\Delta x_+ - \Delta x_-) + O(\Delta x^2)\,. \end{aligned} \tag{12}$$

Using scheme C the approximation becomes

$$\begin{aligned} \Phi_x\big|_{num} &= \Phi_x\big|_A \\ &+ \frac{1}{2}\Phi_{xx}\big|_A(\Delta x_+ - \Delta x_-) + O(\Delta x^2)\,. \end{aligned} \tag{13}$$

No zeroth order errors appear in both schemes. The first order errors are governed by the differences of the cell faces in the x-direction. On a cartesian grid these terms vanish making the schemes second order accurate. Provided the grid is smooth enough the schemes will behave as being second order accurate. Note that the deviation from the exact value of ϕ_x in scheme B is only as half as great as in scheme C. Therefore scheme B is formally more accurate than scheme C.

Skewed grid. The situation for scheme A is depicted in Fig. 5. Scheme A provides for the approximation of ϕ_x at the center of $V_{i,j}$

$$\Phi_x|_{num} = \Phi_x|_{i,j} + \Phi_y|_{i,j}\tan\varphi + \frac{1}{2}\Phi_{xy}|_A\Delta x\tan\varphi + \frac{1}{2}\Phi_{yy}|_A\Delta x\tan^2\varphi + O(\Delta x^2). \quad (14)$$

The zeroth order error depends on the angle φ of the slope-discontinuity. If the angle is not reduced the error remains the same even when the mesh is refined. On smooth meshes the angle φ decreases by mesh refinement which causes the error terms to decrease.
Using the cell vertex schemes to estimate the first derivative at point A (Fig. 6), both schemes lead to the same expressions because all volumes have the same size in the whole mesh and the distribution formulae coincide:

$$\Phi_x|_{num} = \Phi_x|_A + \frac{1}{4}\Phi_{xy}|_A\Delta x\tan\varphi + O(\Delta x^2), \quad (15)$$

In contrast to the cell centered scheme the cell vertex schemes are always at least first order accurate so that even on a mesh with slope-discontinuities the numerical error will be reduced with mesh refinement.

Numerical method

All schemes outlined use central differences and therefore require an amount of artificial viscosity, and here a blend of fourth and second differences as in the work of Jameson [1] is added. The integration of the system of ordinary differential equations is carried out by the use of a five stage Runge-Kutta scheme. Details of the scheme can be found in [5]. In order to drive the solution towards the steady state acceleration techniques such as local time stepping, implicit averaging of residuals, and enthalpy damping are applied. For solid wall boundaries in the cell centered scheme the normal momentum relation is implemented [5]. The cell vertex schemes use a flow tangency condition similar to [3]. The treatment of the farfield boundaries is based on the concept of Riemann invariants. For lifting cases the effect of a single vortex is added to the free stream flow.

Numerical results

The first problem considered is the internal flow over a circular arc of 10% thickness. The onflow Mach-number was chosen to be 0.5 so that the flow field remains subsonic. The grid exhibits a slope-discontinuity caused by the attachment of the circular arc to the wall (Fig. 7). The isobar patterns delivered by Scheme A exhibit disturbances caused by the discontinuity in the mesh. The results of schemes B and C are indistinguishable on this grid due to the fact that the volumes of neighboring cells have almost the same size and therefore the distribution formulae coincide nearly. Here the isobars have smooth slopes and the flow field is symmetric with respect to the inflow and outflow boundaries, as it should be for inviscid subsonic flow. Fig. 8 shows the results for the same problem after a mesh refinement. The disturbances in the cell centered scheme do not vanish, due to the zeroth order errors in the scheme.
The second problem investigated consists of the transonic flow past the NACA 0012 aerofoil. The O-grid was generated by conformal mapping using

160 by 32 cells and all grid lines possess smooth slopes. Fig. 9 shows the results provided by the application of schemes A and C. Both schemes deliver almost the same results, as can be seen from the contour pressure distribution and the lift and drag coefficients. In order to check the order of accuracy a mesh refinement to 320 by 64 cells was performed. The results are given by dashed lines in the same figure. The total pressure losses are almost reduced by a factor of four in both schemes, indicating that the schemes behave as second order accurate schemes.
Fig. 10 shows results gained by the application of scheme B to the same problem. Using this scheme with the same amount of artificial viscosity as in the other schemes the solution did not converge and exhibits oscillations. Only increasing the fourth differences artificial viscosity by a factor of 8 expelled the oscillations at the trailing edge. This behaviour of scheme B when applied on O-grids was already observed by Usab [6]. It seems that in schemes A and C the necessary amount of numerical dissipation is delivered by the higher truncation errors of these schemes on such meshes.

Conclusions

Three finite volume schemes have been investigated. In all cases the same solution method and dissipative terms were employed to achieve a fair comparison. Only one discretization method, namely the cell vertex scheme based on the ideas of Hall, performed well on both smooth and non-smooth grids. The cell centered scheme is not able to provide accurate results on non-smooth grids whereas the cell vertex scheme following Ni needs a large amount of artificial viscosity to ensure a converged solution without shortwave oscillations. The Taylor series analysis for the investigation of model problems proved to be a powerful tool to estimate the accuracy of the discretization schemes theoretically. The results of the theoretical analysis are confirmed by the numerical results.

References

[1] Jameson, A.; Schmidt, W.,; Turkel, E.: Numerical Solutions of the Euler Equations by Finite Volume Methods Using Runge-Kutta Time Stepping Schemes, AIAA-Paper 81-1259 (1981).

[2] Ni, R.: Multiple Grid Scheme for Solving the Euler Equations, AIAA-Paper 81-1025 (1981).

[3] Hall, M.G.: Cell-Vertex Multigrid Scheme for Solution of the Euler Equations, Proceedings of the Conference on Numerical Methods for Fluid Dynamics, Reading, UK (1985).

[4] Hoffmann, J.D.: Relationship Between the Truncation Errors of Centered Finite-Difference Approximations on Uniform and Nonuniform Meshes, Journ. of Comp. Phys. 46, pp 464-474 (1982).

[5] Kroll, N.; Jain, R.K.: Solution of Two-Dimensional Euler Equations - Experience with a Finite Volume Code, DFVLR-interner Bericht 84-19 (1984).

[6] Usab, W.J.: Embedded Mesh Solutions of the Euler Equations Using a Multiple-Grid Method, Ph. D. Thesis, MIT, Cambridge, Massachusetts (1983).

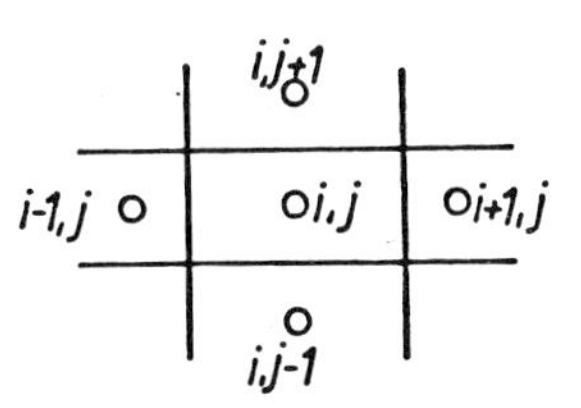

Fig.1: Finite Volume of cell centered scheme

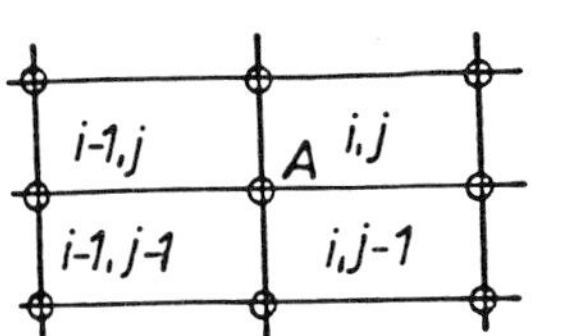

Fig.2: Finite Volume of cell vertex scheme

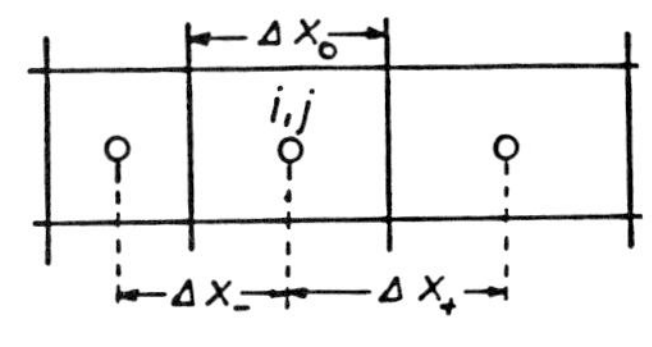

Fig.3: Stretched grid of cell centered schemes

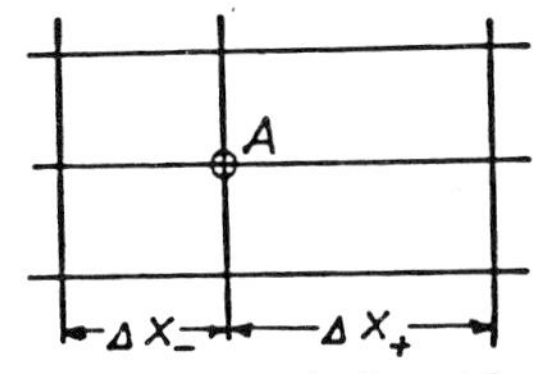

Fig.4: Stretched grid of cell vertex schemes

Fig.5: Skewed grid of cell centered scheme

Fig.6: Skewed grid of cell vertex schemes

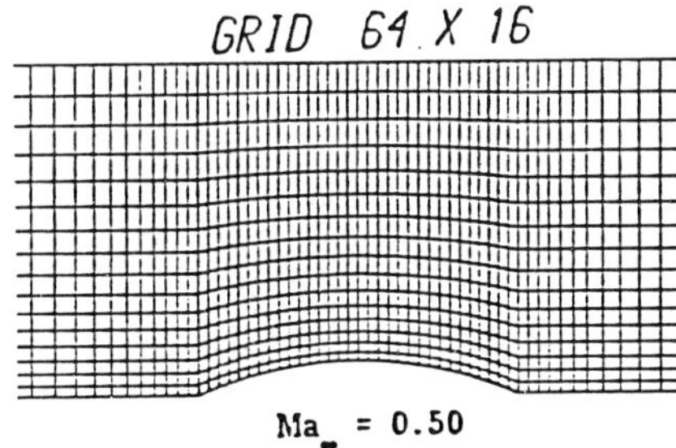

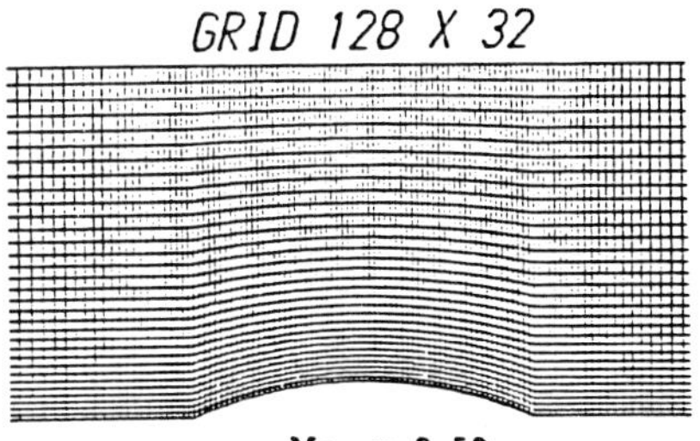

$Ma_\infty = 0.50$

CELL CENTERED SCHEME

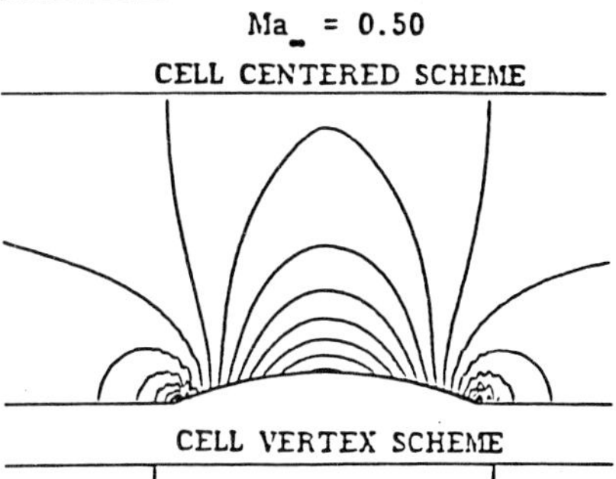

CELL VERTEX SCHEME

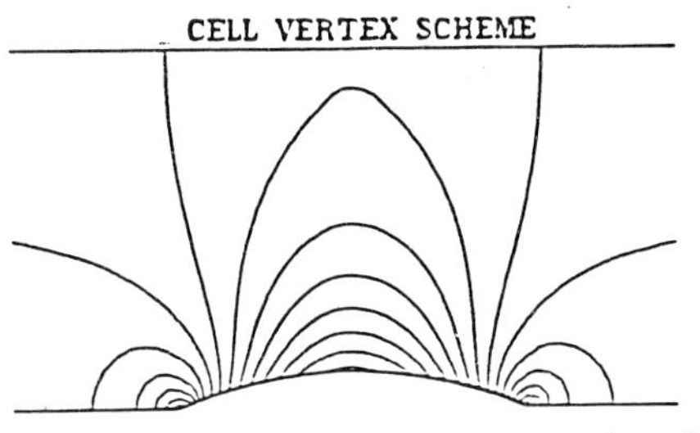

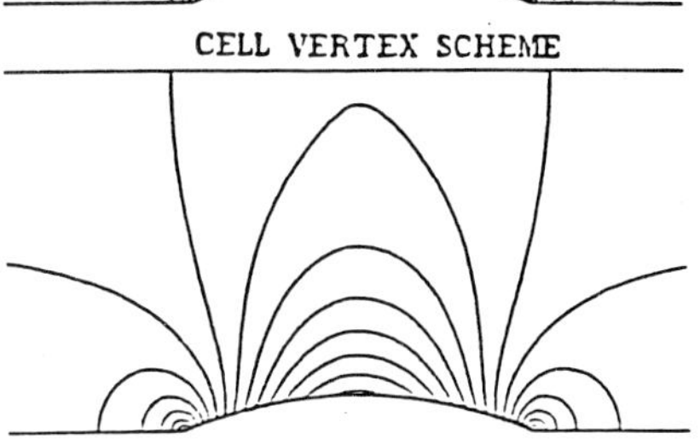

Fig. 7: Internal flow over circular arc

Fig. 8: Effect of grid refinement

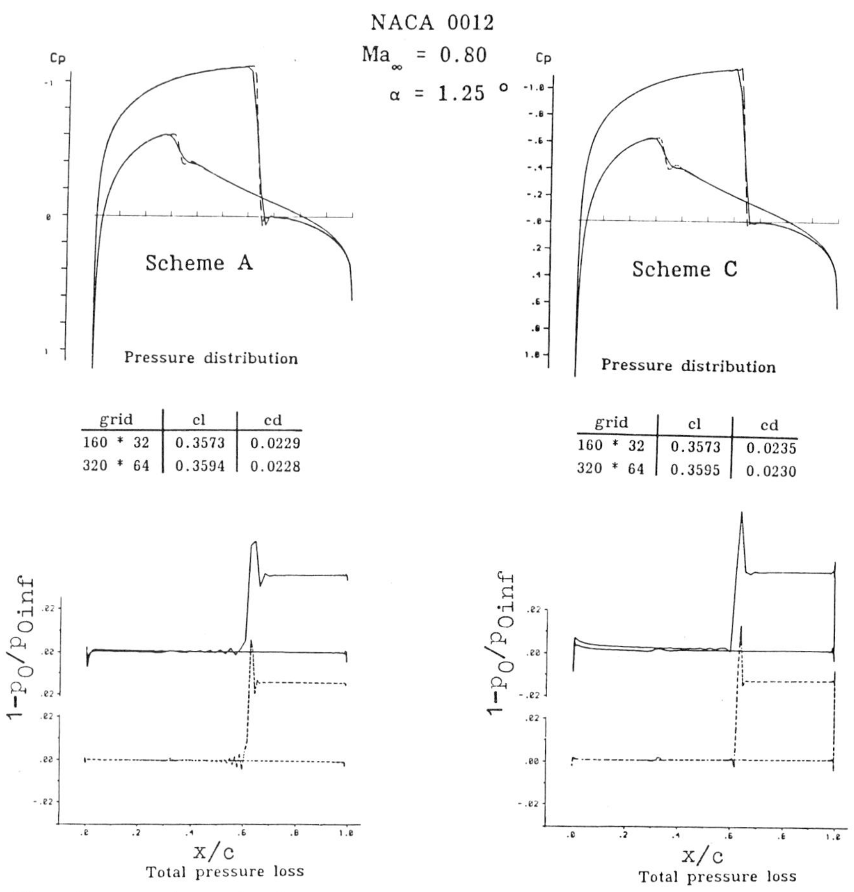

grid	cl	cd
160 * 32	0.3573	0.0229
320 * 64	0.3594	0.0228

grid	cl	cd
160 * 32	0.3573	0.0235
320 * 64	0.3595	0.0230

Fig. 9: Results of schemes A and C on smooth grids for external flow problem

Scheme B

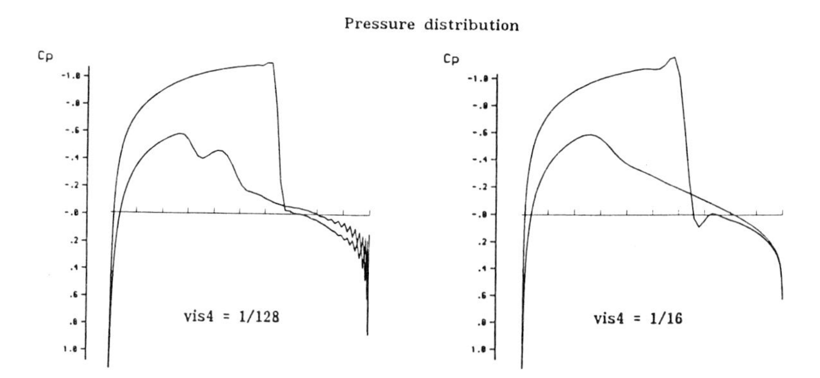

Fig. 10: Results of scheme B with different coefficients of fourth differences artificial viscosity

PANEL CODE SOLVERS

J. Ryan, T.H. Lê, Y. Morchoisne
Office National d'Etudes et de Recherches Aérospatiales
BP 72 - 92322 CHATILLON Cedex, France

SUMMARY

Prediction of 3D incompressible, inviscid flow using panel methods require solving large, full matrices, the cost of which can be very important. A brief review of previously used solvers is presented and a new, highly efficient generalized gradient method is developed and applied to complex configurations.

INTRODUCTION

Panel codes providing pressure, velocity distribution and aerodynamic coefficients over the whole field are increasingly becoming more reliable and precise. Trends of application of these codes are to insert them in iterative procedures such as inviscid-viscid coupling, using boundary layer computations or shape optimisation within artificial intelligence networks. In these cases, problems of cost (time and memory space) are crucial.

All panel codes have in common linear systems with full matrices that have to be solved. When dealing with complex configurations, these systems become very large, and the cost of solving them becomes predominant.

In this paper, we shall first present comparisons between various solvers [1] (direct or iterative) for matrices resulting from discretisation of several aerodynamic configurations (sphere, wing, fuselage, aircraft...).

We then develop a new iterative resolution which generalizes gradient methods, and is shown to be as precise and at least twice as fast as the above mentioned methods (in general, only 2 to 10 iterations are needed for a number of panels going up to 1000).

A SHORT REVIEW OF SOLVERS

Let the linear equation to be solved in matrix form be

$$(*)\quad A x = b$$

where $A = (a_{ij})_{i,j=1,N}$ is a NxN regular matrix, $x = (x_i)_{i=1,N}$ the unknown vector, and b a given vector $b = (bi)_{i=1,N}$.

. The most common solver, a direct method, is the L.U. technique, which consists in decomposing A into the product of a lower triangular (L) and upper triangular matrix (U). Solution of (*) is then found by:

1. solving system $Ly = b$
2. solving $Ux = y$.

Direct methods require around N^3 operations ($\frac{2}{3} N^3$ for L.U.) for a general regular matrix with no particular property (i.e.: symmetric,...).

. Most iterative schemes have the form:

$$x_{i+1} = x_i + B_i r_i$$

where B_i is a matrix function of A and the iteration step i, and $r_i = b - Ax_i$, the residual vector.

In gradient methods, applied to ordinary regular matrices, B is a scalar and determined so as to minimize the residual functional (r_{n+1}, r_{n+1}), ((.,.)) is the inner product in $\mathbb{R}^N$.

Number of operations required for iterative schemes per iteration is about $2 N^2$ (number of operations for the matrix vector product).

When dealing with very large full matrices, three points appear:

a) For an interative method to be as expensive if not more, the total number of iterations must be greater than: $(2/3\ N^3)/(2\ N^2) = N/3$.

b) In direct methods, the solving error is produced by N^3 round-off errors that propagate and increase at each computation step. These are equivalent to an initial error in the elements of A. When N is large, the error can become quite important. In iterative methods, round-off errors are only in N^2, and the solving error is mainly due to truncation error which is a given data.

c) When matrix coefficients of A are cheap to compute (quadrature methods), calculation of these coefficients whenever they are needed can be implemented in iterative schemes instead of storing them in computer memory [2]. Such a disassembled technique is impossible for direct solvers.

The last two points always favour the iterative methods. Concerning the first point, the iteration number is highly dependent on the matrix condition number, though as will be seen below, can be brought down to as low a figure as 2% of panel number.

From this brief review, the following remarks can be made:
- For sparse matrices where coefficient storage is of the same order as that of a vector, two factors have to be taken into account in the choice of a solver: added storage of vectors, which could be a deterrent in front of the size of the original matrix, and convergence speed of the iterative schemes.
Specific properties of the matrix will have a great influence on the choice of the solver.
- For full matrices, extra storage (even up to 10% of matrix space) is of little importance: i.e., for matrix of size N = 10,000, a thousand vector storage is still negligible in terms of memory space and convergence speed will become the main criterion.

MULTIGRADIENT

From the second consideration came the idea of developing an iterative scheme based on storing all former residue vectors and use them to minimize the residual functional (r_{n+1}, r_{n+1}).
Thus, if $r_1, \ldots, r_n$ are the n former residue vectors, x_n, the solution at the nth iteration, we write:

$$x_{n+1} = x_n + \sum_{i=1}^{N} \alpha_{n+1}^i r_i$$

$$(r_{n+1}, r_{n+1}) = (r_n, r_n) - 2 \sum_{i=1}^{N} \alpha^i_{n+1} (r_n, Ar_i)$$

$$+ \sum_{i=j}^{N} \sum_{j=1}^{N} \alpha^i_{n+1} \alpha^j_{n+1} (Ar_i, Ar_j) .$$

In order to minimize this functional $(\alpha^i_{n+1})_{i=1,N}$ must be solution of the following equations:

$$(i=1,N)\left\{ \frac{\partial(r_{n+1}, r_{n+1})}{\partial \alpha_i} = 0 = -2(r_n, Ar_i) + 2\,\alpha^j_{n+1} \sum_{j=1}^{N} (Ar_i, Ar_j) \right\}.$$

Remarks:

- Variation of this iterative scheme consists in just keeping the last q residue vector.
- If $q = 2$, the multigradient is strictly equivalent to the conjugate residue scheme (they both minimize (r_{n+1}, r_{n+1}) in the search plane defined by r_n and Ar_n).

The algorithm of the multigradient is thus:

- Initialization:

x_1, any vector in $\mathbb{R}^N$

$$r_1 = b - Ax_1$$

$$\sigma_1 = (r_1, Ar_1)$$

$$a(1,1) = (Ar_1, Ar_1)$$

- Iterations:

$$\{j = 1,N \quad \sigma_j = (Ar_i, r_n)\}$$

$$\{j = 1,N \quad a(j,n) = (Ar_j, Ar_n)\}$$

Solve $(a\,\alpha = \sigma)$ by Gauss method.

$$x_{n+1} = x_n + \sum_{j=1}^{N} \alpha^j r_j$$

$$r_{n+1} = r_n + \sum_{j=1}^{N} \alpha^j Ar_j .$$

Comparative methods

In order to evaluate the performances of the present method (M.GRAD), we compared it to the L.U. direct method (L.U.) and to two iterative methods (ST.D.) and (C.G.S.).

- Details about L.U. algorithm are given for example in [4].
- ST.D. is the steepest descent technique with the following algorithm [1].

- Initialization:

x_1 any value in $\mathbb{R}^N$

$$r_1 = b - Ax_1$$

- Iterations:

$$\alpha_n = \frac{(r_n, Ar_N)}{(Ar_n, Ar_n)}$$

$$x_{n+1} = x_n + \alpha_n r_n$$

$$r_{n+1} = r_n - \alpha_n Ar_n$$

. C.G.S. (Conjugate Gradient Squared) is an accelerated version of a double conjugate gradient [3].

- Initialization:

x_1 any value in $\mathbb{R}^N$

$$r_1 = b - Ax_1$$

$$g_1 = r_1$$

$$u_1 = r_1$$

- Iterations:

$$\alpha_n = \frac{(r_1, r_n)}{(r_1, Ag_n)}$$

$$h_n = u_n - \alpha_n Ar_n$$

$$x_{n+1} = x_n + \alpha_n(u_n + k_n)$$

$$r_{n+1} = r_n - \alpha_n A(u_n + k_n)$$

$$\beta_n = \frac{(r_1, r_{n+1})}{(r_1, r_n)}$$

$$u_{n+1} = \beta_n h_n + r_{n+1}$$

$$g_{n+1} = u_{n+1} + \beta_n(\beta_n g_n + h_n).$$

RESULTS

Calculations are first performed on test configurations: the sphere and a helicopter fuselage, with 3 sizes of mesh (fig. 1). They are compared on tables below:

- direct L.U. decomposition techniques (L.U.),
- steepest descent (ST.D.),
- conjugate gradient squared (C.G.S.).

Sphere

Method / Grid	L.U.	ST.D.	C.G.S.	M.GRAD
250	0.065 s	0.014 s 4 it	0.014 s 2 it	0.008 s 2 it
500	0.460 s	0.054 s 4 it	0.032 s 1 it	0.032 s 2 it
1000	3.446 s	0.202 s 4 it	0.120 s 1 it	0.119 s 2 it

it = iteration time in seconds

Helicopter fuselage

Method / Grid	L.U.	ST.D.	C.G.S.	M.GRAD
250	0.065 s	0.030 s 9 it	0.020 s 2 it	0.015 s 4 it
500	0.470 s	0.123 s 10 it	0.075 s 3 it	0.067 s 5 it
1000	3.533 s	0.440 s 10 it	0.280 s 3 it	0.242 s 5 it

The realistic complex configuration of a three body lift augmenting device was also used for comparison (fig. 2).

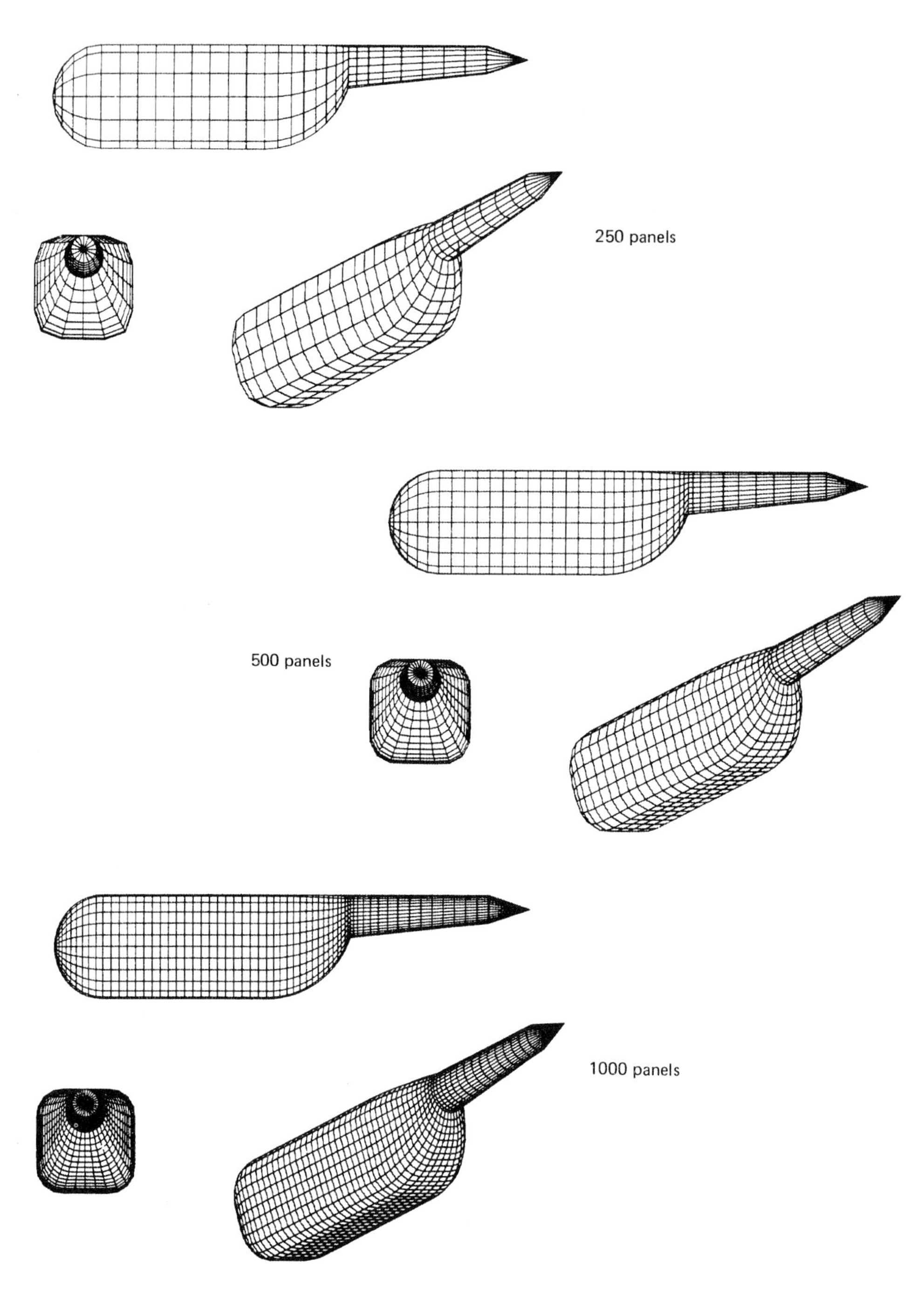

Figure 1

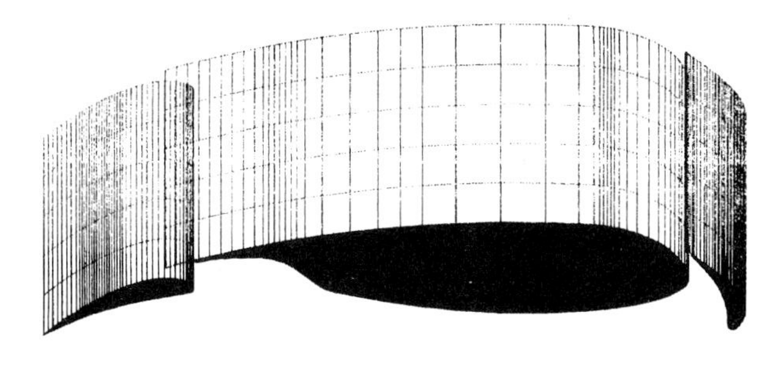

Figure 2

It is meshed with 1837 panels: 682 on the first element, 726 on the main element and 429 on the flap.

Method / Grid	L.U.	ST.D.	C.G.S.	M.GRAD
1837 panels	22.557 s	7.547 s 78 it	3.916 s 20 it	3.436 s 35 it

All computations are done on a CRAY MP with 7 Mega words of central core memory.

In all cases, iterative schemes are cheaper than the L.U. direct method.

In general, C.G.S. and M.GRAD are 50% cheaper than the ST.D. For a 1000 panels, these iterative schemes have reduced the L.U. cost by a ratio of at least 10. For a 10000 panel mesh, ratio will be of a 100.

Another, slightly more complex geometry was then also tested. This consisted of a set of two disconnected bodies: a thick disk and a sphere, both half meshed with 125 panels

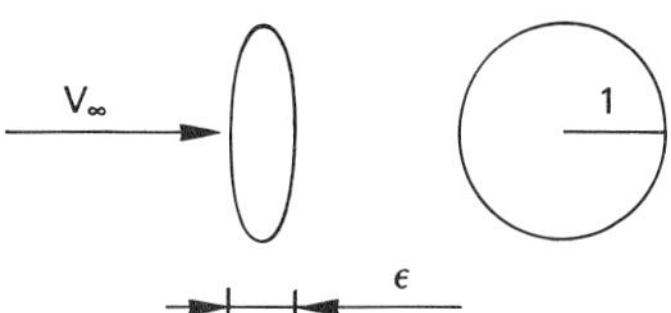

This particular case is representative of configurations where thin and thick elements are combined, such as fuselage + wing + spoilers + flaps.

Two values of the thickness ε of the disk were chosen: as when ε decreases, the matrix has a rising condition number, this allows one to test the stability of the various schemes.

Method ε panels	L.U.	ST.D.	C.G.S.	M.GRAD
10^{-1}	0.065 s	0.055 s 15 it	0.045 s 6 it	0.022 s 5 it
10^{-2}	0.065 s	0.132 s 40 it	0.090 s 12 it	0.045 s 10 it

500 panels

N.B.: A random initialization was also used, and both ST.D. and C.G.S. diverged, while the multigradient still reached the correct solution.

In this more complex case, the multigradient shows itself to be a more stable method, less reliant on initialisation and 50% cheaper than the C.G.S, one of the fastest methods developed up to now.

COMMENTS AND CONCLUSION

The multigradient method is especially attractive for full matrices. For i iterations, the i vectors r_i and the i vectors Ar_i have to be stored. For a large matrix of size N, storing 2 i N is a very small increase of memory space in front of the N x N terms of the matrix.

If one allows 10% of the full matrix space storage for the solving process, if N is around 10,000, the number of iterations i could go up to $N/2\times 10 = 500$. To be cheaper than a direct method, i must be less than $N/3 \simeq 3000$ and as was seen in all ases presented, the number of iterations required by the multigradient never went above 2% of the total panel number. Compared with a fast iterative technique such as C.G.S., the multigradient has been shown to be as fast for smooth configurations, 50% cheaper for ill conditioned cases and more reliable.

REFERENCES

[1] N. Bakhvalow : Méthodes numériques, Editions MIR, Moscou, (1976).

[2] T.H. Lê, Y. Morchoisne, J. Ryan : Techniques numériques nouvelles dans les méthodes de singularités pour l'application à des configurations tridimensionelles complexes. Symp. AGARD/FDP sur "Applications of Computational Fluid Dynamics in Aeronautics". Aix en Provence, 7-10 avril 1986.

[3] P. Joly : Méthodes de gradient . A paraître.Publications du Laboratoire d'Analyse Numérique de Paris VI, (1987).

[4] W.H. Press, B.P. Flannery, S.A. Teukolsky, W.T. Vetterling: Numerical Recipes, (1986).

EULER FLOWS IN HYDRAULIC TURBINES AND DUCTS RELATED TO BOUNDARY CONDITIONS FORMULATION

A. Saxer*, H. Felici*, C. Neury*, I.L. Ryhming**

Hydraulic Machines and Fluid Mechanics Institute
Swiss Federal Institute of Technology
1015 Lausanne, Switzerland

SUMMARY

Internal 3D flow geometries of complex shapes, as found in hydraulic turbomachines (fig. 1), present challenging CFD problems for different reasons. Very often the velocity field in the entrance region to the turbine runner is highly vortical in nature, and the computational region to be considered for the runner presents itself with inhomogeneous boundary conditions over different parts of highly curved bounding surfaces. At high Reynolds numbers, these flows can be ideally simulated by using the Euler equations, whereby the effects of a complex duct geometry on the vortical flow field can be captured. Of particular interest in these problems is the handling of the boundary conditions.

We consider here the flow in a Francis turbine runner (figs. 2 and 3) with a vortical 3D velocity field at the entrance. In order to digest the complexity of this flow, simpler, but related cases are also considered.

GOVERNING EQUATIONS AND NUMERICAL PROCEDURE

Time-marching methods are often used for solving the Euler equations to obtain steady state inviscid solutions. However, for incompressible flow, the time derivative of the density is absent in the continuity equation so that the system of equations to be solved becomes partially hyperbolic and only poorly conditioned for numerical simulation. In fact, the sound waves become much faster than the advection speed and dominate the system.

However, if only the steady incompressible flow is of interest, this drawback can be eliminated by using Chorin's artificial compressibility formulation [1]. An artificial time-dependent term is thus added to the continuity equation so that the governing inviscid equations can be written in the form

$$\partial/\partial t \iiint_{\Omega} \vec{q}\, dvol + \iint_{\partial\Omega} [H(\vec{q}) \cdot \vec{n}]\, ds = \iiint_{\Omega} \vec{f}\, dvol \qquad (1)$$

where $\vec{q} = (p/\rho_o, u, v, w)$ represents the column vector of dependent variables comprising the pressure p referred to the constant density ρ_o, and the components of the velocity vector $\vec{v} = (u, v, w)$ in a Cartesian system (x, y, z).

The quantity $H(\vec{q})\cdot\vec{n} = (c^2\, \vec{v}\cdot\vec{n},\ u\, \vec{v}\cdot\vec{n} + (p/\rho_o)\, \vec{n}\cdot\vec{e}_x,\ v\, \vec{v}\cdot\vec{n} + (p/\rho_o)\, \vec{n}\cdot\vec{e}_y,\ w\, \vec{v}\cdot\vec{n} + (p/\rho_o)\, \vec{n}\cdot\vec{e}_z)$, represents the net flux of $\vec{q}$ transported across the closed surface $\partial\Omega$ surrounding volume Ω of unit vector $\vec{n}$, plus the "pressure" p/ρ_o acting on $\partial\Omega$.

* Research engineer ** Professor

c^2 is the artificial compressibility parameter that can be selected to accelerate the time decay rate in a time-marching procedure. If the coordinate system rotates around the z axis with angular velocity ω, and on the assumption that gravitational forces can be neglected over the hight of the runner, we have to consider the Coriolis, $-2(\vec{\omega} \times \vec{v})$, and the centrifugal, $-\vec{\omega} \times (\vec{\omega} \times \vec{r})$, forces, so that $\vec{f}$ becomes: $\vec{f} = (o, \omega(\omega\, \vec{r} \cdot \vec{e}_x + 2\, \vec{v} \cdot \vec{e}_y), \omega(\omega\, \vec{r} \cdot \vec{e}_y - 2\, \vec{v} \cdot \vec{e}_x), o)$, where $\vec{r}$ is the position vector. The system (1) governs the pseudo-unsteady approach because the solution is only physically meaningful when steady state is reached.

The numerical procedure used has been developed by Rizzi and Eriksson [2] and is briefly outlined as follows. The hyperbolic system (1), discretized in space by the finite-volume technique becomes

$$\Omega_{ijk}\, \partial \vec{q}_{ijk} / \partial t + \delta[H(\vec{q}) \cdot \vec{S}]_{ijk} = \Omega_{ijk}\, \vec{f}_{ijk} \, .$$

$\vec{q}$ and $\vec{f}$ are volumetric averages located at the center of the cell (i, j, k), and $H(\vec{q}) \cdot \vec{S}$ is the non-linear corresponding flux evaluated on the surface S. The desired flux is obtained at cell surfaces using the linear operator μ written in the I, J, K directions as

$$\mu_I\, \psi_{ijk} = 1/2\, (\psi_{i+1/2,j,k} + \psi_{i-1/2,j,k})$$

$$\mu_J\, \psi_{ijk} = 1/2\, (\psi_{i,j+1/2,k} + \psi_{i,j-1/2,k})$$

$$\mu_K\, \psi_{ijk} = 1/2\, (\psi_{i,j,k+1/2} + \psi_{i,j,k-1/2})$$

acting on the cell centered quantity $H(\vec{q})_{ijk}$ and multiplied by the metric terms.

The flux term $H(\vec{q}) \cdot \vec{S}$ is obtained by using the 3D undivided central difference operator δ:

$$\delta\psi_{ijk} = (\delta_I + \delta_J + \delta_K)\, \psi_{ijk} = (\psi_{i+1/2,j,k} - \psi_{i-1/2,j,k}) + (\psi_{i,j+1/2,k} - \psi_{i,j-1/2,k}) + (\psi_{i,j,k+1/2} - \psi_{i,j,k-1/2})$$

$$\psi_{ijk} = [H(\vec{q}) \cdot \vec{S}]_{ijk}$$

which expresses the conservation property in the considered cell.

It has been shown [2] that the truncation error of this scheme is entirely dispersive and not dissipative. To prevent aliasing phenomena, a linear fourth-order operator acting on the dependent variables is added to the advective part.

The resulting system is integrated by an explicit time-marching procedure consisting of a parametrized three-stage Runge-Kutta scheme, and local time stepping is used to accelerate convergence. The scheme has been proven in a number of different applications.

The flow in a Francis runner is analyzed by considering only one inter-blade volume (figs. 4a and 4b). We transform this physical domain onto a computational domain (fig. 5) with an H-H mesh type using the transfinite interpolation technique developed by Eriksson [3] and adapted by Neury [4] to this kind of geometry. Rizzi and Eriksson showed that even in the presence of mesh singularities, for instance a cell surface degenerating into a line or a point, the finite-volume method remains stable but its accuracy diminishes between first and second-order in space.

BOUNDARY CONDITIONS

The surfaces of the computational region may be of fluid or solid kind. Boundary conditions on fluid surfaces can be divided into two types: inlet-outlet and periodical.

Inlet-Outlet

Using the theory of characteristics, it can be shown that, for the hyperbolic system (1) , three conditions have to be imposed at the inlet and one at the outlet. Working with the dependent variables p and $\vec{v}$, we impose the relative vortical velocity distribution (noted $\vec{w}$ in the rotating frame of reference) on the inlet surface and the pressure at the outlet (figs. 6a and 6b). However, the influence of the boundary data location on the computed solution, in terms of distance and number of cells from the profile, must be considered. For that purpose, the flows in two (V_A and V_B) linear NACA8410 cascades were analyzed [5], where the inlet and outlet boundaries were set to one half and one chord length respectively away from the profile. A homogeneous velocity field (inlet angle 22°) at the inlet and a constant downstream pressure for both 0-type meshed configurations were imposed. Each cascade was studied with three grids: 64 x 14, 128 x 28 (figs. 7a and 7b), 256 x 56 cells. In table I, the following integral values for both control volumes and for the corresponding meshes (V_{An}, V_{Bn}, n = 1, 2, 3) are summarized, i.e. M_{II} : outlet flow rate (inlet : M_I = .738606), Γ: circulation, F_S : peripherical momentum ("driving torque"), Fp : profile reaction in the peripherical direction.

In our computations on the finest meshes (V_{A3} and V_{B3}), no visible differences appear for the profile pressure coefficient C_p (fig. 8) in comparison with a panel method. However, the total pressure losses (figs. 9a and 9b) show that upstream and downstream surfaces V_A are too close to the blade and do not ensure constant total pressure at the inlet and outlet surfaces.

Periodicity

The boundary values on these particular fluid surfaces have to be imposed by using a periodicity condition, i.e. averaging the fluxes from both sides of the periodicity surface (same operator μ as in the rest of the field).

Nevertheless, when metric discontinuities appear at a periodic surface, the operator μ may generate oscillations in the computed solution (fig. 10a). These effects can be corrected by using local smoothing of the mesh or by defining locally a new averaging operator μ for the direction in which the oscillations propagate: on each side of the periodic surface, a second-order polynomial operator determine two values on this boundary from the values at the center of the three nearest cells in the curvilinear direction transverse to the periodicity. The metric of the cells is taken into account in this algorithm [5]. To ensure flux conservation, the two values are then averaged. The result is shown in fig. 10b.

Note that, in this case, the μ operator applies directly on the generalized variable $\vec{q}$. In the two other curvilinear directions, the μ operator remains unchanged.

Absorbing boundary conditions on a fluid surface

In order to limit the computational volume we have to introduce artificial boundaries, whereas the actual flow domain is open. On such limiting surfaces (inlet-outlet-periodical), the amplitudes of reflected waves produced by the numerical procedure must be minimized in order to achieve convergence.

For practical applications, Engqvist and Majda [6] presented a theory of local absorbing conditions on these artificial boundaries. The method consists of setting conditions on the characteristic variables by means of the primitive ones $\vec{q}$, rather than on the dependent variables themselves. As output we obtain "new" values $\vec{q}_N$ on artificial boundaries, which are functions of the ingoing and outgoing characteristic variables.

In [5], this technique was applied to external and internal 2D and 3D flows, and the results were compared. As main feature, we found that, although a correct Euler flow was computed, prescribed inflow conditions were modified by the absorbing boundary formulation. In fact, the new values $\vec{q}_N$ are defined by the prescribed ones (imposed or extrapolated from the flow field depending on considered surface), plus a corrective term containing both pressure and velocity. In internal flows, the values prescribed at the inlet and the outlet may not always be the correct ones because they might not be available. By the absorbing boundary corrective term, these values are redefined in order to satisfy the set of characteristic equations. Finally, the computed flow has different inlet and outlet conditions as initially prescribed. Furthermore, if the inlet and outlet surfaces are small compared to the periodic and solid ones, the interest of this technique for internal flows is reduced.

Solid boundary

As described in [7], by introducing the momentum equation in the streamline differentiation of the zero normal flux equation, we obtain a relation between the density ρ, the velocity $\vec{v}$, the body geometry and the normal derivative of pressure, all these quantities beeing located on the body. Introducing then the values of p, $\vec{v}$ on the body as a simple continuation of the quantities at the nearest flow field point (using the first two terms of a Taylor series expansion), and dropping all terms higher than first-order, we obtain a first-order accurate relation which couples the pressure on the body to the density, velocity and pressure in the flow field. However, a first-order boundary condition is consistent with a second-order interior scheme and the convergence is then still of second-order [8].

Consequently, the curvature terms and pressure gradients have to be approximated at the center of the cell in a discrete form. This is done by using a linear average of the curvatures and of the pressures, respectively, at the cell interfaces.

The H-type mesh applied to the rounded leading edge of the blade will introduce a metric discontinuity, which decreases the accuracy of the solution. To reduce this effect, the mesh has to be finer near this singular point, so that the computation can be performed without a special algorithm.

If the leading edge of the blade would have been represented by a cusp, the cells preceeding the leading edge would have been taken into account automatically, since the mesh would then have been "streamline like" near this particular point.

The trailing edge of the Francis runner blade is assumed to be sharp. With a finer H-mesh in this region, this case can be treated with the ordinary linear averaging technique.

Flow computations with a rounded trailing edge were also performed (see NACA8410 cascade example), and showed that if the mesh is sufficiently smooth and without imposing explicitly a Kutta condition, the flow leaves correctly the trailing edge so that the lift is in good agreement with a potential computation (fig. 8). Similar results were obtained for other geometries in [9]. This problem has already been observed and discussed [2] [10], especially for sharp trailing edges. Our computations confirm that with the finite-volume approach, it is possible to obtain correct circulation for flow around highly curved blades with small curvature radii at trailing edge.

In order to evaluate the curvature terms at the center of a cell close to the intersection of two solid surfaces, e.g. band-blade or crown-blade, the two normals on the intersection line are evaluated by a linear extrapolation of the normals at the center of the two nearest cells in the curvilinear direction transverse to the intersection line.

For the parts of the mesh connecting the inlet surface to the leading edge, as well as those connecting the trailing edge to the outlet surface in the Francis runner case, (figs. 4a and 4b), the fluid values have to be computed with a "special periodic" algorithm if the mesh is not smooth enough, so that periodic values are used for computing pressure at the leading edge and the trailing edge of the blade.

RESULTS FOR THE FRANCIS RUNNER

The mesh used for the computation is composed of 32 x 8 x 22 cells (figs. 4a and 4b). For the suction side of the blade, the pressure contours and the relative velocity vectors are represented in figs. 11a and 11b respectively. We observe a global pressure decrease from the leading edge (LE) to the trailing edge (TE), where it is almost constant. However, large gradients appear on LE, along the band and particularly in their intersection region, due to small curvature radii. Moreover, to preserve the radial component of momentum, a pressure gradient appears in this direction, which under the influence of the centrifugal and Coriolis forces induce a flow acceleration from the crown to the band.

On the pressure side, a ridge of high pressure (fig. 12a) running along the stagnation line (fig. 12b), located approximatively 10% chord length away from the LE, indicates that the inlet angle is large so that the flow turns around the nose inducing recirculation zones. The constant pressure line locations (isobars) have been confirmed on model turbine cavitation tests. Moreover, computed integral values such as the torque are within 1% compared to the measurements.

In fig. 13a, the pressure contours are plotted on five interblade mesh surfaces (outlet surface, p = cst). We remark that almost the entire energy is transformed in the first 50% of the blade indicating that the turbine works, but not at the best operational point. In fact, this kind of fixed blade runner has only one maximum efficiency operational point, because it cannot be adapted to other conditions imposed by the wicket-gate.

In fig. 13b, the relative velocity vectors, plotted on the same mesh surfaces, show that the inlet angle increases from the crown to the band and confirm the rapid flow acceleration in the interblade channel and from the center to the maximum radius. In order to avoid the degeneration of crown cells reaching the rotating axis, we define a hollow cylinder with a radius of the order 10^{-6}, referred to the maximum one.

VORTICITY PRODUCTION AND DIFFUSION IN DUCTS

During the last few years, the problem of vorticity production and total pressure loss generation in inviscid compressible external flows has been studied by many authors [11], [12]. In order to investigate how the Euler equations, in the finite-volume formulation described above, generate and transport vorticity in a highly curved geometry such as a Francis runner, the simpler problem of a 2D and 3D S shaped duct has been studied (figs. 14 and 17).

In the 2D configuration, the effect of flow deviation is investigated, moreover in the 3D configuration the behaviour of vorticity near the junction of two solid surfaces is explored.

The 2D S-duct is successively discretized by four H-type meshes (74 x 16, 148 x 32, 296 x 64, 592 x 128 cells) applying a mesh refinement technique. The four computations were performed on the SFIT Cray 1/S until the same degree of convergence was obtained, i.e. a value of 10^{-7} for the pressure variation between two successive time iterations. (A test on the coarse grid showed that using a value of 10^{-12} instead of 10^{-7} did not modify the results).

We imposed a homogeneous inlet velocity field ($\vec{\nabla} \times \vec{v} = 0$) and constant outlet pressure. The same numerical parameters were used for each mesh (CFL = 1.5, c^2 = 5, relaxation parameter in time integration = 0.7, artificial viscosity = 0.03).

Positive and negative vorticity, generated respectively on upper and lower curved walls, is transported by the flow and spread out in the duct (fig. 15a). The numerical capture of vorticity implies total pressure losses shown in figure 15b. The results are controlled by applying the steady form of the Helmholtz equation

$$\left(\vec{v} \cdot \vec{\nabla}\right)\frac{\vec{\Omega}}{\rho} = \left(\frac{\vec{\Omega}}{\rho} \cdot \vec{\nabla}\right)\vec{v} \tag{2}$$

where $\vec{\Omega} = \vec{\nabla} \times \vec{v}$, to the steady computed solution. For the 2D case, the right hand side of (2) falls to zero. The left hand side is presented in figure 15c and shows that in the curved part of the duct, Helmholtz equation is not fulfilled, whereas in the straight part of the duct, the transport of the generated vorticity tends also to zero. Results for the finest mesh (592 x 128) are shown in figures 16a, 16b, 16c. Compared to the coarse grid, the vorticity is located much closer to the walls, i.e. less diffusion is enforced due to smaller cell size. However, the magnitude of the vorticity in this case is reduced by (only) 33%, whereas the total pressure losses are reduced by a factor of 10. This is in contrast to external flows [11], where no significant changes were observed when increasing the number of cells.

Consider next the flow in a 3D S-duct constructed by stacking 2D meshes in the vertical direction. The same solid wall boundary conditions have been applied on all four bounding surfaces of the duct and the same inlet-outlet conditions as in the 2D case.

Near the additional solid boundaries (lower and upper walls), the diffusion due to the discretisation of the Euler equations can now only occur in two directions. Note that the mesh has been refined near the walls, in order to capture wall effects (fig. 17). As can be seen in figs. 18a and 18b, the addition of the two walls leads to a curvature of the vorticity and of the total pressure loss contours. This 3D effect can only be captured if the mesh is sufficiently dense near the walls. With a refined mesh near the walls, the maximum value of the vorticity computed is 0.97 s^{-1} as compared to 1.86 s^{-1} obtained in the Francis runner flow. For the

Francis runner case, however, the inlet flow is rotational with a maximum value of $0.02\ s^{-1}$.

REFERENCES

[1] A.J. Chorin: "**A numerical method for solving incompressible viscous flow problems**", *J. Comp. Phys. 2, 12-26 (1967).*

[2] A. Rizzi, L.E. Eriksson: "**Computation of flow around wings based on the Euler equations**", *JFM 148, 45-71 (1984).*

[3] L.E. Eriksson: "**Practical three-dimensional mesh generation using transfinite interpolation**", *Lecture Series Notes, VKI (1983).*

[4] C. Neury: "**3D mesh generation for calculating flow through radial-axial turbines**", *Proc. Int. Conf. on Numerical Grid Generation in Computational Fluid Dynamics, Pineridge Press, Swansea, 387-398 (1986).*

[5] A. Saxer et H. Felici: "**Etude numérique d'écoulements internes, incompressibles et stationnaires par les équations d'Euler**", *Rapport IMHEF/EPFL T-87-4 (1987).*

[6] B. Engqvist, A. Majda: "**Absorbing boundary conditions for the numerical simulation of waves**", *Math. Comp. 31, 629-651 (1977).*

[7] A.W. Rizzi: "**Numerical implementation of solid-body boundary conditions for the Euler equations**", *Z. angew. Math. Mech. 58, 301-304 (1978).*

[8] R.D. Richtmeyer, K.W. Morton: "**Difference Methods for Initial Value Problems**", 2nd ed., *Interscience, N.Y., 382 (1967).*

[9] L.E. Eriksson, A. Rizzi and J.P. Therre: "**Numerical Solutions of the Steady Incompressible Euler Equations Applied to Water Turbines**", *AIAA-84-2145.*

[10] A. Rizzi, L.E. Eriksson: "**Computation of inviscid incompressible flow with rotation**", *JFM 153, 275-312 (1985).*

[11] K.G. Powell, E.M. Murman, E.S. Perez, J.R. Baron: "**Pressure Loss in Vortical Solutions of the Conical Euler Equations**", *AIAA-85-1701.*

[12] A. Rizzi: "**Euler Solutions of Transonic Vortex Flow Around the Dillner Wing - Compared and Analyzed**", *AIAA-84-2142.*

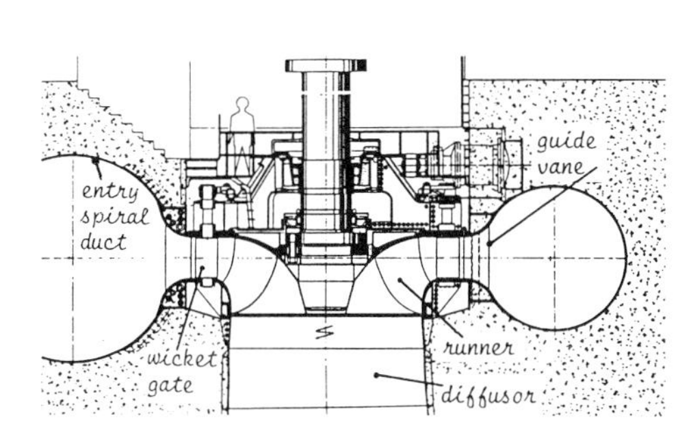

Fig. 1 Axial view of a Francis turbine

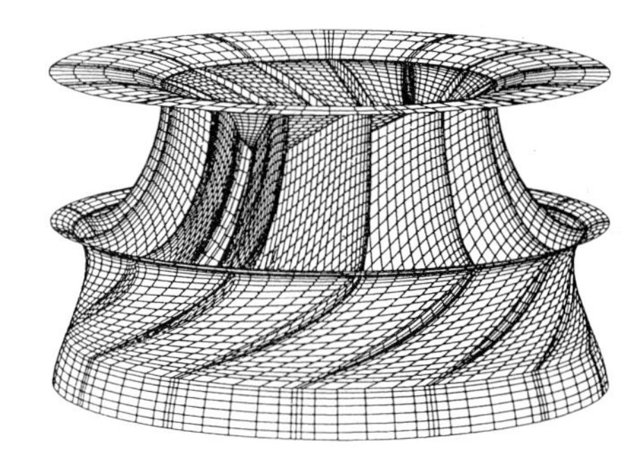

Fig. 2 Francis runner

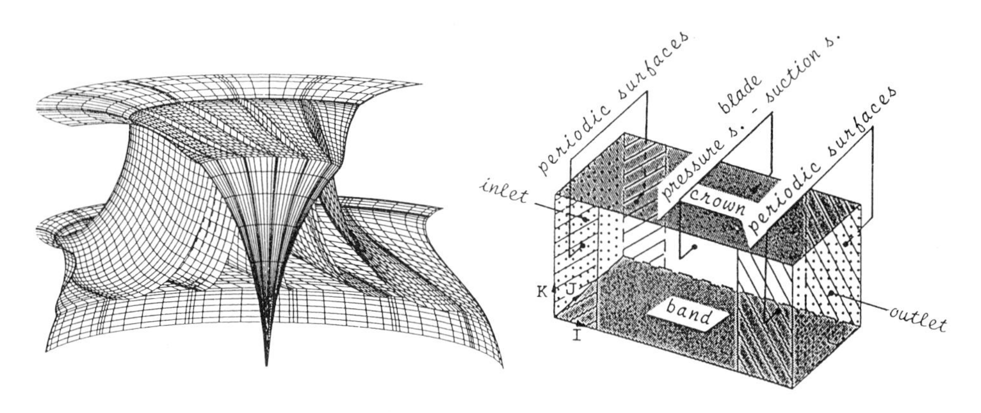

Fig. 3 Francis runner: partial view

Fig. 5 Computational space resulting from the H-H mapping of a Francis runner

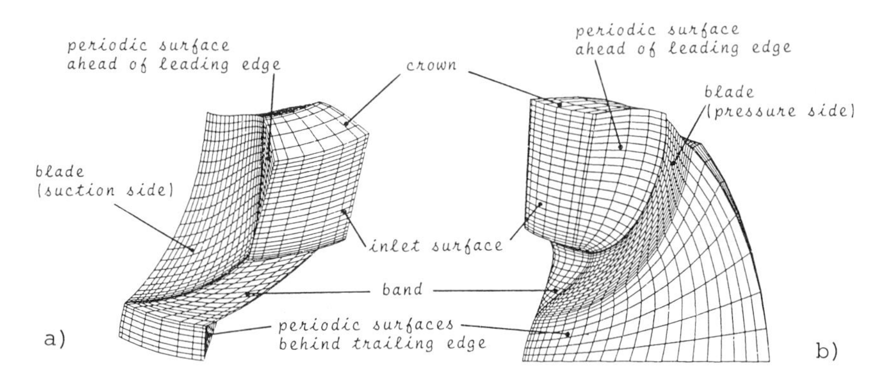

Fig. 4 Francis runner: interblade control volume

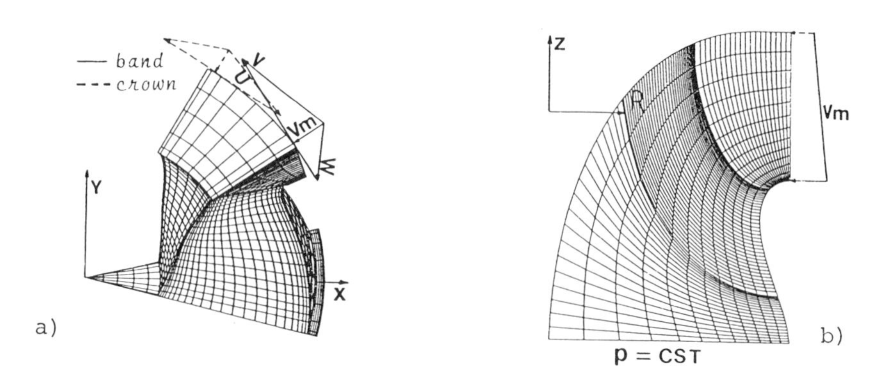

Fig. 6 Inlet-Outlet boundary conditions for a Francis runner

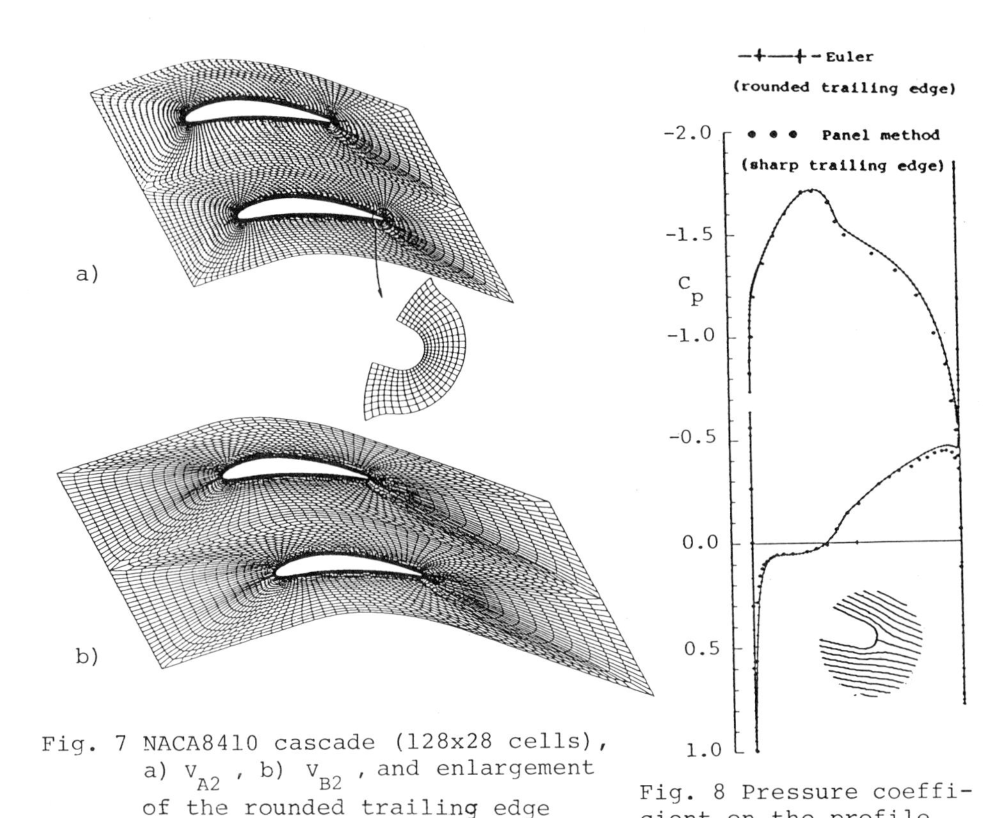

Fig. 7 NACA8410 cascade (128x28 cells), a) v_{A2}, b) v_{B2}, and enlargement of the rounded trailing edge from the 256x56 mesh

Fig. 8 Pressure coefficient on the profile and streamlines near rounded trailing edge

Table I Outlet flowrate (M_{II}), circulation (Γ), driving torque (F_s), and profile reaction (F_p) for NACA8410 cascades

cells	n	M_{II} v_{An}	M_{II} v_{Bn}	Γ v_{An}	Γ v_{Bn}	F_s v_{An}	F_s v_{Bn}	F_p v_{An}	F_p v_{Bn}
64x14	1	0.738503	0.738770	0.549916	0.548155	0.540471	0.537929	0.540954	0.538633
128x28	2	0.738578	0.738557	0.550750	0.550276	0.542577	0.542292	0.542753	0.542416
256x56	3	0.738599	0.738590	0.551306	0.551676	0.542956	0.542953	0.543002	0.542890

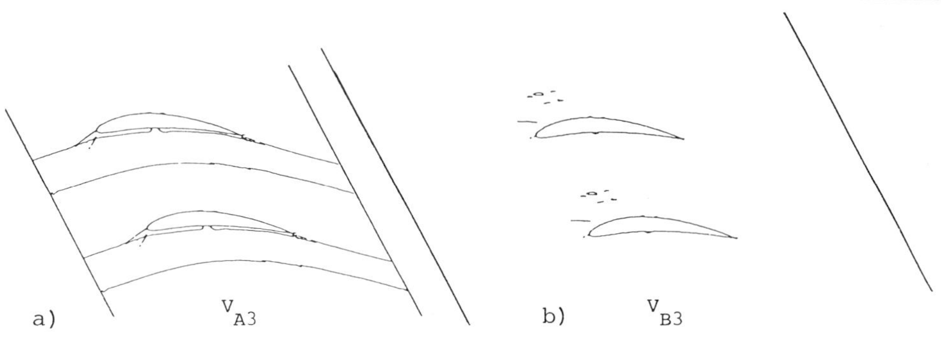

Fig. 9 Total pressure losses (max 6%) influenced by the inlet-outlet data location (NACA8410 cascade)

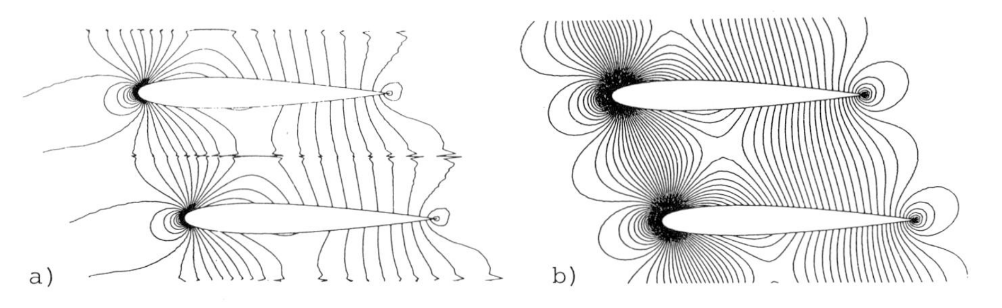

Fig. 10 Computed pressure field, a) with metric discontinuity at periodicity ($\alpha=0^0$, inc: .02), and b) with local modification of the spatial discretisation scheme (8^0,.01)

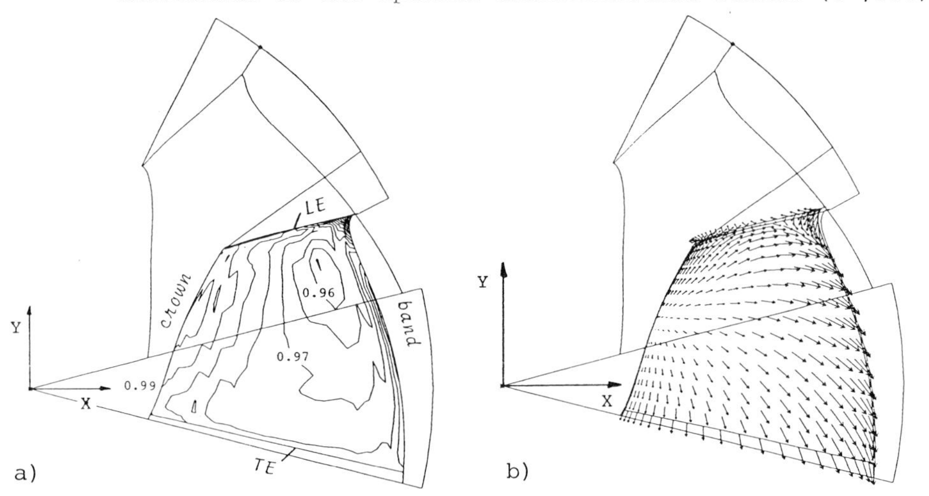

Fig. 11 Pressure field a) and relative velocity vectors b) on the suction side

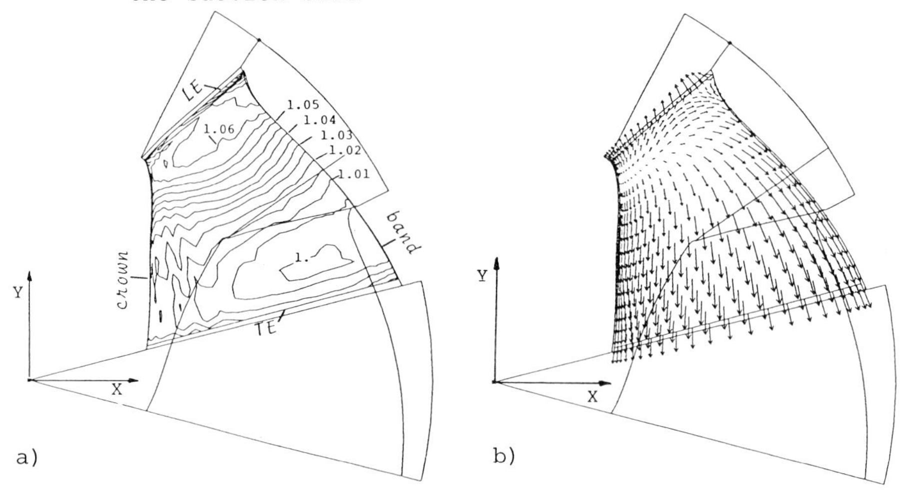

Fig. 12 Pressure field a) and relative velocity vectors b) on the pressure side

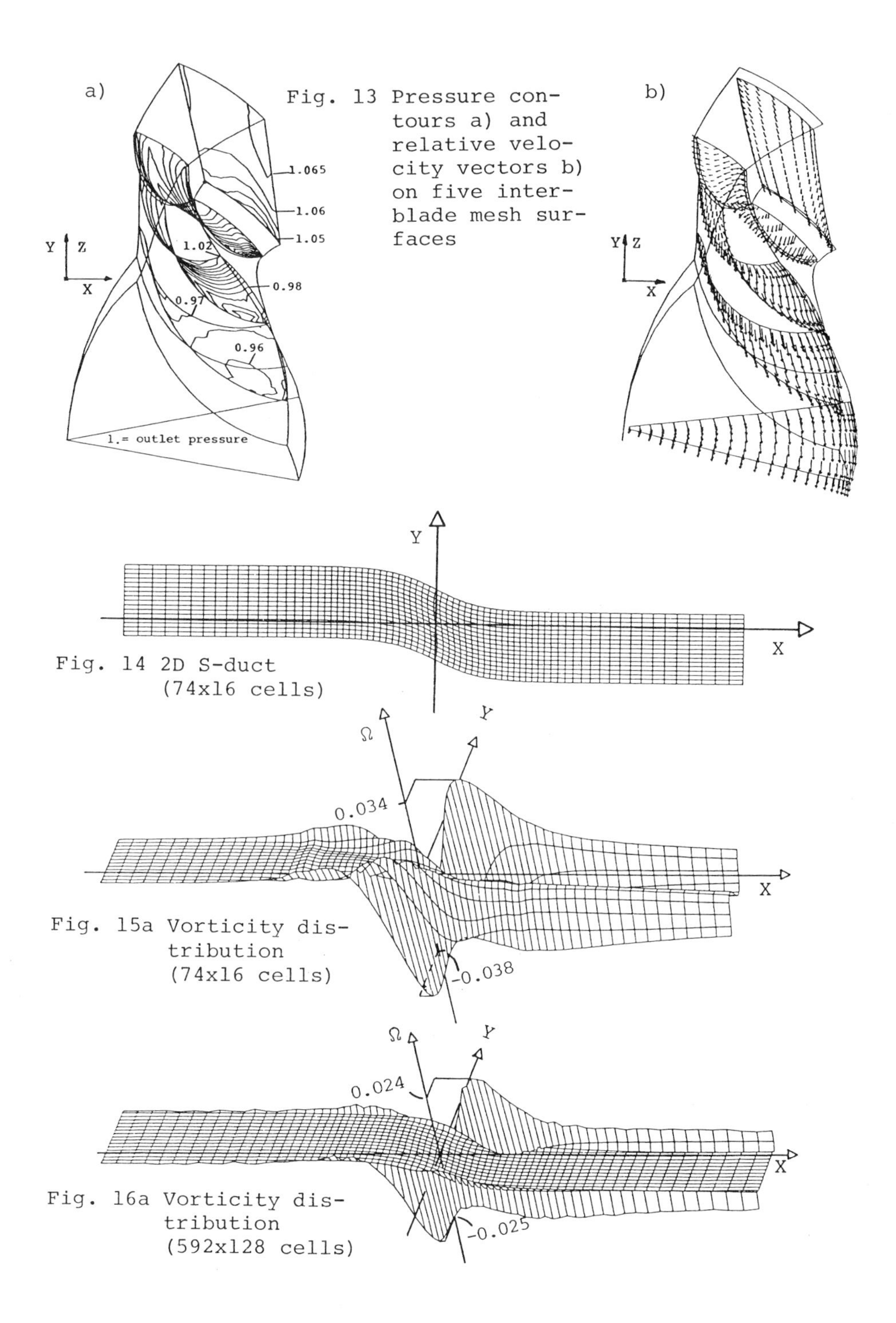

Fig. 13 Pressure contours a) and relative velocity vectors b) on five interblade mesh surfaces

Fig. 14 2D S-duct (74x16 cells)

Fig. 15a Vorticity distribution (74x16 cells)

Fig. 16a Vorticity distribution (592x128 cells)

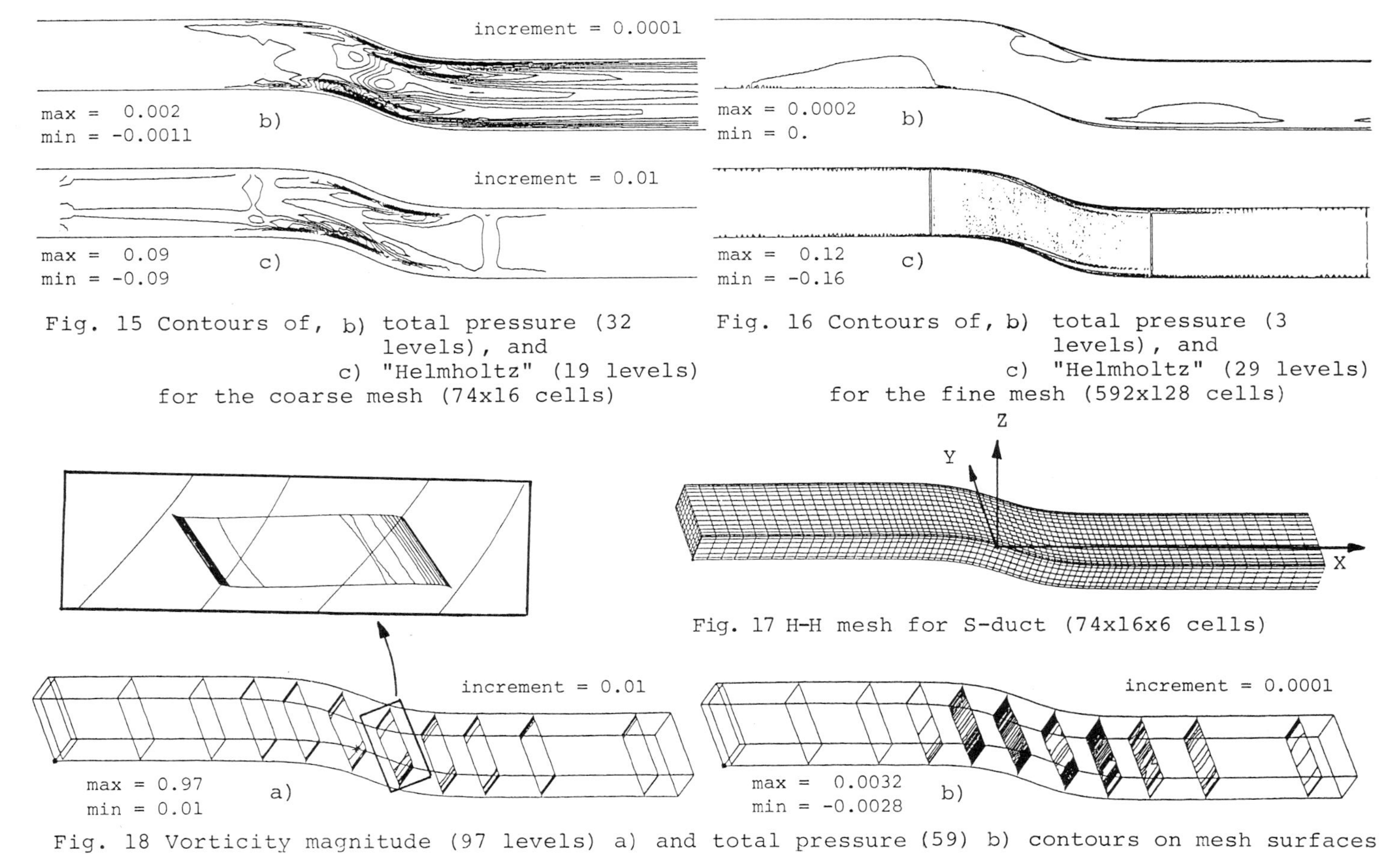

Fig. 15 Contours of, b) total pressure (32 levels), and c) "Helmholtz" (19 levels) for the coarse mesh (74x16 cells)

Fig. 16 Contours of, b) total pressure (3 levels), and c) "Helmholtz" (29 levels) for the fine mesh (592x128 cells)

Fig. 17 H-H mesh for S-duct (74x16x6 cells)

Fig. 18 Vorticity magnitude (97 levels) a) and total pressure (59) b) contours on mesh surfaces

LARGE-EDDY SIMULATION OF TURBULENT BACKWARD FACING STEP FLOW

L.Schmitt and R.Friedrich

Lehrstuhl für Strömungsmechanik
Technische Universität München
Arcisstraße 21, 8000 München 2
Federal Republic of Germany

SUMMARY

A numerical method for computing three-dimensional and time-dependent incompressible turbulent flows in irregular flow domains is presented. The method is tested by calculating laminar recirculating flows and is applied in conjunction with a subgrid-scale model to compute the large-scale structure of the turbulent flow over a backward-facing step.

INTRODUCTION

The large-eddy simulation (LES) technique in which the three-dimensional and time-dependent large scale (grid scale, GS) structure of turbulence is calculated directly and only models for the fine scale (subgrid scale, SGS) structure are introduced, provides detailed information of the flow-field which cannot otherwise be obtained, neither from experiment nor from statistical simulation. In the past we have successfully applied this technique to turbulent flat plate channel (Poiseuille, Couette) and boundary layer flow. Several aspects of these simulations are reported in Schmitt et al. [11], [12] and Richter et al. [9].
In this paper we describe our specific approach to study turbulent backward facing step (bfs) flow. Though the geometry is still simple (fig.1), the flow contains many features of complex flows such as separation, recirculation, reattachment and shear. Besides the interest in understanding these phenomena from a physical as well as engineering viewpoint it is a challenge for numerical methods, for theories and models of turbulence to predict this flow case.

MATHEMATICAL MODEL

The equations for the GS flow quantities,

$$ {}^{\Delta\varphi}\overline{\Phi} = {}^{\Delta\varphi}[\Phi] = 1/\Delta\varphi \int_{\Delta\varphi} \Phi \, d\varphi \quad , \tag{1} $$

defined as averages over grid volumes, ΔV, or their corresponding surfaces, ΔA_j, are derived from the integral conservation equations for mass and momentum applied to a finite grid volume ΔV ('volume balance procedure', Schumann [13]) and read for cartesian coordinates in nondimensionalized form

$$ \delta_j \, {}^{j}\overline{v_j} = 0 \, , \tag{2} $$

$$ \partial_t \, {}^{V}\overline{v_i} + \delta_j (\, {}^{j}\overline{v_j v_i} + {}^{j}\overline{p} \, \delta_{ji} - {}^{j}\overline{\tau_{ji}} \,) = 0 \, , \tag{3} $$

where, $\tau_{ij} = 1/Re \, D_{ij}$, $D_{ij} = \partial_j v_i + \partial_i v_j$, and e.g. $\delta_x \Phi = 1/\Delta x(\Phi(x + \Delta x/2) - \Phi(x - \Delta x/2))$.

The SGS stresses, ${}^{i}\overline{\tau_{ij,SGS}} = -{}^{i}\overline{v_i'v_j'}$, arising from the nonlinear convection terms by introducing the decomposition $\Phi = {}^{\Delta\phi}\overline{\Phi} + \Phi'$ are important if the grid does not resolve the detailed structure of the flow field – they represent the action of the unresolved scales of motion on those that are resolved. While for laminar flows these SGS stresses are usually neglected, for high Reynolds number turbulent flow they must be modelled. In fact there the SGS stresses are needed to supplement the resolvable molecular stresses as a diffusive and as a dissipative agent. Following Schumann and Grötzbach ([13],[4]) we adopt the two-part eddy-viscosity concept,

$$ {}^{i}\overline{\tau_{ij,SGS}} = {}^{ij}\mu_{iso}\,({}^{i}\overline{D_{ij}} - \langle {}^{i}\overline{D_{ij}} \rangle) + {}^{ij}\mu_{inh} \langle {}^{i}\overline{D_{ij}} \rangle \quad , \tag{4} $$

where the first part is to account for locally isotropic SGS stresses and the second for inhomogeneities due to the nonzero components of mean strain ($\langle\ \rangle$ indicates statistical averages). For more details the reader is referred to Schmitt and Friedrich [10].

SOLUTION PROCEDURE

The system of equations is solved in a staggered grid in which ${}^{V}\overline{v_i}$ equals approximately ${}^{i}\overline{v_i}$ for the original grid and ${}^{i}\overline{p}$ is replaced by ${}^{V}\overline{p}$. All GS quantities not defined on the grid are approximated in a second-order manner using algebraic averages, e.g. $\overline{\Phi}^{x} = [\Phi]^{x} = \tfrac{1}{2}(\Phi(x - \Delta x/2) + \Phi(x + \Delta x/2))$. Difference approximations needed for ${}^{i}\overline{D_{ij}}$ and in the SGS model are similar to those given in Schumann [13]. For convenience we give formulae only for an equidistant cartesian grid. The actual code can be applied also for nonequidistant grid spacings in z-direction and (without SGS model) for cylindrical (in y-direction) and natural (in x-direction, with constant curvature) coordinates. There, approximations are based on volume- or surface-weighted averages.
To advance the velocity v_i from time step n to n + 1 the following explicit scheme is used (space averaging bars are omitted, C_i denotes convection, D_i, diffusion):

$$ v_i^* = f_1\, v_i^{n-1} + f_2\, v_i^{n} + f_3\,(C_i^{n} + D_i^{n-1}) \quad , \tag{5} $$

$$ v_i^{n+1} = v_i^* - \delta_i\, p^* \quad , \tag{6} $$

$$ \delta_i\, v_i^{n+1} = 0 \quad . \tag{7} $$

The time integration goes on in cycles. Each cycle starts with an Euler step ($f_1 = 0$, $f_2 = 1$, $f_3 = \Delta t$), followed by approximately 50 leapfrog steps ($f_1 = 1$, $f_2 = 0$, $f_3 = 2\Delta t$) and ends with an averaging step ($f_1 = 0.5$, $f_2 = 0.5$, $f_3 = 1.5\Delta t$). Except for the Euler and averaging steps this scheme is of second order accuracy and in the absence of time-differencing errors and diffusion it is mass-, momentum- and energy-conserving.
To update the velocity v_i^{n+1} using eq.(6), p^* (related to the actual pressure by $p^* = f_3\, p^{n+1}$) has to be determined from a discrete 3d Poisson equation,

$$ \delta_i \delta_i\, p^* = \delta_i\, v_i^* , \tag{8} $$

which results from inserting equation (6) into (7).
The form of (8) is valid for cells not adjacent to boundaries and for those boundary cells where the pressure is given as boundary condition (Dirichlet, periodic b.c. for p^*). Prescription of the velocity component normal to the boundary, $v_i^* \equiv v_i^{n+1}$, yields another type of equation since $\delta_i p^* \equiv 0$ (Neumann b.c. for p^*):

$$ \delta_i[{}^{i}f_v\, \delta_i p^*] = \delta_i[{}^{i}f_v\, v_i^* + (1 - {}^{i}f_v) v_i^{n+1}] \quad , \tag{9} $$

where ${}^{i}f_v = 0$, if the velocity component is prescribed and ${}^{i}f_v = 1$, if it is determined according to equations (5) and (6).

In order to solve the resulting set of equations a direct method is desired, since iterative methods do not conserve mass exactly, which can lead to instabilities. This poses no problem for a box-like computational domain. Here, fast fourier transformation is applied in the y-direction (periodic boundary conditions) and cyclic reduction to the resulting sets of decoupled two-dimensional equations. Dirichlet, periodic or Neumann boundary conditions for p^* at global boundaries can be handled automatically in the solution procedure.
For irregular domains we expand the solution domain to the smallest box that covers the irregular domain, assume for added cells and cells at internal boundaries equations of type (8) and make use of the so-called capacitance matrix technique (Buzbee et al. [2]) to satisfy the original equations of type (9) for cells at internal boundaries. In matrix notation this procedure can be formulated as follows (Schumann [14]). Given are the set of equations $\mathbf{A}\ \mathbf{p}^* = \mathbf{q}$ to be solved, the so-called 'A-problem', where $\mathbf{A}$ is such that no efficient direct solver exists and the matrix $\mathbf{B}$ 'similar' to $\mathbf{A}$ with which a fast solution is possible. The equation $\mathbf{B}\ \mathbf{x} = \mathbf{y}$ defines the 'B-problem'. With 'similar' we mean that the n×n-matrices $\mathbf{A}$ and $\mathbf{B}$ ($n = IM \times JM \times KM$) are equal except for a few $m << n$ rows, representing the equations for the mesh cells at internal boundaries. Decomposing $\mathbf{A}$, $\mathbf{B}$ and $\mathbf{q}$ into (eventually after some reordering)

$$\mathbf{A} = \begin{bmatrix} \mathbf{A}_1 \\ \mathbf{A}_2 \end{bmatrix}, \qquad \mathbf{B} = \begin{bmatrix} \mathbf{B}_1 \\ \mathbf{A}_2 \end{bmatrix}, \qquad \mathbf{q} = \begin{bmatrix} \mathbf{q}_1 \\ \mathbf{q}_2 \end{bmatrix}$$

($\mathbf{A}_1$, $\mathbf{B}_1$: m×n-matrices, $\mathbf{A}_2$: (n-m)×n-matrix, $\mathbf{q}_1$ vector of length m, $\mathbf{q}_2$, of length (n-m)), we get the solution of the A-problem with the help of the m×m-capacitance matrix

$$\mathbf{C} = \mathbf{A}_1\ \mathbf{B}^{-1}\ \mathbf{W}$$

and the n×m-'selection'-matrix

$$\mathbf{W} = \begin{bmatrix} \mathbf{I} \\ \mathbf{0} \end{bmatrix},$$

($\mathbf{I}$: m×m-unit matrix, $\mathbf{0}$: (n-m)×m-zero matrix) applying the steps:

$$\mathbf{B}\ \tilde{\mathbf{p}}^* = \mathbf{q}\ , \tag{10a}$$

$$\hat{\mathbf{w}} = \mathbf{A}_1\ \tilde{\mathbf{p}}^*\ , \tag{10b}$$

$$\mathbf{w} = \mathbf{q}_1 - \hat{\mathbf{w}}\ , \tag{10c}$$

$$\mathbf{C}\ \mathbf{v} = \mathbf{w}\ , \tag{10d}$$

$$\tilde{\mathbf{q}} = \mathbf{q} + \mathbf{W}\ \mathbf{v}\ , \tag{10e}$$

$$\mathbf{B}\ \mathbf{p}^* = \tilde{\mathbf{q}} \tag{10f}$$

($\hat{\mathbf{w}}$,$\mathbf{w}$, $\mathbf{v}$: vectors of length m, $\tilde{\mathbf{p}}^*$ and $\tilde{\mathbf{q}}$, vectors of length n).

That means we obtain the desired solution at the cost of solving two B-problems (10a,f) and a m×m algebraic system of equations (10d). Note that we apply this procedure only to the decoupled 2d equations, which is possible because our flow geometry does not depend on the y-direction. In comparison to a full 3d implementation this is more efficient regarding computation time and storage requirements. The capacity matrices (one for each decoupled plane) depend only on the irregular geometry and are pre-computed at the start of a simulation where solving of m B-problems is necessary to determine one matrix. For the solution of (10d), $\mathbf{C}^{-1}$ is determined by performing its upper and lower triangular matrix (L-U-) decomposition.

BOUNDARY CONDITIONS

For laminar flow boundary conditions cause in general no problem. In the inflow plane a parabolic profile for u is used ($w \equiv 0$). At solid walls no-slip conditions are applied. In the outflow plane different simple extrapolations of the velocity components in combination with Neumann or Dirichlet boundary conditions for p* had practically no influence on the simulation results. As in the turbulent case periodic boundary conditions in y-direction are used. Note that laminar runs have been performed with the 3d code taking only two grid points in y-direction to carry out the fourier transforms.

In the LES of turbulent flows the specification of boundary conditions is a crucial point. In the inflow plane, the instantaneous GS velocity vector and the SGS turbulence energy ${}^{V}\overline{E'}_{iso}$ (needed to determine ${}^{ij}\mu_{iso}$) have to be specified during the whole simulation process. Since no experiment can provide these data we first perform a LES of a fully developed channel flow with periodic boundary conditions in x-direction (Re number, grid, time step, SGS model as in the following bfs simulation), store on magnetic tape all flow quantities from a plane normal to the flow direction for all time steps and then use these data in the inflow plane for the LES of the bfs flow. Thus, on one hand, inlet effects are avoided since not only profiles of one-point correlations like the mean velocity components or the turbulent stresses but also turbulent structures (e.g. expressed in terms of space and time correlation functions) are consistent with what the LES model for the bfs near the inflow region will generate. On the other hand the accordance of the simulation results with experimental experience in the limiting cases of channel flow (upstream of the separation corner and downstream of the reattachment point) can be proved in advance.

In the outflow plane (IR) the statistical longitudinal component $\langle {}^{x}\overline{u} \rangle$ is linearly extrapolated whereas the corresponding fluctuations (needed at time level $n+1$) are taken from the neighbouring cell upstream at time level n. This means for the instantaneous values:

$${}^{x}\overline{u}^{n+1}(IR,J,K) = {}^{x}\overline{u}^{n}(IR-1,J,K) + \langle {}^{x}\overline{u} \rangle (IR-1,K) - \langle {}^{x}\overline{u} \rangle (IR-2,K) \ . \tag{11a}$$

Linear extrapolation of $\langle {}^{x}\overline{u} \rangle$ is necessary to reduce $\langle {}^{z}\overline{w} \rangle$-wiggles near the outflow region which appear in the case of constant extrapolation as a result of continuity, periodicity in y-direction and wall b.c. (these enforce $\langle {}^{z}\overline{w} \rangle (IR,K) = 0$).

For the remaining components ${}^{y}\overline{v}$, ${}^{z}\overline{w}$ and ${}^{V}\overline{E'}_{iso}$ we use:

$$\overline{\Phi}^{n}(IR,J,K) = \overline{\Phi}^{n}(IR-1,J,K) \ . \tag{11b}$$

At solid walls resolution restrictions require special formulations for the instantaneous shear stress components in x- and z-directions. We assume that the tangential velocity value nearest to the wall is in phase with the wall shear stress (Schumann [13]) and that the 'law-of-the-wall' in the form

$$\begin{array}{lll} \langle u \rangle^{+} = z^{+} & & 0 \leqq z^{+} \leqq 5 \\ \langle u \rangle^{+} = -3.05 + 5 \ln(z^{+}) & & 5 < z^{+} \leqq 30 \\ \langle u \rangle^{+} = 5.5 + 2.5 \ln(z^{+}) & & 30 < z^{+} < (2000) \end{array}$$

$$(\langle u \rangle^{+} = \langle u \rangle / u_{\tau}, \ z^{+} = z\, u_{\tau}\, Re, \ u_{\tau} = (|\langle \tau_{w} \rangle|)^{1/2})$$

averaged over the wall nearest grid volume provides the coefficient of proportionality in this relation:

$${}^{z}\overline{\tau}_{zx}|_{w}(t,I,J,K_{1}) = c_{\tau,z}(I)\ {}^{x}\overline{u}|_{1}(t,I,J,K_{1}) \ , \tag{12a}$$

$${}^{x}\overline{\tau}_{xz}|_{w}(t,I_{1},J,K) = c_{\tau,x}(K)\ {}^{z}\overline{w}|_{1}(t,I_{1},J,K) \ . \tag{12b}$$

$c_{\tau,z}$ e.g. is simply

$$c_{\tau,z}(I) = \langle {}^{z}\overline{\tau}_{zx}|_{w}(I,K_1)\rangle / \langle {}^{x}\overline{u}|_1(I,K_1)\rangle \ . \qquad (12c)$$

We are aware of the fact that in recirculating flows the 'law-of-the-wall' (at least in its usual form) is not fully supported by experimental experience. But at least in the mean our model is consistent with what statistical simulations do in order to bridge the near-wall region. There, however it could be shown [7], that variation of the wall functions has no effect on the calculated reattachment length. Moreover, differences ensuing for the flow quantities in the wall region are barely felt away from the wall, indicating that the overall flow field depends only weakly on the wall functions.

APPLICATION

In this chapter we describe the application of our numerical method to laminar and turbulent flow in a 1:2 sudden expansion ($l_{ref} = h$, $u_{ref} = u_{0,1}$, $\Psi_{ref} = u_{b,1}h$, where $u_{0,1}$ is the maximum and $u_{b,1}$ the bulk flow velocity in the inflow plane).
In the laminar range the reattachment length x_{r1} of the primary separation region at the step side wall increases with the Reynolds (Re) number [1] providing an excellent test case for the accuracy of the numerical method. Fig.2 shows our numerical results for different Re numbers ($Re_A = 2u_{b,1}h_1/\nu = 1.415\ Re_{0,1}$, $Re_{0,1} = u_{0,1}h/\nu$) in comparison with the experimental and numerical results of Armaly et al. [1] ($h_2/h_1 = 1.94$) and the numerical results of Kim and Moin [5]. Good agreement is observed between simulation and experiment below $Re_A \approx 400$ and for higher Re number between the present simulations and those in [5]. For $Re_{0,1} = 600$ and 750 we get $x_{r1} = 11.88$ and 13.08. The greatest deviations occur in that Re number range, where a second separation zone appears near the upper wall. We find $x_{s2} = 9.64$, $x_{r2} = 20.59$ ($Re_{0,1} = 600$) and $x_{s2} = 10.55$, $x_{r2} = 24.80$ ($Re_{0,1} = 750$), respectively. As for x_{r1}, these locations deviate from those given in [1] which may be due to effects of three-dimensionality in the experiment ([1],[5]). In figures 3 and 4 the flow patterns for $Re_{0,1} = 150$ ($l_1 = 2.4$, $l_2 = 14.4$, $IM{\times}KM = 84{\times}20$, $\Delta x{\times}\Delta z = .2{\times}.1$) and for $Re_{0,1} = 600$ ($l_1 = 0.$, $l_2 = 30.$, $IM{\times}KM = 100{\times}50$, $\Delta x{\times}\Delta z = .3{\times}.04$) are illustrated by means of contour lines for the streamfunction Ψ and the vorticity component ω_y. We also performed simulations of the flow cases of the GAMM-workshop and found that our results are throughout within the bandwidth of those documented in [6].
In the turbulent case our computational domain extends 16 units in x-direction, 2 in the y- and 2 in the z-direction. An equally spaced grid giving a resolution of $\Delta x{\times}\Delta y{\times}\Delta z = 1/4{\times}1/8{\times}1/8$ was chosen. Scaled with the maximum mean inlet velocity the Re number is about $1.6 \cdot 10^5$. In that Re number range the turbulence is fully developed, the flow can be considered as independent of the Re number [3] and it is two-dimensional and stationary in the mean (allowing to compute statistical quantities by averaging in y-direction and in time). Only the primary separation region exists (see fig.5) for which we obtain $x_{r1} \approx 11.1$ (with a 'standard' code and k-ε-model [8] we get under similar simulation conditions $x_{r1} \approx 6.1$). Though in the LES x_{r1} is about 30% too high in comparison with experiments ([3]: $x_{r1} = 8.5$ for $Re_{0,1} = 1.13 \cdot 10^5$, [15]: $x_{r1} = 8.6$ for $Re_{0,1} = 1.1 \cdot 10^4$) the simulated flow features and statistical flow quantities in general show good agreement with experiment, especially if one compares statistical values on the basis of (nearly) equal positions x/x_{r1} (fig.6,7). A more detailed comparison with experiments (see [10]) suggests that insufficiencies of the SGS model (especially of the inhomogeneous part) and underprediction of the vertical turbulence intensity seem to be mainly responsible for the longer separation region. Figures 8 and 9 give an impression of the simulated turbulent flow. The instantaneous flow patterns of the vorticity component ω_y and of the enstrophy $(\omega_i'')^2$ in the central plane (x,z) are displayed in figures 8b and 8c. In both figures the shear layer is clearly discernible and high vorticity fluctuations can be observed there. The pattern of $\langle \omega_y \rangle$ (fig.8a) should be compared with that of the

laminar case (fig.3b,4b). In figure 9 turbulent mixing is illustrated with the help of a 'hydrogen-bubble' simulation. 32 particles have been released from a 'y-wire' every 15 time steps and were followed over about 500 time steps. Note, that many particles are transported across the mean separating streamline into the recirculation region.
Figure 10 is just to demonstrate the possibilities of our simulation code and to show the direction for future (turbulent) studies. Here, for $Re_{0,1} = 150$ an additional adverse pressure gradient was superimposed on the bfs flow by deflecting the upper wall in a series of small steps. It should be mentioned, that the reattachment length x_{r1} increases in comparison to the case without deflection (see fig.3a).

CONCLUSIONS

A numerical method for solving three-dimensional, time-dependent incompressible flows was applied to laminar and turbulent flow cases. An important feature of the method is the direct solution of the 'pressure' equation for irregular flow domains using the capacitance matrix technique.
Concerning the reattachment lengths good agreement between our simulations, experiments and other computations was obtained in the laminar range where the flow fields are spatially smooth and reach a time-independent stage. In the turbulent case the flow fields vary spatially very strongly, there appear regions with coherent flow structures and stationarity can be observed only for statistical mean values. Here, the numerical method was applied in conjunction with a SGS model and special procedures have been used to generate proper inflow data and to bridge the near-wall regions. Though the reattachment length was overpredicted by about 30%, the simulated (turbulent) flow quantities in general showed good agreement with experiment.
On the whole the results indicate that for the turbulent case not only an increase in resolution and further development of the SGS model seem to be necessary but also more profound investigations of the interactions between numerical method, GS quantities (which must be generated and transported properly) and SGS model (which has to adjust to different characteristic scales). Finally, a good deal of work can still be invested in studying the effect of different boundary conditions.

Acknowledgement. This work is supported by the German Research Society (DFG).

REFERENCES

[1] Armaly,B.F., Durst,F., Pereira,J.C.F., Schönung,B. J.Fluid Mech. 127, 473 - 496, 1983.
[2] Buzbee,B.L., Dorr,F.W., George,J.A., Golub,G.H. SIAM J.Numer.Anal. 8/4, 722 - 736, 1971.
[3] Durst,F., Schmitt,F. Proc. of the 5th Symp. on Turb. Shear Flows. Cornell Uni., Ithaka, N.Y., Aug.7 - 9, 1985.
[4] Grötzbach,G., Schumann,U. Proc. of the Symp. on Turb. Shear Flows. Penn. State Univ., Apr.18 - 20, 1977.
[5] Kim,J., Moin,P. J. Comput. Phys. 54, 308 - 323, 1985.
[6] Morgan,K., Periaux,J., Thomasset,F. (Eds.) A GAMM-workshop. Vieweg, Notes on Numerical Fluid Mechanics 9, 1984.
[7] Obi,S., Peric,M., Scheuerer,G. 2nd Intern. Symp. on Transport Phenomena in Turbulent Flows, Univ. of Tokyo, Tokyo, Oct.1987.
[8] Pun,W.M., Spalding,D.B. Imperial College, HTS/76/2, 1976.
[9] Richter,K., Friedrich,R., Schmitt,L. Proc. of the 6th Symp. on Turb. Shear Flows, Toulouse, France, Sept.7 - 9, 1987.
[10] Schmitt,L., Friedrich,R. Proc. of the 6th Symp. on Turb. Shear Flows, Toulouse, France, Sept.7 - 9, 1987.

[11] Schmitt,L., Richter,K., Friedrich,R. In: Hirschel,E.H. (Ed.): Finite Approximations in Fluid Mechanics. DFG-Priority Research Program, Results 1984-1985. Vieweg, Notes on Numerical Fluid Mechanics 14, 1986.

[12] Schmitt,L., Richter,K., Friedrich,R. In: Schumann,U., Friedrich,R. (Eds.): Direct and Large Eddy Simulation of Turbulence. Proc. of the EUROMECH Colloquium No.199, TU München, Sept.30 – Oct.2, 1985. Vieweg, Notes on Numerical Fluid Mechanics 15, 1986.

[13] Schumann,U. J.Comput. Phys.18, 376 – 404, 1975.

[14] Schumann,U. In: Kollmann,W. (Ed.): Computational Fluid Dynamics. Hemisphere, 402 – 430, 1980.

[15] Tropea,C. Dissertation, Universität Karlsruhe, 1982.

FIGURES

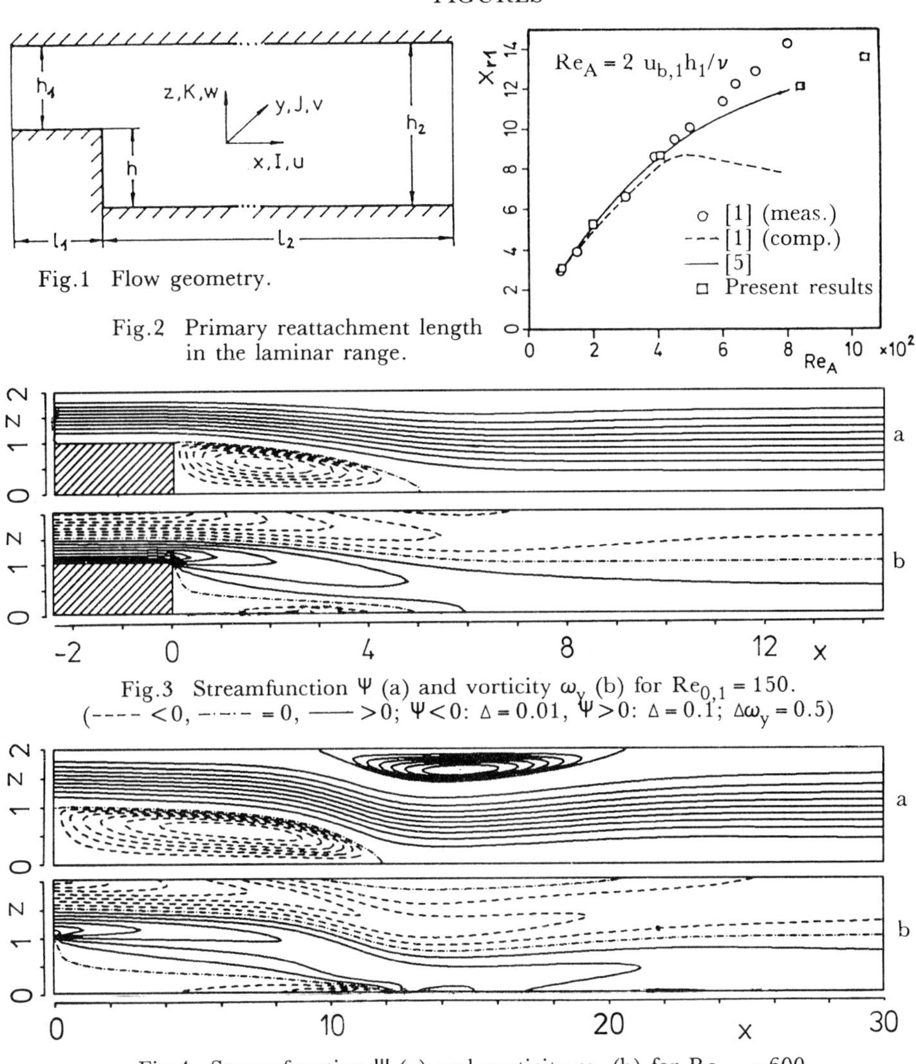

Fig.1 Flow geometry.

Fig.2 Primary reattachment length in the laminar range.

Fig.3 Streamfunction Ψ (a) and vorticity ω_y (b) for $Re_{0,1} = 150$.
(---- <0, —·—·— = 0, —— >0; $\Psi<0$: $\Delta = 0.01$, $\Psi>0$: $\Delta = 0.1$; $\Delta\omega_y = 0.5$)

Fig.4 Streamfunction Ψ (a) and vorticity ω_y (b) for $Re_{0,1} = 600$.
($\Psi<0$: $\Delta = 0.01$, $0<\Psi<1$: $\Delta = 0.1$, $\Psi>1$: $\Delta = 0.002$; $\Delta\omega_y = 0.5$)

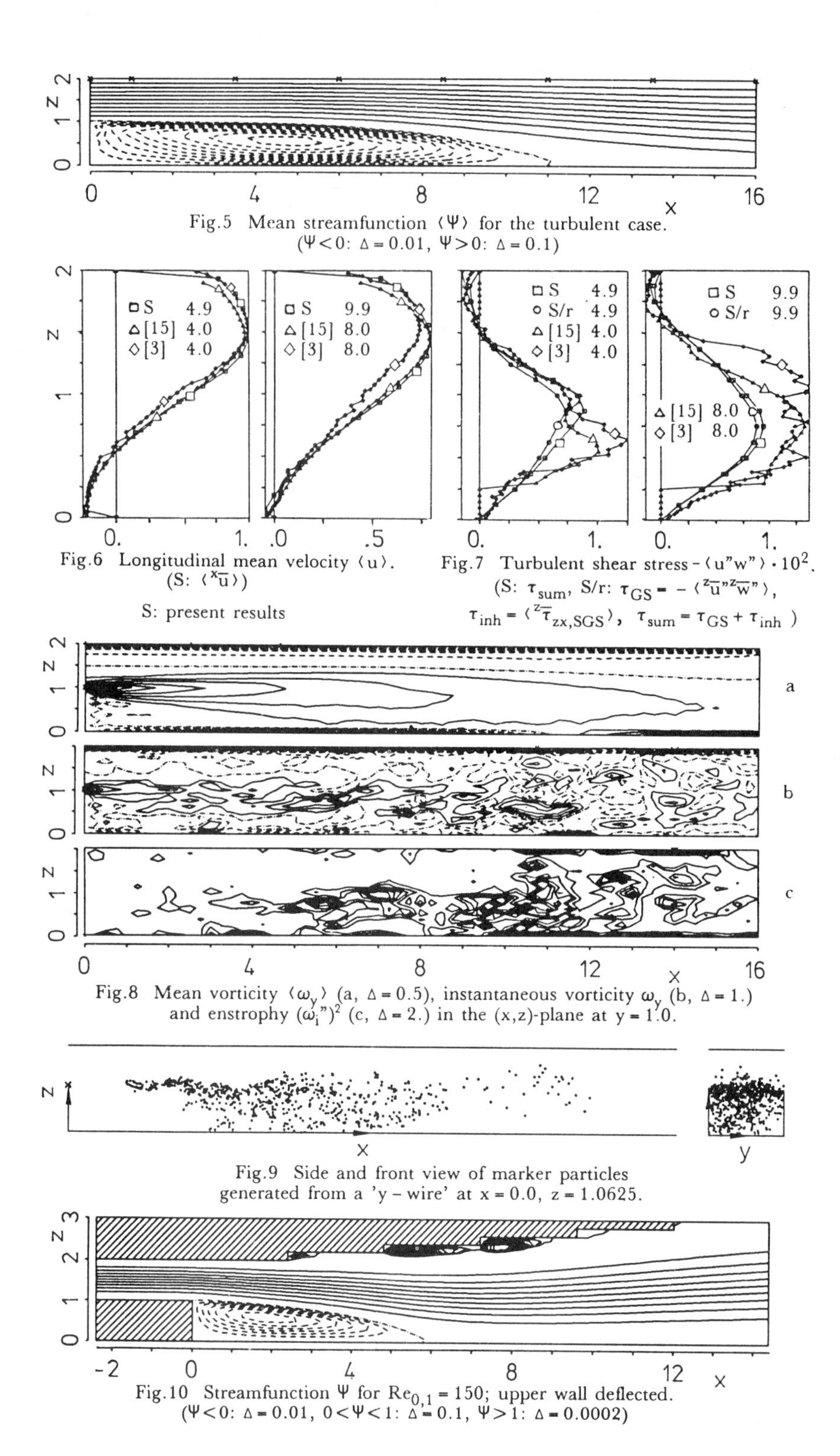

Fig.5 Mean streamfunction $\langle\Psi\rangle$ for the turbulent case. ($\Psi<0$: $\Delta=0.01$, $\Psi>0$: $\Delta=0.1$)

Fig.6 Longitudinal mean velocity $\langle u\rangle$. (S: $\langle {}^{x}\overline{u}\rangle$)

S: present results

Fig.7 Turbulent shear stress $-\langle u''w''\rangle\cdot 10^2$. (S: τ_{sum}, S/r: $\tau_{GS}=-\langle {}^{z}\overline{u}''\,{}^{z}\overline{w}''\rangle$, $\tau_{inh}=\langle {}^{z}\overline{\tau}_{zx,SGS}\rangle$, $\tau_{sum}=\tau_{GS}+\tau_{inh}$)

Fig.8 Mean vorticity $\langle\omega_y\rangle$ (a, $\Delta=0.5$), instantaneous vorticity ω_y (b, $\Delta=1.$) and enstrophy $(\omega_i'')^2$ (c, $\Delta=2.$) in the (x,z)-plane at $y=1.0$.

Fig.9 Side and front view of marker particles generated from a 'y – wire' at $x=0.0$, $z=1.0625$.

Fig.10 Streamfunction Ψ for $Re_{0,1}=150$; upper wall deflected. ($\Psi<0$: $\Delta=0.01$, $0<\Psi<1$: $\Delta=0.1$, $\Psi>1$: $\Delta=0.0002$)

AN ATMOSPHERIC MESOSCALE MODEL: TREATMENT OF HYDROSTATIC FLOWS AND APPLICATION TO FLOWS WITH HYDRAULIC JUMPS

Ulrich Schumann and Hans Volkert
DFVLR, Institut für Physik der Atmosphäre, D-8031 Oberpfaffenhofen

SUMMARY

For numerical simulation of three-dimensional atmospheric flows at micro- and mesoscales a finite-difference method has been developed (programme MESOSCOP). In its non-hydrostatic variant it requires the inversion of an elliptic (Poisson) equation. Here we describe an alternative which applies the hydrostatic approximation and thus avoids the elliptic equation. It has been found, however, that the hydrostatic version requires smaller time steps for stability. The method is applied to investigate the formation of hydraulic jumps in shallow nonlinear fluid flow over a mountain ridge on a rotating plane. By scale analysis and numerical experiments it is shown that the effect of rotation is negligble essentially if F^2/Ro is small, where F is the Froude number and Ro the Rossby number.

INTRODUCTION

A three-dimensional numerical model (MESOSCOP) has been developed to simulate microscale and mesoscale atmospheric flows (horizontal scales less than a few hundred kilometres). The model is implemented in three versions in order to study clouds and precipitation processes over flat terrain, airflow over mountains and turbulent boundary layers. Mesoscale models use either the (complete) non-hydrostatic equations or the hydrostatic approximation [5]. The hydrostatic approximation is valid if the horizontal scales of the flow are much larger than the vertical ones. The approximation results in a possibly more efficient coding since it does not require the solution of an elliptic equation for the pressure as it is necessary for non-hydrostatic incompressible flows. Often, however, the validity of this approximation is not easy to prove. Therefore, it is desirable to have both, the hydrostatic and non-hydrostatic version available, so that the degree of hydrostasy can be determined in concrete flow situations. In a recent paper [8], the numerical version of MESOSCOP for non-hydrostatic flows is described together with five validation examples. Here, we introduce a hydrostatic variant and apply it to investigate the generation of hydraulic jumps in rotating and non-rotating flows. The behaviour of such hydraulic jumps is of interest with respect to cold fronts which are deformed or blocked by high mountains [1, 6].

THE NON-HYDROSTATIC ALGORITHM

MESOSCOP is based on the conservation laws for density ρ , volume specific momentum $\rho\vec{v}$, and several scalars $\rho\psi_k$, $k=1, \ldots ,K$. The number K and the meaning of the scalars depend on the specific applications. The complete model applies for compressible as well as for incompressible fluids. Since the hydrostatic version is restricted to incompressible Boussinesq fluids, we give the equations for this case only; $\bar{\rho}$ is the time-independent reference density:

$$\mathrm{div}(\overrightarrow{\bar{\rho}\mathrm{v}}) \quad = \quad 0\,, \tag{1}$$

$$\frac{\partial \overrightarrow{\bar{\rho}\mathrm{v}}}{\partial t} \;+\; \mathrm{div}(\overrightarrow{\bar{\rho}\mathrm{v}\mathrm{v}}) \;+\; 2\vec{\Omega}\times(\overrightarrow{\bar{\rho}\mathrm{v}}) \;+\; \mathrm{div}(\mathbf{F}) \;=\; -\,\overrightarrow{\mathrm{grad}}(p) \;-\; \rho\,\vec{g}\,, \tag{2}$$

$$\frac{\partial \bar{\rho}\psi_k}{\partial t} \;+\; \mathrm{div}(\overrightarrow{\bar{\rho}\mathrm{v}}\,\psi_k) \;+\; \mathrm{div}(\vec{f_k}) \;=\; \bar{\rho}\,q_k\;; \qquad k = 1,..,K\,. \tag{3}$$

$$\rho \;=\; \rho(p,\,\psi_k)\,. \tag{4}$$

Equation (1) is the continuity equation. Equation (2) describes the nonhydrostatic momentum balance and includes Coriolis forces due to earth's rotation $\vec{\Omega}$, friction by diffusive momentum fluxes $\mathbf{F}$, pressure p, and buoyancy forces by gravity $\vec{g}$. Equation (3) expresses the budget of a scalar ψ_k with nonadvective fluxes f_k, and mass specific sources q_k. Equation (4) represents the equation of state.

For numerical integration, we use a finite difference method based on a staggered grid in space and a combination of the Adams-Bashforth method for momentum with the Smolarkiewicz scheme for scalars. The integration algorithm starts from suitable initial values $\vec{\mathrm{v}}^n$, ψ_k^n, and p^n at time t^n, $n = 0$. For brevity we assume that all sources q_k can be evaluated explicitly. Then the algorithm computes in sequence for $n=0,1,\ldots$:

$$\bar{\rho}\psi_k^{n+1} \;=\; \bar{\rho}\psi_k^n \;-\; \Delta t\,[\mathrm{div}(\,(\overrightarrow{\bar{\rho}\mathrm{v}})^n\;\psi_k^n) + \mathrm{div}(\vec{f}_k^n\,)] \;+\; \Delta t\,\bar{\rho}q_k^n \tag{5}$$

$$\rho^{\sim} = \rho(p^n, \psi_k^{n+1})\,, \tag{6}$$

$$\vec{b}^n \;=\; \mathrm{div}(\,(\overrightarrow{\bar{\rho}\mathrm{v}})^n\;\vec{\mathrm{v}}^n) + \mathrm{div}(\mathbf{F}^n) + 2\vec{\Omega}\times(\overrightarrow{\bar{\rho}\mathrm{v}})^n\,, \tag{7}$$

$$(\overrightarrow{\bar{\rho}\mathrm{v}})^{\sim} \;=\; (\overrightarrow{\bar{\rho}\mathrm{v}})^n \;-\; \Delta t\,[\gamma_0\,\vec{b}^n + \gamma_1\,\vec{b}^{n-1} + \overrightarrow{\mathrm{grad}}(p)^n + \rho^{\sim}\vec{g}\,]\,, \tag{8}$$

($\gamma_0 = 1, \gamma_1 = 0$ for $n = 0$; $\gamma_0 = 1.5, \gamma_1 = -0.5$ for $n > 0$.)

$$r \;=\; \frac{1}{\Delta t}\,\mathrm{div}\,(\overrightarrow{\bar{\rho}\mathrm{v}})^{\sim}\,. \tag{9}$$

Now we invert the Poisson equation

$$\mathrm{div}\,\overrightarrow{\mathrm{grad}}\,(\Delta p) \;=\; r \tag{10}$$

to obtain the mass conserving pressure increments Δp . Finally, pressure and momentum are updated, and time is incremented to the next level:

$$p^{n+1} = p^n + \Delta p\,, \tag{11}$$

$$(\overrightarrow{\bar{\rho}\mathrm{v}})^{n+1} \;=\; (\overrightarrow{\bar{\rho}\mathrm{v}})^{\sim} \;-\; \Delta t\,\overrightarrow{\mathrm{grad}}\,(\Delta p)\,,\ \vec{\mathrm{v}}^{n+1} \;=\; \frac{(\overrightarrow{\bar{\rho}\mathrm{v}})^{n+1}}{\bar{\rho}}\,, \tag{12}$$

$$t^{n+1} = t^n + \Delta t\,. \tag{13}$$

Boundary conditions are described in [8]. The hydrostatic version will be restricted to cases where the boundary condition at the top surface of the computational domain (at $z = z_t$) prescribes the pressure either explicitly or as a function of the vertical velocity at this plane. For flows over mountains with height $z_s(x, y)$, we apply terrain following coordinates [7]. In this case the Poisson equation corresponds to a 25-point operator which is inverted in a block iteration using direct Poisson solvers for constant coefficients. This part of the computation may well require up to 50 % of the total time.

THE HYDROSTATIC ALGORITHM

In the hydrostatic limit, the vertical component of Equation (2) reduces to

$$\frac{\partial p}{\partial z} = -\rho g . \tag{14}$$

Therefore, the hydrostatic version of the algorithm proceeds as follows:

$\bar{\rho}\psi_k^{n+1}$ and $\rho^{\sim}$ are computed as in Equations (5,6).

The horizontal components of $\vec{b}$ and $(\bar{\rho}\vec{v})^{n+1} = (\bar{\rho}\vec{v})^{\sim}$ follow from Equations (7, 8).

Instead of Equation (9) we compute $r = -\rho^{\sim}g$, determine the top pressure from the boundary condition and integrate

$$\frac{\partial p^{n+1}}{\partial z} = r . \tag{15}$$

downwards to optain the new pressure p^{n+1}.

Then the vertical velocity at the surface is determined from the kinematic boundary condition

$$w = u\frac{\partial z_s}{\partial x} + v\frac{\partial z_s}{\partial y} , \tag{16}$$

and finally the vertical component of the momentum $\bar{\rho}w$ follows by integrating the continuity Equation (1) vertically upwards.

In discrete curvilinear coordinates, using the notation defined in [7], the relevant equations read:

$$\delta_3 V G^{33} p = -\overline{V\rho^{\sim}}^3 g, \tag{17}$$

$$G^{33}\bar{\rho}w = \overline{G^{31}\bar{\rho}u}^1 + \overline{G^{32}\bar{\rho}v}^2 , \tag{18}$$

at the ground, and

$$\delta_3 G^{33}\bar{\rho}w = -\delta_1 a - \delta_2 b - \delta_3[\overline{G^{31}\bar{\rho}u}^1 + \overline{G^{32}\bar{\rho}v}^2] \tag{19}$$

in the interior.

Tests have shown, that this algorithm requires smaller time steps than the non-hydrostatic version, in particular if large vertical velocities arise as near

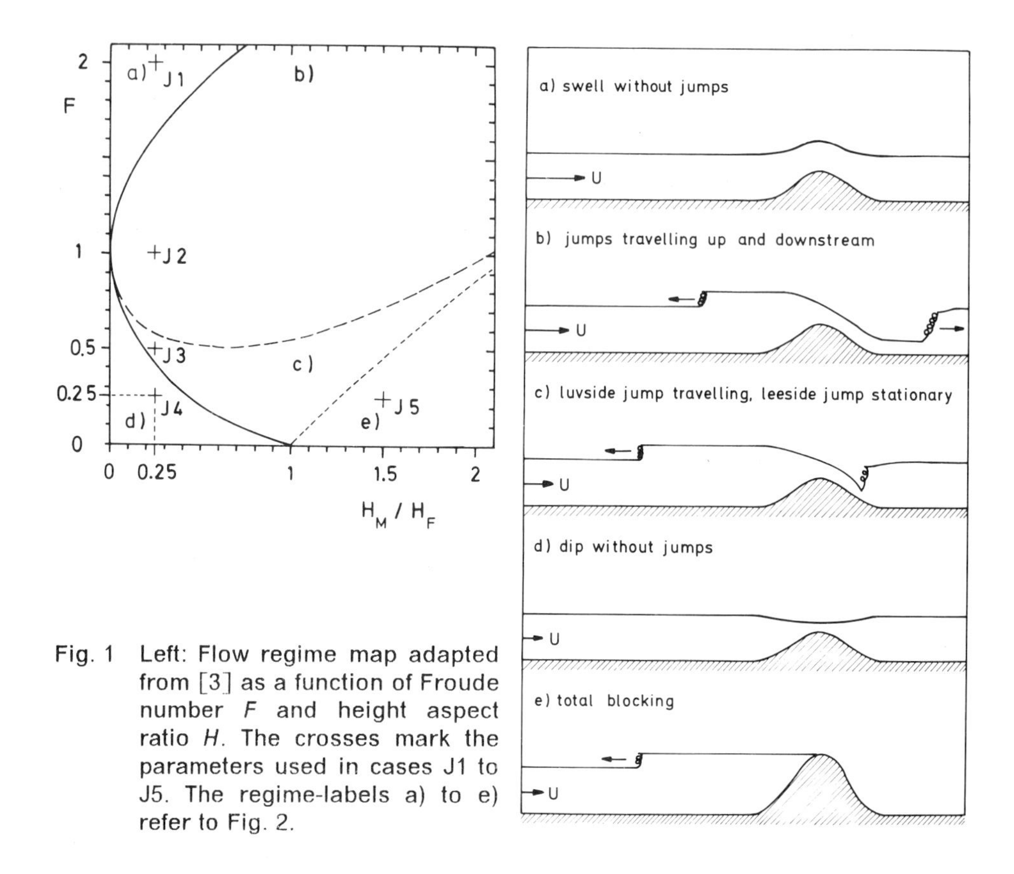

Fig. 1 Left: Flow regime map adapted from [3] as a function of Froude number F and height aspect ratio H. The crosses mark the parameters used in cases J1 to J5. The regime-labels a) to e) refer to Fig. 2.

Fig. 2 Right: Illustration of flow regimes referring to the parameter domains a) to e) indicated in Fig. 1.

hydraulic jumps. Presumably, this stems from the fact that the hydrostatic equations neglect the vertical inertia. Thus, the gain in computational efficiency by avoiding the Poisson equation is at least partly compensated by the larger number of time steps.

INFLUENCE OF ROTATION ON THE FORMATION OF HYDRAULIC JUMPS

We consider a layer of incompressible fluid flowing on a plane rotating at frequency $|\vec{\Omega}| = f/2$ around the vertical axis. The fluid approaches an isolated ridge with uniform speed U. The ridge of height H_M and width L is assumed to have the profile $h_M = H_M \sin^2(\pi x/L)$, for $0 \leq x \leq L$, $h_M = 0$ elsewhere. The fluid is layered in that we have heavy fluid below lighter fluid. The height of the interface between the two layers far upstream the ridge is H_F. The density difference is $\Delta\rho$. Thus the effective gravity amounts to $g' = g\Delta\rho/\bar{\rho}$. Small disturbances in the height of the interface move with the speed of gravity waves $C = (g'H_F)^{1/2}$ in the absence of rotational forces. We assume that the flow can be approximated by shallow water equations, i.e. $H_F \ll L$, $H_M \ll L$. This implies hydrostatic flow. Then the problem is determined by three characteristic numbers:

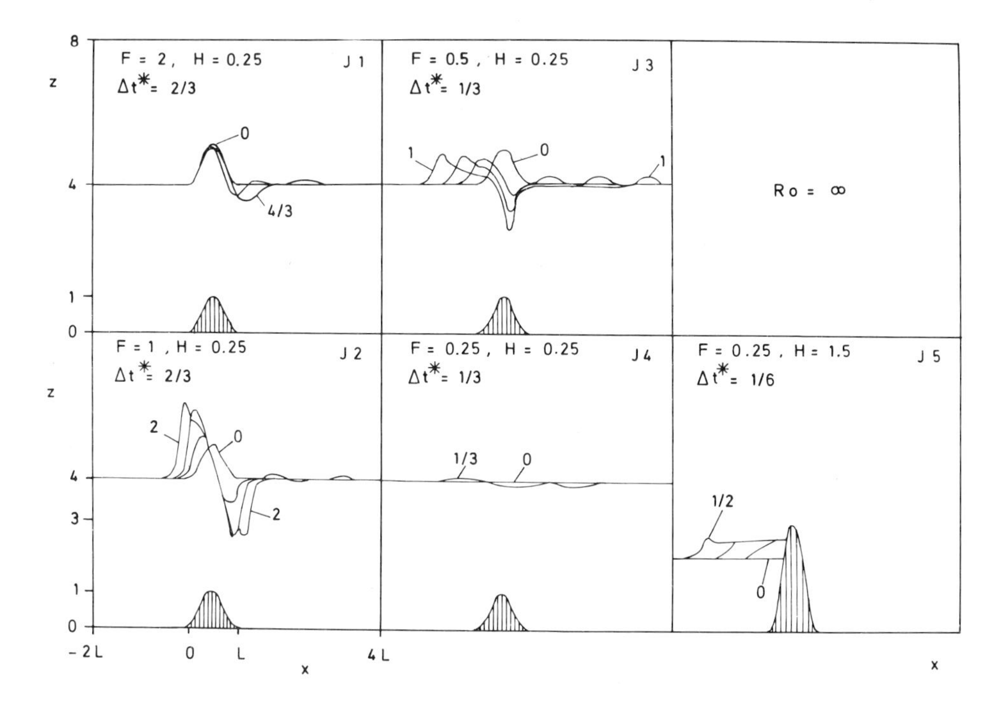

Fig. 3 Development of the top interface of the heavy fluid layer. Each panel shows the results for one of the cases J1 to J5 at a sequence of times with increments Δt^* for $Ro = \infty$ (without rotation). The numbers at the curves denote the respective value of the the normalized time t^*.

$F = U/C$, the Froude number,
$H = H_M/H_F$, the height aspect ratio,
$Ro = U/(fL)$, the Rossby number.

For irrotational flows ($f = 0$, $Ro = \infty$), Houghton and Kasahara [3] and Long [4] have shown that a shallow layer of inviscid fluid flowing at constant speed over a mountain ridge can create hydraulic jumps. Figure 1, maps regimes with and without jumps as a function of the Froude number F and the height ratio H. The various types of flows are illustrated in Fig. 2. The stationary theory predicts the formation of jumps in pairs, one upstream and one downstream of the ridge.

We have applied MESOSCOP to simulate the formation of hydraulic jumps. Five cases are considered (J1 to J5) with Froude-numbers F and height aspect ratios H as indicated by the crosses in Fig. 1. The mountain width is $L = 100$. The computational domain extends over $-2L < x \leq 4L$, $0 < z \leq z_t = 8$, the height of the interface is $H_F = 4$ (except for case J5 where $H_F = 2$), the velocity of the undisturbed flow is $U = 1$. $t^* = tU/L$ defines a non-dimensional time.

Figure 3 shows the development of the flows at a sequence of times for the five cases. As can be seen, the simulation results in the formation of hydraulic jumps in fair agreement with the expectations from Figs. 1 and 2, although the transient solutions are not quite so clear as the idealized picture given in Fig. 2. For J1, the disturbance induced by the mountain cannot propagate upstream because this flow is super-critical in the hydraulic sense. Case J2 shows jumps at the upstream and downstream sides both travelling away from the ridge while for case J3 the downstream jump is stationary. For case J4 the disturbance is small so that the picture is not so clear. Case J5 clearly shows the blocking effect as expected from Fig. 1. The numerical solutions of the hydrostatic and non-hydrostatic algorithms show very small differences. We conclude that both methods are well suited to investigate the formation of hydraulic jumps.

No equivalent theory exists for flows over mountains in a rotating system. Houghton [2] and Williams and Hori [9] consider the transient motion of a shallow water layer on an f-plane without mountains starting from an initial velocity disturbance of magnitude U over a length L. They find that Coriolis forces tend to reduce the amplitude of hydraulic jumps and to delay its formation but apparently do not totally suppress its formation. The reduction of hydraulic jumps by Coriolis forces can be explained as follows: Behind the upstream hydraulic jump the velocity u is less than U. As a consequence, the flow downstream the jump experiences positive acceleration in y-direction due to the Coriolis forces. This in turn causes Coriolis-acceleration in x-direction which tends to reduce the hydraulic jump. The time scale of Coriolis forces is f^{-1}, that of jump creating nonlinear forces is L/U. Hence the delay is large if the Rossby number $Ro = U/(fL)$ is less than a certain limit. No precise information exists on the value of this limiting number.

For the purpose of scale analysis we refer to the shallow water equations where $u(x,t)$ and $v(x,t)$ are the vertically constant velocities in the fluid layer and $h(x,t)$ is the height of the free surface above ground:

$$\frac{du}{dt} = -g'\frac{\partial}{\partial x}(h + h_M) + fv, \tag{20}$$

$$\frac{dv}{dt} = -fu, \tag{21}$$

We introduce non-dimensional variables (denoted by an over-bar) which are of order unity:

$$u = U\bar{u},\ v = U\bar{v},\ x = L\bar{x},\ t = \bar{t}L/U,\ h = H_F\bar{h},\ h_M = H_M\bar{h}_M. \tag{22}$$

This yields the following system:

$$\frac{d\bar{u}}{d\bar{t}} = -F^{-2}\frac{\partial}{\partial\bar{x}}(\bar{h} + H\bar{h}_M) + Ro^{-1}\bar{v}, \tag{23}$$

$$\frac{d\bar{v}}{d\bar{t}} = -Ro^{-1}\bar{u}. \tag{24}$$

From these equations we see that the Coriolis forces (last terms) are small in comparison to the inertia forces if

$$Ro^{-1} \ll 1. \tag{25}$$

In order to have negligble effects of rotation in comparison to the smaller of the two gravity force terms, the following conditions have to be satisfied:

$$F^2/Ro \ll 1, \quad \text{if } H > 1, \tag{26}$$

$$F^2/(RoH) \ll 1, \quad \text{if } H < 1. \tag{27}$$

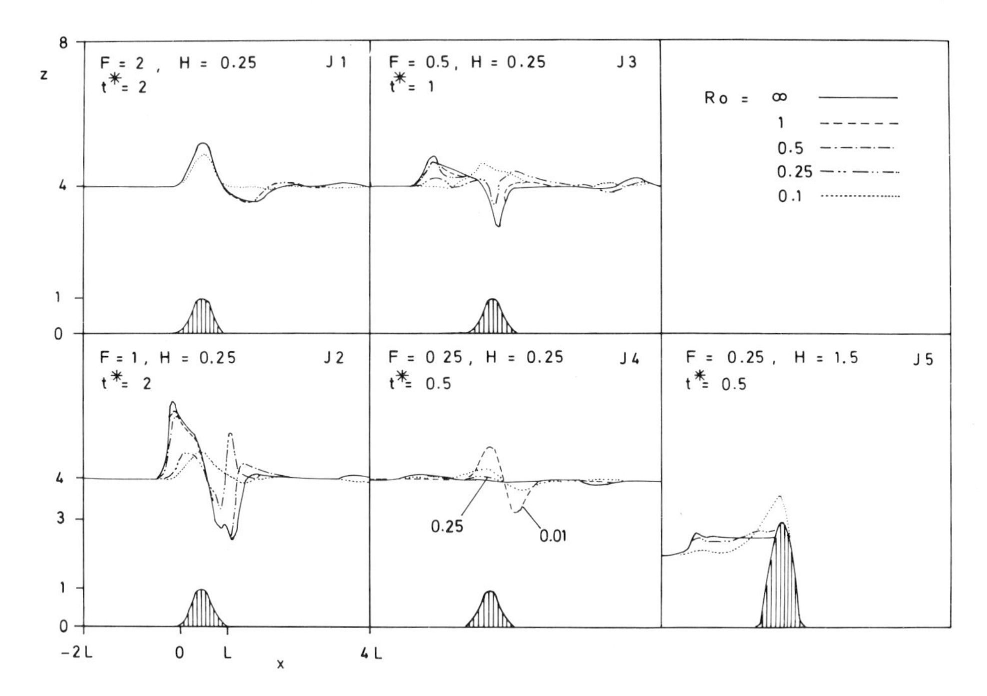

Fig. 4 Influence of rotation on the formation of hydraulic jumps. Each panel shows the results for one of the cases J1 to J5 at a fixed normalized instant of time t^* for various values of the Rossby number *Ro* . For case J4, the labelling of the curves differs and is given explicitly.

In simplified terms, the scale analysis suggests that Coriolis forces are small in comparison to gravity forces if F^2/Ro is small.

Figure 4 shows the solutions for the five cases at fixed times for various values of the Rossby number. For Rossby numbers larger than unity the effect of rotation is negligble throughout all five cases. For very small Rossby numbers, the Coriolis forces cause steady flow over the ridge at constant depth of the fluid layer. We see that the effect of rotation in case J1 with the largest Froude number is negligble for Rossby numbers greater than 1 ($F^2/Ro = 4$). For smaller Froude numbers, e.g. for cases J4 with F = 0.25, the effect of rotation is negligble for the even smaller Rossby number 0.1 ($F^2/Ro = 1.25$). Therefore this case was run also for $Ro = 0.01$. This shows, that F^2/Ro is the controlling characteristic number and that the effect of rotation on the formation of hydraulic jumps is negligble for F^2/Ro less than unity at least if H is of order unity and F is not too large. Otherwise the more complicated conditions given in Eqs. (25-27) apply.

REFERENCES

[1] BAINES, P. G.: "The dynamics of the southerly buster", Australian Met. Mag. **28** (1980) pp. 175-200.

[2] HOUGHTON, D. D.: "Effect of rotation on the formation of hydraulic jumps", J. Geophys. Res. **74** (1969) pp. 1351-1360.

[3] HOUGHTON, D. D., KASAHARA, A.: "Nonlinear shallow fluid flow over an isolated ridge", Comm. Pure Appl. Math. **21** (1968) pp. 1-23.

[4] LONG, R. R.: "Finite amplitude disturbances in the flow of inviscid rotating and stratified fluids over obstacles", Ann. Rev. Fluid Mech. **4** (1972) pp. 69-92.

[5] PIELKE, R. A.: "Mesoscale meteorological modeling", Academic press, Orlando, USA, 612 pp., 1984.

[6] SCHUMANN, U.: "Influence of mesoscale orography on idealized cold fronts", J. Atmos. Sci. (1987) in press.

[7] SCHUMANN, U., VOLKERT, H.: "Three-dimensional mass- and momentum-consistent Helmholtz-equation in terrain-following coordinates", in: W. Hackbusch (Ed.), Notes on Numer. Fluid Mech. Vol. 10, Vieweg, Braunschweig, pp. 109-131, 1984.

[8] SCHUMANN, U., HAUF, T., HÖLLER, H., SCHMIDT, H., VOLKERT, H.: "A Mesoscale Model for the Simulation of Turbulence, Clouds and Flow over Mountains: Formulation and Validation Examples", Beitr. Phys. Atmosph. **60** (1987) in press.

[9] WILLIAMS, R. T., and A. M. HORI, A.M.: "Formation of hydraulic jumps in a rotating system", J. Geophy. Rev. **75** (1970) pp. 2813-2821.

AN IMPLICIT METHOD FOR THE COMPUTATION OF UNSTEADY INCOMPRESSIBLE VISCOUS FLOWS

H. Schütz, F. Thiele
Technische Universität Berlin,
Hermann-Föttinger-Institut für Thermo- und Fluiddynamik,
Straße des 17. Juni 135, D-1000 Berlin 12

SUMMARY

A finite-difference method for the solution of the two-dimensional incompressible, unsteady Navier-Stokes equations in terms of the stream function is presented. The implicit procedure applies backward time and central space approximation to the fourth-order differential equation. Newton linearization together with the direct solution of the linear system of finite-difference equations leads to a successive correct solution to the unsteady Navier-Stokes equations. For the circular cylinder and the symmetric airfoil the unsteady flow is numerically calculated and compared with available data.

INTRODUCTION

Most methods for the numerical solution of the two-dimensional Navier-Stokes equations are based on formulations which result in coupled systems of second-order differential equations. Lecointe and Piquet [1] have solved the vorticity-stream function formulation numerically by means of compact schemes for the incompressible flow around a circular cylinder. The procedure proposed consists of a "Mehrstellen" spatial discretization and an optimized ADI method to solve the Poisson equation. Using the Beam-Warming method the vorticity equation is time-differenced so that this technique allows the splitting of the space operator. The same formulation in generalized curvilinear coordinates has been applied by Ghia et al. [2] in order to investigate the unsteady flow past airfoils. Their numerical method solves the discretized equations by applying an ADI method for the vorticity transport equation and a block Gaussian elimination technique for the stream function equation. Fasel [3] has developed a numerical procedure based upon the Navier-Stokes equations in terms of vorticity and velocity, for the investigation of the stability of boundary layers. By contrast with the previous formulation two Poisson equations for the velocity components have to be solved. Backward time and central space differencing are employed to maintain an overall second order accurate scheme. The simultaneous solution of the three systems of finite-difference equations has been obtained by a coupled algorithm with line-iteration.

In recent years the stream function formulation of the Navier-Stokes equations has successfully been used to study steady flow problems in two dimensions [4-7]. For two-dimensional flows this formulation has several advantages in comparision with other approaches as only one fourth-order differential equation has to be solved. Since the solution of higher-order differential equations is much more difficult numerical procedures have only been developed in the last ten years. In 1975 Roache and Ellis [4] suggested the Biharmonic Driver method which makes use of a direct linear (BID) biharmonic solver. For flows in simple geometries this technique converged within a few iterations but only at low Reynolds numbers. To overcome this restriction Walter and Larsen [5] improved the BID method by applying an iterative Newton scheme to deal with the nonlinear terms of the stream function equation. The finite-element method of Tuann and Olsen [6] replaced the fourth-order differential equation and the boundary conditions by an equivalent variational principle. Numerical solutions for the steady flow past a circular cylinder were presented for Reynolds numbers ranging from 1 to 100. Thiele et al. [7] extended the finite-difference method of Hermitian type to solve the Navier-Stokes equations in terms of a stream function. The main features of the procedure consist of Newton-Raphson linearization and direct solution of the system of finite-difference equations. Using a combination of well-known numerical methods Schreiber and Keller [8] developed an efficient and reliable stream function solver. The techniques applied include central differences on a uniform net, Richardson extrapolation, sequences of Newton and chord iterations together with the LU-factorization of the Jacobian matrix. Accurate results for the driven cavity problem were obtained up the Reynolds numbers 10,000.

For the prediction of unsteady incompressible viscous flows a fully implicit method will be developed which uses the stream function formulation of the Navier-Stokes equations. The stream function equation is first discretized by backward finite-differences in time such that at each time step a nonlinear fourth-order differential equation must be solved. By using Newton linearization the derivatives in space are approximated by a combination of the finite-difference method of Hermitian type and the compact differencing technique [9]. The linear system of algebraic equations for the unknowns, the grid point values of the function and its first derivatives is directly solved by applying a block LU-decomposition of the coefficient matrix. The procedure developed is then applied to calculate the unsteady flow around a circular cylinder as well as the symmetric airfoil.

STREAM FUNCTION EQUATION AND BOUNDARY CONDITIONS

The incompressible two-dimensional Navier-Stokes equations, in conjunction with the continuity equation, are written in terms of the scalar stream function ψ as a nonlinear fourth-order differential equation

$$\frac{\partial\Delta\psi}{\partial t} + \frac{\partial\psi}{\partial y}\frac{\partial\Delta\psi}{\partial x} - \frac{\partial\psi}{\partial x}\frac{\partial\Delta\psi}{\partial y} - \frac{1}{Re}\Delta\Delta\psi = 0 \qquad (1)$$

where x,y are non-dimensional Cartesian coordinates. Here, Re is Reynolds number defined as $Re = U_\infty L_R/\nu$. The stream function

$$u = \frac{\partial\psi}{\partial y}, \qquad v = -\frac{\partial\psi}{\partial x} \qquad (2)$$

satisfies the continuity equation by definition.

The application to flows around an arbitrary body requires a transformation to general curvilinear coordinates. Here, we consider the transformation to orthogonal coordinates and the flow around a circular cylinder. The advantage of an orthogonal transformation consists in much simpler metric coefficients without any restriction in regard of the physical solution. As general transformations to the circular cylinder always exist in two dimensions this flow geometry will be the basic problem which has to be considered for the numerical solution procedure.

For the circular cylinder the transformation of the coordinates is given by the following relations

$$x = e^{\eta} \cos \xi \; ; \qquad y = e^{\eta} \sin \xi \tag{3}$$

where the function e^{η} assures an appropriately clustered grid point distribution close to the cylinder surface. In the transformed ξ,η-plane the stream function equation reads

$$\begin{aligned} &\frac{\partial}{\partial t}\left[e^{2\eta}\left(\frac{\partial^2\psi}{\partial\xi^2} + \frac{\partial^2\psi}{\partial\eta^2}\right)\right] + 2\frac{\partial\psi}{\partial\xi}\left(\frac{\partial^2\psi}{\partial\xi^2} + \frac{\partial^2\psi}{\partial\eta^2}\right) + \frac{\partial\psi}{\partial\eta}\left(\frac{\partial^2\psi}{\partial\xi\,\partial\eta^2} + \frac{\partial^3\psi}{\partial\xi^3}\right) \\ &- \frac{\partial\psi}{\partial\xi}\left(\frac{\partial^3\psi}{\partial\eta^3} + \frac{\partial^3\psi}{\partial\xi^2\,\partial\eta}\right) - \frac{1}{Re}\left\{ 4\left(\frac{\partial^2\psi}{\partial\xi^2} + \frac{\partial^2\psi}{\partial\eta^2}\right) - 4\left(\frac{\partial^3\psi}{\partial\xi^2\,\partial\eta} + \frac{\partial\psi}{\partial\eta^3}\right) + \right. \\ &\left. + \frac{\partial^4\psi}{\partial\xi^4} + 2\frac{\partial^4\psi}{\partial\xi^2\,\partial\eta^2} + \frac{\partial^4\psi}{\partial\eta^4} \right\} = 0 \;. \end{aligned} \tag{4}$$

The boundary conditions follow from the definition of the stream function and can be determined straightforwardly. The no-slip condition at the solid wall is

$$\eta = 0: \qquad \psi = \text{const.} \quad \frac{\partial\psi}{\partial\eta} = 0 \;. \tag{5}$$

In the far field of the cylinder the boundary conditions are derived from the inviscid flow

$$\psi = \left(e^{\eta}-e^{-\eta}\right) \sin \xi, \qquad \frac{\partial\psi}{\partial\xi} = \left(e^{\eta}-e^{-\eta}\right) \cos \xi, \qquad \frac{\partial\psi}{\partial\eta} = \left[e^{\eta}+e^{-\eta}\right] \sin \xi \;. \tag{6}$$

In order to start the solution procedure initial conditions have to be specified. We assume that the potential flow is valid within the whole region whereas equation (5) holds for the cylinder surface.

FINITE-DIFFERENCE APPROXIMATION AND SOLUTION PROCEDURE

Due to the parabolic character of the stream function equation with respect to the time derivative the finite-difference approximation can be performed separately in time and space. We first rewrite equation (4) as

$$Re\,\frac{\partial \mathbf{E}(\psi)}{\partial t} + Re\,\mathbf{F}(\psi) + \mathbf{G}(\psi) = 0 \;. \tag{7}$$

Here, $\mathbf{F}(\psi)$ and $\mathbf{G}(\psi)$ represent the nonlinear convection term and the biharmonic diffusion term, respectively.

The time discretization of the first term in equation (7) applies three-point backward finite differences with constant step size except for the first time step where two-point formulae are used. This approximation results in a fourth-order differential equation

$$\frac{Re}{\Delta t}\left(1 + \frac{\gamma}{2}\right)\mathbf{E}^n + Re\ \mathbf{F}^n + \mathbf{G}^n = \frac{Re}{\Delta t}\left(1 + \gamma\right)\mathbf{E}^{n-1} - \frac{Re}{\Delta t}\frac{\gamma}{2}\mathbf{E}^{n-2} \equiv RHS \qquad (8)$$

which has to be solved for each time level n. The parameter γ allows a choice between first and second order accurate approximations

$$\begin{aligned} \gamma = 0: &\quad O(\Delta t) \quad \text{two-point backward} \\ \gamma = 1: &\quad O(\Delta t^2) \quad \text{three-point backward}\ . \end{aligned} \qquad (9)$$

The right hand side of equation (8) depends only on the values of the previous time step.

The nonlinear convection term $\mathbf{F}(\psi)$ implies a Newton iteration procedure of the form

$$\psi^n = \bar{\psi}^n + d^n \qquad (10)$$

where the bar represents the solution of the previous iteration and d^n is the corresponding correction. By applying this linearization procedure equation (8) reduces to the linear differential equation in terms of the correction d

$$\frac{Re}{\Delta t}\left(1+ \frac{\gamma}{2}\right)\mathbf{E}^n(d) + Re\ \mathbf{F}_1^n(d,\bar{\psi}) + \mathbf{G}^n(d) = RHS - \frac{Re}{\Delta t}\left(1+ \frac{\gamma}{2}\right)\mathbf{E}^n(\bar{\psi}) - Re\ \mathbf{F}_2^n(\bar{\psi}) - \mathbf{G}^n(\bar{\psi}) \qquad (11)$$

where quadratic terms of the correction have been neglected. $\mathbf{F}_1^n(d,\bar{\psi})$ and $\mathbf{F}_2^n(\bar{\psi})$ contain the linearized part of the convection terms.

A main difficulty in solving equation (11) results from the derivatives of third and fourth order. An accurate approximation of Hermitian type has been proposed by Thiele et al. [7]. This method relies on five-point formulae using the function and its first derivatives including the mixed one. Hence, at each grid point four unknowns appear in the finite-difference representation of equation (11). Wagner [9] improved this technique considerably by eliminating the mixed derivative so that only three finite-difference equations are required to determine the grid point values of the function $d_{i,j}$ and the first derivatives $(d^\xi)_{i,j}$ and $(d^\eta)_{i,j}$. Using "Mehrstellen" formulas in each space direction together with the central collocation of the differential equation (11) this technique is an optimal combination of the finite-difference method of Hermitian type and the compact differencing procedure. For constant grid spacing the "Mehrstellen" formulas are given by

$$d^\eta_{i,j-1} + 4\, d^\eta_{i,j} + d^\eta_{i,j+1} + \frac{3}{\Delta\eta}\left(d_{i,j-1} - d_{i,j+1}\right) = O(\Delta\eta^4) \qquad (12)$$

$$d^\xi_{i-1,j} + 4\, d^\xi_{i,j} + d^\xi_{i+1,j} + \frac{3}{\Delta\xi}\left(d_{i-1,j} - d_{i+1,j}\right) = O(\Delta\xi^4)\,. \qquad (13)$$

The collocation method which employs finite-difference approximations such as

$$d^{\xi\xi\xi}_{i,j} = \frac{15}{2(\Delta\xi)^3}\left(d_{i+1,j} - d_{i-1,j}\right) - \frac{3}{2(\Delta\xi)^2}\left(f^\xi_{i-1,j} + 8f^\xi_{i,j} + f^\xi_{i+1,j}\right) + O(\Delta\xi^4) \qquad (14)$$

$$d^{\eta\eta\eta\eta}_{i,j} = \frac{12}{(\Delta\eta)^4}\left(d_{i,j+1} - 2d_{i,j} + d_{i,j-1}\right) + \frac{6}{(\Delta\eta)^2}\left(f^\eta_{i,j+1} - f^\eta_{i,j-1}\right) + O(\Delta\eta^2) \qquad (15)$$

leads to the third finite-difference equation required to determine the grid point values of $d_{i,j}$, $(d\xi)_{i,j}$ and $(d\eta)_{i,j}$. As the first derivatives of the stream function are in fact the velocity components, most boundary conditions can be directly incorporated into the finite-difference scheme.

The technique described is applied to all interior grid points. This method results in a 3×3 block-tridiagonal system of finite-difference equations where the submatrices have a tridiagonal structure. Wagner [9] developed a solution procedure which is based upon the LU-decomposition of the matrix. Following Schreiber and Keller [8] this direct solver allows achoice between a Newton and chord iteration. The chord method updates the whole right hand side of equation (11) while the Newton method requires in addition the decomposition of the matrix at each iteration. At each time level the direct stream function solver is applied to the solution of equation (7) which yields a fully implicit method. Due to the Newton linearization the procedure converges within few iterations at each time step. In this way a successive correct solution to the unsteady viscous flow equations is always assured.

RESULTS

In order to investigate the features of the numerical solution procedure the flow around a circular cylinder is considered. This unsteady flow problem has been investigated frequently by several authors. Thus the numerical results can be compared with experimental as well as theoretical data.

A typical grid applied in the numerical calculations was 61 meridional points and 99 radial points. The average number of iterations required at each time step consists of one Newton and five chord iterations. Here, it should be pointed out that no artificial viscosity or upwind approximation had to be introduced into the numerical scheme at higher Reynolds numbers to get a converging solution procedure. This may be due to the fact that the convection terms are approximated by fourth-order accurate formulae. As only central finite-difference approximations are used a small disturbance such as an impulsive rotation of the cylinder was necessary to obtain the unsteady flow behaviour. Fig. 1 shows the flow field around a circular cylinder for Re = 200 at various times. The Strouhal number for the free vortex shedding case calculated by the Fourier analysis is $S_0 = 0.165$. This value is in good agreement with available data.

The application of the method developed to calculate the flow around a symmetric airfoil requires a transformation to a circle. The transformation used consists of the mapping of the airfoil onto a near-circle by means of a complex function and the mapping of the near-circle onto the circle which is described by two Fourier series expansions.

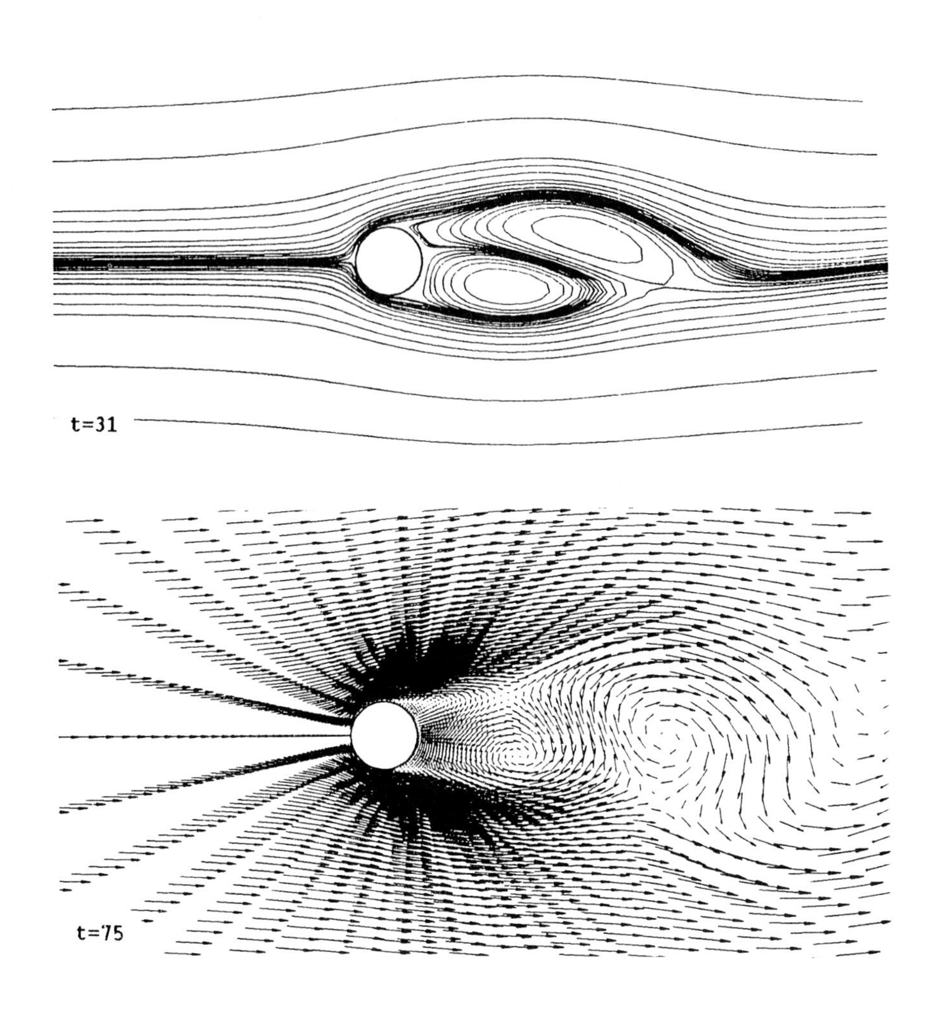

Figure 1: Stream function and velocity distribution for Re = 200 at t = 31 and t = 75, respectively.

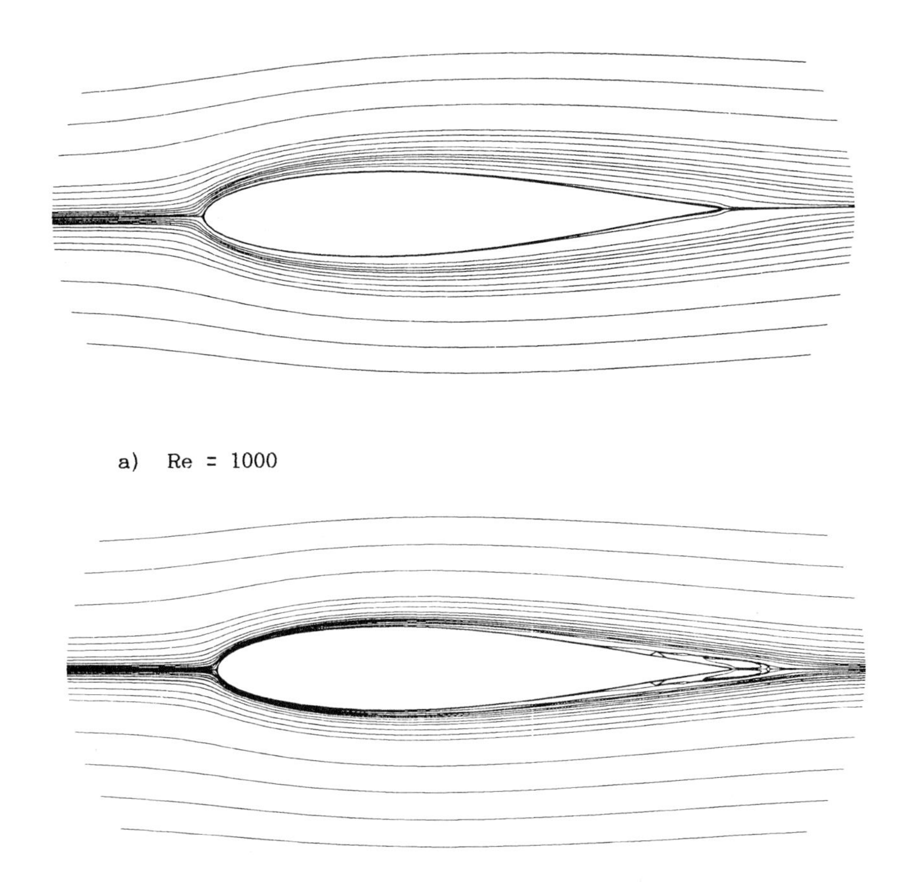

Figure 2: Symmetric airfoil flow at two different Reynolds numbers

In Fig. 2 the numerical results are presented for the flow around an airfoil. The stream function distribution indicates an attached flow at Re = 1000 whereas the flow separates at the trailing edge for Re = 10,000. For both cases an imposed disturbance did not lead to an unsteady motion. Oscillations were damped out so that the time-dependent solution approached to the steady flow field.

CONCLUDING REMARKS

A fully implicit finite-difference method has been derived to solve the unsteady Navier-Stokes equations for incompressible two-dimensional flows. Due to the stream function formulation the Newton method applied to the nonlinear convection terms converged within a few iterations. The results presented for the circular cylinder and the symmetric airfoil demonstrate that the solution procedure developed is able to simulate unsteady flows quite accurately.

ACKNOWLEDGMENT

This work was supported by the Deutsche Forschungsgemeinschaft - Schwerpunktprogramm: Finite Approximationen in der Strömungsmechanik.

REFERENCES

[1] Lecointe, Y., Piquet, I.: "On the use of several compact methods for the study of unsteady incompressible viscous flow round a circular cylinder". Computers and Fluids, 12 (1984) pp. 281-292.

[2] Ghia, K.N., Osswald, G.A., Ghia, U.: "Analysis of two-dimensional incompressible flow past airfoils using unsteady Navier-Stokes equations" In: Numerical and Physical Aspects of Aerodynamic Flows III (T. Cebeci, Ed.), Springer-Verlag, New York (1985) pp. 318-338.

[3] Fasel, H.: "Investigation of the stability of boundary layers by a finite-difference model of the Navier-Stokes equations". J. Fluid Mech., 78 (1976) pp. 355-383.

[4] Roache, P.J., Ellis,M.A.: "The BID method for the steady-state Navier-Stokes equations". Computers and Fluids, 3 (1975) pp. 305-320.

[5] Walter, K.T., Larsen, P.S.: "The FON method for the steady two-dimensional Navier-Stokes equations". Computers and Fluids, 9 (1981) pp. 365-376.

[6] Tuann, S.Y., Olson, M.D.: "Numerical studies of the flow around a circular cylinder by a finite element method". Computers and Fluids, 6 (1978) pp. 219-240.

[7] Thiele, F., Wagner, H., Wepler, J.: "Solution of the Navier-Stokes equations using the finite-difference method of Hermitian type". Proc. Fourth GAMM-Conference on Numerical Fluid Mechanics; Notes on Numerical Fluid Mechanics, 5 pp. 326-336, Vieweg Verlag, Braunschweig, 1982.

[8] Schreiber, R., Keller, H.B.: Driven cavity flows by efficient numerical techniques. J. Comput. Phys., 49 (1983) pp. 310-333.

[9] Wagner, H.: Ein Differenzenverfahren für die Navier-Stokes Gleichungen in mehrfach zusammenhängenden Gebieten. PhD Thesis, TU Berlin 1985.

COMPUTATION OF VISCOUS SUPERSONIC FLOW AROUND BLUNT BODIES

R. Schwane, D. Hänel
Aerodynamisches Institut der RWTH Aachen
Wüllnerstr. zw. 5 und 7, 5100 Aachen, West Germany

SUMMARY

The three-dimensional Navier-Stokes equations are solved with an upwind relaxation scheme combined with a shock fitting procedure. The three-dimensional grid generation is carried out by an optimization procedure. Computational results are presented for supersonic flow around two blunt body configurations.

INTRODUCTION

The computation of 3-D, supersonic flow around blunt bodies is of increasing interest for the future development of spacecrafts and supersonic airplanes. The flow around blunt bodies is characterized by strong shock waves, embedded subsonic regions, and shock-boundary layer interactions with separation. To attack such flow problems the Navier-Stokes equations must be solved. The numerical method required for such a problem should be sufficient accurate in viscous flow regions and should have the properties of a high resolution scheme in the nearly inviscid flow portion.

In previous works /1, 2 / a computational method for the two-dimensional equations has been developed which meets essentially these requirements. This method is based on an upwind high resolution scheme using van Leer's flux splitting. A favourable rate of convergence and robustness was proved in a number of flow calculations.

The present paper deals with the extension of this method to three dimensions. Computed results for a hemisphere-cylinder body and for a double-ellipsoidal body are presented and compared with other results.

GOVERNING EQUATIONS AND METHOD OF SOLUTION

The viscous three-dimensional flow of a compressible gas is described by the time-dependent Navier-Stokes equations. The conservative non-dimensional equations are formulated in a curvilinear body fitted time-dependent coordinate system with

$$\xi = \xi(t, x, y, z) \quad , \qquad \eta = \eta(t, x, y, z) \quad , \qquad \zeta = \zeta(t, x, y, z) \; . \tag{1}$$

The coordinate system is chosen nearly orthogonal with the η -coordinate normal to the body surface in order to apply the thin-layer approximation of the Navier-Stokes equations. Then the governing equations read

$$Q_t + F_\xi + G_\zeta + \left(H - \frac{1}{Re} S \right)_\eta = 0 \tag{2}$$

where Q represents the vector of the conservative variables, F, G, H are the Euler fluxes and S contains the viscous terms and heat conduction terms in η -direction. The detailed formulation can be found elsewhere, e.g. in /3/.

In the present paper the gas is assumed to be thermally and calorically perfect, the viscosity is described by Sutherland's law. The conservation equations are approximated by an implicit conservative difference scheme of first order accuracy in time and second order in space.

An upwind high resolution scheme is used for the Euler part and a central discretization for the viscous terms. For the Euler terms the flux vector splitting by van Leer /4/ is adopted. According the MUSCL principle /5/ the conservative variables in the split Euler fluxes at the cell interfaces are extrapolated choosing piecewise linear distribution over the cell resulting in a second-order accurate approximation. To avoid numerical oscillations the higher order extrapolation is limited by a limiter function of van Albada et al. /6/.

The time derivative is descretized by an Euler backward difference. This results in an implicit unfactored scheme, which is solved iteratively for each time step.

$$LHS(Q^n)\cdot(Q^{n+1,\nu}-Q^n)=RHS(Q^n)\,. \tag{3}$$

The operator $LHS(Q^n)$ consists of the Jacobians of the split fluxes and of that of the viscous terms, the $RHS(Q^n)$ represents the steady-state difference scheme. A block line Gauß-Seidel method in alternating directions is applied as the relaxation method. For steady state solutions first order upwinding for the LHS is used and the time-step is controled by the residual to accelerate the convergence.

MESH GENERATION

In three-dimensional calculations the mesh arrangement becomes very important for accurate results since in general the number of grid points is restricted by the computer capacity. Especially for viscous flow problems, the mesh must be sufficient dense close to solid surfaces and almost orthogonal in regions, where viscous flow effects dominate. In the present work the mesh is generated numerically with an iterative method similar to that in /7/. The distribution of the grid points is calculated by a decision function F which must be minimized

$$F=\sum_i\sum_j\sum_k(\alpha\,\theta_{i,j,k}+S_{i,j,k})=Min. \tag{4}$$

where θ is a measure for orthogonality and S for smoothness and volume control. The measures θ and S are defined by the local vectors $\vec{r}$ between the neighbouring mesh points, e.g.

$$\theta_{i,j,k}=\sum_{m=1}^{6}\sum_{n=1}^{6}\frac{(\vec{r}_m\cdot\vec{r}_n)^2}{|r_m|^2\cdot|r_n|^2}\quad,\qquad m\neq n \tag{5}$$

$$S_{i,j,k}=\sum_{m=1}^{6}|r_m|^2\,\beta_m\,. \tag{6}$$

The solution is carried out by an iterative relaxation method. The applications have shown that this mesh generation method is robust and can be easily adapted to different flow problems. Typical C-O mesh configurations as used for the present blunt-body calculations will be shown below.

BOUNDARY CONDITIONS

The domain of integration for computing the supersonic flow around a blunt body is sketched in Fig. 1. From this three typical boundaries are formed by the bow shock, by the body contour and by the outflow boundary. On the body the boundary conditions are given by the no-slip and thermal conditions at the wall, at the outflow boundary the variables are extrapolated from the interior flow field assuming supersonic and boundary layer flow, respectively.

A shock-fitting procedure is applied at the outer boundary, so that the position of the detached bow shock is a result of the calculation. The state behind the shock is determined by the jump conditons over the shock.

Starting from a guessed shock position the variables behind the shock are calculated from the Rankine-Hugoniot relations and used as the boundary conditions. After one time step the solution is advanced and a corrected value of the pressure can be extrapolated from the interior flow field to the shock. From this pressure value and from the local shock slope a transient free stream Mach number Ma_∞^ν can be calculated. According to its difference to the prescribed Mach number Ma_∞ the shock can be shifted locally by:

$$\Delta \vec{r}^{\,\nu} = a_\infty \cdot \Delta t \cdot (Ma_\infty^\nu - Ma_\infty). \tag{7}$$

The vector Δ r is developed on the shock surface in a triangulated grid of regular cell size, and then redistributed to the original computational grid by interpolation. This fitting procedure avoids the problems near grid singularities and allows time steps as large as used in the inner scheme.

RESULTS

The present method of solution was applied to different viscous, two-dimensional flow problems /1, 2 /and was found to be a robust and efficient solver. However, the use of an upwind scheme for the solution of the Navier-Stokes equations requires a careful checking of the viscous results to avoid undesirable influences of the upwind discretization and flux splitting on the spatial accuracy. Several aspects of that problem were investigated in /8, 9/.

The present paper deals with the application of the three-dimensional algorithm for supersonic flow around blunt bodies. For testing the three-dimensional algorithm, the mesh generator, as well as the shock fitting procedure, the supersonic flow around a hemisphere-cylinder body was calculated and the results were compared with that of axisymmetric calculations in /10, 11/.

The C-O mesh arrangement for the hemisphere-cylinder body with 50x27x14 grid points is shown in Fig. 2. The calculations were carried out for an inflow Mach number of 2.94 and a Reynolds number of $2.2 \cdot 10^5$.

The first calculations with the original flux splitting by van Leer resulted in an unphysical high wall temperature in regions of strong Mach number changes (e. g. stagnation point), which was mainly caused by the non-preservation of the total enthalpy using the split fluxes. The present study has shown that the deviation can be removed substantially by a modification of the original split energy flux /4/ which reads now:

$$F^{\pm}_{energy} = \pm \frac{1}{4} \rho \cdot a \cdot (Ma \pm 1)^2 \cdot h_{tot}(Q^{\pm}). \tag{8}$$

By this, the total enthalpy h_{tot} is preserved and transported to the cell interfaces through the split mass fluxes. The Fig. 3 shows the improvement of the accuracy for the wall temperature distribution, related to the inflow stagnation temperature. The full and dashed lines represent the results with the modified and the original split fluxes, respectively, whereas the circles mark the axisymmetric results from /10/.

Further comparisons were made with the results by Müller /11/. The isotherms and the shock contour from the present calculation and from /11/ are plotted in Fig. 4a and 4b. Although the isothermes do not represent the same values the results are in close agreement.

The isobars are plotted in the symmetry plane and in the outflow plane in Fig. 5. It shows a typical disturbance near the polar singularity of the C-O grid (see Fig. 2). To avoid this mesh singularity, which impairs the accuracy and the convergence as well, a subgrid was inserted to spread the grid spacing in that region.

Fig. 6 shows the mesh arrangement with a subgrid for a double-ellipsoidal body, as defined geometrically in the experimental work /12/. The surface of the body is formed by two ellipsoidals with the same center but different half axis. The nearly orthogonal grid distribution was obtained after 30 iterations of the present grid optimization procedure, starting from a rough algebraic estimation. The flow around this double-ellipsoidal body was calculated for an inflow Mach number of 8.15 and a Reynolds number of $2\ 10^6$ with a mesh of 76x16x14 grid points. This sparse mesh used for the development and test of the present scheme requires careful clustering of the grid points in critical regions, like near the intersection of the two ellipsoidals where the boundary layer interacts with the embedded shock. Fig. 7 shows the calculated isobars at the surface of the body and in the symmetry plane. The isobars indicate an embedded shock near the intersection of the ellipsoidals which is interacting with the outer bow shock. The corresponding lines of constant density are plotted in Fig. 8. A comparison with the first experimental results /12/, given as a Schlieren picture for this case in Fig. 9, shows qualitative agreement.

CONCLUSIONS

An implicit relaxation method was developed for the solution of the time-dependent three-dimensional Navier-Stokes equations. The method is based on higher order upwinding and flux splitting for the Euler terms and central discretization for the viscous terms. This method is coupled with a shock fitting procedure which allows the same large time steps as the implicit scheme. A robust and flexible optimization procedure was developed for the generation of nearly orthogonal meshes with controled clustering of the grid points. Calculations for three-dimensional supersonic flow over blunt bodies and comparison with other results have confirmed the spatial accuracy of the present method. For more complex configurations and more detailed flow informations the grid point density has to be increased and for higher Mach numbers real gas effect has to be taken into account. Corresponding work is planed.

REFERENCES

[1]. SCHRÖDER, W., HÄNEL, D.: "An unfactored implicit scheme with multigrid acceleration for the solution of the Navier-Stokes equations", Computers and Fluids, 15 (1987) pp. 313-336.

[2]. SCHRÖDER, W.: "Numerische Integration der Navier-Stokes-Gleichungen unter Verwendung des Mehrgitter-Konzeptes", Thesis, RWTH Aachen (1987).

[3]. PULLIAM, T. H., STEGER, J. L,: "Implicit finite-difference simulations of three-dimensional compressible flow", AIAA J., 18 (1980) pp. 159-167.

[4]. VAN LEER, B.: "Flux-vector splitting for the Euler equations", Lecture Notes in Physics 170 (1982) pp. 507-512.

[5]. VAN LEER, B.: "Towards the ultimate conservative difference scheme V. A second-order sequel to Godunov's method", J. Comp. Phys. 32 (1979) pp. 101-136.

[6]. VAN ALBADA, G. D., VAN LEER, B., and ROBERTS, W. W.: "A comparative study of computational methods in cosmic gas dynamics", Astron. Astrophys., 108 (1982) pp. 76-84.

[7]. CARCAILLET, R., KENNON, S. R., DULIKRAVITCH, G. S.: "Generation of optimum three-dimensional computational grids", Notes on Numerical Fluid Mechanics, 13 (1985) pp. 38-46, Vieweg.

[8]. VAN LEER, B., THOMAS, J. L., ROE, P. L., NEWSOME, R. W.: " A comparison of numerical flux formulas for the Euler and Navier-Stokes equations", AIAA paper 87-1104 CP (1987).

[9]. HÄNEL, D.: On the accuracy of upwind schemes in solutions of the Navier-Stokes equations", AIAA paper 87-1105 CP (1987).

[10]. VIVIAND, H., GHAZZI, W.: "Numerical solution of the Navier-Stokes equations at high Reynolds numbers with application to the blunt body problem", Proceedings of the Fifth International Conference on Numerical Methods in Fluid Dynamics, Lecture Notes in Physics, 59 (1976) pp. 434-439, Springer Berlin.

[11]. MÜLLER, B.: "Calculation of axisymmetric laminar supersonic flow over blunt bodies", DFVLR report, IB-84 A02 (1984).

[12]. DE COSTA, J. L., AYMER DE LA CHEVALERIE, D., ALZIARY DE ROQUEFORT, T.: "Ecoulement tridimensionnel hypersonique sur une combinaison d'ellipsoides", Rapport N.4RDMF86, Université de Poitieres, France, (1987).

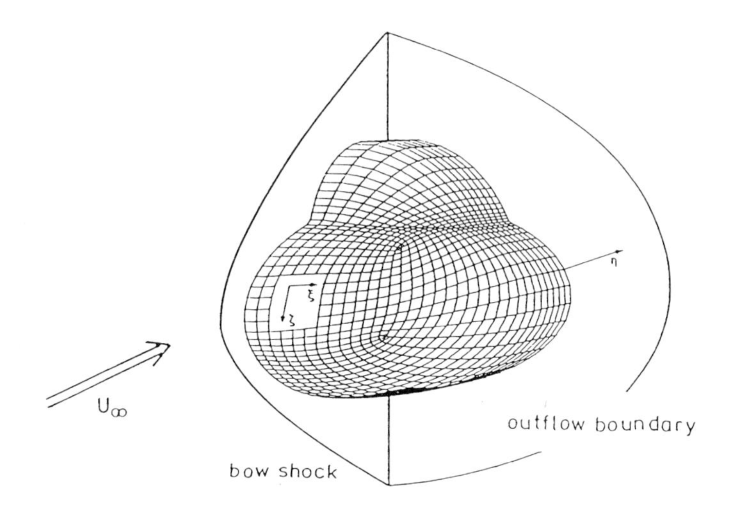

Fig. 1
Domain of integration for the calculation of supersonic flow around blunt bodies.

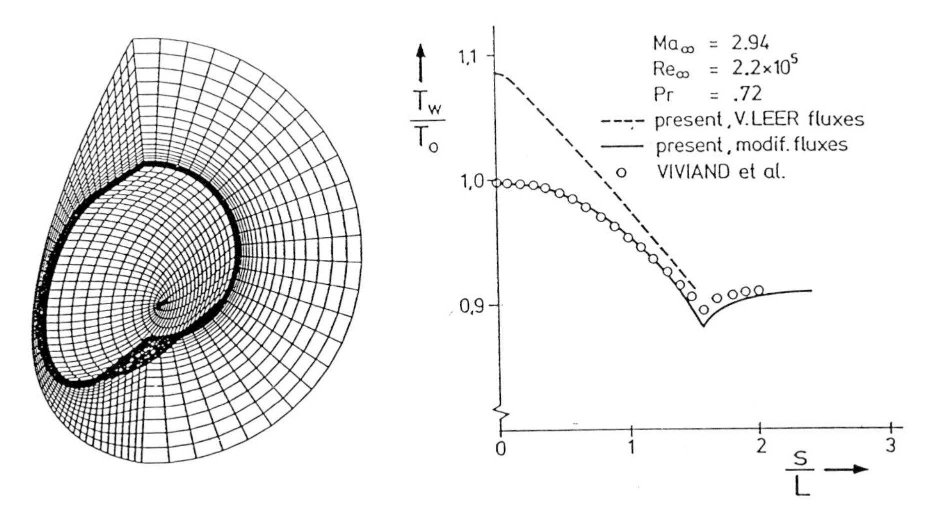

Fig. 2
Computational mesh for a hemisphere-cylinder body with 54x27x14 grid points.

Fig. 3
Wall temperature distribution related to the inflow stagnation temperature, over the arc length in the mid-plane of a hemisphere-cylinder body.

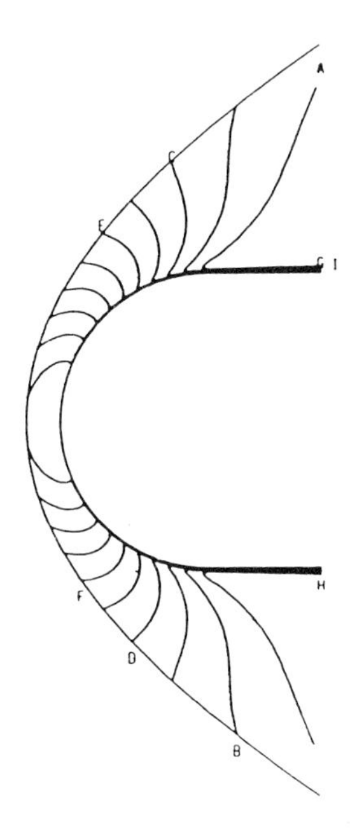

Fig. 4a
Lines of constant temperature in the mid-plane of a hemisphere-cylinder body (Ma = 2.94, Re = $2.2 \; 10^5$) present 3-D results.

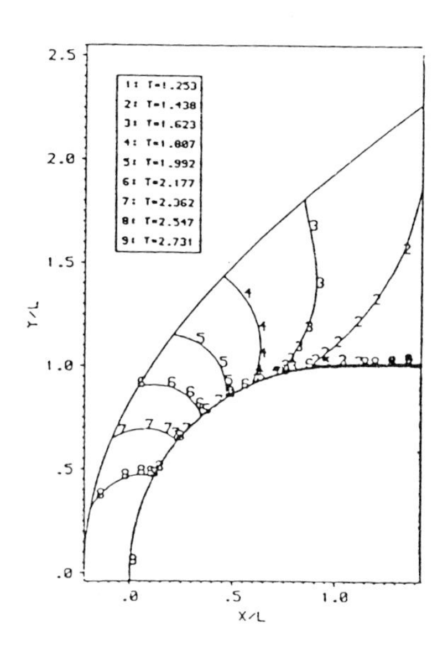

Fig. 4b
Legend see Fig. 4a, axisymmetric calculation /11/.

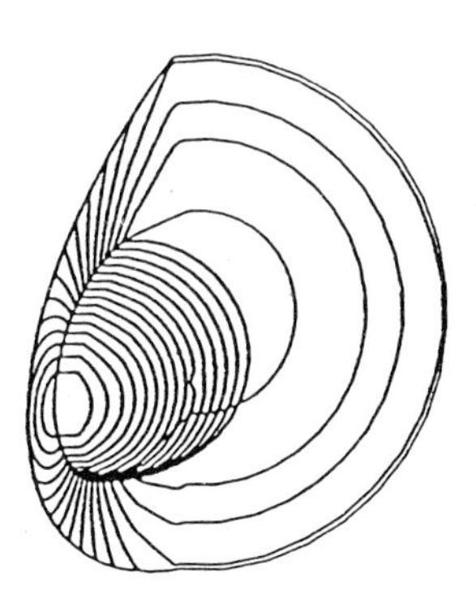

Fig. 5
Lines of constant pressure for the hemisphere-cylinder body.

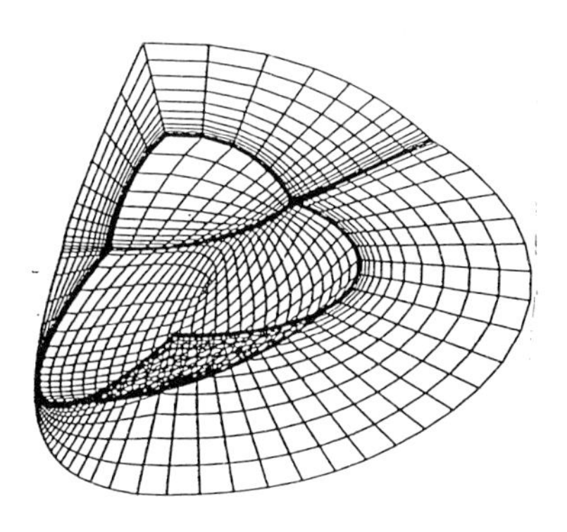

Fig. 6
Computational mesh for double-ellipsoidal body with 79x16x14 grid points.

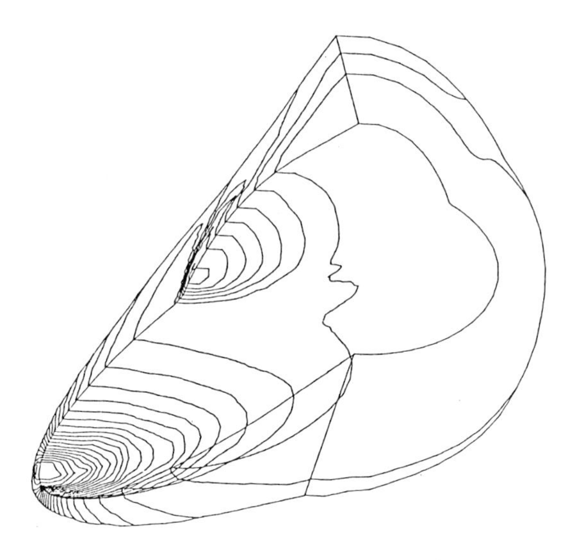

Fig. 7
Lines of constant pressure for the flow around the double-ellipsoidal body ($Ma = 8.15$, $Re = 2\ 10^6$, $\alpha = 0$).

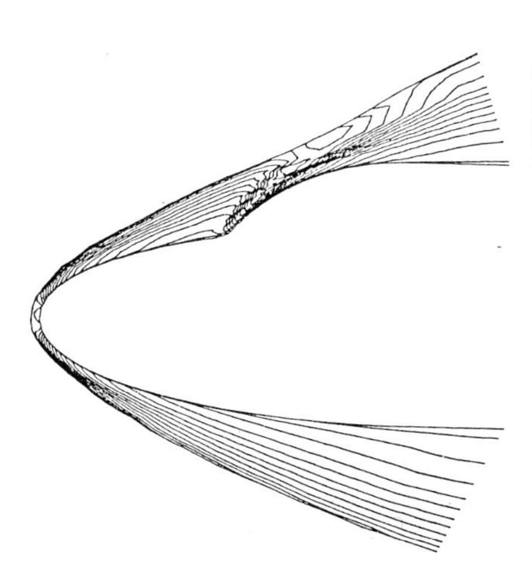

Fig. 8
Lines of constant density in the symmetry plane. (Legend see Fig. 7).

Fig. 9
Schlieren picture of the flow around the double-ellipsoidal body (taken from /12/). (Legend see Fig. 7).

NUMERICAL ALGORITHM FOR THE INVESTIGATION OF STRATIFIED FLOWS IN THE VARIABLES "VECTOR POTENTIAL-VORTICITY"

Yu.Shokin, V.Belolipetsky, V.Kostyuk

Computing Center, Siberian Division Academy of Sciences, Akademgorodok,660036,Krasnoyarsk,36,USSR

SUMMARY

The numerical algorithm for the solving of stratified fluid flows problems in variables "vector potential-vorticity" is considered. Numerical solution is constructed on the staggered mesh. The examples of computations for two- and three-dimensional flows are presented.

INTRODUCTION

The present work is devoted to a simulation of velocity and temperature fields in stratified reservoirs under various conditions of the intake. The used mathematical model of a heavy non-homogeneous fluid flow is based on the following assumptions:

- a temperature regime of the reservoir is known;
- the influence of heat-exchange processes may be neglected;
- equations of viscous fluid movement are taken in Boussinesq approximation;
- flows are laminar;
- in assumption that surface waves do not affect the flow, the rigid lid boundary condition is used to describe the water surface.

The system of equations is written in the variables "vector potential-vorticity". A numerical solution is constructed on the grid with divided velocity components using the method of splitting by physical factors.

EQUATIONS OF NON-HOMOGENEOUS HEAVY FLUID FLOW

A numerical study of viscous incompressible stratified by density fluid flows in the field of gravitational force is based on Navier-Stokes equations in Boussinesq approximation. The system of Navier-Stokes equations may be written in two equivalent forms [1, 2].

In natural variables "velocity-pressure".

$$\frac{\partial \rho}{\partial t} + (\bar{V} \cdot \nabla)\rho = 0, \tag{1}$$

$$\frac{\partial \bar{V}}{\partial t} + (\bar{V} \cdot \nabla)\bar{V} = \frac{1}{Fr^2}(-\nabla p + \bar{F}) + \frac{1}{Re}\Delta \bar{V}, \tag{2}$$

$$\mathrm{div}\,\bar{V} = 0. \tag{3}$$

Here (x,y,z) is a rectangular Cartesian coordinate system (axis y is directed downward); t is the time; the coordinates are related to the characteristic length H, time is related to H/U_o, U_o is a characterictic velocity value; ρ is the deviation of density from the characteristic value ρ_o related to $\Delta\rho$ $(\Delta\rho = \rho_{max} - \rho_{min})$; p is the deviation of pressure from the hydrostatic value p_o related to $\Delta\rho g H$; $\bar{V} = (u,v,w)$ is the velocity vector, u,v,w are related to the characteristic velocity U_o; $Re = \frac{U_o H}{\nu}$ is the Reynolds number; $Fr = \frac{U_o}{\sqrt{gH\Delta\rho/\rho_o}}$ is the density Frude number; ν is the kinematic viscosity coefficient; g is the acceleration due to gravity; $\bar{F} = (0, \rho, 0)$.

Pass on to the variables vorticity $\bar{\omega}$, vector potential $\bar{\psi}$. To simplify the establishment of boundary conditions for the vector potential components a scalar potential φ is introduced [3]

$$\bar{V} = \mathrm{rot}\,\bar{\psi} + \nabla\varphi. \tag{4}$$

A vorticity vector is defined as follows

$$\bar{\omega} = \mathrm{rot}\bar{V}. \tag{5}$$

In the variables vector potential-vorticity system of equations (2), (3) has the form

$$\frac{\partial \bar{\omega}}{\partial t} + (\bar{V} \cdot \nabla)\bar{\omega} = (\bar{\omega} \cdot \nabla)\bar{V} + \frac{1}{Re}\Delta\bar{\omega} + \frac{1}{Fr^2}\,\mathrm{rot}\,\bar{F}, \tag{6}$$

$$\nabla(\mathrm{div}\,\bar{\psi}) - \Delta\bar{\psi} = \bar{\omega}, \tag{7}$$

$$\Delta \varphi = 0. \tag{8}$$

The system of equations in variables of the stream function ψ - vorticity ω for the two-dimensional flows has the form:

$$\frac{\partial \rho}{\partial t} + (\bar{V} \cdot \nabla) \rho = 0,$$

$$\frac{\partial \omega}{\partial t} + (\bar{V} \cdot \nabla) \omega = \frac{1}{Fr^2} \frac{\partial \rho}{\partial x} + \frac{1}{Re} \Delta \omega , \tag{9}$$

$$\Delta \psi = -\omega ,$$

$$u = \frac{\partial \psi}{\partial y}, \quad v = -\frac{\partial \psi}{\partial x}.$$

A large part of numerical methods of solution of two-dimensional tasks was suggested for equations written in the variables stream function-vorticity [1,2]. For solving three-dimensional problems the Navier-Stokes equations are mainly used in natural variables velocity-pressure.However there are works [3] in which for solution of three-dimensional problems use is made equations in the variables vector potential-vorticity.

NUMERICAL SIMULATION OF TWO-DIMENSIONAL FLOWS OF STRATIFIED FLUID

Consider a problem on the flow of viscous incompressible non-homogeneous by density fluid in a two-dimensional region. Confine ourselves to a study of flows in a rectangular region. In general case for a non-rectangular region, if there exists an unsingular sufficiently smooth transformation of coordinates representing the studied region into a rectangular, for solving equations (9) written in new coordinates, one can use a numerical algorithm constructed for a rectangular region.

Let at some moment for time $t_n = n \cdot \Delta t$ (Δt is the quantity of step in time) fields of vorticity ω^n, stream function ψ^n and density ρ^n be known. To define sought functions at the temporal layer t_{n+1} apply an explicit scheme based on the method of splitting by physical processes [4].
At the I stage the diffusity equation for ω is solved:

$$\frac{\omega^* - \omega^n}{\Delta t} = \frac{1}{Re} \Delta \omega^n. \tag{10}$$

At the II stage the transfer equations for ρ and ω are solved:

$$\frac{\rho^{n+1} - \rho^n}{\Delta t} + \frac{\partial u\rho}{\partial x} + \frac{\partial v\rho}{\partial y} = 0,$$

$$\frac{\omega^{**} - \omega^*}{\Delta t} + \frac{\partial u\omega}{\partial x} + \frac{\partial v\omega}{\partial y} = 0. \tag{11}$$

At the III stage final value of vorticity and stream function are define:

$$\frac{\omega^{n+1} - \omega^{**}}{\Delta t} = \frac{1}{Fr^2} \frac{\partial \rho^{n+1}}{\partial x},$$

$$\Delta \psi^{n+1} = - \omega^{n+1}. \tag{12}$$

Consider a finite-difference scheme using a staggered grid. Values of the stream function ψ are determined at the knots of the mesh, values of density ρ and vorticity ω are at the centres of cells, values of horizontal component of the velocity vector are in the middles of side faces, values of vertical component of velocity are in the middles of upper and lower faces. Further we restrict ourselves to a consideration of the grid uniform along the y axis and non-uniform along the x axis, Δx_i, Δy are the quantity of steps in space.

A numerical analog of equation (10) has the form:

$$\omega^*_{i,j} = \omega^n_{i,j} + \frac{\Delta t}{Re}\left[2\frac{\Delta x_{i-1}(\omega^n_{i+1,j} - \omega^n_{i,j}) - \Delta x_i(\omega^n_{i,j} - \omega^n_{i-1,j})}{\Delta x_i \cdot \Delta x_{i-1}(\Delta x_i + \Delta x_{i-1})} + \right.$$

$$\left. + \frac{\omega^n_{i,j+1} - 2\omega^n_{i,j} + \omega^n_{i,j-1}}{\Delta y^2}\right] \tag{13}$$

(with a stability condition $\Delta t_{dif} \leq \frac{Re\, \Delta x^2 \cdot \Delta y^2}{2(\Delta x^2 + \Delta y^2)}$ at $\Delta x_i = \Delta x$).

Difference scheme for solving equations (11) is constructed on the basis of the integral conservative law written for each cell. Applying the approximation of the first order of accuracy obtain the upstream scheme (the scheme is stabilized for

$$\Delta t_{adv} \leq \left(\frac{|u|_{max}}{\Delta x} + \frac{|v|_{max}}{\Delta y} \right)^{-1}).$$

At the final stage ω^{n+1}, ψ^{n+1} are defined from the solution of the equations (12). The time step for the entire algorithm was estimated from the condition

$$\Delta t \leq \min(\Delta t_{dif}, \Delta t_{adv}). \qquad (14)$$

The above stated approach with the use of a grid with divided components of velocity has the following properties:

a) since a boundary value of the vorticity ω is determined by known velocity projections, then disappears a necessity to require the condition $\frac{\partial \psi}{\partial n} = 0$ to be satisfied which at difference approximation results in a zero discharge between a rigin boundary and the nearest grid point or it imposes rigid conditions on the profile of the boundary layer in boundary cells.

b) in the work [5] it is noted that application of approximated boundary conditions for a vorticity may lead to diminishing a time step in comparison with the step obtained from an analysis of a linear equation. The boundary condition for the vorticity in the presented method allows one to conduct stable calculations with the step Δt determined from condition (14).

NUMERICAL ALGORITHM FOR A CALCULATION OF THREE-DIMENSIONAL STRATIFIED FLOWS

Consider Navier-Stokes equations written in the variables vector potential-vorticity. Note some properties of equation for the vector potential (7):

1) for the existence of equation (7) it is necessary for the condition $\operatorname{div} \bar{\omega} = 0$ [6] to be satisfied;

2) without violating generality, one can estimate that $\operatorname{div} \bar{\psi} = 0$ [6]. Therefore instead of (7) one can consider the equation

$$\Delta \bar{\psi} = -\bar{\omega} \qquad (15)$$

on condition $\operatorname{div} \bar{\psi} = 0$. If it is required on the boundary Γ of region R $\operatorname{div} \bar{\psi}/_{\Gamma} = 0$, a sufficiently smooth solution of equation (15) with the right side satisfying the condition $\operatorname{div}\overline{\omega} = 0$ will satisfy the equation $\operatorname{div}\overline{\psi} = 0$ in R.

Consider an algorithm of solution of system (1),(6),(8), (15) that provides at every step in time solenoidality of the vorticity vector and does not require setting boundary conditions for the vorticity at rigid surfaces. The algorithm consists of three stages:

1) The intermediate vector velocity field

$$\frac{\overline{V}^{*}-\overline{V}^{n}}{\Delta t} = -\operatorname{rot}\overline{V}^{n} \times \overline{V}^{n} + \frac{1}{Re}\Delta\overline{V}^{n} + \frac{1}{F_r^2}\overline{F}^{n} \qquad (16)$$

and final density magnitudes

$$\frac{\rho^{n+1}-\rho^{n}}{\Delta t} = -(\overline{V}^{n}\cdot\nabla)\rho^{n} \qquad (17)$$

are determined. The intermediate velocity field $\overline{V}^{*}$ governs correct vorticity characteristics within the flow region under investigation.

2) From the equation $\overline{\omega}^{n+1} = \operatorname{rot}\overline{V}^{*}$ vorticity is found.

3) Scalar potential and vector potential are determined from the solving of the corresponding elliptic boundary problems. For solution equations (16),(17) the explicit difference schemes are used. Solutions of discrete analogs of elliptical equations (8),(15) with corresponding boundary conditions are found by successive over-relaxation method. Numerical schemes are constructed on the stagered mesh. For a flow region in the form of a rectangular parallelepiped the scalar potential, density and pressure are determined at the centres of cells, projections of vorticity vector and vector potential are determined in the middles of corresponding edges, projections of velicities - at the centres of faces. At every section of the boundary Γ magnitudes of a normal component of the velocity vector are known. From this boundary condition and relation $\operatorname{div}\overline{\psi}/_{\Gamma} = 0$ conditions on Γ are found for scalar and vector potentials [4].

DISCUSSION OF COMPUTATIONAL RESULTS

According to a numerical algorithm investigations were made of stratified fluid flows in rectangular reservoir for different magnitudes of Reynolds and Frude numbers [4]. The calculations have shown that the main parameter governing the flow picture is the density Frude number. Numerical experiment permitted one to determine critical magnitudes of Frude number Fr^*. With the numbers $Fr < Fr^*$ for the case of surface positioning of a discharge aperture the flow splits into two characteristic regions: the upper is involved into the discharge aperture, the lower is circulated in the bottom region of the reservoir with low velocities.

The flow picture of a non-homogeneous fluid depends on the type of a density (temperature) stratification. If there is a prononced thermocline the fluid splitting occurs in the region of the most density gradient. If the temperature of water smoothly changes with depth the thickness of the layer involved into the aperture is continuously decreasing with decreasing the Frude number.

With the constructed algorithm calculations of spatial flows of non-viscous stratified fluid in the reservoir of a rectangular shape have been carried out. A case of the position of the intake at the surface has been considered. For finite magnitudes of Frude numbers by the character of involvement of fluid into the aperture flows can be divided into two types:

a) at $Fr > 0.35$ into the discharge aperture fluid is involved from all the layers with depth but at the bottom region velocity is the less the closer the magnitude of the Frude number is to 0.35;

b) at $Fr < 0.3$ fluid splits into two regions: the upper one is involved into a discharge aperture, the lower circulates with low velocities. When an aperture is considerably narrower than the reservoir width and is situated symmetrically to the plane $z = 1/2$, then at $Fr < 0.3$ the fluid splitting in the vicinity of this plane is more singular than in the periphery. At the side walls thickness of the involved layer is bigger but flow velocities are less. At the bottom part of reservoir an unsteady circulation flow of a complex structure is formed.

Calculations of the three-dimensional lid-driven cavity flow are made. For large values of the Reynolds number (Re $\geqslant$ 2000) the typical three-dimensional structures-corner vortices and Taylor-Görtler-like vortices are found.

The developped algorithm is in a natural way adapted to parallel calculations.

REFERENCES

[1] Roach P.:"Computational fluid dynamics", Hermosa Publishers, Albuquerque, 1976.

[2] Belotserkovskii O.M.:"Chislennoe modelirovanie v mekhanike sploshnykh sred", Nauka, Moskow, 1984.-520 s.

[3] Aregbesola Y.A.S., Burley D.M.:"The vector and scalar potential method for the numerical solution of two-and three-dimensional Navier-Stokes equations", J.Comput. Phys., 1977, 24, N°4, p.398-415.

[4] Belolipetskii V.M., Kostyuk V.Yu.:"Chislennoe reshenie zadachi protekaniya dlya sistemy uravnenii neodnorodnoi zhidkosti", Chisl.metody mekhaniki sploshn.sredy, Novosibirsk, 1984, t.17, N°2, s.3-9.

[5] Paskonov V.M., Polezhayev V.I., Chudov L.A.:"Chislennoe modelirovanie protsessov teplo- i massoobmena," Nauka, Moskow, 1984.-288 s.

[6] Kochin N.Ye., Kibel I.A., Roze N.V.:"Teoreticheskaya gidrodinamika.Ch.1", Nauka, Moskow, 1963.-583 s.

AN EXPLICIT RUNGE-KUTTA METHOD FOR 3D TURBULENT INCOMPRESSIBLE FLOWS

Chao-Ho Sung, Naval Ship R&D Center, Bethesda, MD 20084-5000, U.S.A.

Cheng-Wen Lin, ORI, Inc., 1375 Piccard Dr., Rockville, MD 20850, U.S.A.

C.M. Hung, NASA Ames Research Center, Moffett Field, CA 94035, U.S.A.

ABSTRACT

A computer code has been developed to solve for the steady-state solution of the 3D incompressible Reynolds-averaged Navier-Stokes equations. The approach is based on the cell-center central-difference finite-volume formulation and an explicit one-step multistage Runge-Kutta time stepping scheme. The Baldwin-Lomax turbulence model is used. Techniques to accelerate the rate of convergence to a steady-state solution include the preconditioned method, the local time stepping and the implicit residual smoothing. Improvements in computational efficiency have been demonstrated in several areas including a four-stage Runge-Kutta scheme, the estimate of a local time step size with the viscous effect included and the implementation of the boundary conditions at the far field, the solid wall and the symmetric plane. This numerical procedure has been used to simulate the turbulent horseshoe vortex flow around an airfoil/flat-plate juncture. The comparison between prediction and experiment indicates a good agreement.

1. INTRODUCTION

The use of explicit Runge-Kutta methods as the time-stepping schemes for the solutions of the compressible Euler equations has become popular since the appearance of the papers by Jameson, Schmidt and Turkel [1] and Rizzi and Eriksson [2]. There are two main reasons for the popularity. The first reason is that explicit methods are easy to vectorize and to code. The ease of vectorization is a particularly important advantage when supercomputers are used. The second reason is that the severe stability restriction of the conventional explicit methods has been relaxed by the use of Runge-Kutta methods with an enlarged stability region. This approach has been extended to the compressible Navier-Stokes equations by Swanson and Turkel [3] and Vatsa [4]. The extension to the imcompressible flows appears in the paper by Rizzi and Eriksson [5] but is limited to the Euler equations. This approach will be extended to solve for the steady-state solution of the imcompressible 3D Reynolds-averaged Navier-Stokes equations in this paper. The numerical procedure developed will be applied to predict the turbulent incompressible flow around an airfoil/flat-plate juncture. The prediction will be compared with the experimental measurement by Dickinson [6].

2. ANALYSIS AND NUMERICAL PROCEDURE

The preconditioned incompressible Reynolds-averaged Navier-Stokes equations are

$$Eq_t + F_x + G_y + H_z = 0, \quad \text{or } q_t + E^{-1}F_x + E^{-1}G_y + E^{-1}H_z = 0 \qquad (1)$$

where the subscriptes are partial derivatives with respect to the time t and the three Cartesian coordinates x, y and z. The preconditioned matrix E and the column vectors of the variable q and of the three components of fluxes F, G and H are defined as

$$E = \begin{bmatrix} \beta^{-2} & 0 & 0 & 0 \\ (1+\alpha)\beta^{-2}u & 1 & 0 & 0 \\ (1+\alpha)\beta^{-2}v & 0 & 1 & 0 \\ (1+\alpha)\beta^{-2}w & 0 & 0 & 1 \end{bmatrix}, \quad q = \begin{bmatrix} p^* \equiv p/\rho \\ u \\ v \\ w \end{bmatrix}, \quad F = \begin{bmatrix} u \\ u^2+p^*-\tau_{xx} \\ uv-\tau_{xy} \\ uw-\tau_{xz} \end{bmatrix}$$

$$G = \begin{bmatrix} v \\ uv-\tau_{yx} \\ v^2+p^*-\tau_{yy} \\ vw-\tau_{yz} \end{bmatrix}, \quad H = \begin{bmatrix} w \\ uw-\tau_{zx} \\ vw-\tau_{zy} \\ w^2+p^*-\tau_{zz} \end{bmatrix} \tag{2}$$

where p is the pressure, ρ is the constant density, u, v and w are the three Cartesian components of the velocity and the Reynolds stresses are defined as

$$\tau_{ij} = R_e^{-1}\,\nu\,\left(\frac{\partial u_i}{\partial x_j} + \frac{\partial u_j}{\partial x_i}\right), \qquad i,j = 1,\ 2,\ 3 \tag{3}$$

where $(u_1,\ u_2,\ u_3) = (u,\ v,\ w)$, $(x_1,\ x_2,\ x_3) = (x,\ y,\ z)$, ν is the sum of the kinematic and the eddy viscosity and Re is the Reynolds number. The variables are nondimensionalized by the free stream condition at infinity in the following manner: u, v and w by u_∞, p and τ by $\rho_\infty u_\infty^2$, x, y and z by a characteristic length L chosen as the chordlenth and $Re = u_\infty L/\nu_\infty$. α and β^{-2} are the preconditioned parameters.

The idea is to choose a β for a given α such that the disparity in propagating speeds will be reduced during the transient state. The fact that the transient-state solution may not be physically valid is not relevant since only the steady-state solution is of interest. Neglecting the viscous terms, equations (1) is rewritten as

$$q_t + Aq_x + Bq_y + Cq_z = 0, \tag{4}$$

In order to find the wave propagating speeds, equation (4) is Fourier transformed. Then the wave propagating speeds are given by the eigenvalues λ's of $D = \omega_1 A + \omega_2 B + \omega_3 C$, where the absolute values of Fourier components ω_1, ω_2 and ω_3 are bounded by 1. It is found that

$$\lambda_1 = U,\ \lambda_2 = U,$$

$$\lambda_{3,4} = \tfrac{1}{2}[(1-\alpha)U \pm \sqrt{(1-\alpha)^2U^2 + 4\beta^2(\omega_1{}^2 + \omega_2{}^2 + \omega_3{}^2)}], \tag{5}$$

where $U \equiv \omega_1 u + \omega_2 v + \omega_3 w$. Minimization of the maximum of the ratios of the eigenvalues in (5) gives the condition for the choice of β [7]. Numerical experiments on the Euler solutions of the M1 wing indicate that the choice $\alpha = -1$ and $\beta^2 = 1$ gives a reasonably good convergence rate.

Explicit one-step time-stepping schemes of the set of ordinary differential equations can be expressed as

$$\frac{dq}{dt} = Q(q) \longrightarrow q^{n+1} = R(\Delta t J) q^n \qquad (6), (7)$$

where J is the Jacobian of Q and R is a polynomial of degree m in the matrix $z = \Delta t J$. It is noted that m is the number of functional evaluations and is also called the number of Runge-Kutta stages. The stability region of (7) is defined as $S = \{z: |R(z)| \leq 1\}$. For J having imaginary eigenvalues λ, i.e. hyperbolic systems, the maximum stability limit is at the boundary S_1 at the intersection of the stability region S and the imaginary axis. Then the time-stepping scheme (7) is stable if the time step size is chosen as

$$\Delta t \leq \frac{S_1}{\max(\lambda)} \equiv \frac{CFL}{\max(\lambda)} = \frac{CFL}{\lambda_0} . \qquad (8)$$

Despite possible ambiguity, S_1 will be identified as the Courant-Friedrichs-Lewy (CFL) number.

The scheme used by Jameson, Schmidt and Turkel [1]

$$R(z) = 1 + z + \frac{1}{2!} z^2 + \frac{1}{3!} z^3 + \frac{1}{4!} z^4 \qquad (9)$$

is fourth-order accurate with CFL = $2\sqrt{2}$, the scheme used by Rizzi and Eriksson [2]

$$R(z) = 1 + z + \frac{1}{2} z^2 + \frac{1}{2} z^3 \qquad (10)$$

is second-order accurate with CFL = 2 and the scheme to be used in this paper

$$R(z) = 1 + z + \frac{5}{9} z^2 + \frac{4}{27} z^3 + \frac{4}{81} z^4 \qquad (11)$$

is first-order accurate with CFL = 3.

Since CFL is proportional to m, one is tempted to think that schemes with a higher number of stages may be more efficient since a larger time step can be used. This turns out not to be true because a larger m also implies that more functional evaluations are required to forward one time step. Numerical experiments indicate that 4 to 6 stages seem to be the most efficient.

In order to use a Runge-Kutta scheme, equation (1) must be written in the semidiscrete form such as (6). This is done by a finite-volume formulation with the dependent variables defined at the center of a computational cell and the fluxes evaluated at the midpoint of each cell surface using central difference. A generalized thin-layer approximation suitable for more than one solid wall is adopted to simplify computation. The Baldwin-Lomax turbulence model will be used.

The estimate of the local time step size given by (8) is based on hyperbolic equations. A direct application to the Navier-Stokes equations leads to an overestimate of the time step size near the wall and may result in oscillation or even divergence. Swanson and Turkel [8] analyzed an unconditionally stable difference scheme for a simple diffusion equation and extended the scheme to the 2D Navier-Stokes equations. They gave an estimate of Δt based on the inviscid model but the time step size can automatically be reduced in the region where the viscous effect is important. In the 3D version, it can be expressed as

$$\Delta t \leq \frac{1}{\Delta t_H^{-1} + 2 R_e^{-1} \nu (g^{11} + g^{22} + g^{33})}, \qquad (12)$$

where Δt_H is based on the inviscid model such as given by (8). An alternative approach used in this paper is the following. Applying Fourier transform, the time step size is obtained as

$$\Delta t \leq \frac{CFL}{\lambda_0+4R_e^{-1}\nu(g^{11}+g^{22}+g^{33})}, \tag{13}$$

where λ_0 is given by (8). It is interesting to note that (12) and (13) are similar and are exactly the same when CFL = 2. Numerical experiments performed using three different time step sizes given by (8), (12), and (13) will be discussed later.

Although there is some inherent dissipation in the Runge-Kutta scheme used, an additional dissipation is needed to damp high-frequency oscillations. A fourth-difference artificial dissipation model will be adapted.

The mathematical motivation for using the residual smoothing technique is to reduce the spectral radius of the function on the right-hand side of the equation. This is equivalent to using a higher CFL number in (8) than is allowed in the absence of residual smoothing. This is legitimate only when the time variation is small or when only the steady-state solution is of interest. An implicit residual smoothing scheme by Jameson [9] is used. As a result, CFL = 5 or higher, instead of 3, can be used. The Dirichlet boundary condition is used.

Boundary conditions at the solid wall, the symmetric plane and the far field will be considered. Far field boundary conditions are particularly difficult to handle in incompressible viscous flows. Despite the obvious deficiency, methods based on hyperbolic equations will be applied to the Navier-Stokes equations. Numerical experiments by Rudy and Strikwerda [10] on the non-reflecting boundary condition for the boundary layer flow over a flat-plate indicate that this approach is plausible. A 3D version of the non-reflecting boundary condition will be used.

The most commonly used boundary conditions (called Method I) for the solid wall and the symmetric plane are essentially based on one-dimensional consideration and are only adequate for simple geometry. For complex geometries with considerable secondary or reversed flows, a more careful treatment (called Method II) is needed to account for the curvature effect. When this is done, the rate of convergence is found to be improved.

3. TURBULENT FLOW PAST AIRFOIL/FLAT-PLATE JUNCTURE

The numerical procedure for incompressible flows described will be used to simulate the turbulent flow past an airfoil/flat-plate juncture shown in Figure 1a. The airfoil is a NACA 0020 with the nose replaced by a 1.5:1 ellipse. The chordlength is 25.9 cm with a maximum thickness of 6.1 cm. The freestream velocity is 34 m/s corresponding to a Reynolds number of 5.5×10^5 based on the chordlength and a Mach number of 0.1. The flat-plate boundary layer thickness is 4.7 cm, or equivalently 0.18 chordlength, at the leading edge location without the airfoil.

The computational domain is the following. The inflow boundary is 1.7 chordlength upstream of the leading edge of the airfoil and the outflow boundary is 1.8 chordlength downstream of the trailing edge. The outer boundary is 1.9 chordlength from the symmetric plane and the top boundary is 0.7 chordlength from the flat-plate. The computational grid is constructed by stacking the 2D grid shown in Figure 1b. Three different grids are used. For production runs, CPU time can be saved by first running in the coarse grid 33 x 25 x 25 for 1000 time steps and then continuing in the fine grid 65 x 49 x 49 for 500 time steps. The root-mean-square residual of pressure is reduced by more than three orders of magnitude after 1000 time steps in the coarse grid. Computations were performed on CRAY 2 of the Numerical Aerodynamic Simulation (NAS) Program at NASA Ames Research Center. The storage required is approximately 5.5 MW. It took approximately 0.6 CPU seconds to forward one time step in the coarse grid and 3.5 CPU seconds in the fine grid. All the numerical results to be discussed are based on this airfoil/flat-plate juncture. The only exception is the comparison of three explicit Runge-Kutta schemes when the Euler solution of the M1 wing is used.

4. DISCUSSION OF RESULTS

The comparison with respect to computational efficiency of three Runge-Kutta schemes given by (9), (10) and (11) is shown in Figure 2. The test case is the Euler solution of the M1 wing at an angle of attack of 10 degrees in a 65 x 17 x 17 grid. It is shown that scheme (11) used in this paper required approximately 15 percent less CPU time than scheme (9) used by Jameson, Schmidt and Turkel [1] and approximately 30 percent less CPU time than scheme (10) used by Rizzi and Eriksson [2].

Convergence histories using three different estimates (8), (12) and (13) of the local time step are compared. When CFL is raised to 6, the root-mean-square residual of the pressure of the local time step based on hyperbolic equations oscillates violently as shown in Figures 3. However, at this CFL number, solutions converge using either Swanson and Turkel's local time step or the present method. The rate of convergence is slightly faster using the local time step derived in this paper.

In the numerical simulation of the turbulent flow past an airfoil/flat-plate juncture, the region in front of the airfoil where horseshoe vortices form is a place where the solution converges to the steady-state slower than other places. As the number of time steps increases, the horseshoe vortex gradually moves upstream and finally settles down at about 0.08 chordlength upstream of the leading edge of the airfoil. Thus the rate of the forward movement of the horseshoe vortex is a good criterion to judge the efficiency of a numerical method. Figure 4 shows the comparison using the solid wall and symmetric plane boundary conditions of Method I and Method II. After 1000 time steps, the vortex core moves to slightly over 0.6 chordlength upstream of the leading edge using Method I while it reaches the steady-state location of 0.8 chordlength upstream using Method II.

Comparison between prediction and experiment [6] of the pressure on the flat-plate along the Y = constant lines are shown in Figure 5a to 5f. In general, the agreement is good although the computation tends to overpredict in the vicinity of the maximum thickness of the airfoil. This is most evident at the maximum thickness location X/C = 0.18 of Figure 5d where data were taken just outside the airfoil at Y/C = 0.2. It will be noted later in Figure 6b that the overprediction of pressure coincides with the underpredication of the streamwise velocity in the vicinity of the maximum thickness of the airfoil. Thus, one possible source of the discrepancy between the prediction and the experiment is the blockage effect of the wind tunnel near the location of the maximum thickness of the airfoil.

Comparison between the prediction and the experiment of the three components of the mean velocity are shown in Figure 6a to 6c. Comparisons are made on four planes from 0.75 chordlength upstream of the leading edge of the airfoil to 0.18, 0.75 and 1.50 chordlengths downstream of the leading edge. Again, the general agreement is good.

ACKNOWLEDGEMENT

The authors wish to thank Dr. Lars-Erik Eriksson for his assistance in initiating this investigation. The CRAY 2 computer time is generously provided by the Numerical Aerodynamic Simulation (NAS) Program at NASA Ames Research Center. This work was sponsored partially by Office of Naval Technology (OCNR 233) under the ship and Submarine Advanced Performance Quiet Hull Design Technique, DTNSRDC Work Unit Number 1-1506-730 and partially by Office of Naval Research, Special Focus Program on Numerical and Experimental Investigation of Appendage Flow, Drag and Wake, Program Element 61153N, DTNSRDC Work Unit Number 1-1542-106.

REFERENCES

[1] Jameson, A., W. Schmidt and E. Turkel, "Numerical Solutions of the Euler Equations by Finite Volume Methods Using Runge-Kutta Time-Stepping Schemes", AIAA Paper 81-1259 (June 1981).

[2] Rizzi, A. and L.-E. Eriksson, "Computation of Flow Around Wings Based on the Euler Equations", J. Fluid Mech. Vol. 148, pp. 45-71 (1984).

[3] Swanson, R.C. and E. Turkel, "A Multistage Time-Stepping Scheme for the Navier-Stokes Equations", AIAA Paper 85-35 (January 1985).

[4] Vatsa, V.N. "Accurate Solutions for Transonic Viscous Flow Over Finite Wings", AIAA Paper 86-1052 (May 1986).

[5] Rizzi, A. and L.-E. Eriksson, "Computation of Inviscid Incompressible Flow With Rotation", J. Fluid Mech. Vol. 153, pp. 275-312 (1985).

[6] Dickinson, S.C., "Flow Visualization and Velocity Measurements in the Separated Region of an Appendage-Flat Plate Junctions", Proceedings of the Ninth Biennial Symposium on Turbulence, University of Missouri-Rolla (October 1-3, 1984); also, "Velocity Measurements in the Separated Region of an Appendage-Flat Plate Junction", the American Physical Society Division of Fluid Dynamics Annual Meeting, Tucson, Arizona (November 24-26, 1985).

[7] Turkel, E. "Preconditioned Methods for Solving the Incompressible and Low Speed Compressible Equations", NASA ICASE Report 86-14 (March 1986).

[8] Swanson, R.C. and E. Turkel, "Pseudo-Time Algorithm for the Navier-Stokes Equations", NASA ICASE Report 86-37 (May 1986).

[9] Jameson, A. "Numerical Solution of the Euler Equation for Compressible Inviscid Fluids" in Numerical Methods for the Euler Equations of Fluid Dynamics (1985).

[10] Rudy, D.H. and J.C. Strikwerda, "Boundary Conditions for Subsonic Compressible Navier-Stokes Calculations", Computers and Fluids, Vol. 9, pp. 327-338 (1981).

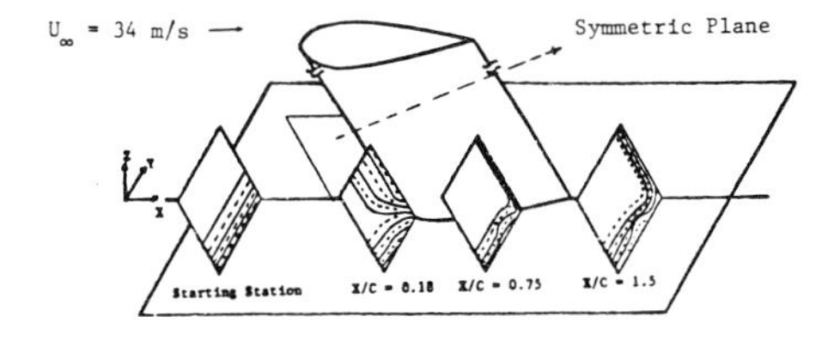

Fig. 1a - Airfoil/Flat-Plate Juncture and Measurement Stations

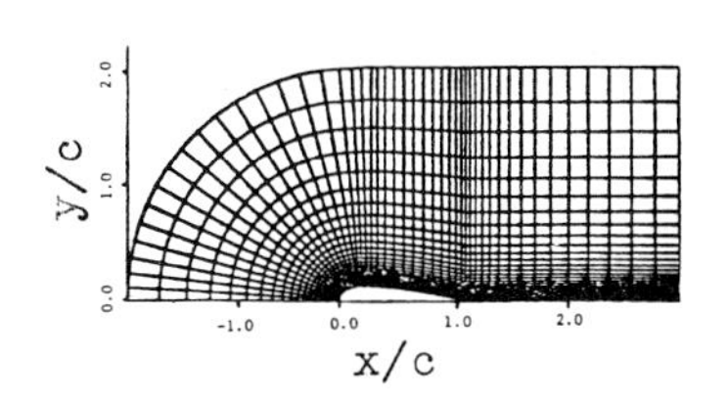

Fig. 1b - 2D Section of Computational Grid

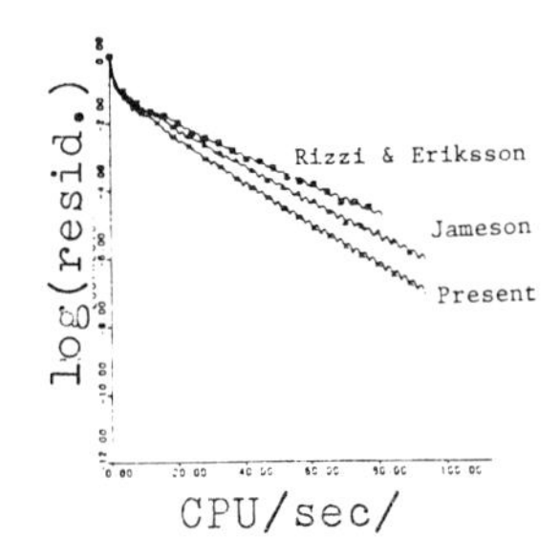

Fig. 2 - Comparison of Three Runge-Kutta Schemes

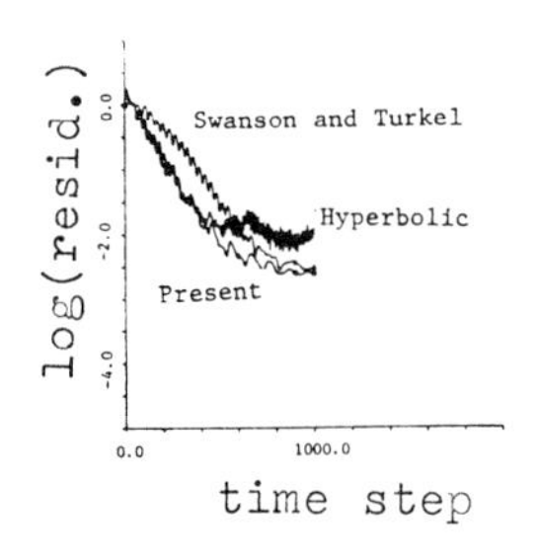

Fig. 3 - Comparison of Three Local Time Step Estimates

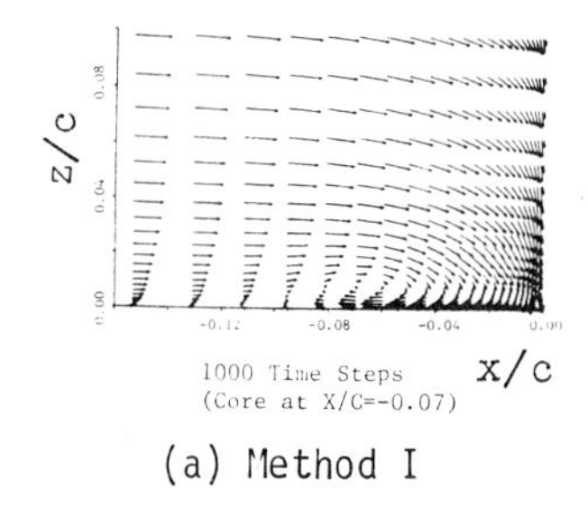

(a) Method I

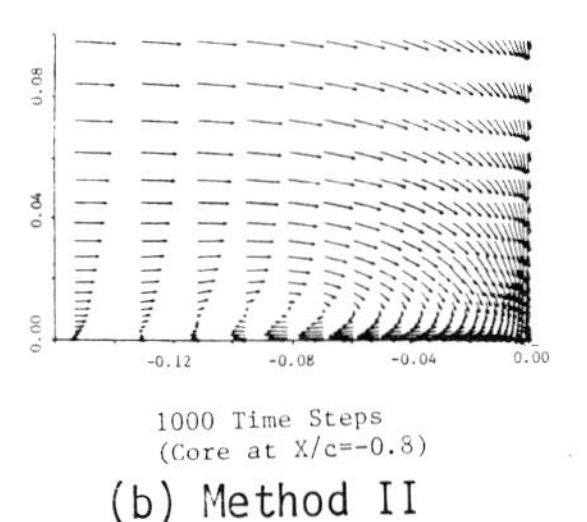

(b) Method II

Fig. 4 - Comparison of Methods for Solid Wall and Symmetric Plane Boundary Conditions

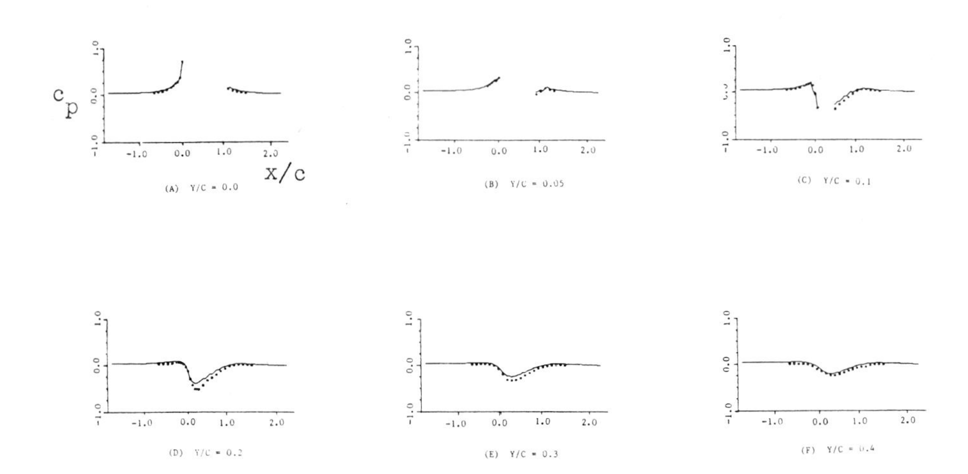

Fig. 5 - Comparison of Pressure Coefficient on Flat-Plate

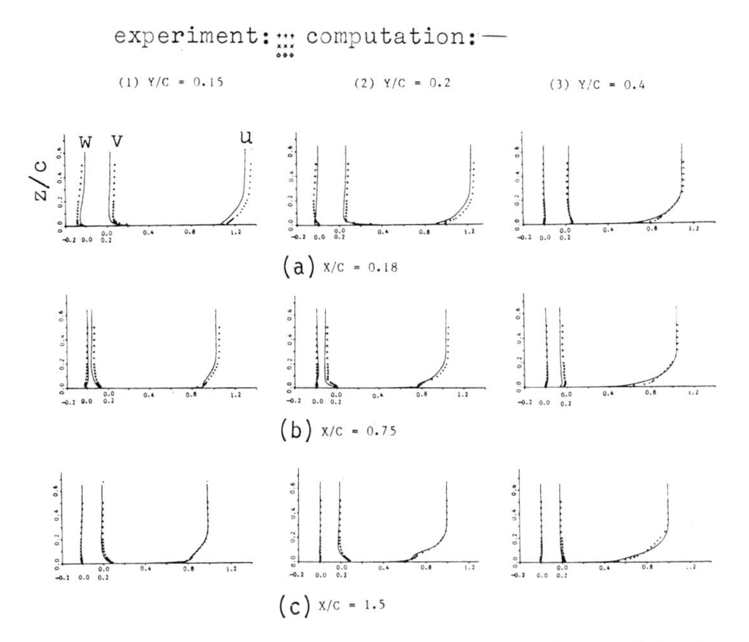

Fig. 6 - Comparison of Three Components of Mean Velocity

STEADY LAMINAR ENTRANCE FLOW IN A CURVED CIRCULAR PIPE

F.N. Van de Vosse, A.A. Van Steenhoven, A. Segal* and J.D. Janssen

Eindhoven University of Technology, Department of Mechanical Engineering
P.O.BOX 513, 5600 MB Eindhoven
The Netherlands

*Delft University of Technology, Department of Mathematics and Informatics
P.O.BOX 356, 2600 AJ Delft
The Netherlands

SUMMARY

A standard Galerkin finite element method is used to approximate the solution of the three-dimensional Navier-Stokes equations for steady incompressible Newtonian entrance flow in a 90-degree curved tube (curvature ratio $\delta = 1/6$) for a Dean number $\kappa = 122$. The computational results are compared with the results of laser-Doppler velocity measurements in an equivalent experimental model. For both the axial and secondary velocity components, a fair agreement between the computational and experimental results is found.

INTRODUCTION

If the Navier-Stokes equations are rewritten in an orthogonal curvilinear toroidal co-ordinate system, two important dimensionless parameters are found [16] : the curvature ratio $\delta = a/R$ and the Dean number $\kappa = Re\sqrt{\delta}$, with a the radius of the tube, R the curvature radius and Re the Reynolds number based on the diameter of the tube and the mean axial velocity ($Re = 2aW/\nu$) with W the mean axial velocity and ν the kinematic viscosity. The Dean number can be interpreted as the ratio of the square root of the product of convective inertial forces and centrifugal forces and the viscous forces.

In this study a curvature ratio $\delta=1/6$ with Reynolds numbers of 300 is used, which leads to a Dean number $\kappa = 122$. Furthermore, a cartesian co-ordinate system is used in order to incorparate straight in- and outflow tubes in the numerical model and to enable a generalization of the method for other 3D gemetries. Similar numerical flow investigations have been carried out by Singh [10], Yao and Berger [15], Liu [8], Stewartson et al. [12] and Soh and Berger [11], who, however, used a uniform inlet flow profile instead of the parabolic flow profile used here. Besides, these authors used finite difference or analytical approximations, transforming the problem into toroidal co-ordinates.

NUMERICAL METHOD

The Navier-Stokes equations for steady incompressible flow are given by the momentum equations together with the continuity equation. For a bounded domain $\Omega \in R^3$ with boundary Γ they read:

$$\begin{cases} \rho(\vec{u}\cdot\nabla)\vec{u} = \rho\vec{f} + \nabla\cdot\underline{\underline{\sigma}} \\ \nabla\cdot\vec{u} = 0 \end{cases} \tag{1}$$

with ρ the density, $\vec{u}$ the velocity, $\underline{\underline{\sigma}}$ the Cauchy stress tensor and $\vec{f}$ the body force. For Newtonian fluids the Cauchy stress tensor can be written as:

$$\underline{\underline{\sigma}} = - p\,\underline{\underline{I}} + 2\eta\nabla_s\vec{u} \tag{2}$$

with p the pressure, $\underline{\underline{I}}$ the unity tensor, η the dynamic viscosity and $\nabla_s\vec{u}$ the rate of strain tensor defined as the symmetrical part of $\nabla\vec{u}$. The following boundary conditions are considered:

$$\begin{cases} \vec{u} = \vec{g} & \text{at } \Gamma_1 \quad \text{(inflow or no-slip condition)} \\ (\underline{\underline{\sigma}}\cdot\vec{n}) = \vec{0} & \text{at } \Gamma_2 \quad \text{(stress-free outflow condition)} \\ \left.\begin{array}{l} (\underline{\underline{\sigma}}\cdot\vec{n})\cdot\vec{t}_1 = 0 \\ (\underline{\underline{\sigma}}\cdot\vec{n})\cdot\vec{t}_2 = 0 \\ (\vec{u}\cdot\vec{n}) = 0 \end{array}\right\} & \text{at } \Gamma_3 \quad \text{(symmetry condition)} \end{cases} \tag{3}$$

with $\Gamma_1 \cup \Gamma_2 \cup \Gamma_3 = \Gamma$, $\Gamma_i \subset \Gamma_{j\neq i} = \emptyset$ (i=1,2,3), $\vec{n}$ the outer normal on Γ and $\vec{t}_j$ (j=1,2) the orthonormal tangential vectors on Γ.
Using a standard Galerkin finite element approximation and substitution of:

$$u_i^h = \sum_{n=1}^{N} u_{in}\,\varphi_{in} \qquad i=1,2,3$$

$$p^h = \sum_{m=1}^{M} p_m\,\psi_m \tag{4}$$

yields the following system of non-linear equations (Cuvelier et al. [4]):

$$\begin{cases} S\underline{u} + N(\underline{u})\,\underline{u} + L^T\underline{p} = \underline{f} \\ L\underline{u} = \underline{0} \end{cases} \tag{5}$$

with:

$$S = \begin{bmatrix} S^{11} & S^{12} & S^{13} \\ S^{21} & S^{22} & S^{23} \\ S^{31} & S^{32} & S^{33} \end{bmatrix} \quad , \quad S^{ij}(k,l) = \int_\Omega \eta\Big[\sum_{\alpha=1}^{3}\Big(\frac{\partial\varphi_{ik}}{\partial x_\alpha}\frac{\partial\varphi_{jl}}{\partial x_\alpha}\Big)\delta_{ij} + \frac{\partial\varphi_{ik}}{\partial x_j}\frac{\partial\varphi_{jl}}{\partial x_i}\Big]\,d\Omega$$

$$[N(\underline{u})\underline{u}]_{ik} = \sum_{j=1}^{3} \int_\Omega \varrho \left[\sum_{n=1}^{N} u_{jn}\varphi_{jn} \sum_{l=1}^{N} u_{il} \frac{\partial\varphi_{il}}{\partial x_j}\right] \varphi_{ik}\, d\Omega$$

$$L = [L^1\ L^2\ L^3], \qquad L^i(m,l) = -\int_\Omega \psi_m \frac{\partial\varphi_{il}}{\partial x_i}\, d\Omega$$

$$\underline{f} = [\underline{f}^1\ \underline{f}^2\ \underline{f}^3]^T, \qquad \underline{f}^i(k) = \int_\Omega \varrho\, f_i\, \varphi_{ik}\, d\Omega$$

$$\underline{u} = [\underline{u}^1\ \underline{u}^2\ \underline{u}^3]^T, \qquad \underline{u}^i(k) = u_{ik}\,, \qquad \underline{p}(m) = p_m\,.$$

Direct solution of the set of equations (5) is time and memory consuming owing to the fact that zero components appear on the principal diagonal of the coefficient matrix, due to the absence of the pressure in the continuity equation. In general this requires a partial pivoting procedure which disturbes the band structure of the matrix. To overcome this difficulty, the penalty function method is applied by solving instead of (5) :

$$[S + N(\underline{u}_\varepsilon) + \tfrac{1}{\varepsilon} L^T M_p^{-1} L\,]\underline{u}_\varepsilon = \underline{f}$$

$$\underline{p}_\varepsilon = \tfrac{1}{\varepsilon} M_p^{-1} L \underline{u}_\varepsilon \qquad (6)$$

with :

$$M_p(k,l) = \int_\Omega \psi_k \psi_l\, d\Omega \qquad k=1,\dots,M \quad ,l=1,\dots,M\,. \qquad (7)$$

Caray and Krishnan [2] showed that for the steady Navier-Stokes equations the solution of (6) converges to the solution of (5). These results are confirmed by numerical experiments of Reddy [9]. The main advantage of the penalty function method over the direct solution of equations (5) is, that the pressure is eliminated from the momentum equations resulting in a smaller set of equations that can be solved without the demand of pivoting procedures.

A survey of finite elements for the three-dimensional Navier-Stokes equations is given by Fortin et al.[5,6,7]. The Brezzi-Babuska condition and the applicability of a penalization formulation (i.e. a discontinuous pressure approximation) are of importance for the choice of an element. The simplest element, which is at least second order accurate and which satisfies the above mentioned conditions, is the full quadratic velocity - linear pressure ($Q_2^{(27)}$-P_1) element (see Fig. 1).

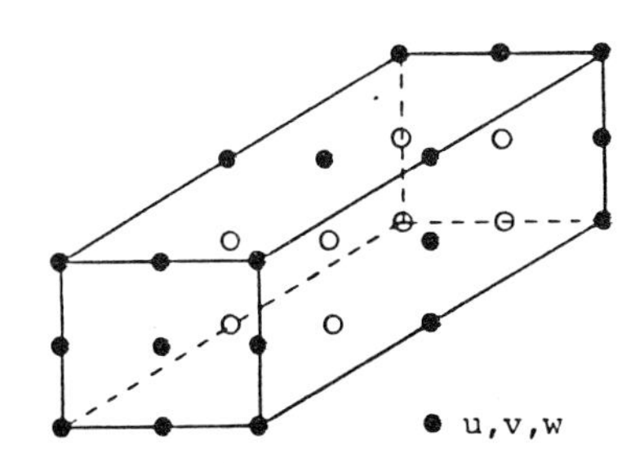

Fig. 1 : The 27-noded hexahedral (Q_2-P_1) element.

Similar to the (P_2^+-P_1) element [3], which has been proved to be successful for the two-dimensional Navier-Stokes equations [13,14], the pressure is approximated by its value and its derivatives in the centroid of the element and thereby is discontinuous over the element boundary. The number of unknowns per element is 81 for the velocity and 4 for the pressure. In the next section the (Q_2-P_1) element will be applied to the flow in a 90-degree curved tube and the results will be compared with data obtained from laser-Doppler experiments. Although no predictions about the order of accuracy are made, this flow configuration is thought to be complex enough for evaluation of the applicability of the element to complex three-dimensional flow problems.

RESULTS

The entrance flow in a curved tube with curvature ratio δ=1/6 was calculated for Re=300 ($\kappa\approx$122). To this end, a finite element mesh was generated as depicted in Fig. 2. Upstream of the entrance (θ=0) and downstream of the exit ($\theta=\pi/2$) of the 90°-bend, an inlet and an outlet section with a length of twice the tube diameter were used. A parabolic axial velocity distribution and zero secondary velocities were imposed as inflow condition, whereas stress-free outflow conditions were used.

Experiments were performed in a perspex model with an internal radius of 4mm and a curvature radius of 24mm (δ=1/6). The three velocity components were measured with a laser-Doppler anemometer in a way described by Bovendeerd et al. [1]. A long inlet section of 0.4m ensured a fully developed parabolic flow at the entrance of the bend. Also downstream the bend a long straight glass pipe was present. The Reynolds number of the flow was kept at Re=300 (κ=122). The velocity components were measured at 5 cross-sections in the bend (θ=0 (π/8) π/2) corresponding with the element boundaries in the finite element mesh (see Fig. 2).

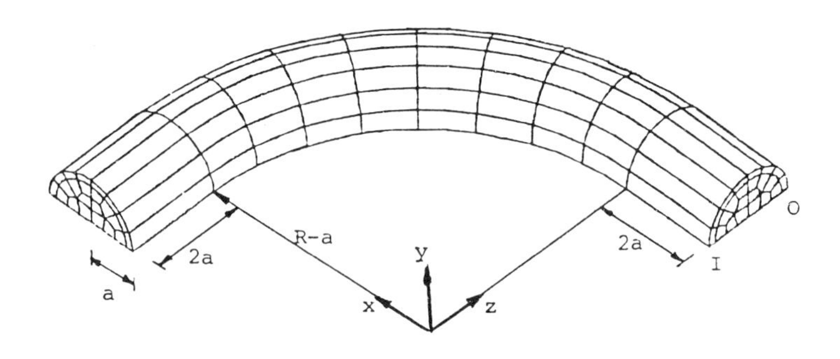

Fig. 2 : Geometry and finite element mesh as used in the calculations (220 elements, 2205 nodes).

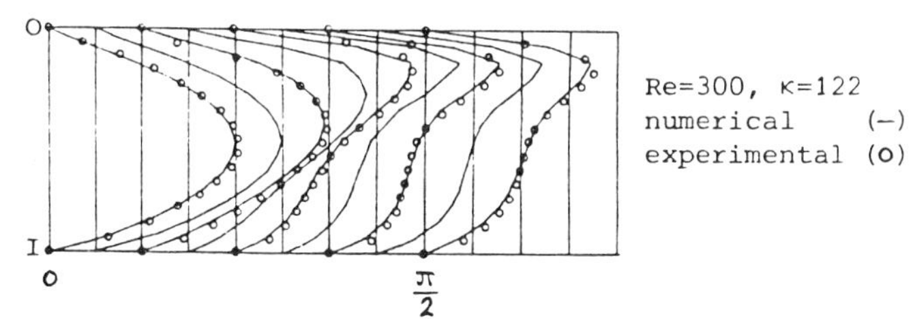

Fig. 3 : Axial velocity in the plane of symmetry of the curved tube (δ=1/6) as calculated (-) and measured (o) for Re=300 (κ=122). 0,I denote outer and inner bend, respectively.

In Fig. 3 the development of the axial velocity in the plane of symmetry is given. Moreover the measured values of the axial velocity are given and indicated by circles. At the first two velocity profiles (θ = 0 and θ = $\pi/16$) hardly any influence of the curvature is visible, although a slight shift towards the inner bend is observable at θ = 0. Further downstream, the maximum of the axial velocity profile is shifted towards the outer bend. Despite an overall over-estimation of the axial velocity (probably due to a measurement inaccuracy of the imposed flow value) the calculations agree well with the experiments at all cross-sections.

In Fig. 4, a more complete picture of the axial flow development is given. Here the axial velocity distribution is given by isovelocity contours at three cross-sections in the tube. An almost parabolic axial velocity distribution is observed at θ = 0. This paraboloid, however, is shifted slightly towards the inner bend. The experimental iso-velocity contours of the axial velocity component at this axial position show a shift from the upper wall, probably due to the inaccuracy in the determination of the wall position. Halfway the bend ($\theta=\pi/4$) the shift towards the outer bend has continued and large wall shear rates occur at the outer bend of the tube. A 'C-shaped' isovelocity contour is found for the high velocity region. At the cross-section $\theta=\pi/2$, the 'C-shaped' contours further develop. The agreement between the experiments and the calculations is fair.

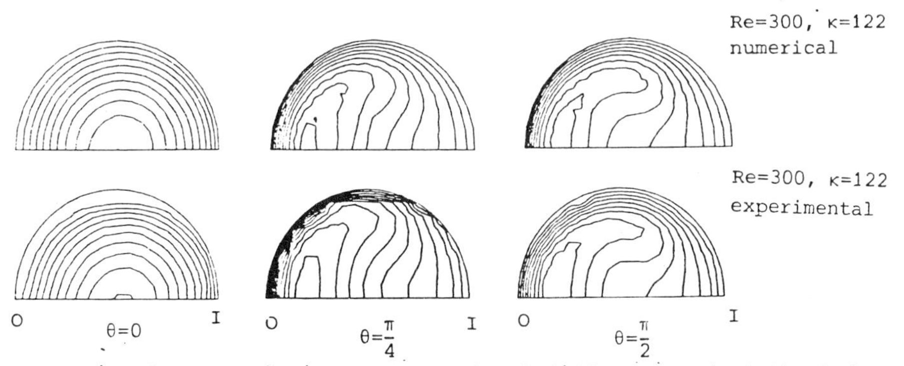

Fig. 4 : Isovelocity contours (level difference $\Delta w/W$=0.2) of the axial velocity in the curved tube (δ=1/6) as calculated and measured for Re=300 (κ=122). 0,I denote outer and inner bend, respectively.

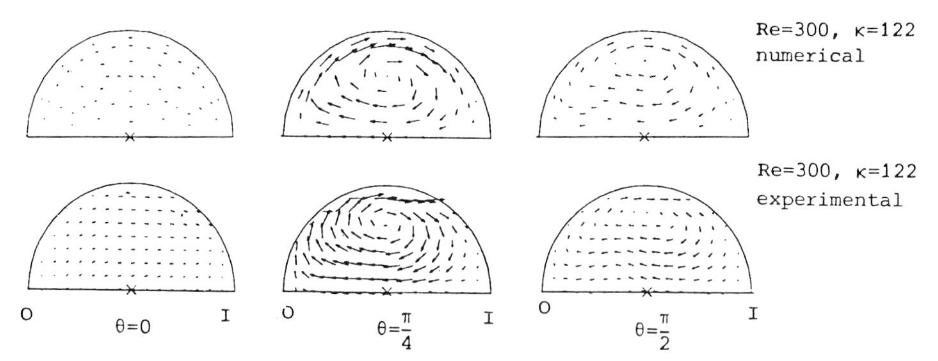

Fig. 5 : Normalized secondary velocity vectors in the curved tube ($\delta=1/6$) as calculated and measured for Re=300 ($\kappa=122$). O,I denote outer and inner bend, respectively.

In Fig. 5 the development of the secondary velocity field is given. At the entrance ($\theta=0$) the upstream influence of the bend is visible in a unidirectional secondary flow from the outer bend towards the inner bend, which amounts to be about 5% of the mean axial flow. The same upstream influence of the bend was found in the experiments. At $\theta=\pi/4$ a vortex has developed which near the plane of symmetry is directed from the inner bend towards the outer bend and at the upper wall from the outer bend back to the inner bend. Although direct comparison with the experiments is difficult because of the different locations at which the velocities are determined, good agreement is found at this axial station. Finally, at $\theta=\pi/2$, the secondary velocities are significantly lower, especially near the plane of symmetry. The center of the vortex has moved further in the inner bend direction while, moreover, a deformation of the shape of the vortex has taken place. A similar behaviour is found in the experiments.

DISCUSSION

In the preceding section the results of calculations of the entrance flow in a curved tube have been presented by means of the axial and secondary velocity distributions. With respect to the applicability of the numerical model to three-dimensional flow problems, the following remarks are made. The agreement between the experimental and numerical results indicates that the finite element method applied offers an accurate tool for three-dimensional flow simulations. Although a relatively coarse mesh was used, the results seem to be quite accurate. For the finite element mesh applied here, about 25 hours CPU and about 2 days I/O were needed per Newton iteration (7 Newton iterations were needed to reach Re=300) on a mini-computer (APOLLO DSP90). Therefore, application of this solution procedure for more complex three-dimensional geometries demands more sophisticated computer capacity, like array processing or super-computers.

REFERENCES

[1] BOVENDEERD, P. H. M., VAN STEENHOVEN, A. A., VAN DE VOSSE, F. N., VOSSERS, G.: " Steady entry flow in a curved pipe ", J. Fluid Mech. ,177 (1987) pp. 233-246.

[2] CARAY, G. F., KRISHNAN, R.: " Penalty finite element method for the Navier-Stokes equations ", Comp. Meth. Appl. Mech. Eng., 47 (1984) pp. 183-224.

[3] CROUZEIX, M., RAVIART, P.-A.: " Conforming and nonconforming finite element methods for solving the stationary Stokes equations I ", RAIRO Anal. Numer., 7 (1973) pp. 33-76.

[4] CUVELIER, C., SEGAL, A., VAN STEENHOVEN, A. A.: " Finite element methods and Navier-Stokes Equations ", D.Reidel Publishing Comp., Dordrecht /Boston /Lancaster /Tokyo, (1986).

[5] FORTIN, M.: " Old and new finite elements for incompressible flows ", Int. J. Num. Meth. Fluids., 1 (1981) pp. 347-364.

[6] FORTIN, M., FORTIN, A.: " Newer and newer elements for incompressible flow ", in : "Finite elements in Fluids vol. 6", Eds. Gallagher R.H., Caray G.F., Oden J.T. and Zienkiewicz O.C., John Wiley & Sons Ltd. (1985).

[7] FORTIN, M., FORTIN, A.: " Experiments with several elements for viscous incompressible flows ", Int. J. Num. Meth. Fluids., 5 (1985) pp. 911-928.

[8] LIU, N. S.: " Developing flow in a curved pipe ", INSERM-Euromech 92, 71 (1977) pp. 53-64.

[9] REDDY, J. N.: " On penalty function methods in finite element analysis of flow problems ", Int. J. Num. Meth. Fluids., 2 (1982) pp. 151-171.

[10] SINGH, M. P.: " Entry flow in a curved pipe ", J. Fluid Mech. , 65 (1974) pp. 517-539.

[11] SOH, W. Y., BERGER, S. A.: " Laminar entrance flow in a curved tube ", J. Fluid Mech. , 148 (1984) pp. 109-135.

[12] STEWARTSON, K., CEBECI, T., CHANG, K. C.: " A boundary-layer collision in a curved duct ", Q. J. Mech. Appl. Math., 33 (1980) pp. 59-75.

[13] VAN DE VOSSE, F. N., SEGAL, A., VAN STEENHOVEN, A. A., JANSSEN, J. D.: " A finite element approximation of the unsteady two-dimensional Navier-Stokes equations ", Int. J. Num. Meth. Fluids, 6 (1986) pp. 427-443.

[14] VAN DE VOSSE, F. N.: " Numerical analysis of carotid artery flow " Ph.D.Thesis University of Technology, Eindhoven (1987).

[15] YAO, L. S., BERGER, S. A.: " Entry flow in a curved pipe ", J. Fluid Mech., 67 (1975) pp. 177-196.

[16] WARD-SMITH, A. J.: " Internal fluid flow ", Clarendon Press, Oxford (1982).

Acknowledgement.

We wish to thank H.J. de Heus for the work he performed obtaining the experimental data.

A numerical technique to calculate reacting flows using finite rate chemical kinetics

J.B. Vos*
Hydraulic Machines and Fluid Mechanics Institute
Swiss Federal Institute of Technology Lausanne
CH-1015 Lausanne, Switzerland

Summary

Within the framework of a research project carried out in the Netherlands, studying turbulent flows with combustion, a computer code has been developed. A technique developed to enable the calculation of flows with combustion using a combustion model based on finite rate chemical kinetics is described. Several calculation results are presented.

Introduction

The Faculty of Aerospace Engineering of the Delft University of Technology and the Prins Maurits Laboratory TNO, the Netherlands, carry out a joint research project studying the flow and combustion processes in Solid Fuel Combustion Chambers (SFCCs). An SFCC consists of a solid fuel grain with an inner bore. Air is fed into this bore, and at the interface between the pyrolyzing solid fuel and the air, combustion will take place. A recirculation zone created by a sudden expansion at the inlet of the channel serves as a flame stabilizer, see Fig. 1. The sudden expansion decreases the flow velocity and hence increases the residence time of the gas in the channel. Furthermore, the sudden expansion increases the turbulence level, yielding a better mixing of fuel and oxidizer. Possible applications of SFCCs are, for example, waste combustion or aerospace propulsion systems. Because combustion temperatures in an SFCC are high (ranging from 1500 K to 2500 K) SFCCs may be used for the combustion of highly toxic materials like PCBs, which are burnt completely at these high temperatures. However, the greatest interest in SFCCs is as aerospace propulsion system (Solid Fuel Ramjet). Many future projects for aerospace propulsion systems are based on the use of Solid Fuel Ramjets for the flight through the atmosphere. However, before SFCCs can be applied, basic research into the flow and combustion processes is required. The SFCC-research project aims at both an experimental and theoretical approach to increase the understanding of turbulent flows with combustion. An experimental facility has been built allowing testing of SFCCs at various pressures and inlet air temperatures. Spectroscopic diagnostic techniques are used to determine the instantaneous temperature and the chemical composition of the burning gas. Fuels burnt in the experimental facility are Plexiglas (PMMA), which allows for the visual observation of the combustion process, and Polyethylene (PE).

---

* This work was carried out when the author was working at the Faculty of Aerospace Engineering of the Delft University of Technology, the Netherlands.

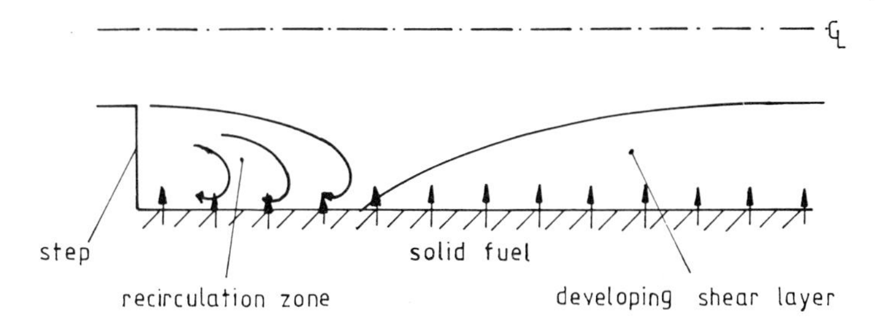

Fig. 1: Flow pattern in an SFCC

The theoretical research into SFCCs focuses at developing computer models which describe turbulent flows with combustion. A computer code, called COPPEF (Computer Program for Parabolic and Elliptic Flows), has been developed which calculates 2-Dimensional turbulent flows in channels with and without a sudden expansion. Heat and mass transfer effects at the solid fuel boundary are accounted for, and three different combustion models have been implemented [1], [2]. In this paper the numerical techniques used to solve the partial differential equations and the numerical problems encountered with the implementation of a finite rate chemical kinetics combustion model are discussed. Results of calculations for H_2-O_2 and C_2H_4-O_2 flames will be presented.

Governing Equations

The flow in an SFCC can be considered as highly turbulent. Reynoldsnumbers based on the stepheight and inlet velocity vary between 10^5 and 10^7. Making detailed predictions of turbulent flows with combustion is not possible yet for several reasons. First of all, combustion is a process in which many species and many chemical reactions are involved. A detailed description of the combustion processes requires the knowledge about the intermediate species formed, and about the chemical reactions which are important. Presently, only for the combustion of simple fuels like H_2 and hydrocarbon fuels upto C_2H_6 reaction schemes are available, but even for these simple fuels, the number of reactions and species to be considered is large. Secondly, turbulence and the effect of turbulence on the combustion process are extremely complex phenomena, in which a spectrum of length and time scales is involved. Statistical techniques are often used to describe turbulent flows, which however, introduces the "turbulence closure problem". There are many turbulence closure models available, but they only work satisfactorily for relatively simple flows. Furthermore, all these models are derived for non-reacting flows, and it may be questioned whether they can be used for flows with combustion. The effect of turbulence on the combustion process is even more complicated. Combustion is a molecular process of collisions between atoms and/or molecules. If the collision energy of the colliding particles is high enough, new atoms or molecules may be formed. On a macroscopic scale, the mixing of fuel and oxidizer is controlled by the turbulence, while on a molecular scale the mixing is controlled by molecular diffusion. Turbulence affects the macroscopic mixing of fuel and oxidizer, resulting in regions with only oxidizer or only fuel, prohibiting combustion in these regions. Furthermore, since the temperature is a measure of the collision energy, fluctuations in temperature may result in regions with fast proceeding chemical reactions, but may also prohibit combustion in other regions.

From this discussion it is clear that simplifications have to be made in order to obtain a model which can be solved numerically. To describe the flow and combustion processes in an SFCC it is assumed that the flow can be regarded as 2-Dimensional axisymmetric and steady. Turbulence is taken into account by Favre- or mass-weighted averaging of the conservation equations, and it is assumed that the k-ε turbulence closure model can be used in flows with chemical reactions to model the terms containing products of fluctuating variables. To obtain a coupling between data obtained from spectroscopic measurements and numerical predictions, a combustion model based on finite rate chemistry is used to describe the combustion process. The effect of turbulence on the combustion process, however, is neglected.

All governing partial differential equations can be cast in the general form

$$\frac{\partial}{\partial x}(\rho u\varphi) + \frac{1}{r}\frac{\partial}{\partial r}(r\rho v\varphi) - \frac{\partial}{\partial x}\left(\Gamma_\varphi \frac{\partial\varphi}{\partial x}\right) - \frac{1}{r}\frac{\partial}{\partial r}\left(r\Gamma_\varphi \frac{\partial\varphi}{\partial r}\right) = S_\varphi \quad ,(1)$$

where φ stands for the velocities u and v, the enthalpy h, the turbulence quantities k and ε, and the mass fraction for species s, Y_s. Γ_φ is the turbulent diffusion coefficient, and S_φ the source term for the variable φ [1]. This system of equations has to be completed by an equation of state and by providing a thermodynamic relation between the enthalpy and the temperature.
The net production rate of mass fraction of species s by chemical reactions can be calculated as follows. Consider the general system of chemical reactions

$$\sum_s \nu^a_{s,k} A_s \underset{k_{f,k}}{\overset{k_{b,k}}{\rightleftarrows}} \sum_s \nu^b_{s,k} A_s \quad , \quad (2)$$

where $\nu_{s,k}$ are the stoichiometric coefficients and where A_s denotes the chemical symbol of species s. The net chemical production rate of mass fraction of species s may be calculated from [3]

$$\dot{\omega}_s = M_s \sum_k (\nu^b_{s,k} - \nu^a_{s,k})\, k_{f,k}\, \rho^{m_1} \prod_s \left(\frac{Y_s}{M_s}\right)^{\nu^a_{s,k}} \cdot \left[1 - \frac{1}{K_{c,k}} \rho^{m_2} \prod_s \left(\frac{Y_s}{M_s}\right)^{(\nu^b_{s,k}-\nu^a_{s,k})}\right] \quad , \quad (3)$$

where

$$m_1 = \sum_s \nu^a_{s,k} \quad , \quad m_2 = \sum_s (\nu^b_{s,k} - \nu^a_{s,k}) \quad .$$

The forward reaction rate constant is calculated from an Arrhenius expression,

$$k_{f,k} = B_k \, T^{\alpha_k} \, e^{-E_k/T} \quad , \qquad (4)$$

where B_k and α_k are constants, and E_k the activation energy. The equilibrium constant, $K_{c,k}$, defined as the ratio of the forward and backward reaction rates, is calculated from

$$K_{c,k} = (R^*T)^{-m_2} \exp((\sum_s \nu^a_{s,k} F^o_s - \sum_s \nu^b_{s,k} F^o_s) / RT) \quad , \qquad (5)$$

where R^* is the universal gas constant divided by 1 atmosphere, and F^o_s the Gibbs free energy of species s.

The system of equations can be solved if all the necessary boundary conditions are specified. At the inlet of the channel, each variable φ is specified. At the centerline and at the outlet, the velocity in radial direction is set to zero, while a no-gradient condition is used for the other variables. At the solid fuel wall, the velocity in axial direction is set to zero (no-slip), while the velocity in radial direction is specified to simulate the pyrolizing fuel. The wall temperature is assumed to be constant. The values of k and ε near the wall are calculated using wall functions. The effects of heat and mass transfer are incorporated in the wall functions.

Solution Procedure

The system of coupled partial differential equations (1) is solved by means of a segregated solution approach. Each dependent variable φ is solved from an algebraic difference equation, and the non-linear couplings between the equations are accounted for by means of an iteration procedure. A finite volume integration scheme is used to reduce the partial differential equations to algebraic difference equations. A staggered grid system is employed in which the axial and radial velocities are stored at the boundaries of the control volumes for the other variables. The total fluxes (convective + diffusive) across the control volume interfaces are approximated by means of the Power-law scheme [4]. Due to the segregated solution approach, problems arise with the strong coupling between the pressure and velocity field. If the pressure field is prescribed correctly, the velocity field obtained by solving the momentum equations will satisfy the continuity equation. There is however no equation from which the pressure field can be solved directly. It is clear that the continuity equation acts as a constraint on the velocity field obtained by solving the momentum equations, and it is in fact this constraint which determines the pressure field. To calculate the pressure field from the momentum and continuity equations, the SIMPLER (Semi Implicit Method for

Pressure Linked Equations Revised), algorithm [4] is adopted. Besides the SIMPLER-algorithm, a line continuity correction method [1] and a block correction procedure [5] are incorporated in the COPPEF computer code to ensure that at each iteration the continuity equation is completely satisfied. Initial test calculations showed that complete satisfaction of the continuity equation enhances the convergence rate of the iteration procedure.

The pseudo-time-splitting technique

Solving the mass fraction equations with the integration procedure as described in the previous section lead to problems caused by the source dominated character of these equations. The net chemical production rate term in these equations is much larger than the convection and diffusion terms. Only by strongly limiting the maximum allowable change of Y_S in one iteration cycle, it was possible to keep the value of Y_S between zero and one. However, it was estimated that in this situation approximately 10^8 iteration cycles were required to obtain a converged solution. A closer look on the mass fraction equations shows that these equations are ill posed since there is no term in these equations which balances the chemical production rate term.

Chemical reactions may occur if fuel and oxidizer are mixed, if the temperature is high enough, and if the residence time of the gasmixture in the considered geometry is large enough. The latter condition is used to overcome the problems with the solution of the mass fraction equations. The solution of the mass fraction equations is split into two steps. In the first step convection and diffusion of mass fractions are taken into account, while in the second step chemical reactions are taken into account simultaneously for each species by integrating the chemical source term to the residence time of the gas in the considered control volume. The second step of this technique can be looked upon as an iteration process one level lower than the overall iteration procedure. The two steps of the pseudo-time-splitting technique are coupled by the overall iteration procedure. The justification of splitting the mass fraction equations into a convection/diffusion part and a chemical production part is that convection and diffusion are connected to the time and length scales of the mean motion, while chemical reactions have time and length scales which are several orders of magnitude smaller. It is thus possible to "freeze" the gas mixture in a grid cell and let it react during the time it is present in this volume. In fact, each grid cell can be seen as a chemical reactor, and all these grid cells are coupled since they exchange mass by convection and diffusion.

The integration method of Gear [6] suitable for systems of stiff ordinary differential equations, is used to integrate the chemical source term.

Calculation Results

Calculations were carried out in which the pyrolizing solid fuel wall was simulated by a porous wall through which gaseous fuel was injected. Initial calculations showed that only for flows with high inlet temperatures, T>1100 K, sustained combustion was obtained. Application of COPPEF to flows with low inlet temperatures led to the solution that no combustion was taking place. Since there was no interest in such a solution, an ignition pulse was simulated during the iteration process for flows with low inlet temperatures [7].

Calculations were carried out studying the flow and combustion of O_2/N_2 mixtures with H_2 or C_2H_4 (ethylene gas) injected at the wall. Ten different species and 24 different reactions were used to describe the combustion of H_2 with O_2 [1], while 27 species and 87 reactions were used in the description of the combustion of C_2H_4 [8]. The overall oxidizer/fuel ratio is defined as

$$(O/F) = (\rho u\, Y_{O_2}\, A)_{inlet} \,/\, (\rho v_w Y_{Fuel} A)_{wall}. \tag{6}$$

Fig. 2 shows the isotherms for the combustion of H_2 with O_2/N_2 with an overall O/F ratio of 8, which is equal to the stoichiometric O/F ratio. (The stoichiometric O/F ratio is the O/F ratio at which both the oxidizer and fuel are burnt completely). It may be seen from Fig. 2 that there is a region of maximum temperatures along the entire channel. Above this region, the mixture is oxidizer rich, and below it is fuel rich. Maximum temperatures are ~ 2650 K. Fig. 3 shows the isotherms of the combustion of H_2 with O_2/N_2 with an O/F ratio of 80. As the inlet temperature was low, 300 K, an ignition pulse was simulated during the iteration process. It was observed that, after igniting the gasmixture, the burning mixture extinguished during the iteration process. Only in a small region close to the inlet sustained combustion was obtained. It is believed that this flame extinction was caused by the combination of a large O/F ratio and a low inlet temperature. The heat released in the combustion process was insufficient to heat the cold inlet gas, and the flame extinguished. Only in the region where the mixture was locally fuel rich sustained combustion was obtained. This partial flame extinction has also been observed in calculations of C_2H_4 - air flames, and the O/F ratio's for which it occurred correspond close to the O/F ratio's for which it is difficult to obtain combustion in the experimental facility. Fig. 4 shows the isotherms for the combustion of C_2H_4 with O_2/N_2, by using the geometry and inlet conditions of the experimental facility. Maximum temperatures are located in the recirculation zone, where the mixture is locally fuel rich. In this region large amounts of CO are formed, indicating that here the combustion is not complete [7]. By increasing the amount of C_2H_4 blown into the flow, the size of the fuel rich zone is increased, but it does not affect the value of the maximum temperature or maximal amount of CO formed. Only the location where these maximum values occur are affected. This is in agreement with experimental results [9], where it was observed that maximum temperatures were found in regions where the local O/F ratio was close to the stoichiometric O/F ratio, and that, for the O/F ratio range used in the experiments, the value of the maximum temperature was independent of the overall O/F ratio.

Conclusions

The developed pseudo-time-splitting technique, which enables the calculation of flows with combustion using a finite rate chemical kinetics combustion model, works satisfactorily. By simulating an ignition pulse during the iteration process, the finite rate chemical kinetics combustion model can be applied for flows with low inlet gas temperatures. With the finite rate chemical kinetics model, it is possible to predict (partially) extinguishing flames.

Acknowledgements

The work described in this paper was sponsored by the Technology Foundation (STW), the Project Office for Energy Research (PEO), the Netherlands Organization for the Advancement of Pure Research, Working Group Supercomputers (ZWO-WGS), the Prins Maurits Laboratory TNO and the Faculty of Aerospace Engineering of the Delft University of Technology. The Swiss Federal Institute of Technology Lausanne is acknowledged for giving the opportunity to present this paper.

References

[1] VOS, J.B.: "The calculation of turbulent reacting flows with a combustion model based on finite chemical kinetics", PhD-thesis, Faculty of Aerospace Engineering Delft University of Technology, April 1987

[2] ELANDS, P.J.M.: "The prediction of the flow and combustion in an SFCC by means of two combustion models based on the diffusion flame concept", AIAA 87-1803, 23rd Joint Propulsion Conference, San Diego 1987

[3] WILLIAMS, F.A.: "Combustion Theory, 2nd Edition", Benjamin/Cummings Publishing Corporation, Menlo Park 1985

[4] PATANKAR, S.V.: "Numerical Heat Transfer and Fluid Flow", Hemisphere Publishing Corporation, Washington 1980

[5] SETTARI, A and AZIZ, K.: "A generalization of the additive correction methods for the iterative solution of matrix equations", SIAM J. of Numerical Analysis, Vol. 10, no. 3, 1973, p. 506

[6] GEAR, C.W.: "Numerical Initial Value Problems in Ordinary Differential Equations", Prentice Hall, Englewood Cliffs 1971

[7] VOS, J.B.: "Simulating an ignition pulse in turbulent reacting flows calculated with a finite chemical kinetics combustion model", AIAA 87-1979, 23rd Joint Propulsion Conference, San Diego 1987

[8] WESTBROOK, C.K.: "Chemical Kinetics of Hydrocarbon Oxidation in Gaseous Detonations", Combustion and Flame, Vol. 46, 1982, p. 191.

[9] GELD, C.W.M. van der, KORTING, P.A.O.G. and WIJCHERS, T.: "Combustion of PMMA, PE and PS in a ramjet", Fac. of Aerospace Eng. Delft Univ. of Techn./Prins Maurits Laboratory TNO, Report LR 514/PML 1987-C18, 1987

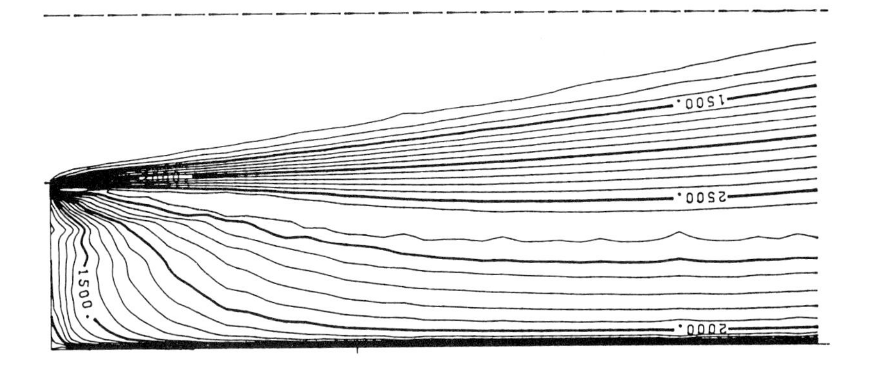

Fig. 2: Isotherms for the combustion of H_2 with O_2/N_2, T_{in} = 1100 K, T_w = 300 K, O/F = 8, radial length scale enlarged

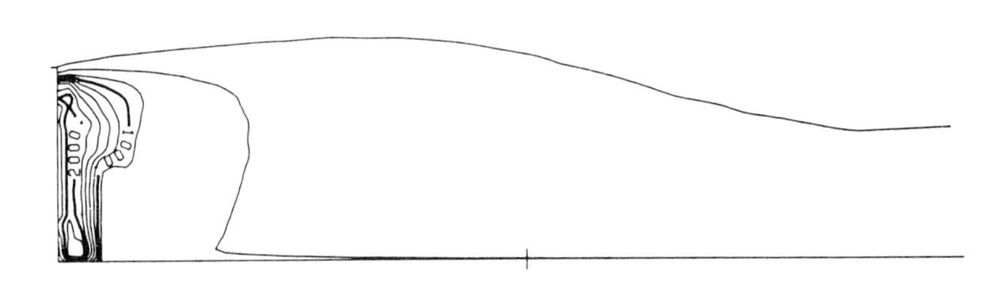

Fig. 3: Isotherms for the combustion of H_2 with O_2/N_2, T_{in} = 300 K, T_w = 800 K, O/F = 80, radial length scale enlarged

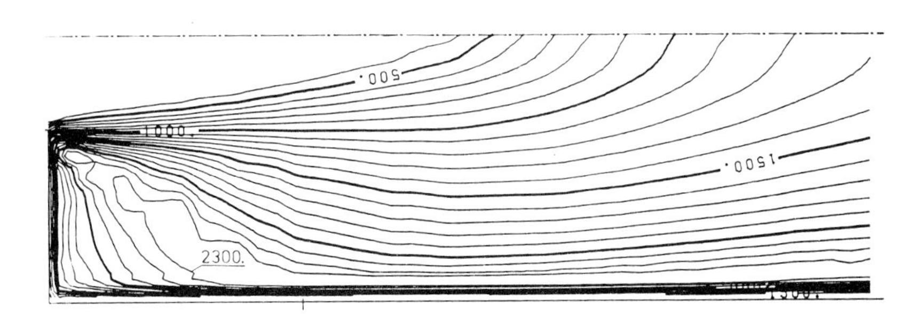

Fig. 4: Isotherms for the combustion of C_2H_4 with O_2/N_2, T_{in} = 300 K, T_w = 800 K, O/F = 4.29, radial length scale enlarged

A METHOD FOR THE SOLUTION OF THE REYNOLDS-AVERAGED NAVIER-STOKES EQUATIONS ON TRIANGULAR GRIDS

by

N.P.Weatherill, L.J.Johnston, A.J.Peace and J.A.Shaw

Aircraft Research Association Limited, Manton Lane, Bedford, England

SUMMARY

A numerical method for the solution of the Reynolds-averaged Navier-Stokes equations is presented. The approach, which utilises the concept of a triangular mesh and a finite volume algorithm for the governing flow equations, is demonstrated by simulating turbulent flow over a single aerofoil and laminar flow over a two-element high lift aerofoil/flap configuration.

INTRODUCTION

The long term aim of the research described in this paper is to provide a computational method to predict the maximum lift conditions for the high lift devices used on modern aircraft. The complex geometries of such configurations, together with the complicated flow structure make this a particularly challenging problem. Traditionally, the problem has been attempted by simplifying the physics of the flow in different regions with some appropriate patching between the subsequent zonal solutions. Such approaches usually require a rather ad hoc treatment of flow separation and, in consequence, are unreliable for flows close to maximum lift. An alternative approach, and the one adopted here, is to simulate the flow using the Reynolds-averaged Navier-Stokes equations. These equations represent a more accurate description of the physics of the flow, and with a suitable turbulence model for the viscous stresses and heat fluxes produce a means by which accurate flow simulation at maximum lift conditions can, potentially, be achieved. There are obvious difficulties in following such an approach; high computational costs, possible inadequacies in turbulence models and technical problems of ensuring numerical errors do not swamp real flow features. However, we feel that with the availability of high speed computers and the recent developments in computational methods, this approach is now worthy of investigation.

Multi-element aerofoil geometries represent complex configurations in two dimensions. As such the generation of a suitable mesh on which to discretise the equations is not straightforward. Although conformal mapping techniques have been developed, these are presently restricted to a maximum of three elements. The approach adopted here is to use triangular meshes. In view of the importance of the accurate simulation of the viscous flow near the solid boundaries, a regular structured mesh has been constructed around each element. These regular sub-meshes are connected by unstructured triangular meshes generated using an algorithm based on the Delaunay criterion [1].

The algorithm used to solve the Reynolds-averaged Navier-Stokes equations is based on the finite volume method developed by Jameson for the solution of the Euler equations on unstructured triangular meshes [2]. This approach utilises a polygonal control volume which surrounds each node and fluxes are summed over each edge in the triangulation. In the present implementation, these inviscid fluxes have been augmented with a suitable discretisation of the viscous stresses and heat fluxes, and an algebraic turbulence model [3].

The present approach has been used to simulate the viscous turbulent flow over single aerofoils. It is necessary to consider an aerofoil in isolation both to enable our results to be compared with those obtained by other workers and to ensure that the method is performing adequately on simple geometries. Our results for such cases compare favourably with experiment and other theoretical predictions. To highlight the flexibility of our approach we present results for the laminar flow over a two-element high lift aerofoil/flap configuration.

MESH GENERATION

The mesh generation technique is a combination of an analytic curvilinear mesh point generator and an algorithm based on the Delaunay criterion which connects points to form a triangulation. Mesh points $P_{i,j}$, $[1 \le i \le i_{max(k)}, 1 \le j \le j_{max(k)}$, k=1,number of elements] which possess a polar topology are generated local to each aerofoil using a simple analytic mapping technique. This approach provides a means by which grid spacing near the solid boundaries can be explicitly controlled. In the near wall regions around each element a structured regular mesh is obtained by performing a direct triangulation to the points $P_{i,j}$ $[1 \le i \le i_{max(k)}, 1 \le j \le j_{min(k)}]$. The imposed point connectivity within each quadrilateral in these regions is defined as $P_{i,j} \to P_{i-i,j}, P_{i,j} \to P_{i-i,j+1}$ and $P_{i,j} \to P_{i,j+1}$. It is necessary to ensure that these structured regions associated with different aerofoils do not overlap.

The remaining grid points $P_{i,j}$ $[1 \le i \le i_{max(k)}, j_{min(k)} < j \le j_{max(k)}]$ are used to discretise the remaining flow domain and are connected to conform to the Delaunay criterion. This connectivity is based on the polygonal regions of the Voronoi diagram, defined by

$$V_m = \{p : \|p-P_m\| < \|p-P_n\|, \forall n \neq m\}$$

where P_m is the set of points to be connected. If all point pairs which have a segment of boundary of the Voronoi diagram in common are joined by a straight line the result is a triangulation of the data points. Such a triangulation, which is called a Delaunay triangulation, optimises the regularity of the triangles.

A mesh with spacing appropriate to a simulation of laminar flow for an aerofoil/flap configuration is shown in Figure 1. The global view, Figure 1a, shows how the grid around the main element is regular and only in the region where the two component grids overlap is the mesh clearly unstructured. Figures 1b and 1c show the mesh in the gap region, in which in Figure 1b the inner-near wall structured region has been omitted. In this example, 5 and 4 grid lines are used in the structured region for the main element and flap, respectively. The mesh points have been smoothed using a Laplacian filter with particular care taken at the boundaries between the structured and unstructured regions.

FLOW SOLUTION

Governing equations

The time dependent Navier-Stokes equations for two-dimensional flow can be written in integral form as

$$\frac{d}{dt}\int_V \underline{w} dV + \int_S \underset{\sim}{H}\, \underline{n}\, dS = 0, \tag{1}$$

where $\underline{w} = (\rho,\rho u,\rho v,\rho E)^T$ and $\underset{\sim}{H} = \underline{F}\ \underline{i} + \underline{G}\ \underline{j} = (\underline{F}^I + \underline{F}^V)\underline{i} + (\underline{G}^I + \underline{G}^V)\underline{j}$. For convenience, the flux tensor $\underset{\sim}{H}$ has been split into the inviscid and viscous contributions, denoted by superscripts I and V, respectively. $\underline{F}^I = (\rho u,\rho u^2+p,\rho uv,\rho uH)^T$, $\underline{F}^V = [0,-\sigma_{xx},-\sigma_{xy},(-u\sigma_{xx}-v\sigma_{xy}+q_x)]^T$, $\underline{G}^I = (\rho v,\rho uv,\rho v^2+p,\rho vH)^T$ and $\underline{G}^V = [0,-\sigma_{xy},-\sigma_{yy},(-u\sigma_{xy}-v\sigma_{yy}+q_y)]^T$. The stress and heat flux elements are denoted by σ and q, respectively. The integration in equation (1) is taken over a volume V, whose surface S has outward unit normal $\underline{n}$.

Using the divergence theorem, equation (1) can be rewritten as

$$\frac{d}{dt}\int_V \underline{w} dV + \oint_C (\underline{F}dy - \underline{G}dx) = 0 \tag{2}$$

where C is the boundary of V, and the contour integral is carried out in an anticlockwise sense.

For laminar flows, the effective viscosity is equal to the laminar viscosity which is given by Sutherland's law. For turbulent flow, the equations retain the same form as for laminar flow, provided that Reynolds stresses are modelled using the eddy viscosity hypothesis. The effective viscosity is then the sum of the laminar and turbulent viscosities. However, the variables must now be interpreted as Reynolds-averaged, time mean quantities. An algebraic turbulence model has been implemented [3] for single element configurations.

The boundary conditions for the Navier-Stokes equations applied at solid surfaces are zero normal flow, no slip and zero normal temperature gradient at the wall. Non-reflecting boundary conditions are applied at the farfield, based on the introduction of Riemann invariants for a one-dimensional flow normal to the outer boundary of the flow domain.

Numerical formulation

The numerical formulation for equation (2) with the viscous terms $\underline{F}^V$ and $\underline{G}^V$ set to zero was proposed by Jameson[2]. It involves performing flux balances for each polygonal domain which encloses each node within the triangulation. Since flow variables are known at nodes the inviscid contribution to the contour integral in equation (2) is given by

$$\oint_C (\underline{F}^I dy - \underline{G}^I dx) \approx \frac{1}{2V}(\underline{F}^I_{k+1} + \underline{F}^I_k)(y_{k+1} - y_k) - \frac{1}{2V}(\underline{G}^I_{k+1} + \underline{G}^I_k)(x_{k+1} - x_k) \tag{3}$$

where $k = 1,2\ldots, K+1$ denote the nodes with position (x_k,y_k) which define the polygonal domain. This summation process over all polygonal domains is implemented by computing fluxes across every edge in the triangulation and sending the contributions to the nodes whose polygonal boundary contains the edge.

The viscous components of the contour integral in equation (2) involve the evaluation of derivatives of the primitive variables on each polygonal contour. To be consistent with the overall contour integral method of evaluating derivatives, it remains to define a suitable contour for obtaining first derivatives on edges. There is no unique way to perform the integration, but perhaps the simplest is to use the contour shown in Figure 2, so that the derivative at E is the sum of four terms. For example,

$$\left.\frac{\partial u}{\partial x}\right|_E \approx \frac{1}{2A}\,[(u_2+u_3)(y_3-y_2) + (u_1+u_3)(y_1-y_3) + (u_4+u_1)(y_4-y_1) + (u_2+u_4)(y_2-y_4)]$$

where A is the sum of the areas of the two adjacent triangles which define the contour. Such an approach is applicable everywhere except on boundaries where a suitable modification is made.

Although equation (2) contains dissipative terms, it is found that for stability of the numerical formulation these must be augmented by artificial dissipation terms, of the form used in the solution of the Euler equations [2]. The basic idea of accumulated edge differences has been used. This involves a term generated from a weighted difference between variables at a given node and its nearest neighbours. The weighting is adapted to gradients in the flow variables.

For application to the Navier-Stokes equations, it has been found necessary to perform a simple, but what proves to be an effective modification to the accumulated edge differencing for the artificial dissipation. In the viscous regions adjacent to aerofoil surfaces it is important to ensure that the artificial dissipation does not dominate the real viscous effects. In these regions, a regular structured grid has been constructed thus ensuring that directional properties of each edge are known. Utilising this property, we have investigated three formulations of the accumulated edge differencing for the artificial viscosity. First, a formulation which is referred to as the 3/3 dissipation model involves accumulating contributions from all edges emanating from a node. With reference to Figure 3, this involves evaluating differences along the edges from node O to nodes ABCDEF. An alternative formulation is to neglect contributions from the edges OF and OC; this is referred to as the 2/3 dissipation model. Finally, continuing in this vein, contributions along OF, OC, OA and OD can be neglected and this is called the 1/3 dissipation model. The relevance of these three dissipation models will become apparent in the following discussions. However, it can be noted here that in the 1/3 model, artificial dissipation terms computed from differences in a direction normal to the surface are neglected. This is important, since it is such differences that have the dominant effect in the real viscous flow. Therefore, the 1/3 model is an attempt to ensure that the artificial dissipation introduced into the solution scheme does not swamp the physical diffusion of the flow. We note that outside the structured regions, the dissipation is implemented consistent with the 3/3 model.

The discretisation procedure outlined leads to a set of coupled differential equations which are integrated using a 3-stage time stepping scheme. Local time stepping and residual smoothing are utilised to accelerate convergence to a steady state.

RESULTS

As a demonstration of the techniques described in the previous sections, we now present results for single and multi-element aerofoil geometries.

Firstly, to demonstrate the effect of the artificial dissipation models described previously, Figure 4 shows boundary layer profiles in the form of semi-logarithmic plots, using the 1/3 and 3/3 models for a subcritical flow case on a single aerofoil. The 1/3 model gives results which conform to the expected analytic profile for turbulent flow, whilst the 3/3 model shows profiles which are obviously affected by artificial dissipation. This confirms the views expressed earlier, that the normal component of the artificial dissipation should be omitted in shear layers to eliminate swamping of the real viscous effects. Figure 5 shows the turbulent flow pressure distribution as predicted by the present method, using a reduced dissipation model, for the RAE 2822 aerofoil, flow case 9. Comparison is made with experimental data. Agreement is generally favourable, with the present result giving a shock which is slightly too smeared. This may be a result of the grid being too coarse (4920 nodes, 120 on surface).

Turning now to a two-element aerofoil case, Figure 6 shows results on the Williams test case A [4], where Euler and laminar solutions are compared with the exact incompressible flow solution. The grid used for the laminar case is that depicted in Figure 1; 4503 nodes are used in total with 100 and 70 nodes on the surfaces of the main element and flap, respectively. The pressure distributions indicate that the Euler result (at low Mach number) is close to the exact result. The laminar solution is indicative of large viscous effects and, in fact, examination of the skin-friction distribution shows that substantial regions of separated flow are present on both the main element and the flap. Although this laminar result has perhaps little meaning in itself, it does show that viscous solutions can be obtained on complex geometries by the present method.

Finally, to further demonstrate the generality of the approach we have followed, a result is shown in Figure 7 on the three-element test case of Suddhoo and Hall [5]. Only an Euler solution is presented here, the grid containing 4400 nodes in total with each element having 95 surface nodes. Comparison with the exact solution is again favourable.

CONCLUSIONS

In this paper, we have presented a method for the solution of the Reynolds-averaged Navier-Stokes equations around multi-element aerofoils. The method involves the generation of a triangular mesh and the solution of the flow equations using a finite volume algorithm. The demonstration of turbulent flow on a single-element aerofoil and laminar flow on a multi-element configuration has indicated the status of our research at the present time. Based on these findings, we believe that progression to turbulent flow on high-lift configurations is possible, and this is our next aim. To model the physics of these flows accurately, we anticipate that this will involve a higher order turbulence model than the simple algebraic model used for single aerofoils.

ACKNOWLEDGEMENT

The authors are indebted to the Aircraft Research Association for financial support of this work, and to Professor A.Jameson for his development work on the numerical scheme for the Euler equations.

REFERENCES

[1] WEATHERILL, N.P.: "A method for generating irregular computational grids in multiply connected planar domains", to appear Int.J. Numerical Methods in Fluids, 1988.

[2] JAMESON,A., BAKER,T.J., and WEATHERILL,N.P.: "Calculation of inviscid transonic flow over a complete aircaft", AIAA paper 86-103,1986.

[3] BALDWIN,B., and LOMAX,H.: "Thin-layer approximation and algebraic model for separated turbulent flows", AIAA paper 78-257,1978.

[4] WILLIAMS,B.R.: "An exact test case for the plane potential flow about two adjacent lifting aerofoils", Aeronautical Research Council R & M No.3717,1973.

[5] SUDDHOO,A., and HALL,I.M.: "Test cases for the plane potential flow past multi-element aerofoils", Aero.J., 89,p403,1985

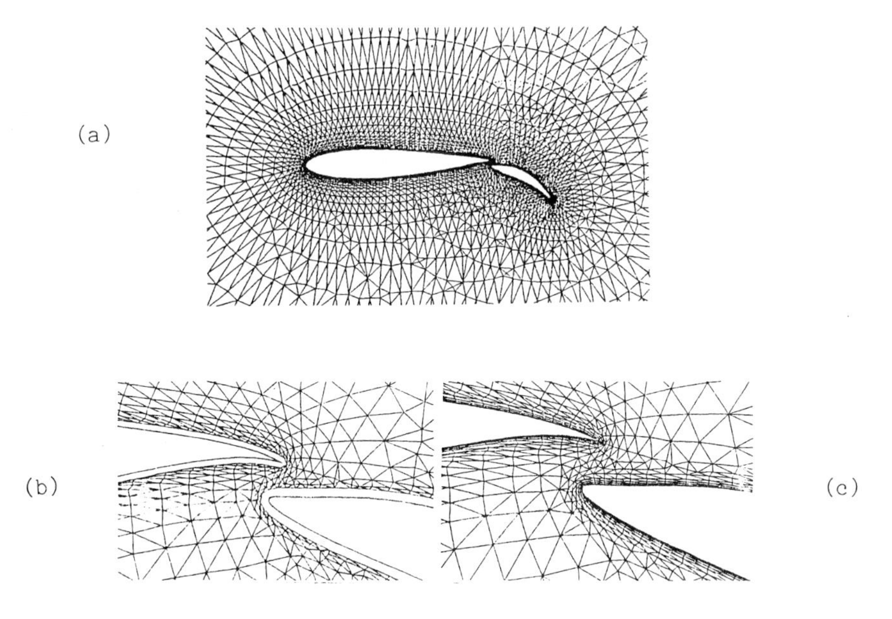

Figure 1

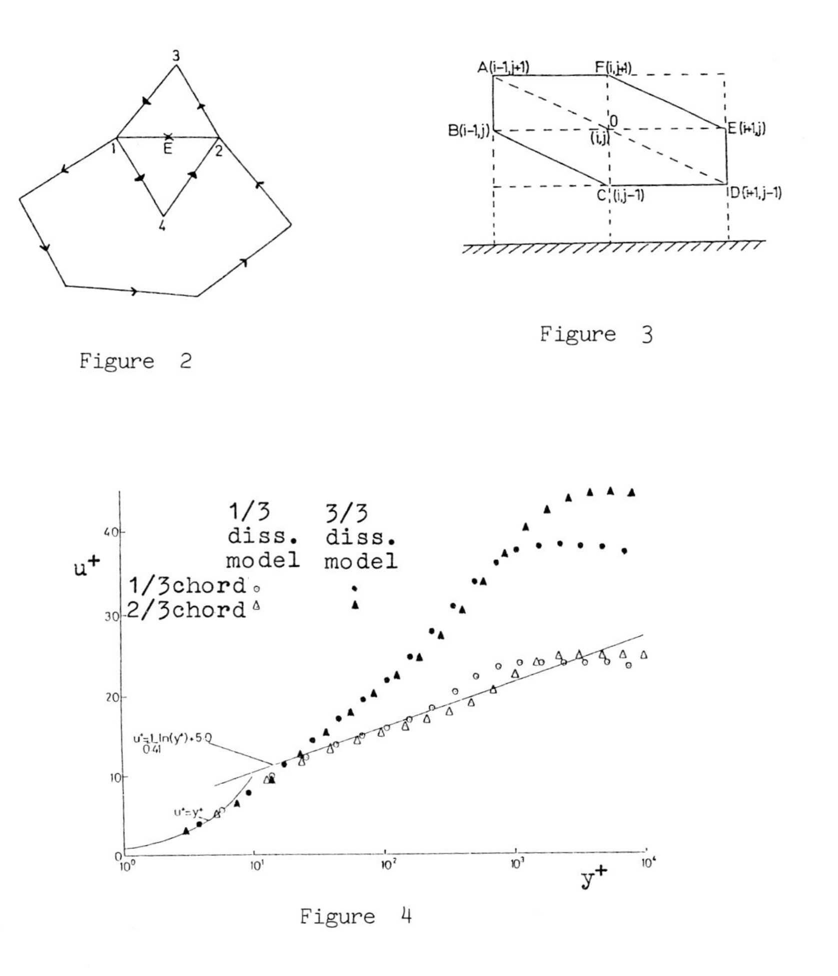

Figure 2

Figure 3

Figure 4

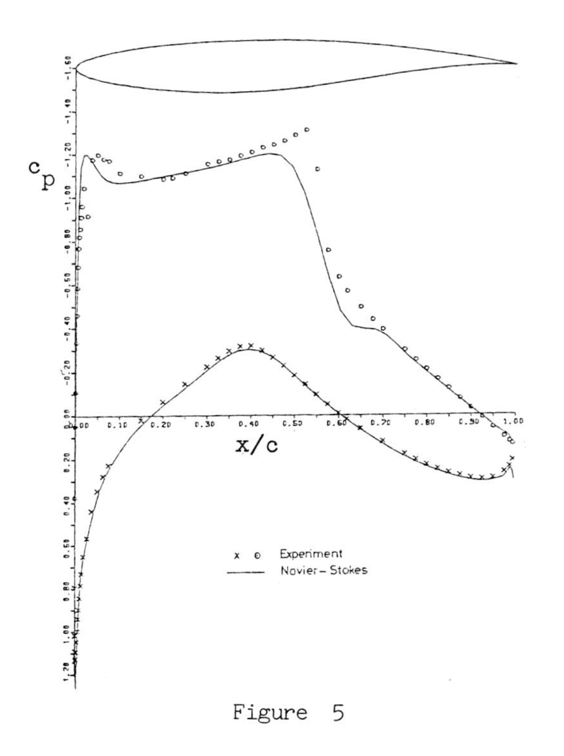

Figure 5

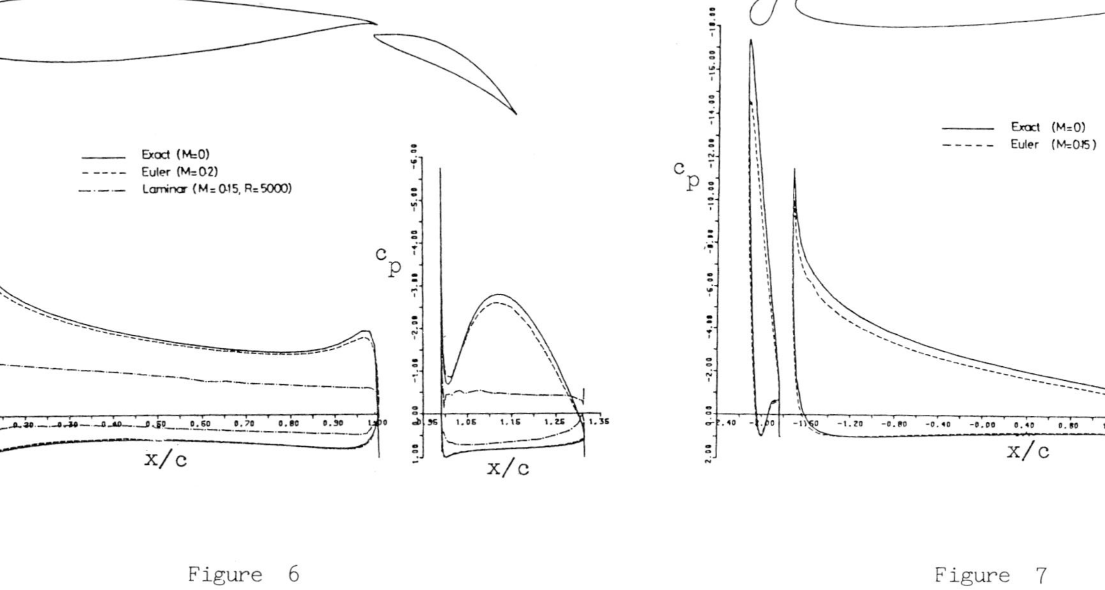

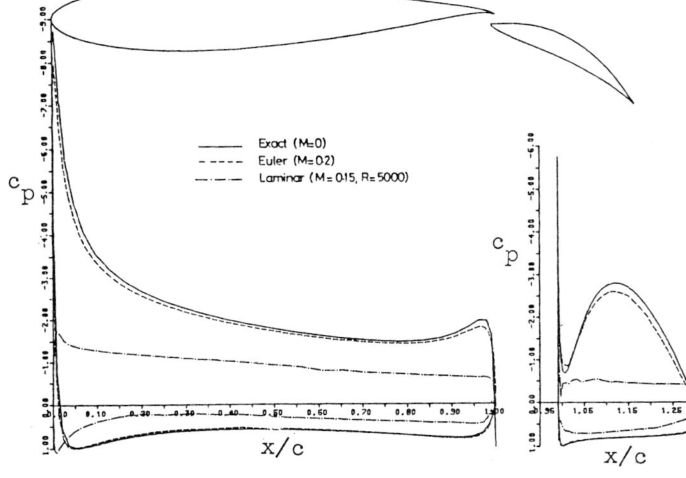

Figure 6

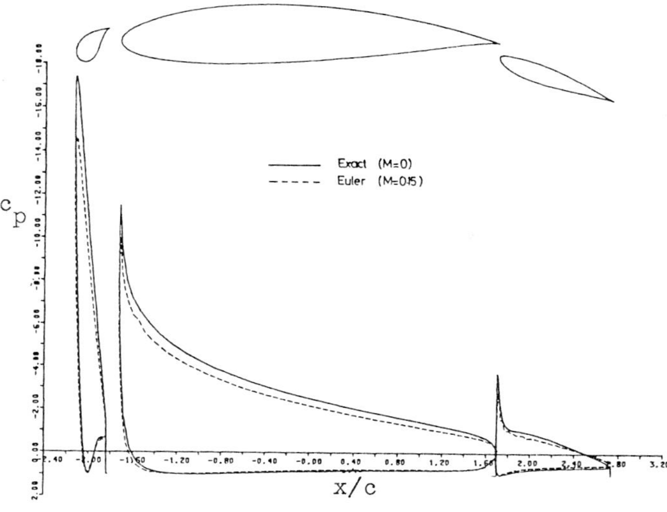

Figure 7

EULER SOLVERS FOR
HYPERSONIC AEROTHERMODYNAMIC PROBLEMS

BY

C.WEILAND, M.PFITZNER, G.HARTMANN

MESSERSCHMITT-BÖLKOW-BLOHM GMBH

SPACE SYSTEMS GROUP

8000 MUNICH 80, FRG

SUMMARY

Euler solvers for the prediction of hypersonic flow fields are considered.Time-marching and space-marching methods are developed including equilibrium real gas effects.The equations of state used are discussed.For the space-marching method some aspects of the equations and the solution algorithm are presented in detail.Calculations of the flow fields around and through real configurations are given.

1.INTRODUCTION

Future European projects regarding space transportation systems,hypersonic transport aircrafts and advanced rocket motors require the control of the aerothermodynamical problems.To that end it is necessary to adjust the available numerical methods for the integration of the various systems of governing equations to real gases in thermal and chemical equilibrium and non-equilibrium as well.

This paper deals with the Euler equations.

The theoretical tools have to be developed for a consistent formulation of these equations with equilibrium real gases.All the methods using the Jacobian of the flux vectors (i.e.:flux vector splitting and split matrix methods,and quasi-conservatively formulated Euler equations as well) need the derivation of the Jacobian with general equations of state. For embedded subsonic flow areas especially in the stagnation point region the time-dependent method and for pure super- and hypersonic flow areas the space-marching method will be applied.For higher freestream Machnumbers the bow shock fitting method instead of the shock capturing method is preferred. Therefore the bow shock fitting procedure for the space-marching and the time-marching method have to be derived for general equilibrium real gases. The theory of the time-dependent scheme is reported in /1/ using the split-matrix method for the integration of the quasi-conservatively formulated Euler-equations/2/.

In the following some aspects of the used space-marching technique will be presented.Finally,the discussion of complete Euler solutions for outer and inner realistic flow situations concludes the paper.

2. EQUATIONS OF STATE FOR EQUILIBRIUM REAL GASES

The system of Euler equations contains three thermodynamic variables,for instance p (pressure) , ρ (density) and ε (internal energy per unit mass) with $\varepsilon = \frac{e}{\rho} - \frac{1}{2}(u^2+v^2+w^2)$,e (total energy per unit volume).

The caloric equation of state combines the internal energy ε with the density ρ and the temperature T.

$$\varepsilon = \varepsilon(\rho, T). \tag{1}$$

In the case of thermally ideal gas the internal energy ε is only a function of T, $\varepsilon=\varepsilon(T)=\int c_v dT$,which means, that for moderate p no dissociation and no ionization occurs ($T \leqslant 2000K$ for air).For constant specific heat c_v ,the gas is called calorically ideal ($T \leqslant 500K$ for air) and one finds

$$\varepsilon=c_v T.$$

The thermal equation of state for real gas is given by

$$p=Z(\rho,T)\rho RT \qquad \begin{array}{ll} Z(\rho,T) & \text{:compressibility factor} \\ R & \text{:special gas constant.} \end{array} \tag{2}$$

If no dissociation,no ionization and no van-der-Waals effects are present the compressibility factor is Z=1 and the thermally ideal gas equation is valid

$$p=\rho RT.$$

By combining the two equations of state (1),(2) the pressure p can be expressed as a function of the density ρ and the internal energy ε .

One gets

$$p(\rho,\varepsilon)=Z(\rho,\varepsilon)\rho RT(\rho,\varepsilon). \tag{3}$$

For a thermally and calorically ideal gas this leads to the well-known relation

$$p=(\gamma-1)\rho\varepsilon \qquad \gamma=\frac{c_p}{c_v} \text{ :ratio of specific heats.}$$

The numerical methods employing the characteristic eigenvalue decomposition (i.e.:split-matrix and flux vector splitting methods) make use of the speed of sound

$$c^2=\left(\frac{\partial p}{\partial \rho}\right)_s.$$

In the selected thermodynamic variables ρ, ε it reads

$$c^2(\rho,\varepsilon) = \frac{p}{\rho^2}\left(\frac{\partial p}{\partial \varepsilon}\right)_\rho + \left(\frac{\partial p}{\partial \rho}\right)_\varepsilon . \tag{4}$$

A consistent equilibrum real gas formulation of the quasi-conservative Euler-equations requires the knowledge of the pressure $p=p(\rho,\varepsilon)$ and its derivatives

$$\left(\frac{\partial p}{\partial \rho}\right)_\varepsilon \quad \text{and} \quad \left(\frac{\partial p}{\partial \varepsilon}\right)_\rho \quad \text{,respectively.}$$

These values can be taken from tables, or Mollier-fit routines for air,for instance the one of Tannehill and Mugge/5/, or more generally from the program code of Gordon & McBride/6/ which contains also the equilibrium chemical kinetics of combustions.

3.SPACE MARCHING FORMULATION

The steady three-dimensional Euler equations are formulated in cylindrical coordinates,as the frame of reference,because the finite difference method derived is often applied to fuselage-like configurations.
It holds

$$J(U)\frac{\partial U}{\partial z} + K(U)\frac{\partial U}{\partial r} + L(U)\frac{\partial U}{\partial \Phi} + H(U)=0 \tag{5}$$

where the matrices J(U),K(U),L(U) are the Jacobians defined by

$$J(U) = \frac{\partial E(U)}{\partial U}$$

and

$E^T = (\rho u, p+\rho u^2, \rho uv, \rho uw, (e+p)u)$ flux vector

$U^T = (\rho, \rho u, \rho v, \rho w, e)$ solution vector

$H(U)^T = \frac{1}{r}(\rho v, \rho vu, \rho(v^2-w^2), 2\rho vw, (e+p)v)$ curvature terms.

The transformation into the computational space is carried out by using a generalized coordinate system $(\bar{z},\xi,\Phi)$ where the marching coordinate is given by $z=\bar{z}$

$$\begin{aligned} z &= \bar{z} \\ \xi &= \xi(z,r,\Phi) \\ \Theta &= \Theta(z,r,\Phi). \end{aligned} \tag{6}$$

Then the equ.(5) transforms to

$$J(U)\frac{\partial U}{\partial z} + \overline{K(U)}\frac{\partial U}{\partial \xi} + \overline{L(U)}\frac{\partial U}{\partial \Theta} + H(U) = 0 \tag{7}$$

with

$$\begin{aligned} \overline{K(U)} &= J\xi_z + K\xi_r + L\xi_\Phi \\ \overline{L(U)} &= J\Theta_z + K\Theta_r + L\Theta_\Phi . \end{aligned}$$

Equ.(7) is approximated by finite differences and solved with a semi-implicit sweep method for an ideal gas assumption in /5,6/. For real gas the Jacobian J(U) looks

$$J(U)=\begin{vmatrix} 0 & 1 & 0 & 0 & 0 \\ \psi-u & u(2-\bar{p}_\varepsilon) & -v\bar{p}_\varepsilon & -w\bar{p}_\varepsilon & \bar{p}_\varepsilon \\ -uv & v & u & 0 & 0 \\ -uw & w & 0 & u & 0 \\ u(\psi-\frac{e+p}{\rho}) & \left(\frac{e+p}{\rho}-u^2\bar{p}_\varepsilon\right) & -uv\bar{p}_\varepsilon & -uw\bar{p}_\varepsilon & u(\bar{p}_\varepsilon+1) \end{vmatrix}$$

with $\psi=\bar{p}_\varepsilon(u^2+v^2+w^2-\frac{e}{p})+p_\rho=\bar{p}_\varepsilon(u^2+v^2+w^2-\frac{e+p}{\rho})+c^2$

$\frac{e+p}{\rho}=h_0$:total enthalpy $\bar{p}_\varepsilon=\frac{\partial p}{\partial \varepsilon}\cdot\frac{1}{\rho}$

and similar for K(U) and L(U).

A bow shock fitting procedure is used because in hypersonic flows the strength of bow shocks is large and the application of a shock capturing procedure may lead to some difficulties regarding the conservation properties of the governing equations. Applying the bow shock fitting procedure means transforming the physical space into the computational space such ,that the bow shock contour coincides with a coordinate surface ($\xi(z,r,\Phi)=1$). Thus at the bow shock one has six unknowns , the five flow variables of the solution vector and the contour function of the shock.On the other hand with the five Rankine-Hugoniot equations and the one scalar equation of the sweep result/6/ these quantities can be determined.

$\rho V_\nu=\rho_\infty V_{\nu\infty}$	continuity
$p+\rho V_\nu^2=p_\infty+\rho_\infty V_{\nu\infty}^2$	normal momentum
$V_\tau=V_{\tau\infty}$	tangential momentum (two independent relations)
$\frac{e+p}{\rho}=\frac{e_\infty+p_\infty}{\rho_\infty}$	energy
$g=\mu_1\rho+\mu_2\rho u+\mu_3\rho v+\mu_4\rho w+\mu_5 e$	sweep equation

V_ν = shock normal velocity

V_τ = shock tangential velocity.

The evaluation of these equations results in

$$\begin{vmatrix} \rho u \\ \rho v \\ \rho w \end{vmatrix}_{sh} = \frac{\rho}{\rho_\infty} \begin{vmatrix} \rho u \\ \rho v \\ \rho w \end{vmatrix}_{\infty} + \frac{\alpha_\infty}{\beta^2} \left(1 - \frac{\rho}{\rho_\infty}\right) \begin{vmatrix} \xi_z \\ \xi_r \\ \frac{1}{r}\xi_\phi \end{vmatrix}$$

$$\left|\frac{\rho}{\rho_\infty}\right|_{sh} = \frac{1}{2A}\left(-B + \sqrt{B^2 - 4AC}\right)$$

$$e_{sh} = e_\infty \frac{\rho}{\rho_\infty} + \left(1 - \frac{\rho_\infty}{\rho}\right)\left(p_\infty \frac{\rho}{\rho_\infty} - \frac{\alpha_\infty^2}{\beta^2 \rho_\infty}\right)$$

$$p(\rho,\varepsilon)_{sh} - p_\infty - \left(\frac{\alpha_\infty(\xi_z)}{\beta(\xi_z)}\right)^2 \frac{1}{\rho_\infty}\left(1 - \frac{\rho_\infty}{\rho}\right) = 0 \tag{8}$$

$$A = \frac{g_\infty - \alpha_\infty \mu_\nu/\beta^2}{\mu_5} + p_\infty \qquad \alpha_\infty = (\rho\vec{v})_\infty \cdot \text{grad}\,\xi$$

$$B = -\left(\frac{g - g_\infty}{\mu_5} + A + C\right) \qquad \beta = |\text{grad}\,\xi|$$

$$C = \frac{\alpha_\infty^2}{\beta^2 \rho_\infty} \qquad \xi_z|_{shock} \equiv \frac{\partial F}{\partial z}$$

$$\mu_\nu = \mu_2 \xi_z + \mu_3 \xi_r + \mu_4 \frac{1}{r}\xi_\phi \qquad F: \text{shock contour function.}$$

The equ.(8) is solved iteratively for $\partial F/\partial z$ by means of a Newton method with the condition that $p(\rho,\varepsilon)$ is given by an equilibrium real gas routine.

4.RESULTS

The consideration of equilibrium real gas effects influences only weakly the static pressure distribution,while the density increases and the temperature diminishes correspondingly.As long as the vibrational degrees of freedom are excited,-i.e.:the specific heats are only a function of the temperature and the thermally ideal gas equation holds-,the decrease of the temperature is inversely proportional to the increase of the density.If the molecules of the gas dissociate, the compressibility factor $Z(\rho,T)$ grows and the reduction of the temperature will be faster than the rise of the density. Fig.1 shows the calculation of the flow field around an ogive-cylinder configuration with $M_\infty = 8$ and $\alpha = 6°$ using the space-marching procedure for perfect gas (fig.1a) and real gas (fig.1b).Lines of constant temperatures are plotted on the body surface,in the upper and lower planes of symmetry and in the last computed crossflow plane (equal numbers in the plots refer to same levels of temperatures).The figures are qualitatively very similar but the levels are slightly shifted.

For the same configuration fig.2 shows the lines of constant temperature in the case $M_\infty = 8$ and $\alpha = 10^0$. The results of an ideal gas flow field calculation ($M_\infty = 8$, $\alpha = 30^0$) around a simplified shuttle configuration are presented in fig.3. In the nose region the time-dependent method /1/ was applied. The inviscid streamlines on the body surface are plotted (fig.3a). In fig.3b the isobars on the body surface are displayed calculated with the space-marching technique. Fig.4 concerns with flow field calculations past the Hermes configuration. A comparison of the temperature distribution in the nose region for ideal gas (fig.4b , $\Delta T = 250K$) and real gas (fig.4c , $\Delta T = 50K$) is given. Note that on the leeward side (plane of symmetry) the temperature is lower in case of ideal gas than real gas (fig.4d). The flow field in a rocket motor nozzle is computed by using a 2-D rot. variant of the time-dependent method /2,7/, where the equations of state for a hydrogen/oxygen combustion process were determined by means of the computer code of Gordon & McBride /4/ (figs.5).

REFERENCES:

/1/ **Pfitzner,M.;Weiland,C.:** AGARD-CP-428,Paper 22,Bristol,(1987).

/2/ **Weiland,C.:** Notes on Numerical Fluid Mechanics,Vol.13,Vieweg Verlag,(1986),pp.383-390

/3/ **Tannehill,J.C.;Mugge,P.H.:** NASA-CP-2470 (1974).

/4/ **Gordon , McBride :** NASA SP-273 (1976).

/5/ **Babenko,K.I.;Voskresenskii,G.P.;Lyubimow,A.N.;Rusanow,V.V.:** NASA TT F-380,(1966).

/6/ **Weiland,C.:** AGARD-CP-342,Paper 19,Rotterdam,(1883).

/7/ **Hartmann,G.:** Berechnung von Düsenströmungen unter Berücksichtigung von Realgaseffekten,Diploma Thesis, Universität Stuttgart ,(Juli 1987).

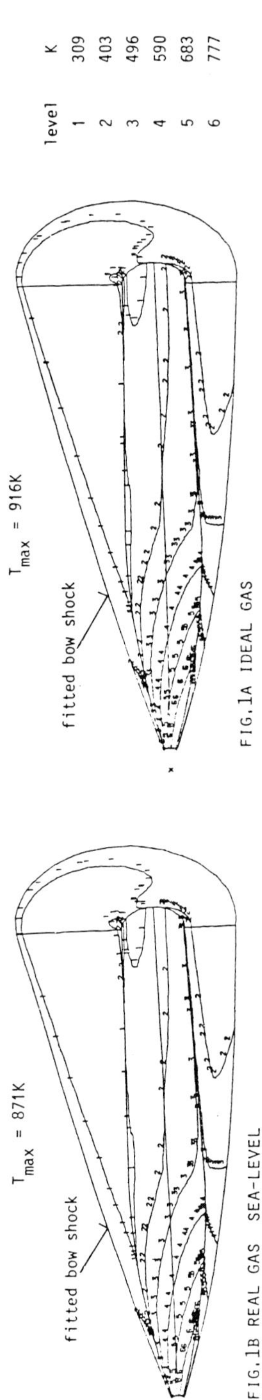

FIG.1 ISOLINES OF TEMPERATURE ON THE BODY SURFACE, IN PLANES OF SYMMETRY AND IN THE EXIT CROSSFLOW PLANE; OGIVE-CYLINDER CONFIGURATION $M_\infty = 8$, $\alpha = 6^0$

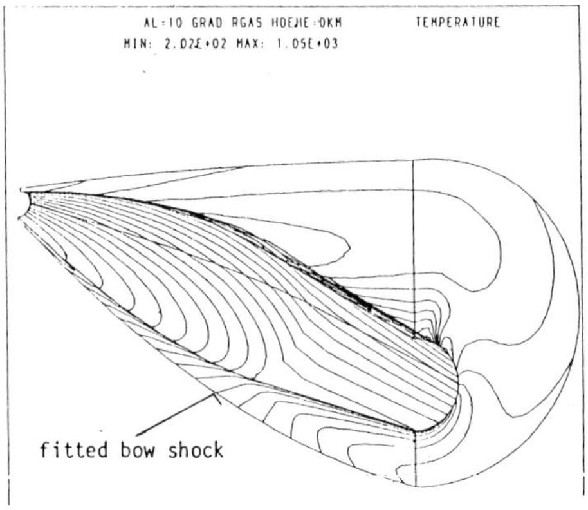

AL=10 GRAD RGAS HOEHE=0KM TEMPERATURE
MIN: 2.02E+02 MAX: 1.05E+03

fitted bow shock

FIG.2 ISOLINES OF TEMPERATURE; OGIVE-CYLINDER $M_\infty = 8$; $\alpha = 10^0$

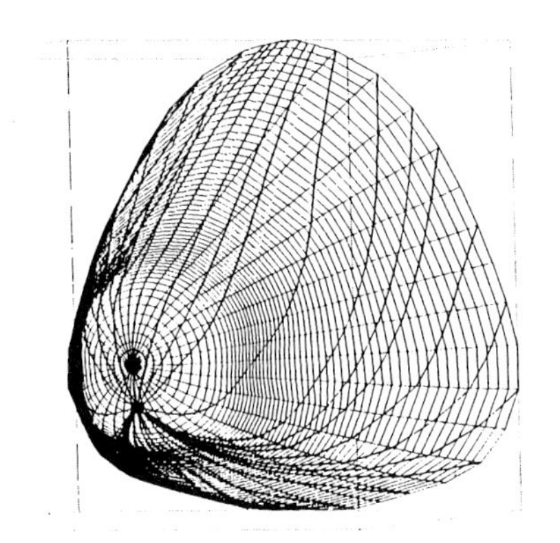

FIG.3A SIMPLIFIED SHUTTLE NOSE: INVISCID STREAMLINES ON SURFACE $M_\infty = 8$; $\alpha = 30^0$

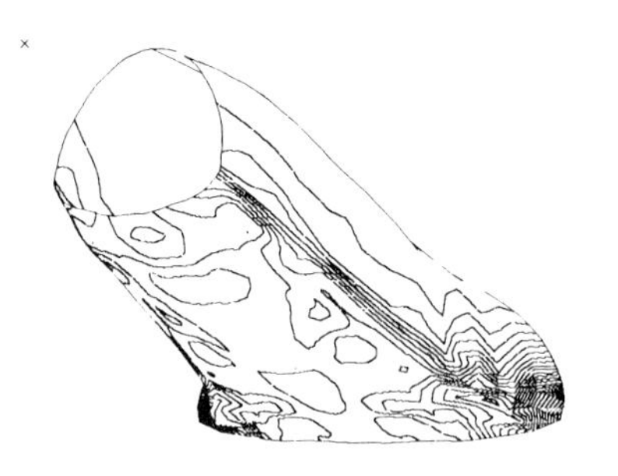

FIG.3B SIMPLIFIED SHUTTLE: ISOBARS ON BODY SURFACE $M_\infty = 8$; $\alpha = 30^0$

FIG. 4 HERMES-CONFIGURATION $M_\infty = 10$; $\alpha = 30^0$

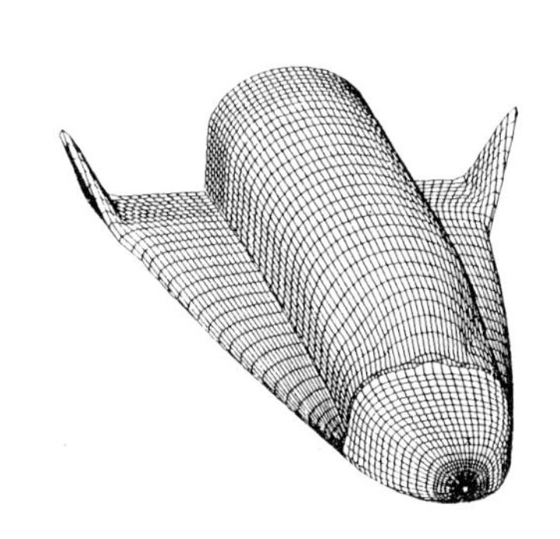

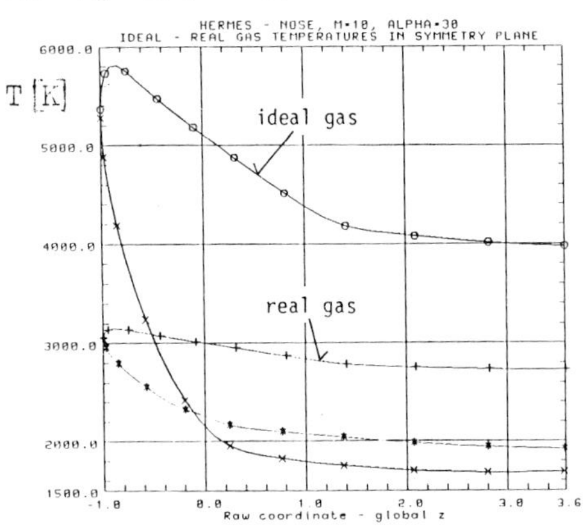

FIG. 4D NOSE-REGION: TEMPERATURE DISTRIBUTION IN PLANES OF SYMMETRY

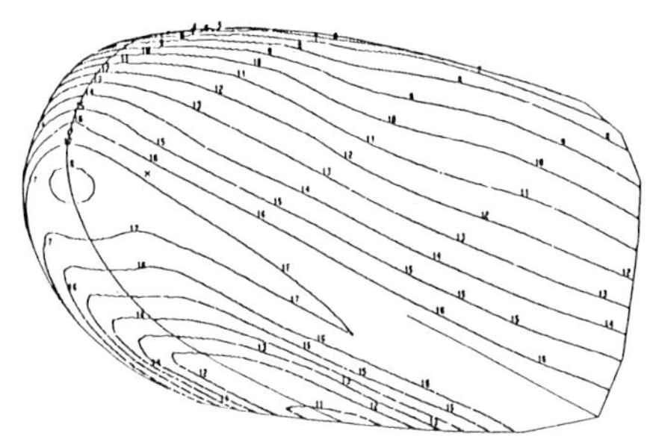

FIG. 4B NOSE-REGION: ISOLINES OF TEMPERATURE IDEAL GAS (T_{MIN} =1680K; T_{MAX} = 5810K; ΔT = 250K) T_∞=271K

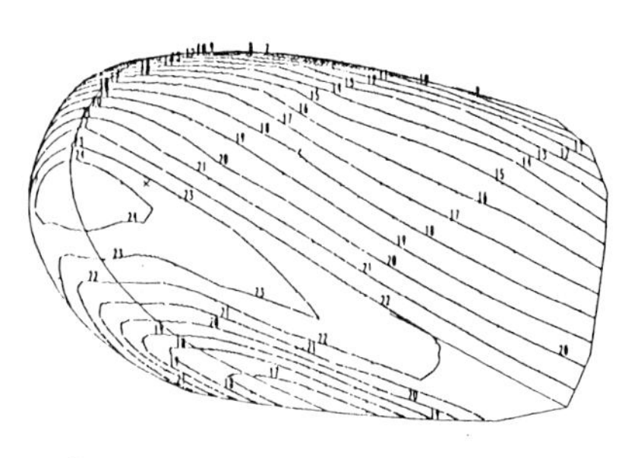

FIG. 4C NOSE-REGION: ISOLINES OF TEMPERATURE; REAL GAS, ALTITUDE 50KM , T_∞=271K (T_{MIN}=1910K, T_{MAX}=3150K, ΔT=50 K)

FIG. 5 NOZZLE FLOW; HYDROGEN/OXYGEN COMBUSTION

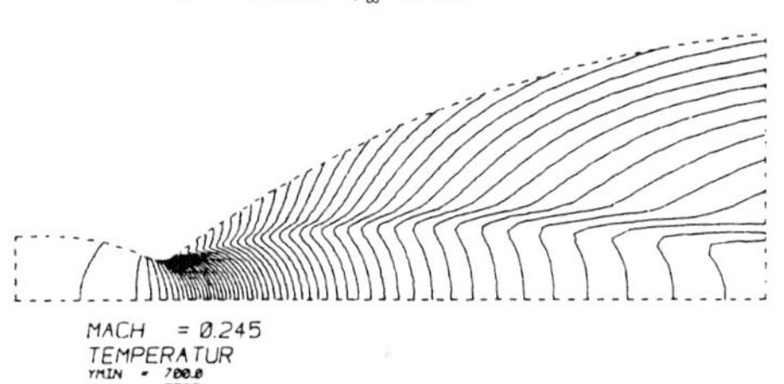

LINES OF CONSTANT TEMPERATURE

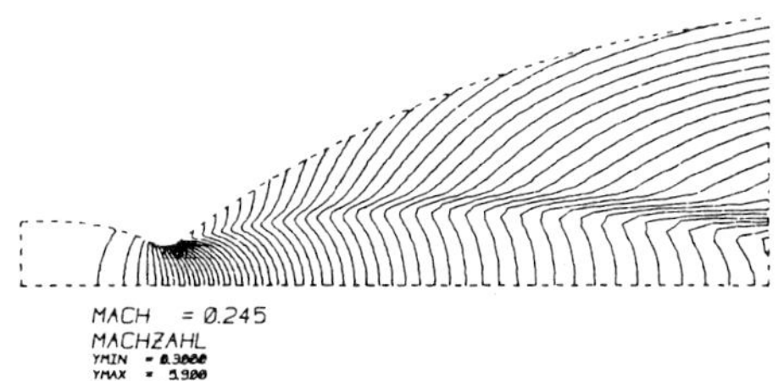

LINES OF CONSTANT MACHNUMBERS

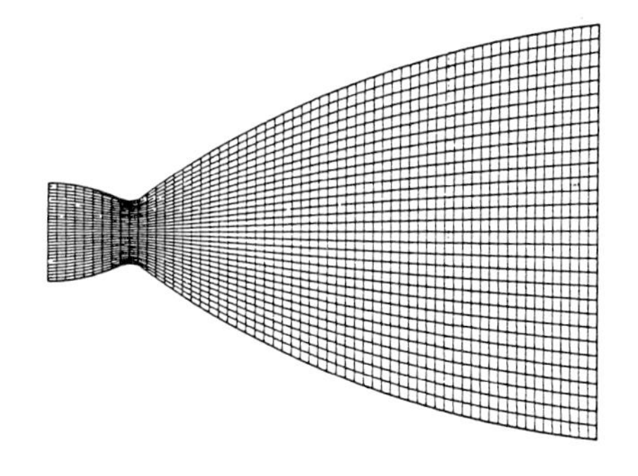

NOZZLE CONTOUR

COUPLED EULERIAN AND LANGRANGIAN NUMERICAL METHODS FOR THE COMPUTATION OF THE FLOWFIELD AROUND AN AIRFOIL

A.Zervos and S. Voutsinas

Laboratory of Aerodynamics, National Technical University of Athens
42, 28th Octovriou Str., Athens 106 82, Greece

SUMMARY

A numerical method is proposed for the prediction of the incompressible flowfield around an airfoil. The flowfield is decomposed into two regions. The Eulerian-Finite Element Method is used in the inner region and the Langrangian-Vortex Particle Method in the outer region. The two regions are connected with an integral equation on a common smooth boundary which is solved by the Boundary Element Method. The non-slip condition is fulfilled producing vortex particles along the boundary of the airfoil which are, afterwords, distributed on the nodes of the inner's region computational grid by the Cloud-in-Cell technique.In order to have a grid free formulation in the outer region vorticity is approximated there as a sum of vortex particles. All vortex particles are displaced by using the transport equations. A local diffusion process is incorporated in the numerical procedure. The first results for the starting flow of an airfoil show that the method is promissing.

INTRODUCTION

The numerical simulation of the viscous flowfield past aerodynamic bodies demands large amount of computer time and storage. The decomposition of the flowfield into different regions and the application of different methods in each region according to the specific physical characteristics can be an attractive alternative.

The aim of this work is the coupling of an Eulerian and a Langrangian methodology for the prediction of the flowfield around an airfoil at high Reynolds numbers. Near the body surface where the viscous effects (vorticity production-diffusion) are very important an Eulerian-differential methodology is the most appropriate choice. In the region relatively far from the body, where the viscous effects are of different scale and the displacement phenomena become dominant the Langrangian-vortex particle method (VPM) can simulate adequately the flowfield. As differential methodology the Finite Element Method (FEM) was chosen mainly because of its grid generation simplicity. The two regions are connected with an integral equation on a common boundary which is derived from Green's theorem and solved by the Boundary Element Method (BEM).

Both methods (BEM-VPM,FEM) have been alrady used individually for similar problems [1,2,3,4,5]. The BEM-VPM in order to simulate adequately the viscous flow near a boundary demands a large number of vortex particles due to the fact that the flowfield induced by a single particle is of local validity. On the other hand Eulerian differential methods (including FEM) must use a fine computational grid in order that the numerical length scale is compatible with the viscous diffusion length scale.

FORMULATION AND METHOD OF SOLUTION

In our approach the flowfield is decomposed into two parts (Figure 1): Ω_I, Ω_O with common boundary a closed smooth curve s_O, within which the airfoil is included. Let $\psi(x)$ denote the perturbation stream function and $\omega(\vec{x})$ the vorticity field, whereas indices I,O denote restrictions on Ω_I, Ω_O respectively. The equations of the problem take the following form:

- kinematic equations
  In the inner region Ω_I

$$\nabla^2 \psi_I = -\omega_I \tag{1}$$

with the inviscid boundary condition on the surface of the airfoil s_a

$$\psi_I = u_\infty y - v_\infty x \quad \text{(see Figure 1)}. \tag{2}$$

In the outer region Ω_O

$$\nabla^2 \psi_O = -\omega_O \tag{3}$$

with the condition at infinity

$$\psi_O(\vec{x}) = \mathcal{O}\left(\frac{1}{\vec{x}}\right), \quad |\nabla\psi_O(\vec{x})| = \mathcal{O}\left(\frac{1}{|\vec{x}|^2}\right), \qquad x \to \infty . \tag{4}$$

The coupling conditions are:

$$\psi_I|_{s_O} = \psi_O|_{s_O} = \psi_{s_O} \tag{5}$$

$$\frac{\partial \psi_I}{\partial \eta}\Big|_{s_O} = \frac{\partial \psi_O}{\partial \eta}\Big|_{s_O} = \partial_\eta \psi_{s_O} \tag{6}$$

-dynamic equations
The transport equations take the form:

$$\frac{d\omega_I}{dt} = \nu \nabla^2 \omega_I, \ \Omega_I \tag{7}$$

$$\frac{d\omega_O}{dt} = \nu \nabla^2 \omega_O, \ \Omega_O \tag{8}$$

and the no-slip condition on s_a is

$$\frac{\partial \psi_I}{\partial \eta} = u_\infty \eta_y - v_\infty \eta_x . \tag{9}$$

In region Ω_I a continuous distribution of vorticity $\omega_I(x)$ is assumed and an Eulerian methodology is applied. In region Ω_O a discrete distribution of vorticity $\omega_o(x)$ is assumed and a Langrangian methodology is applied.

The problem in Ω_I is formulated in variational form [6] resulting in:

$$\int_{\Omega_I} \nabla\psi\nabla\beta d\Omega = \int_{\Omega_I} \omega_I \beta d\Omega + \int_{s_O} \frac{\partial\psi}{\partial\eta}\beta ds \qquad \forall \beta \in W \tag{10}$$

where β is a test function which for Dirichlet conditions on s_a belongs to

$$W = \{\upsilon \in H^1(\Omega_I) : \upsilon|_{s_a} = 0\}.$$

The problem in Ω_O is solved by using an integral equation on the common boundary s_o. Green's theorem for Ω_O takes the form [7]:

$$\psi_O(x_o) = \int_{s_O} \frac{\partial\psi_{O(x)}}{\partial\eta} \varphi(x-x_o) ds(x) - \int_{s_O} \psi_O(x) \frac{\partial\varphi}{\partial\eta}(x-x_o)\, ds(x)$$

$$+ \iint \nabla^2\psi_O(x)\varphi(x-x_o) d\Omega(x), \quad x_o \in \Omega_O . \tag{11}$$

By taking into account the coupling conditions (5) and (6), the equation (3) and the values of $\varphi(x-x_o)$ and $\partial\varphi/\partial n(x-x_o)$

$$\varphi(x-x_o) = \frac{1}{2\pi} \ln|x-x_o|, \quad \frac{\partial\varphi}{\partial\eta}(x-x_o) = \frac{1}{2\pi} \frac{\vec{x}-\vec{x}_o}{|\vec{x}-\vec{x}_o|} \vec{\eta} \tag{12}$$

the integral equation on the common boundary becomes [8]:

$$\pi\psi_I(\vec{x}_O) = \int_{s_O} \frac{\partial\psi}{\partial\eta}(\vec{x}) \ln|\vec{x}-\vec{x}_O| ds(\vec{x}) - \int_{s_O} \psi_I(\vec{x}) \frac{\vec{\eta}(\vec{x})(\vec{x}-\vec{x}_O)}{|\vec{x}-\vec{x}_O|^2} ds(\vec{x})$$

$$- \int_{\Omega_O} \omega_O(\vec{x}) \ln|\vec{x}-\vec{x}_O| d\Omega, \qquad \vec{x}_O \in s_O . \tag{13}$$

NUMERICAL PROCEDURE

The inviscid flowfield

The variational equation (10) is approximated by employing the Finite Element Method and the integral equation (13) by employing the Boundary Element Method. Both approximations use linear interpolation functions.

For a grid-free formulation in Ω_O, ω_O is approximated as a sum of vortex particles [9]:

$$\omega_O(x) = \Sigma\, \Gamma_j \zeta_{\varepsilon_j}(x-x_j) \quad , \; x_j \in \Omega_O \tag{14}$$

where ζ_ε is a cut-off function with radial symmetry and compact support [10].

A typical FE triangular discretization of the region Ω_I is used. If N is the number of nodes in Ω_I and M the number of nodes on s_o then, by considering as unknowns the values of the streamfunction ψ on the nodes, the variational equation (10)is discretized in N+M equations.

The integral equation (13) is easily discretized if we consider as unknowns the values of $\psi s_o(x)$ on the nodes of s_o and $\partial\psi s_o/\partial n\ (x)$ on the segments of s_o. Equation (13) is then written at the M centers of the segments of the boundary s_o. The final system is, then, constituted by N+2M equations and unknowns. ψ_o is calculated using Green's theorem.

The solution of the kinematic equations produces the inviscid flowfield.

Vorticity generation

The no-slip condition is fullfiled producing vortex particles along the boundary of the airfoil. If the ψ distribution on each side of the grid triangles is linear then the velocity on each triangle is constant. A vortex sheet is introduced on each element in such a manner that its intensity γ will cancel locally the tangential velocity C_u on the surface of the airfoil (Figure 2).

$$\gamma = -2C_u \tag{15}$$

Considering the γ distribution as constant, the intensity of each vortex particle produced is:

$$\Gamma = \gamma \Delta l = -2C_u \Delta l \tag{16}$$

Vorticity distribution

The vorticity distribution on each element of Ω_I is considered continuous. The vortex particles produced are distributed on the nodes of the computational grid. The Cloud in Cell technique [11], adopted for triangular elements is used. If a vortex particle with intensity Γ is inside a triangular element $e_{1.2.3}$ then the value of vorticity on each node is (Figure 3):

$$\omega_i^e = \frac{\Gamma}{A}\,\frac{A_i}{A} = \frac{\Gamma A_i}{A^2}, \qquad i=1\div 3 \tag{17}$$

where $A=A_1+A_2+A_3$ is the area of the triangle. The continuous distribution of vorticity on element e is:

$$\omega^e(x,y) = \sum_{i=1}^{3} \omega_i^e\, \beta_i^e\,(x,y) \tag{18}$$

where $\beta_i^e\,(x,y)$ are interpolation functions.

A vortex particle in region Ω_O remains as vortex particle and its influence in Ω_I is performed through the integral equation on s_o.

Vorticity transport

In order to complete a time step the transport equations are used for the displacement of all vortex particles.

For the inner region Ω_I each vortex particle is displaced with the velocity of the element to which it belongs. The contribution of other possible vortex particles included on the same element is also computed, accounting for the inner structure of the large scale.

For the outer region Ω_O, each vortex particle is displaced with the local velocity as computed by the representation theorem and the contribution of all other particles in Ω_O.

In all cases the Adams-Bashford scheme is used:

$$\vec{x}_{K+1} = \vec{x}_K + \Delta t(1.5\vec{u}_K - 0.5\vec{u}_{K-1}) . \quad (19)$$

Vorticity diffusion

At a first approach vorticity diffusion has been taken into account locally, i.e. for each particle separately, using the singular solution of the heat equation.

The solution of the equation

$$\frac{\partial \omega}{\partial t} = \nu \nabla^2_\omega \quad (20)$$

is $\omega(x;t) = \frac{1}{4\pi\nu t} \int_{R^2} \omega(x') e^{-\frac{|x-x'|^2}{4\nu t}} dx'$.

For discrete vortex particles it becomes

$$\omega(x;t) = \Sigma \, \Gamma_j \, \zeta_\varepsilon(x-x_j) \quad (21)$$

where $j_\varepsilon (x-x_j) = \frac{1}{\pi\varepsilon^2} e^{\frac{|x-x_j|^2}{\varepsilon^2}}$ is the cut-off function and $\varepsilon=2\sqrt{\nu t}$ is the cut-off length.

By letting at each time step the value of the cut-off length to change with time as

$$\varepsilon^2_{K+1} = \varepsilon^2_K + 4\nu\Delta t \quad (22)$$

we introduce the local diffusion process.

RESULTS

Results are presented for the steady case and for the starting flow of a NACA 0012 airfoil. Figure 4 shows the computational grid used for the FE calculations. In Figure 5 the flowfield and the pressure distribution around the airfoil at a 6^0 angle of attack are presented. The results are identical to those calculated through classical numerical schemes and very close to experimental measurements for high Reynolds numbers. Figure 6 shows the position of the vortex particles and Figure 7 the flowfield at an early stage of the flow development of the starting motion around the airfoil at 0^0 angle attack. These preliminary results were produced with a coarse grid and show the flow retardation, simulating the boundary layer, near the body surface.

CONCLUSIONS

The proposed approach seems to be promissing. It is actually exploited in order to investigate the influence of different parameters such as limiting size of the inner region, grid density, local diffusion process etc. Grids of the type presented in Figure 8 seem to be suitable. More elaborate schemes for the vorticity production (such as the introduction of several vortex sheets on each element) and viscous diffusion (random walk prediction schene for the inner region) are currently tested.

REFERENCES

[1] Basu B.C. and Hancock G.J., "The unsteady motion of a two-dimensional aerofoil in incompressible, inviscid flow", J.Fluid Mech., vol. 87 (1978) pp 159-178.

[2] Vezza M. and Galbraith R.A.McD., "A method for predicting unsteady potential flow about an aerofoil", Int.Journal for Num. Methods in Fluids, vol.5, pp 347-356, 1985.

[3] Glowinski, "Numerical methods for non-linear problems", Springer-Verlag, (1984), 2nd edition.

[4] Wu J.C. and Thomson J.F., "Numerical solutions of time-dependent incompressible. Navier-Stokes equations using an integro-differential formulation", Computers and Fluids, vol.1, pp 197-215, 1973.

[5] Thomasset, F. "Implementation of Finite Element Methods for Navier Stokes Equations", Springer Series Comp. Physics, Springer-Verlag 1981.

[6] Oden, J.T., Reddy, J.N. "An Introduction to the Mathematical Theory of Finite Elements". Wiley 1976.

[7] Günter, N.M., "Potential Theory and its Applications to Basic Problems of Mathematical Physics", Frederick Ungar Publishing Co. N.Y.1967.

[8] Jonson, C., Nedelec, J.C.,"On the Coupling of Boundary Integral and Finite Element Methods". Mathematics of Computation, vol. 35, 1980, p.1063-1079.

[9] Leonard, A. "Vortex Methods for Flow Simulation", J.Comp. Physics, vol. 37, p.289 (1980).

[10] Zervos A., Voutsinas S. and Liakea E., "On the numerical and physical consequences of the use of cut-off functions in Vortex Methods" Z.A.M.M., vol.67(1987)5,pp T343-T344.

[11] Baker G.R., "The Cloud in Cell technique applied to the roll up of vortex sheets", J.of Comp. Physics 31, pp76-95 (1979).

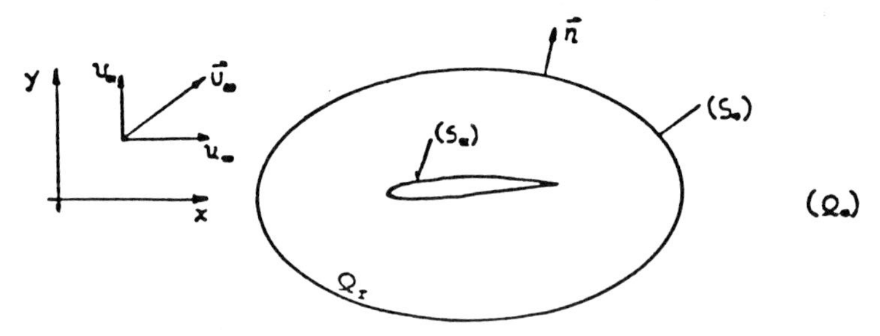

Fig.1 Problem configuration

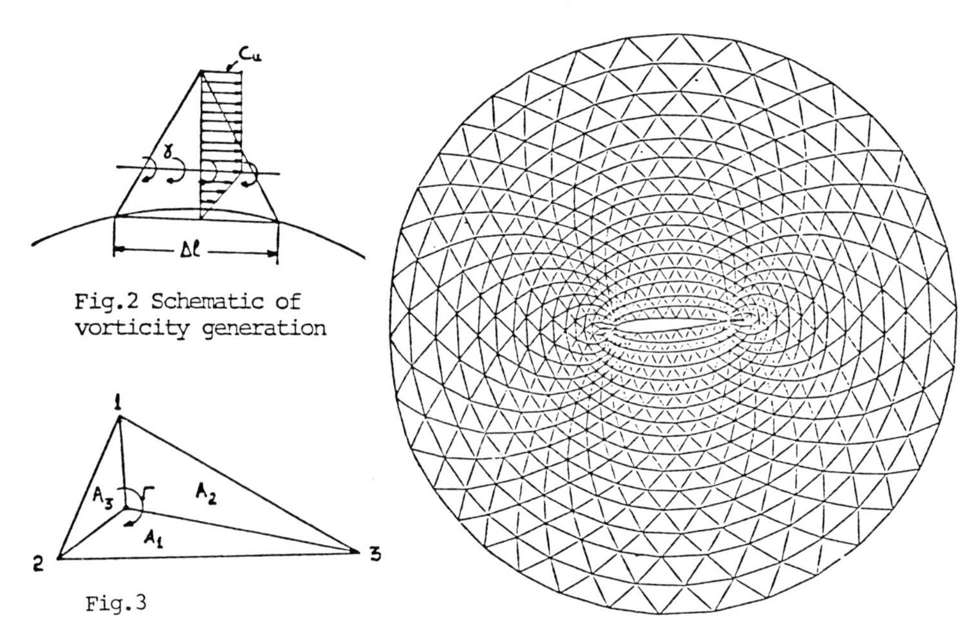

Fig.2 Schematic of vorticity generation

Fig.3

Fig.4 Computational grid gor the FE calculation

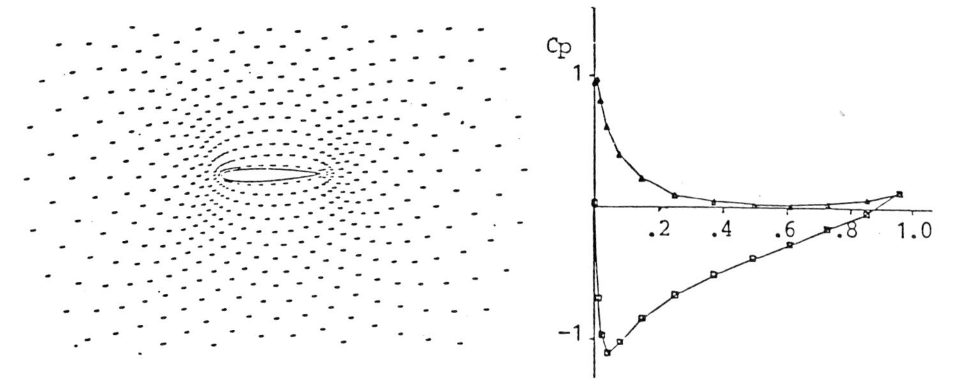

Fig.5 Flow field and pressure distribution for NACA 0012 airfoil (a=6°)

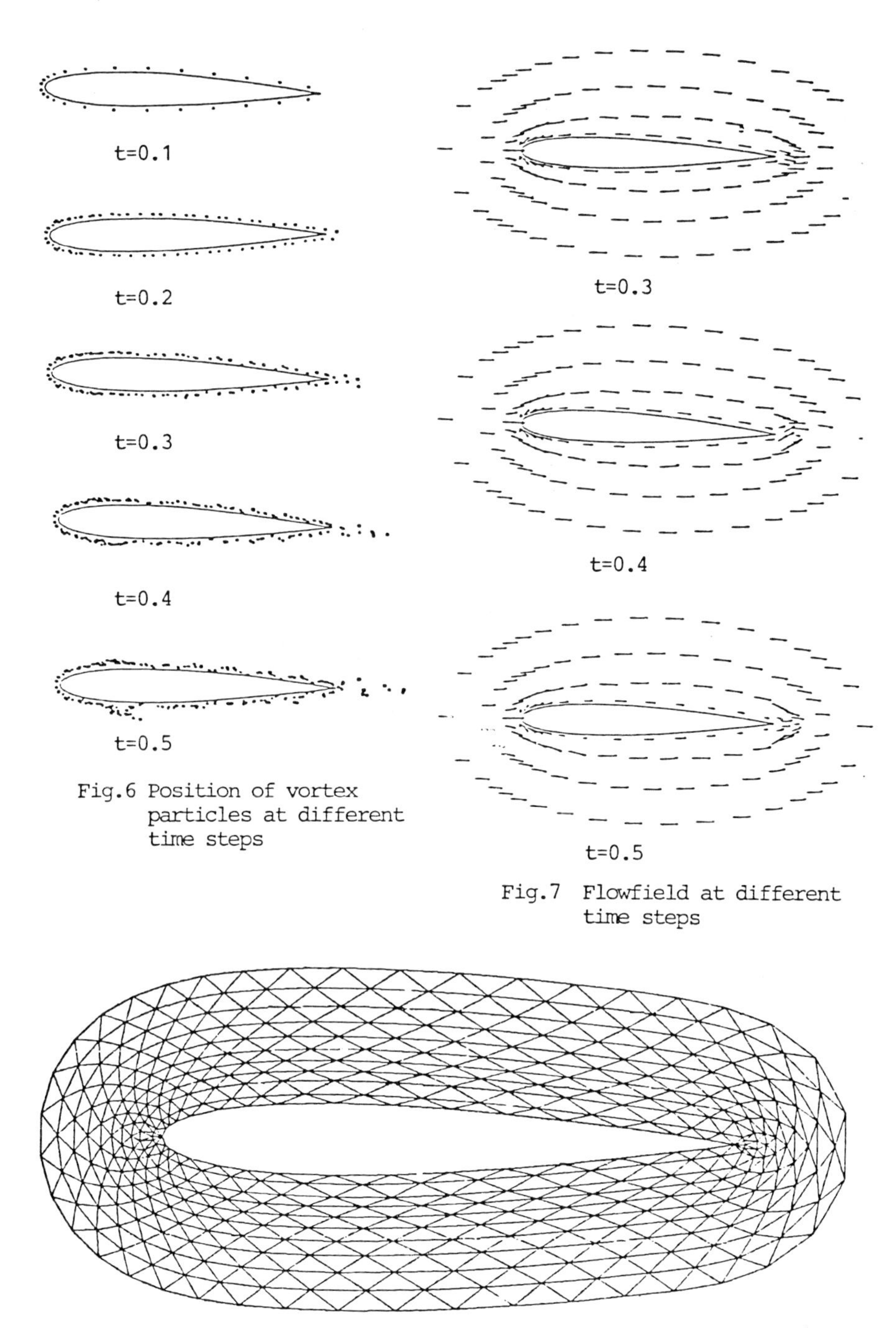

Fig.6 Position of vortex particles at different time steps

Fig.7 Flowfield at different time steps

Fig.8 Computational grid for the coupled problem

GAMM-WORKSHOP :
NUMERICAL SIMULATION OF COMPRESSIBLE NAVIER-STOKES FLOWS
PRESENTATION OF PROBLEMS AND DISCUSSION OF RESULTS

M.O. Bristeau
INRIA, B.P. 105, 78153 Le Chesnay Cedex, France
R. Glowinski
University of Houston, Texas, U.S.A. and INRIA
J. Periaux
AMD/BA, B.P. 300, 92214 Saint-Cloud Cedex,France
H. Viviand
ONERA, 92320 Chatillon, France.

1. INTRODUCTION

A workshop on the Numerical Simulation of Compressible Navier-Stokes Flows was organized by INRIA in December 1985 with the sponsorship of the GAMM Committee on Numerical Methods in Fluid Mechanics. The aim of this workshop was to compare accuracy and efficiency of Navier-Stokes solvers on selected external and internal flow problems using different numerical approaches. The test problems were defined on simple analytical geometries and correspond to 2D laminar flows with moderate gradients in order to avoid the introduction of turbulent models and also large instabilities due to strong shock-boundary layer interactions. In practice, the proposed test cases are transonic or supersonic steady flows at low or moderate Reynolds numbers. In this paper, we present an overview of the test-problems, of the comparisons made during this workshop, and of the main conclusions. The detailed contributions are published in [1] .

2. PROBLEMS FOR ANALYSIS

Two test problems were proposed :
- external flows past a NACA0012 airfoil (Problem A),
- internal flows in a double throat nozzle (Problem B).

2.1. Problem A :

For the external flows, seven test cases were proposed :

A1 :	$M_\infty = .8$	$\alpha = 10°$	Re = 73
A2 :	$M_\infty = .8$	$\alpha = 10°$	Re = 500
A3 :	$M_\infty = 2.$	$\alpha = 10°$	Re = 106
A4 :	$M_\infty = 2.$	$\alpha = 10°$	Re = 1000
A5 :	$M_\infty = .85$	$\alpha = 0°$	Re = 500
A6 :	$M_\infty = .85$	$\alpha = 0°$	Re = 2000
A7 :	$M_\infty = .85$	$\alpha = 0°$	Re = 10000

with

M_∞ = Mach number at infinity,
α = angle of attack,
Re = Reynolds number.

The cases A1 and A3 were prescribed because experimental results where available for the data of the problem (see Fig. 2).

The Reynolds numbers of case A5, A6, A7 associated with the same Mach number were chosen in order to allow the comparison of the thickness of the boundary layers and to check that they vary as $1/\sqrt{Re}$ (Fig. 3).

2.2. Problem B

The double throat nozzle (Fig. 1) of test problem (B) was designed with the aim of generating strong viscous interaction phenomena in steady, laminar, compressible flows in a well-bounded domain (see Fig. 4).

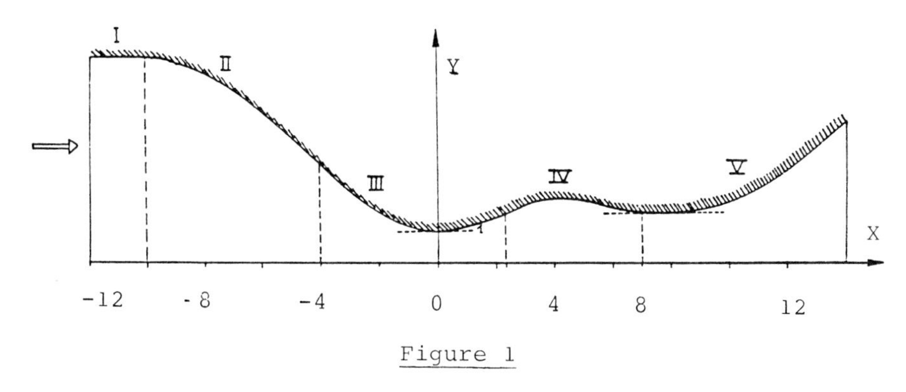

Figure 1

The flow in this nozzle was computed for the following three values of the Reynolds number :

B_1 : $Re_0 = 100$
B_2 : $Re_0 = 400$
B_3 : $Re_0 = 1600$

with

$$Re_0 = a_0 \, L \, \rho_0/\mu$$

where the subscript 0 refers to reservoir conditions, and L is the nozzle half-height at the first throat x = 0.
For both problems (A) and (B) the main challenge was the computation of the pressure and skin friction coefficients on the wall.

3. RESULTS

This meeting brought together sixteen numerical and one experimental contributions. Different approaches for solving the Navier-Stokes equations have been compared : finite differences, finite elements and finite volumes , generally unsteady formulation was used in order to capture steady solutions.

3.1. Comparison of results of Problem (A)

At first sight the results appear quite similar on contour maps of density, Mach number, pressure and on the pressure coefficients on the airfoil. However significant differences appear in the computed skin friction and heat flux coefficients and also integral values of lift, drag and moment coefficients.

The following features of the outputs of the different contributors can be compared :

i) stability of the solution at different Reynolds numbers via the smoothness of the isolines which depends strongly on the number of nodes and on the numerical viscosity.

ii) symmetry of the solution, in particular of the heat flux and skin friction coefficients on the airfoil at angle of attack $\alpha = 0°$.

iii) possibility to handle supersonic flows.

iv) quality of the mesh : density and regularity in the vicinity of the leading and trailing edges.

v) maximum and minimum values of the skin friction and heat flux coefficients which depend highly on the discretization in the vicinity of the leading edge.

vi) influence of the Reynolds number on the thickness of the boundary layer via the closeness of the Mach lines near the wall.

vii) position of the separation point for the flow of test case A3.

viii) integral values of lift, drag and moment coefficients.

3.2. Comparison of results of Problem (B)

All the solutions agree quite well about these features which are revealed by contour maps of Mach number, pressure and entropy. Cross-plots of results pertaining to five numerical solutions of Problem B (see Table 1) are also published in [1] .

Table 1

Solution Nr	Authors	Mesh size
1..........	F.Bassi, F.Grasso, A.Jameson L.Martinelli and M.Savini	152 x 32
2..........	O.Labbé	208 x 29
3..........	J.L.Thomas, R.W.Walters, B.Van Leer and C.L.Rumsey	121 x 81 for Re_0 = 100 246 x 81 for Re_0 = 400 and 1600
4..........	J.N.Scott and M.R.Visbal	218 x 61
5..........	D.Schwamborn	201 x 32

Table 1

The general conclusion which can be drawn from these cross-plots is that there is a very good agreement between all five solutions despite the differences in grids and in numerical methods. It is clear that these numerical solutions relate to the same flow field, and this fact supports the initial assumption that the problem is well-posed at least in a global manner (it cannot be excluded that small differences observed in particular

near the outflow boundary reflect the need for some additional boundary condition).
Two examples of cross-plots are shown on Figures 5 and 6.
The plots of C_f along the wall (Fig. 5) for Re_0 = 400 show appreciable differences over a short distance upstream of the outflow boundary. These differences denote a local sensitivity of the solution to the numerical treatment at this boundary, especially in the highly viscous part.

The distributions of pressure (Fig. 6), Mach number and entropy along the symmetry axis generally show very good agreement between all five solutions. At Re_0 = 400, some differences can be seen around the sharp maximum of pressure (Fig. 6), at the downstream end of the compression region, and also in the weak recompression which occurs around x = 8.

4. CONCLUSIONS

From a first comparison of the results, it appears that the scatter of computed solutions is much larger for Problem (A) than for Problem (B). Since the methods used are the same or very similar, a possible source of this difference can be that problem (A) deals with external flows on non bounded domains while Problem (B) is defined on bounded ones.

Concerning Problem (A) even if the results look quite similar on the pressure and Mach number in the fluid and also on the pressure distribution on the wall, significant differences appear in the computed skin friction and heat flux coefficients. This discrepancy can be explained by the fact that these outputs involve first derivatives of the primitive variables (velocities and temperature) making of importance the choice of the formulation and also the number and the location of nodes in the spatial discretization.

The good agreement of results for Problem (B) seems to indicate that the chosen boundary conditions are appropriate with the small restriction mentioned concerning the outflow condition.

Problem (A) appears to be more sensitive to numerical features (mesh, location of outer boundary) than Problem (B) and the trailing-edge singularity may play a role in this behaviour.

It is clear that much effort remains to be done to achieve good accuracy at acceptable computing cost for practical 3D applications. The volume [1] should provide valuable basic reference cases for the assessment of the progress in numerical methods and techniques for viscous fluid flow simulation.

REFERENCES

[1] Numerical Simulation of Compressible Navier-Stokes Flows. A GAMM-Workshop, M.O. Bristeau, R. Glowinski, J. Périaux, H. Viviand eds., Notes on Numerical Fluid Mechanics, Vieweg, Vol. 18.

[2] J. Allègre, M. Raffin, J.C. Lengrand, Experimental flowfields around NACA0012 airfoils located in subsonic and supersonic rarefied air streams (see [1]).

[3] W. Haase, Solutions of the Navier-Stokes Equations for Sub- and Supersonic Flows in Rarefied Gases (see [1]).

[4] L. Cambier, Computation of viscous transonic flows using an unsteady type method and a zonal grid refinement technique (see [1]).

[5] J.L. Thomas, R.W. Walters, B. Van Leer, C.L. Rumsey, An Implicit Flux split algorithm for the Compressible Navier-Stokes equations (see [1]).

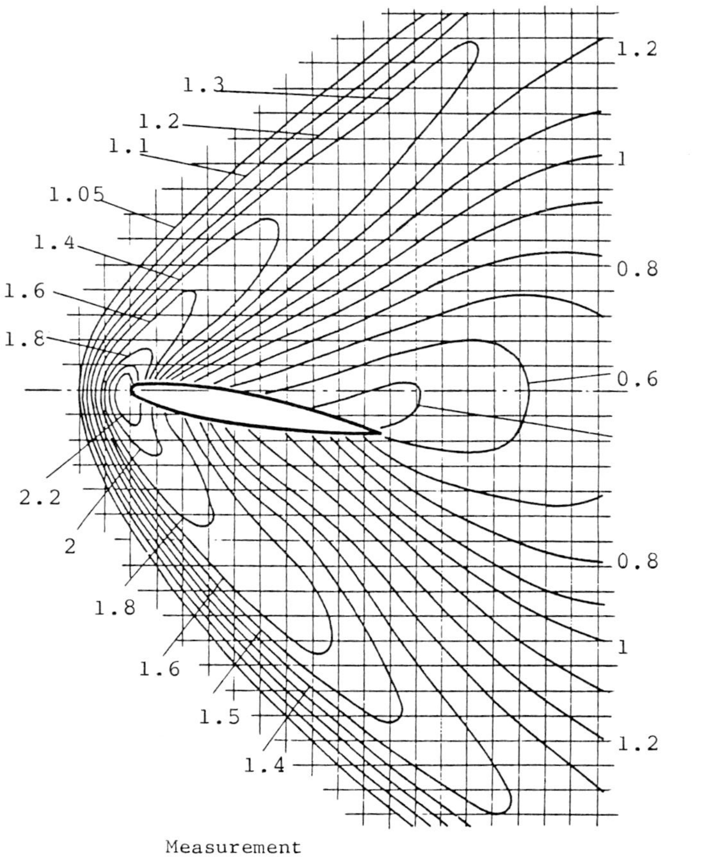

Measurement

by J. Allègre, M. Raffin, J.C. Lengrand [2]

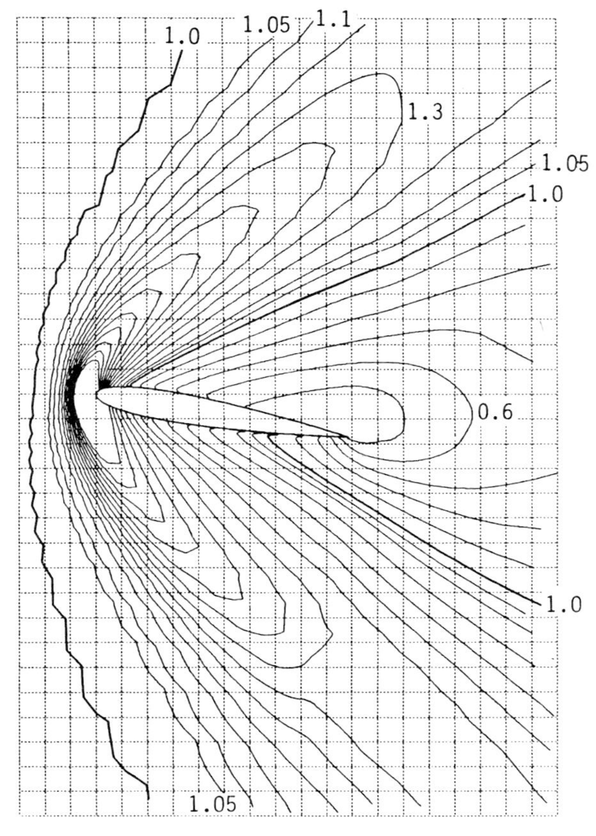

Numerical result

by W. Haase [3]

Figure 2 : Density contours around the NACA 0012 airfoil

$M_\infty = 2.$, Re = 106, $\alpha = 10°$.

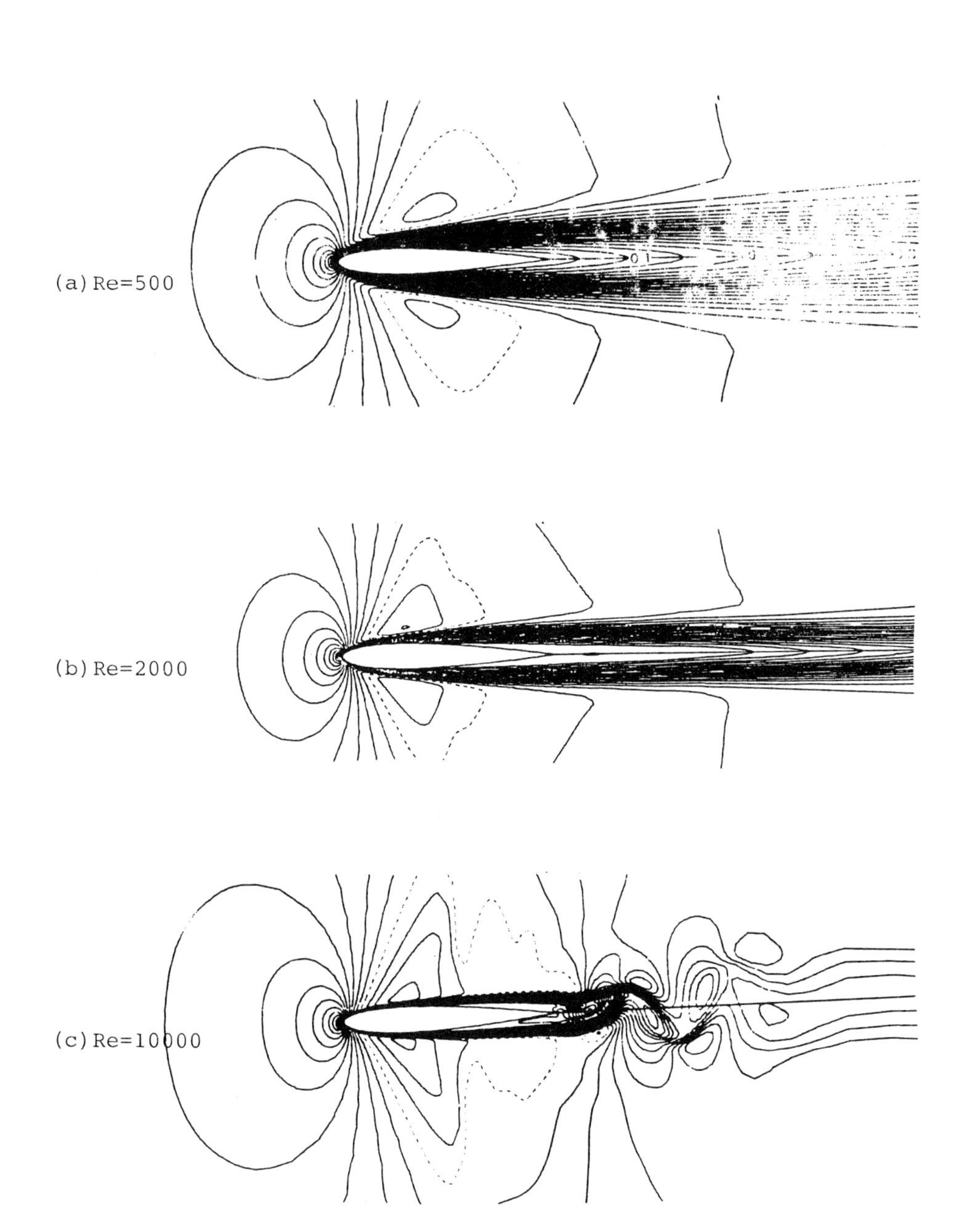

Figure 3 : Mach contours around the NACA0012 airfoil
$M_\infty = 0.85$, $\alpha = 0°$
by L. Cambier [4]

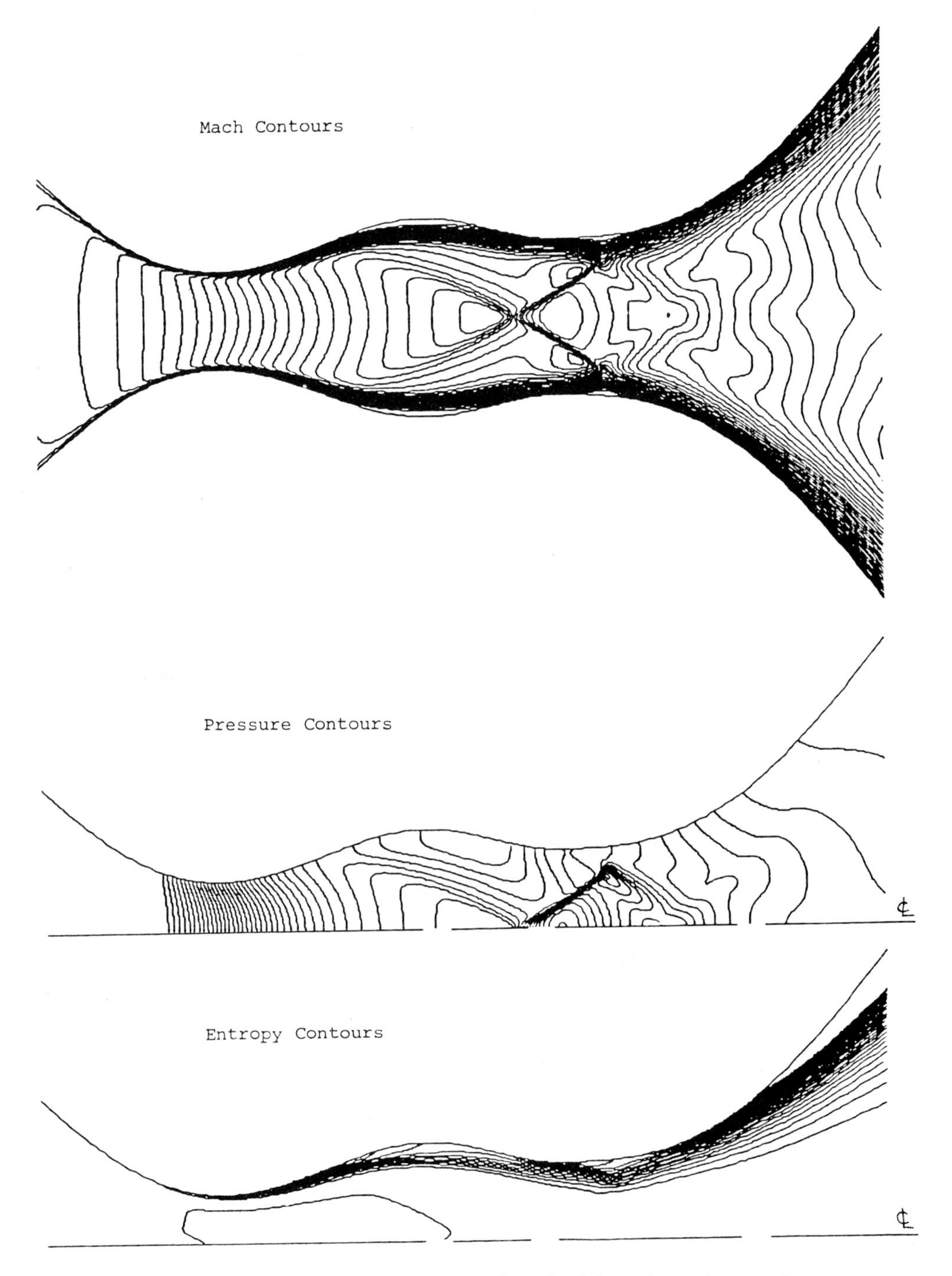

Figure 4 : Flowfield contours in the double throat nozzle ; Re = 1600 ; 246 x 81 grid. Computed by J.L.Thomas R.W.Walters, B.Van Leer, C.L.Rumsey [4] .

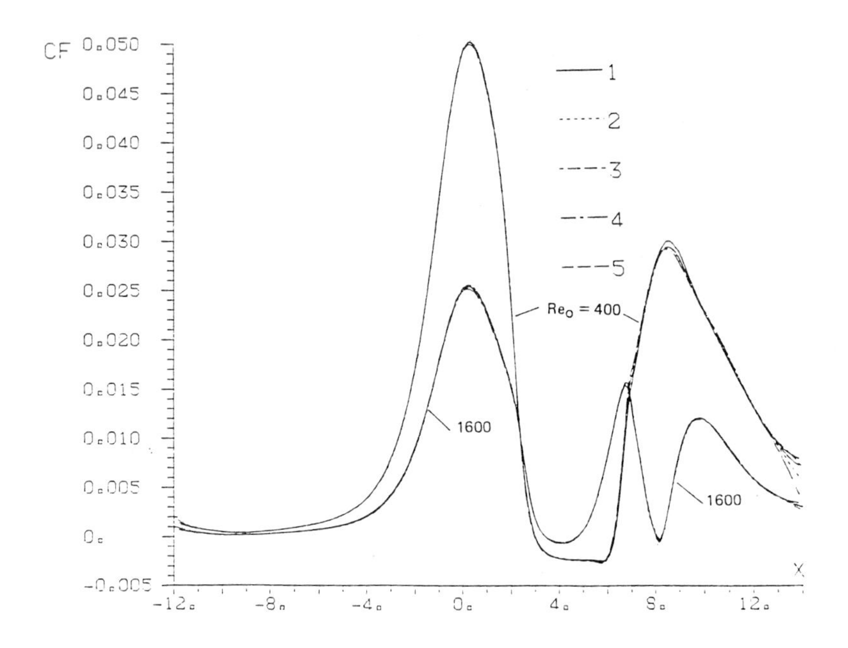

Figure 5 : Skin friction coefficient distributions Re_0 = 400 and 1600.

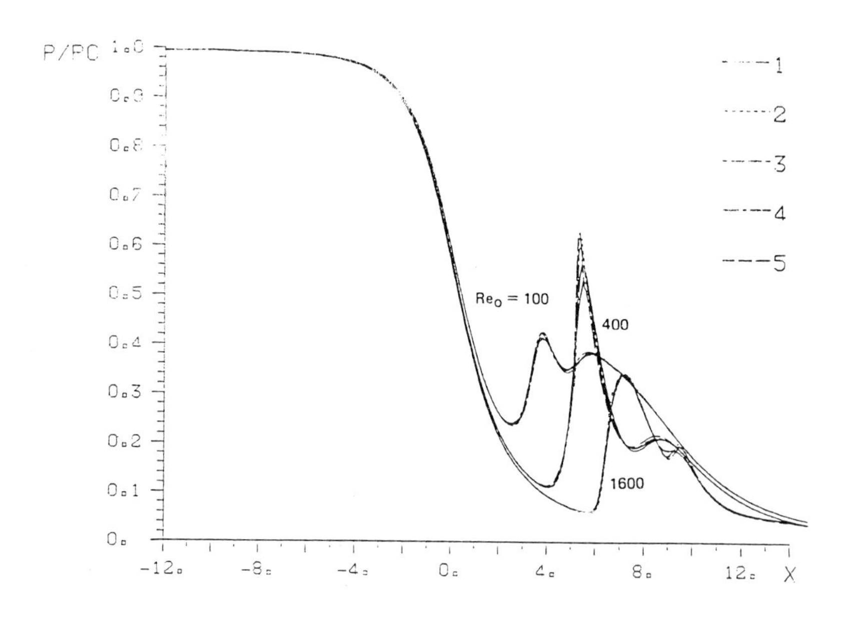

Figure 6 : Pressure distributions on symmetry axis.

Report on the GAMM Workshop on
THE NUMERICAL SIMULATION OF COMPRESSIBLE EULER FLOWS
10–13 June 1986, INRIA–Le Chesnay France

A. DERVIEUX*, J. PERIAUX, A. RIZZI***, B. VAN LEER******

* A. DERVIEUX, INRIA-Sophia Antipolis, 06565 VALBONNE CEDEX, France

** J. PERIAUX, AMD-BA, 78 quai Carnot 92214 SAINT CLOUD CEDEX, France

*** A. RIZZI, FFA, P.O. Box 11021, S-161 11 BROMMA, Sweden

**** B. VAN LEER, Delft Univ. P.O. Box 356, NL-2600 AJ DELFT, The Netherlands and University of Michigan, U.S.A.

ACKNOWLEDGEMENTS

The organizers wish to express their thanks to Drs. F. Angrand, J-A. Desideri, L. Fezoui and B. Stoufflet whose continuous scientific help contributed for a large part during the preparation and the synthesis of the Workshop to the success of this event.

1. INTRODUCTION

The purpose of this event was to compare in terms of accuracy and efficiency the codes solving compressible inviscid, mainly steady, Euler flows. This Workshop was a sequel of the GAMM Workshop held in 1979 in Stockholm and also of the SIAM theoretical Workshop held in 1983 in Rocquencourt, although, this time because of the fast expansion of researches concerning Euler numerical models, the full potential approaches were not included in the present Workshop.
Since 1979, many other Euler workshops have been held, in particular several focussing on airfoil calculations. However, we think that many recently derived methods were not presented at these workshops because, among other reasons, the methods were not enough developed, or had not been applied to the proper flow problems.

2. PROBLEMS FOR ANALYSIS

In order to permit a first analysis of recent methods, we have chosen a sample of test cases which cover :
- a scale of difficulty in implementation, starting from easier cases, with simple geometrical data
- a various range of applications, including both external and internal flows.

2.1. Geometrical data

The geometries of the test cases were of two kinds :

i) Several simplified geometries were proposed on which many new methods could be applied more easily ; a cylinder and several airfoils - NACA0012 and KORN - in order to allow rather precise comparisons during the Workshop, and to permit further comparisons, since their simplicity favorize a greater diffusion of the test-case. For the cylinder, a mandatory mesh was given, so that the respective accuracy of approximations presented by the contributors could be evaluated.

ii) Several slightly more complicated geometries of industrial interest introduced some realistic difficulties : a bi-naca airfoil, a cascade, an 2-D air intake and a 3-D delta wing. The purpose was to permit a more global evaluation of the methods, including basic accuracy, mesh generation and adaptation, efficiency of solution algorithms, among other points.

2.2. Test-cases

The participation in the Workshop was possible according the following rule : each contributor had to compute three of the seven mandatory (M) test cases.

Problem number	M_∞	α	Flow features	Mandatory
1) 2-D external flow over a full circular cylinder				
1.1	.38		subcritical	
1.2	.50		shock ; separation	M
1.3	.60		steady shocks, unsteady separation	
1.4	3.0		supersonic	M
1.5	8.0		hypersonic	
2) 2-D external flow over a NACA0012 airfoil				
2.1	.63	2°	subcritical	
2.2	.30	20°	subcritical and unsteady (?)	
2.3	.85	1°	sensitive lift	M
2.4	.81	0° Perturb!	3 conservative potential solutions	
2.5	.82	0°	5 conservative potential solutions	

3) 2-D external flow over a Korn airfoil					
3.1	.75	0°		shockless supercritical	M
4) 2-D external/internal flow over a bi-NACA0012 double airfoil					
4.1	.30	6°		subcritical ; potential solution known	
4.2	.55	0°		internal shock	
4.3	.55	6°		internal and upper shock	M

5) 2-D flow through a cascade (proposed by C. Hirsch).

Problem number	M	α	T_{in}	P_{in}	Flow features	Mandatory
5.1	.89	30°	287 K	1013 bar	shockless critical	
5.2	1.001	30°	287 K	1013 bar	internal shock structure, oblique shock impuiguig on upper trailing edge, experimental data available	M

6) 2-D flow through an engine inlet.

Problem number	M	M_{engine}		Flow features	Mandatory
6.1	.70	.3	0	subcritical	
6.2	1.2	.3	0	single shock at entrance	
6.3	2.0	.27	0	lambda shock at entrance	M

Problem number	M_∞	α	Flow features	
7.1	.40	10°	subcritical, vorticity	experimental data available
7.2	.85	10°	transonic, vorticity	
7.3	1.2	10°	supersonic, vorticity	

3. PARTICIPATION

The meeting brought together 20/30 attendees coming from France (7), Italy (3), USA (3), Great-Britain, Sweden, the Netherlands and Japan and about 20 invited contributions were presented including a survey paper by Prof. B. van Leer on the numerical solutions of the Euler equations, a lecture by Prof. C. Hirsch and a lecture by Prof. M. El Sennar on experimental results.

3.1. Contributions

We shall refer in the sequel to the following list of contributors :

1. F. ANGRAND, V. BILLEY, J. PERIAUX, J-P. ROSENBLUM, B. STOUFFLET (INRIA, AMD-BA, France)
2. M. BORREL, J.L. MONTAGNE (ONERA, France)
3. M. BREDIF, J.J. CHATTOT, J.L. DELAYE (Matra, France)
4. F. CHALOT, L.P. FRANCA, I. HARARI, T.J.R. HUGHES, F. SHAKIB, M. MALLET, J. PERIAUX, B. STOUFFLET (Stanford Univ., USA, AMD-BA, France)
5. V. COUAILLER, J.P. VEUILLOT (ONERA, France)
6. A. DADONE (Bari, Italy)
7. A. DERVIEUX, J.A DESIDERI, F. FEZOUI, B. PALMERIO, J.P. ROSENBLUM (INRIA, AMD-BA, France)
8. A. ECER, H.U. AKAY (Purdue Univ. Indianapolis, USA)
9. F. EL DABAGHI, J. PERIAUX, O. PIRONNEAU, G. POIRIER (INRIA, AMD-BA, France)
10. M. ELSENNAR (Amsterdam, The Netherlands)
11. P.W. HEMKER, B. KOREN (Amsterdam, The Netherlands)
12. C. HIRSCH (Bruxelles, Belgium)
13. B. KROUTHEN, A. RIZZI (FFA, Sweden)
14. A. LERAT, J. SIDES (ENSAM, ONERA, France)
15. G. MORETTI, A. LIPPOLIS (GMAF, New-York, USA, Bari, Italy)
16. K. MORGAN, J. PERAIRE, J. PEIRO, M. VAHDATI, O.C. ZIENKIEWICZ (Swansea, U.K)

17. M. PANDOLFI, F. LAROCCA, T.T. AYELE (Torino, Italy)
18. N. SATOFUKA, K. MORINISHI (Kyoto, Japan)
19. B. VAN LEER (Delft, The Netherlands)
20. B. VAN LEER, J.L. THOMAS, WALTERS (NASA Langley, USA)
21. L. ZANNETTI (Torino, Italy)

3.2. Methods used

Many classical and recent numerical methods were presented in the contributions :

- various forms of the Euler equations (conservative, lambda, Clebsch, stream function, ...)
- finite-difference, finite-volume and finite-element approximations
- centered and upwind schemes
- shock capturing, shock fitting
- unsteady methods : (explicit, implicit, multigrid algorithms) and steady methods

Although **steady calculations** was mainly addressed, **unsteady** computations were presented to describe the unstable behaviours that arose in some test cases (1.2, 1.3, 2.2).

4. RESULTS

An analysis of each test case will be presented in the book [1]. We present here general comments and a partial analysis of the classical test case 2.3.

4.1. General comments

Although several calculations were standard ones, the meeting brought many interesting informations concerning :

- the capture of unsteadiness (cylinders)
- the arising of separation, dead waters ...
- the internal viscosity involved in old and new schemes
- the fact that 2-D mesh generation was not yet a trivial task
- what can be done by mesh adaptation
- gains in efficiency when using sophisticated methods
- speed up rates on supercomputer architectures
- the local production of numerical entropy by different schemes in distinct areas of the flow (mainly occuring at stagnation points, sharp edges and along highly curved bodies).

It appears that 3-D computations around a sharp wing was a rather difficult test case out of range for many contributors.

4.2. Analysis of results for test case 2.3

It is a flow past a NACA0012 airfoil, M=.85,α=1^0. This is a classical test case which was studied in the 1979 workshop and in the AGARD WG 07 ; the present workshop brought many new answers. However, we think that a flow solution accurate within 1% is **not** yet clearly **obtained** , neither in the conditions of a mandatory number of nodes (it was 4000), nor, even, when a free mesh, with possibly many nodes and adaption were chosen. However, we shall try to establish a hierarchy in the results.

Method of comparison

For a first evaluation, we shall consider the dispersion of three computed global coefficients, namely the lift coefficient C_L, the drag coefficient C_D and the moment coefficient C_M ; for a precise definition of these numbers, we refer to [1].

Analysis

The distribution of results concerning C_L (TABLE 1) indicates striking discrepancies. We turn to correlations between these coefficients, that can be evaluated in Figures 1 (C_L *versus* C_D) and Figure 2 (C_L *versus* C_M). We have tried to draw some zones in which results seem to gather. For comparison, we added two results obtained a few years ago at the occasion of the AGARD W 07 which were judged among the best with the following numbers :
22. M.D. SALAS and J. SOUTH
23. W. SCHMIDT and A. JAMESON.
Lastly, we want to mention that, in a private communication, Pr. A. Jameson indicated that he made recent calculations giving a lift of 0.38.

For this test case, we can bring the following additional comments :
- the infinite boundary condition played an important role ;
- the number of nodes has not appeared to be so critical since rather good results were obtained using meshes of 3000-4000 nodes
- (moreover) the spurious entropy level, as long as it is not too large, was not strongly related to global accuracy ; is 0.1% accuracy for entropy necessary to reach 1% accuracy in lift coefficient ?
- the "concentrated" results essentially contain conservative Euler discretizations ; as a consequence, we cannot consider these results to be method-independent.

CONCLUSION

A large sample of cases were proposed. They were solved by using a variety of methods. Many "reference" calculations were proposed by the contributors, that we would like to thank for the large amount of work and results they provided ; for illustration we present a few figures selected from these contributions. A detailed analysis of each case will be jointed to the final contributions in [1].

REFERENCE

[1] Report on the workshop, to be published by Vieweg.

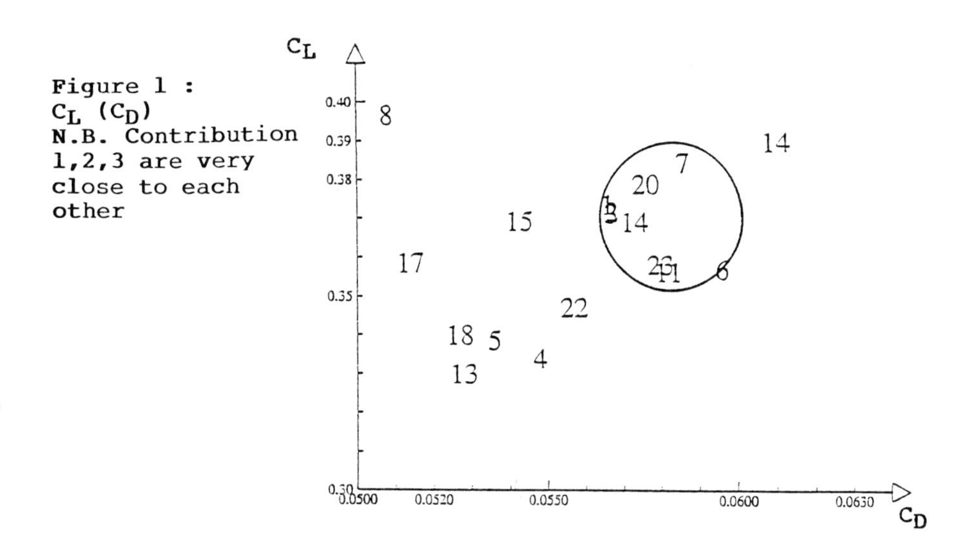

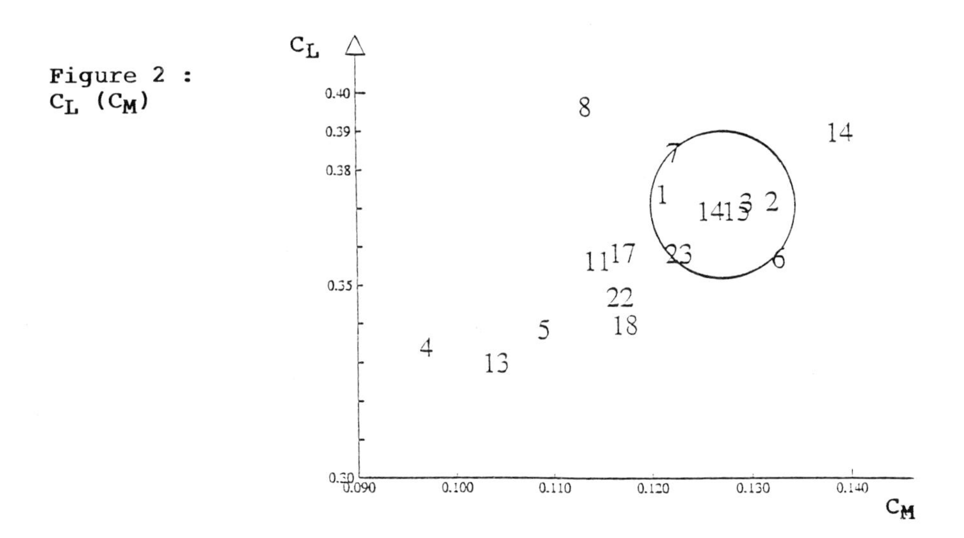

Test case 2.3
FLOW PAST A NACA0012
M = .85, $\alpha = 1°$

TABLE 1 (TEST CASE 2.3)

AUTHORS	MESH SIZE	C_L	C_D	C_M
1	2025	.374	.0565	-0.1208
2	8481	.3721	.0566	-0.1317
3	6369	.3717	.0566	-0.12928
4	2492	.334	.0547	-0.0965
5	4725	.3386	.0535	-0.1086
6	4096	.3572	.0595	-0.1323
7	2025	.3847	.0585	-0.122
8	3312	.3965	.0507	-0.113
11	4096	.3565	.0582	-0.1144
13	3584	.330	.0528	-0.104
14	6144	.3901	.0610	-0.139
14'	8192	.3696	.0573	-0.126
15	4096	.3696	.05426	-0.1286
17	5400	.3586	.0514	-0.117
18	4257	.3399	.0527	-0.1172
20	16705	.3793	.0576	

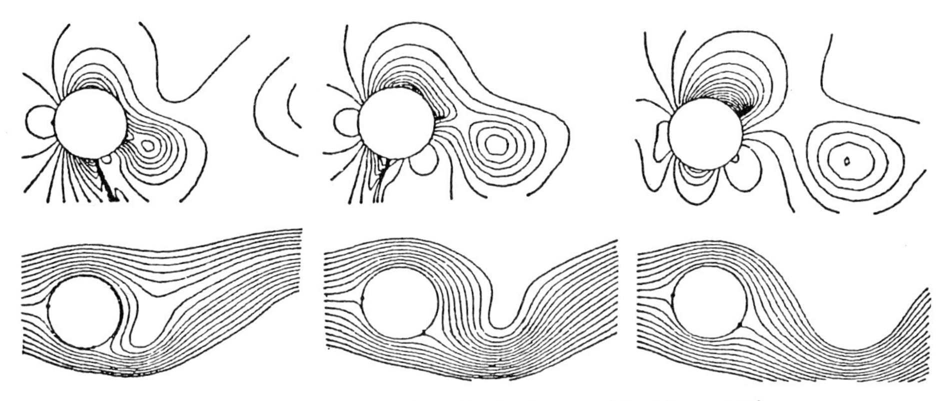

Figure 3 : TEST CASE 1.2 (contribution 17)

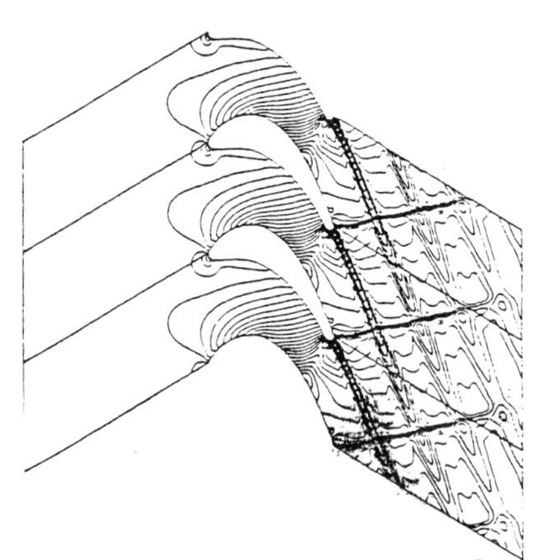

Figure 4 : TEST CASE 5.2 (contribution 20)

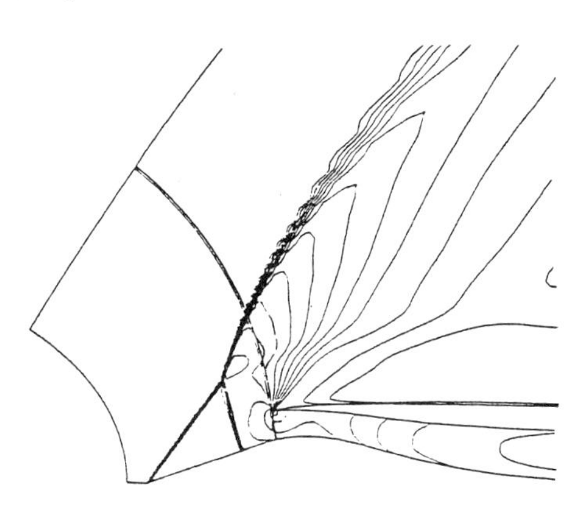

Figure 5 : TEST CASE 6.3 (contribution 2)